Mathematical Statistics with Applications

Statistical Distributions
Professor N. Balakrishnan
McMaster University

Statistical Process Improvement
Professor G. Geoffrey Vining
Virginia Polytechnic Institute

Stochastic Processes
Professor V. Lakshmikantham
Florida Institute of Technology

Survey Sampling
Professor Lynne Stokes
Southern Methodist University

Time Series
Sastry G. Pantula
North Carolina State University

Mathematical Statistics with Applications

Asha Seth Kapadia
The University of Texas, Health Science Center at Houston

Wenyaw Chan
The University of Texas, Health Science Center at Houston

Lemuel Moyé
The University of Texas, Health Science Center at Houston

CRC Press
Taylor & Francis Group
Boca Raton   London   New York

CRC Press is an imprint of the
Taylor & Francis Group, an **informa** business

A CHAPMAN & HALL BOOK

Published in 2005 by
CRC Press
Taylor & Francis Group
6000 Broken Sound Parkway NW, Suite 300
Boca Raton, FL 33487-2742

First issued in paperback 2022

© 2005 by Taylor & Francis Group, LLC
CRC Press is an imprint of Taylor & Francis Group, an Informa business

No claim to original U.S. Government works

ISBN 13: 978-1-03-247789-3 (pbk)
ISBN 13: 978-0-8247-5400-6 (hbk)

DOI: 10.1201/9781315275864

Library of Congress Card Number 2004061863

Library of Congress Cataloging-in-Publication Data

Kapadia, Asha Seth, 1937-
    Mathematical statistics with applications / Asha Seth Kapadia, Wenyaw Chan, Lemuel A. Moyé.
        p. cm. -- (Statistics, textbooks and monographs ; v. 179)
    Includes bibliographical references and index.
    ISBN 0-8247-5400-X
    1. Mathematical statistics. I. Chan, Wenyaw. II. Moyé, Lemuel A. III. Title. IV. Series.

QA276.K218 2005
519.5--dc22   2004061863

**Visit the Taylor & Francis Web site at
http://www.taylorandfrancis.com**

**and the CRC Press Web site at
http://www.crcpress.com**

Mathematical Statistics with Applications

Asha Seth Kapadia
Wenyaw Chan
Lemuel A. Moyé
The University of Texas-Houston
School of Public Health
Houston, Texas

To: *My mother Sushila Seth, my sisters Mira, Gita, Shobha, and Sweety, my son Dev, and my daughter-in-law Sonia*

A.S.K.

To: *Alice, Stephanie, and Jeffrey*

W.C.

To: *Dixie and the DELTS*

L.A.M.

Preface

Each of us has taught courses in mathematical statistics or advanced probability to students who are interested in becoming applied workers, and each of us has faced the same frustration. Students who require grounding in theoretical statistical structures are commonly unprepared for the theoretical onslaught that awaits them. The course is an exercise in frustration for them as they labor hard to understand mathematical material that they have been told that they must know, yet they do not see the direct relevance between the theoretical discussions and the application of statistics to their applied field.

Our experience has motivated and equipped us to write a textbook that will provide all of the introduction that applied students need to mathematical statistics while keeping useful applications in clear view. We have felt the need for this type of textbook from students, and have finally assembled the resources that we needed to complete this task.

Asha S. Kapadia
Wenyaw Chan
Lemuel A. Moyé

Acknowledgments

We owe special thanks to many people who assisted us in completing this task, particularly, Dr. Dung-Tsa Chen from University of Alabama at Birmingham, Dr. Nan Fu Peng of National Chiao Tung University, Taiwan. Also, Dr. Elaine Symanski, Mr. Yen-Peng Li, Mr. Hung-Wen Yeh, Ms. Regina Fisher and Mr. Guido Piquet from the University of Texas School of Public Health at Houston who helped us with either editing, programming, or copying.

This book has used several data sets as illustrations for the concepts introduced in this text. These data sets were generated from three NIH funded grants R01 NR04678-03: (PI: Lederman), R01 NR03052: (PI: Meininger), and R29 HL47659-01: (PI: Taylor). Additionally, the authors wish to thank Dr. Regina P. Lederman, Professor, School of Nursing, UTMB; Dr. Janet C. Meininger, the Lee and Joseph D. Jamail Distinguished Professor at the School of Nursing UT-Houston; and Dr. Wendell C. Taylor, Associate Professor, School of Public Health, UT-Houston for permission to use their data sets.

Special thanks go to Michele P. Mocco for aiding with the manipulation and setting of many of the figures included in our text.

We would like to thank Richard Tressider and Kevin Sequeira for their help in bringing this final product to market. Finally, we express our appreciation to Ms. Maria Allegra for her encouragement and patience with our persistence in developing this book's concept.

Introduction

Mathematical statistics is a fundamental course in the graduate training of statistics in both theoretical and applied training programs. This course provides the background and technical sophistication to students as they prepare to take more advanced courses in multivariate analysis, decision theory, stochastic processes, or computational statistics.

Yet this foundation course in mathematical statistics is usually one of the most difficult challenges in the statistical curriculum. One reason for the difficulty is the heavy mathematical prerequisite of standard mathematical statistics curricula. While many graduate students are well prepared in background mathematics, other students with a less focused and more applied background have not had recent experience with the required material. The standard texts in mathematical statistics provide either no review, or at best, a cursory review of the topics on which a heavy reliance is placed in the body of the text. This is insufficient for the student whose exposure to the topics in advanced calculus is remote.

A second difficulty presented by courses in mathematical statistics is the applied student's sudden exposure to material that is wholly theoretical. Mathematical theory devoid of direct application can present unanticipated challenges to the graduate student with a focused interest in applying advanced statistics to a specific area of application, e.g., communications, economics, agronomy, demography or public health.

Typically, there are two levels of textbooks in statistics. At the higher level is the textbook on mathematical statistics, devoting itself to the abstract, measure-theoretic treatment of probability and inference. At a second, lower level are application-centric textbooks that do not concern themselves so much with theory but with the implementation of statistics in data collection and analysis. Neither of these levels meets the requirements of the researcher with a field of application who needs to learn mathematical statistics. *Mathematical Statistics with Applications* is a textbook for these scientists. Specifically, this textbook provides in-depth, non-measure theoretic treatment of mathematical statistics for the researcher with a field of application. It is not a textbook for theoreticians who will go into a career of mathematical statistics, but for applied specialists who need to learn mathematical statistical theory in order to deepen their understanding of its implications for their chosen field. The group of scientists and students who need to understand the results of mathematical statistics but who are interested in areas of practical application is distinct from the traditional group of mathematical specialists and sub-specialists who have been the traditional consumers of mathematical statistics textbooks.

There is a larger, available audience of students whose needs are not met by the available textbooks in mathematical statistics, but who nevertheless must gain an introduction to this complicated topic. Educators, environmental researchers, behavioral and biological scientists, applied mathematicians, astronomers, physical scientists, public health researchers, quantitative scientists, engineers, and computer scientists, among others, desire to learn mathematical statistics. In our view, this larger audience is not well served by the available books on mathematical statistics with their traditional approach.

Mathematical Statistics with Applications provides graduate and advanced undergraduate students exposure to mathematical statistics only after a thorough discussion of the subject's prerequisites. The first chapter concentrates on the development of the basic mathematics that must be mastered before the subject can be absorbed. There is a discussion of partial differentiation, differentiation under the integral sign, multiple integration and measure theory. These concepts are redeveloped not with the understanding that the student needs only a brief re-acquaintance with the concepts to refresh their memory, but are instead presented assuming that the student knows or remembers little about the important details of these topics. Students who complete this chapter will be sufficiently well versed in the background prerequisite material and therefore able to enter the classic statistical material unencumbered by crippling prerequisite weaknesses.

The second, unique feature of this text is the degree to which useful applications of the theory are discussed. These applications are provided to enrich, and not supplant the underlying theoretical methodology. By providing appropriate examples, the students can see and gauge for themselves the utility of the theory in a useful and relevant field of application.

The combination of thorough discussions and rich applications has produced a large volume of work which in all likelihood could not be completed in its entirety in a two semester course. In order to guide topic selection for students, the optional sections of the chapters that might be skipped are marked with asterisks in the section heading.

The textbook is the foundation of a two semester sequence in mathematical statistics. Students are required to have as prerequisite a background in calculus, a standard prerequisite for graduate level courses in statistics. Commonly experienced class sizes permit the applied material to be discussed in detail. The first semester of such a course would cover Chapters 1 through 6, and the second semester would encompass the remaining five chapters.

It is also possible that the combination of severe time constraints and strong mathematical backgrounds would permit an accelerated one semester course in mathematical statistics. Under these circumstances, this textbook could be the basis of an aggressively scheduled fourteen week sequence.

Week	Chapter
1	1–2
2	3
3–4	4
5–6	5
7	6
8–9	7
10	8
11–12	9
13	10
14	11

Asha S. Kapadia, Ph.D.
Wenyaw Chan, Ph.D.
Lemuel A. Moyé, M.D., Ph.D.
University of Texas School of Public Health

Contents

1

Review Of Mathematics

1.1 Introduction

This chapter provides a review of the mathematics that is required for a good understanding of probability and mathematical statistics. Our intention is to provide a survey of the fundamental concepts in calculus and finite mathematics that will facilitate the concepts introduced in Chapters 2-11. For more details the student is referred to other texts in mathematics.

1.2 Combinatorics

This section will review a few rules of enumeration that are helpful in our study of probability theory.

1.2.1 Examples in Enumeration

If we have a total of n_1 individuals from Population A and n_2 individuals from Population B then there are a total of $n_1 n_2$ pairs containing one member from each population. If we have k non-overlapping populations $P_1, P_2, ...P_k$ with $n_1, n_2, ...n_k$ individuals respectively, then, there are $n_1 n_2 ...n_k$ different groups such that each group of k elements has exactly one member from each of the populations. In this example we are not concerned with specific order of individuals, but simply enumerating total number of possible groups. Continuing, if there are n individuals in a population and we choose a sample of r individuals in such a way that after choosing an individual from the population we send the individual back to the population, so that the size of the population stays constant, then there are n^r such samples.

If however we do not send the chosen individuals back to the population then there are a total of $\dfrac{n!}{r!(n-r)!}$ different samples or *combinations*. As we will learn in Chapter 2, these alternative methodologies of sample selection are referred to as *sampling with replacement* and *sampling without replacement* and will be the basis of important probability laws that will be introduced in Chapter 2.

1.2.2 Permutations and Combinations

If we want to choose r objects out of k distinct objects in such a way that order matters, then we have a total of $k(k-1)(k-2)\cdots(k-r+1) = \dfrac{k!}{(k-r)!}$ ways of choosing these k objects. For example, there are $\dfrac{3!}{(3-2)!} = 3!$ ways or *permutations* of choosing two alphabets out of A,B,C. The choice may result in AB, BA, AC, CA, BC, CB. Since order matters, the choices AC and CA are considered different choices and we must count each choice. If we were to choose all three in a way that order of choosing mattered, then we would have 6 possible ways of choosing them (ABC, ACB, BCA, BAC, CAB, CBA). The total number of possible permutations is represented by

$$_k P_r = \frac{k!}{(k-r)!}. \tag{1.1}$$

Since we know that there are $k!$ different arrangement of k distinct objects taken all at a time we can write

$$_k P_k = k!.$$

In these examples we assumed that both k and r are integers and by definition $_k P_r$ is defined as the number of permutations of k objects taken r at a time. The term "permutations" implies ordering in that AB is different from BA as opposed to the use of the term "combinations" which is used when order is ignored. The number of ways of choosing two individuals out of three (A,B,C) is simply the number of combinations that can be formed when each individual is combined with the other two in this example (AB,BC,AC) while treating AB and BA as one combination. In general the number of combinations of k objects taken r at a time is given by

$$_k C_r = \binom{k}{r} = \frac{k!}{r!(k-r)!} \tag{1.2}$$

where

$$_k C_k = 1. \tag{1.3}$$

We also define $_k C_r = 0$ for $r > k$.

Formal mathematical proof of formulae $(1.1)-(1.3)$ may be obtained from any intermediate level book in mathematics (Draper and Kingman 1967).

    The problem of distinguishable permutations of elements that are not all different may also arise.

Suppose there are n letters among which there are n_1 A's, n_2 B's....n_{26} Z's

($\sum\limits_{i=1}^{26} n_i = n$).Then the total number of distinguishable permutations taking all the

objects together is given by

$$\frac{n!}{n_1!n_2!...n_{26}!} \qquad (1.4)$$

Example 1.1 Assume that $2A$'s and $3B$'s are available to us and we need to arrange all of them in such a way that no two arrangements are the same. Then the different arrangements possible taking all the letters together from formula (1.4) will be

$$\frac{5!}{2!3!} = 10 .$$

In this case the arrangements are

ABBBA, ABBAB, ABABB, AABBB, BABAB, BABBA, BAABB, BBABA, BBBAA, BBAAB.

∎

1.2.3 Gamma Function

The term $n!$ can also be expressed as the gamma function that is represented by

$$\Gamma(\alpha) = \int_0^\infty x^{\alpha-1}e^{-x}dx, \, \alpha > 0$$

For $\alpha > 0$

$$\Gamma(\alpha+1) = \alpha\Gamma(\alpha) ,$$

where $\Gamma\left(\dfrac{1}{2}\right)$ is defined as $\Gamma\left(\dfrac{1}{2}\right) = \sqrt{\pi}$

and for n a positive integer

$$\Gamma(n+1) = n!$$

∎

1.3 Pascal's Triangle

Algebraic manipulation of combinations often provides important relationships. For example the following relationship often arises in the context of combinations

$$\binom{n}{k} + \binom{n}{k-1} = \binom{n+1}{k} \tag{1.5}$$

In order to demonstrate this we first write

$$\frac{n!}{k!(n-k)!} + \frac{n!}{(k-1)!(n-k+1)!} = \frac{n!}{(k-1)!(n-k)!}\left[\frac{1}{k} + \frac{1}{n-k+1}\right]$$

$$= \frac{n!}{(k-1)!(n-k)!}\frac{(n+1)}{k(n-k+1)}$$

$$= \binom{n+1}{k}$$

The name Pascal's Triangle is associated with the following representation of combinations in a triangular form. Note that in the "triangle" below, any combination in a row is the sum of the combinations, to its left and to its right in the preceding row.

$$\binom{0}{0}$$

$$\binom{1}{0} \qquad \binom{1}{1}$$

$$\binom{2}{0} \qquad \binom{2}{1} \qquad \binom{2}{2}$$

$$\binom{3}{0} \qquad \binom{3}{1} \qquad \binom{3}{2} \qquad \binom{3}{3}$$

$$\cdots\cdots\cdots\cdots\cdots\cdots\cdots\cdots\cdots\cdots\cdots\cdots\cdots\cdots\cdots\cdots\cdots$$

$$\binom{n}{0} \qquad \binom{n}{1} \qquad \binom{n}{2}\cdots\cdots\cdots\cdots\cdots\binom{n}{n}$$

Note that the representation above concurs with formula (1.5).

1.4 Newton's Binomial Formula

It could be very helpful if we could represent the powers of sums as the sums of combinations. In order to do this we will examine the expansion of $(x+y)^n$ for different values of n.

(a) If n is a non-negative integer, then $(x+y)^n = \sum_{k=0}^{n} \binom{n}{k} x^k y^{n-k}$

(b) If n is negative or fractional, then

$$(x+y)^n = x^n \left[1 + \frac{y}{x}\right]^n = x^n \sum_{k=0}^{\infty} \binom{n}{k}\left[\frac{y}{x}\right]^k, \text{ for } \left|\frac{y}{x}\right| < 1$$

Here, $\binom{n}{k} = \dfrac{n(n-1)....(n-k+1)}{k!}$

Furthermore using the above definition of $\binom{n}{k}$, it can be shown that

$$\binom{-t}{k} = (-1)^k \binom{t+k-1}{k} \qquad (1.6)$$

Formula (1.6) is an important component of *the negative binomial distribution* that is discussed in Chapter 4 section 4.8 and is worthy of elaboration. We begin by recognizing that the left hand side in the above equation may be expanded as

$$\frac{-t(-t-1)(-t-2)....(-t-k+1)}{k!}$$

$$= \frac{(-1)^k t(t+1)....(t+k-1)}{k!} = \frac{(-1)^k (t+k-1)...t(t-1)!}{k!(t-1)!}$$

$$= \frac{(-1)^k (t+k-1)!}{k!(t-1)!} = (-1)^k \binom{t+k-1}{k}.$$

Use of the Binomial formula and (1.6) produces useful and fascinating equalities. Several of these are:

(i) $\binom{n}{0} + \binom{n}{1} + ... + \binom{n}{n} = 2^n$

(ii) $\binom{n}{0} - \binom{n}{1} + \binom{n}{2} + ... + (-1)^n \binom{n}{n} = 0$

(iii) $\binom{n}{1} + 2\binom{n}{2} + 3\binom{n}{3} + ... + n\binom{n}{n} = n2^{n-1}$

(iv) $\binom{n}{1} - 2\binom{n}{2} + 3\binom{n}{3} - .... + (-1)^{n-1}n\binom{n}{n} = 0$

(v) $\begin{pmatrix} a \\ 0 \end{pmatrix}\begin{pmatrix} b \\ n \end{pmatrix} + \begin{pmatrix} a \\ 1 \end{pmatrix}\begin{pmatrix} b \\ n-1 \end{pmatrix} + ... \begin{pmatrix} a \\ n \end{pmatrix}\begin{pmatrix} b \\ 0 \end{pmatrix} = \begin{pmatrix} a+b \\ n \end{pmatrix}$

(vi) $\begin{pmatrix} n \\ 0 \end{pmatrix}^2 + \begin{pmatrix} n \\ 1 \end{pmatrix}^2 + \begin{pmatrix} n \\ 2 \end{pmatrix}^2 .... \begin{pmatrix} n \\ n \end{pmatrix}^2 = \begin{pmatrix} 2n \\ n \end{pmatrix}$

(vii) $\displaystyle\sum_{j=0}^{k} \begin{pmatrix} a+k-j-1 \\ k-j \end{pmatrix}\begin{pmatrix} b+j-1 \\ j \end{pmatrix} = \begin{pmatrix} a+b+k-1 \\ k \end{pmatrix}$

$(viii)$ $\begin{pmatrix} 2n \\ n \end{pmatrix} 2^{-2n} = (-1)^n \begin{pmatrix} -\dfrac{1}{2} \\ n \end{pmatrix}$

In the above equalities $a, b, n,$ and k are positive integers. Equality (v) is known as the Vandermonde Convolution. ∎

1.5 Exponential Function

A useful result that we will now derive is

$$\lim_{n \to \infty} \left(1 + \frac{x}{n}\right)^n = e^x.$$

(1.7)

We can demonstrate the accuracy of expression (1.7) by recalling two familiar findings. The first is $\displaystyle\sum_{n=0}^{\infty} \frac{x^n}{n!} = e^x$, and the second helpful relationship is

$(a+b)^n = \displaystyle\sum_{k=0}^{n} \begin{pmatrix} n \\ k \end{pmatrix} a^k b^{n-k} = \sum_{k=0}^{n} \begin{pmatrix} n \\ k \end{pmatrix} b^k a^{n-k}$. We start with

$$(1+a)^n = \sum_{k=0}^{n} \begin{pmatrix} n \\ k \end{pmatrix} a^k$$

(1.8)

Let $a = \dfrac{x}{n}$ then the expression (1.8) may be written as

$$\left(1 + \frac{x}{n}\right)^n = \sum_{k=0}^{n} \begin{pmatrix} n \\ k \end{pmatrix} \frac{x^k}{n^k} = \sum_{k=0}^{n} \frac{n!}{k!(n-k)!} \frac{x^k}{n^k} = \sum_{k=0}^{n} \frac{x^k}{k!} \left(\frac{n(n-1)(n-2)..(n-k+1)}{n^k} \right)$$

$$= \sum_{k=0}^{n} \frac{x^k}{k!} \left(1 - \frac{1}{n}\right)\left(1 - \frac{2}{n}\right)\left(1 - \frac{3}{n}\right)...\left(1 - \frac{k+1}{n}\right)$$

Taking limits as $n \to \infty$ we can now write

$$\lim_{n \to \infty} \left(1 + \frac{x}{n}\right)^n = \lim_{n \to \infty} \sum_{k=0}^{n} \frac{x^k}{k!} \left(1 - \frac{1}{n}\right)\left(1 - \frac{2}{n}\right)\left(1 - \frac{3}{n}\right)\cdots\left(1 - \frac{k+1}{n}\right)$$

$$= \lim_{n \to \infty} \sum_{k=0}^{n} \frac{x^k}{k!} \lim_{n \to \infty}\left(1 - \frac{1}{n}\right)\left(1 - \frac{2}{n}\right)\left(1 - \frac{3}{n}\right)\cdots\left(1 - \frac{k+1}{n}\right)$$

$$= \lim_{n \to \infty} \sum_{k=0}^{n} \frac{x^k}{k!} = e^x.$$

(1.9)

There are other forms of this equality, namely

$$\lim_{n \to \infty} \left(1 - \frac{x}{n}\right)^n = e^{-x}$$

(1.10)

and, more generally, if $\{x_n\}$ is a sequence of real numbers that converges to x, then

$$\lim_{n \to \infty} \left(1 + \frac{x_n}{n}\right)^n = e^{\lim_{n \to \infty} x_n}.$$

(1.11)

It is advised that one be acquainted with these concepts as we will be assuming familiarity with them in all subsequent chapters. ∎

1.6 Stirling's Formula

Stirling's formula is attributed to the eighteenth century Englishman James Stirling, who in his *Methodus Differentialis* (1730), derived the following approximation

$$n! \approx \sqrt{2\pi} n^{n+\frac{1}{2}} e^{-n}.$$

An alternative form of this is presented in Feller (1957) as

$$n! \approx \sqrt{2\pi} \left(n + \frac{1}{2}\right)^{n+\frac{1}{2}} e^{-\left(n+\frac{1}{2}\right)}$$

(1.12)

where the sign $\approx$ in (1.12) is used to indicate that the ratio of the two sides tends to one as $n \to \infty$. ∎

1.7 Multinomial Theorem

The multinomial theorem is a generalization of the binomial theorem (section 1.4) and may be stated as follows:

$$(x_1 + x_2 + \ldots + x_n)^k = \sum_{k_1,k_2,\ldots k_n} \sum \ldots \sum x_1^{k_1} x_2^{k_2} \ldots x_n^{k_n} \binom{k}{k_1 k_2 \ldots k_n}$$

in such a way that $k_1 + k_2 + \ldots + k_n = k$. Here the combinatorial function is expanded from including multiple objects to including multiple types of object. We denote this as

$$\binom{k}{k_1 k_2 \ldots k_n} = \frac{k!}{k_1! k_2! \ldots k_n!} \tag{1.13}$$

Formula (1.13) implies the total number of ways of selecting a sample of k objects from among n types of objects such that we choose exactly k_1 objects of the first type, k_2 objects of the second type and so on and k_n objects of the nth type. In this case, even though objects within type i are assumed indistinguishable the order of appearance in the sample by type of object is important.

Example 1.2 If we have 3 different types of objects and we are going to select exactly five out of these in such a way that we select two objects of the first and third type and only one object of the second type, then the total number of different ways that this can be achieved is

$$\binom{5}{2,1,2} = \frac{5!}{2!1!2!} = 30 \qquad\blacksquare$$

Formula (1.13) is extremely useful in statistical theory when considering the occurrence of several competing events, and is the foundation of the *multinomial distribution* discussed in Chapter 4, section 4.4.

Example 1.3 Assume that three letters are to be chosen from the 26 letters in the alphabet $(A,B,C,\ldots Z)$. We are going to determine the number of ways in which exactly one A, one B and one C shows up. Using the formula above, there are

$$\frac{3!}{1!1!1!0!\ldots 0!} = 3!$$

different ways of selecting this sample. Note that in this derivation the term $0!$ appears 23 times in the denominator corresponding to the 23 letters that are not among the letters chosen. $\blacksquare$

1.8 Monotonic Functions

Assume that the functional relationship between two variables x and y is described by $y = f(x)$. Suppose that for each value of x the corresponding value of y may be obtained such that as x increases y stays the same or increases then y is called a monotonically increasing function of x. Also, if as x increases, y stays the same or decreases then y is called a monotonically decreasing function of x. In either case the relationship between x and y is called strictly monotonically increasing (or monotonically decreasing) assuming that y does not stay the same for x increasing (or decreasing).

Example 1.4 (a) The functions $y = x^3 + 3x^2 + 2x + 1$ and $y = e^x, x > 0$ are both strictly monotonically increasing functions of x.
(b) The function $y = e^{-x}, x > 0$, is an example of a strictly monotonically decreasing function of x. ∎

1.9 Convergence and Divergence

The concept of convergence and divergence is usually associated with infinite series. Consider an infinite series

$$f(x) = a_1 + a_2 + ...$$

Define the sum of the first n terms $\mathbf{S}$ as such that

$$S_n = \sum_{i=1}^{n} a_i$$

Then if $\lim_{n \to \infty} S_n$ approaches a finite quantity, the infinite series is said to be convergent otherwise it is referred to as a divergent series.

Example 1.5 Let $S_n = 1 + r + r^2 + r^3 + ... + r^{n-1} = \dfrac{1 - r^n}{1 - r}$

For $|r| < 1$ the series is clearly convergent. If $|r| \geq 1$ then $\lim_{n \to \infty} S_n \to \infty$ and the series is divergent. Notice that for $r = -1$, $S_n = 1 - 1 + 1 - 1 + ...$ and $\lim_{n \to \infty} S_n$ does not exist, because for $n = $ an even number, $S_n = 0$ and for $n = $ an odd number, $S_n = 1$.

1.9.1 Tests for convergence

In the absence of an expression for S_n there are useful tools available to determine if an infinite series is convergent or divergent:

(1) In an infinite series $a_0 + a_1 + a_2 + ...$, the necessary condition for convergence is $\lim_{n \to \infty} a_n = 0$

(2) A series of the form $a_0 + a_1 + a_2 + ...$ where terms are alternately positive and negative is called an alternating series. It converges only if $\lim_{n \to \infty} a_n = 0$ as well as when $|a_{n+1}| < |a_n|$ for all $n = 1, 2, ...$

(3) An alternating series that converges after taking absolute values of all its terms is said to be *absolutely convergent*.

(4) If $a_0 + a_1 + a_2 + ...$ is an infinite series such that $\lim_{n \to \infty} \dfrac{a_{n+1}}{a_n} = A$ then if $A < 1$, and all $a_i > 0$ $(i = 0, 1, ...)$ the series is convergent and if $A > 1$ the series is referred to as divergent series. This test is called the *Cauchy's test-ratio* test and it fails when $A = 1$.

Example 1.6 Consider the following infinite series

$$S = 1 - \frac{1}{\sqrt{3}} + \frac{1}{\sqrt{5}} - \frac{1}{\sqrt{7}} + ...$$

The nth term of this series is

$$a_n = (-1)^{n+1} \frac{1}{\sqrt{2n-1}}$$

$$\lim_{n \to \infty} a_n = 0.$$

Also,

$$\left| (-1)^{n+1} \frac{1}{\sqrt{2n-1}} \right| > \left| (-1)^{n+2} \frac{1}{\sqrt{2(n+1)-1}} \right|$$

The given series is therefore convergent using (2) above.   ■

1.10 Taylor's Theorem

The need to represent a function $y = f(x)$ by a power series often arises in mathematics. Taylor's Theorem provides us with such a series for expressing many functions frequently encountered in statistics (see for example Chapter 7). These functions may be expressed as finite or infinite series. For the infinite series, the convergence properties in the interval of interest need to be determined before such a representation is meaningful. According to Taylor's Theorem a function $f(x)$ may be approximated by the following series

$$f(x) = f(x_0) + f'(x_0)\frac{(x-x_0)}{1!} + f''(x_0)\frac{(x-x_0)^2}{2!} + \dots$$

$$+ f^{(n-1)}(x_0)\frac{(x-x_0)^{(n-1)}}{(n-1)!} + A_n$$

$$\text{where } f^{(i)}(x_0) = \left.\frac{\partial^i f(x)}{\partial x^i}\right|_{x=x_0} \quad i = 1, 2, \dots n-1$$

and A_n represents the remainder or the difference between the function and its approximation by the infinite series and is given by

$$f^{(n)}(\eta)\frac{(x-x_0)^n}{n!} \text{ where } x_0 < \eta < x$$

If $A_n \to 0$ as $n \to \infty$, $f(x)$ may be expanded in a Taylor series about $x = x_0$. A_n may also be used to determine the error involved when an infinite series is approximated by a series containing only n terms. For $x_0 = 0$ the expansion is known as the Maclaurin's series.

Example 1.7 Let $g(x) = \sin x$

Then, the Taylor series for the exponential function at 0 (Maclaurin's series) is calculated as follows

$$g(0) = 0$$
$$g'(0) = \cos 0 = 1$$
$$g''(0) = -\sin 0 = 0$$
$$.$$
$$.$$
$$.$$
$$g^{(4k)}(0) = 0$$
$$g^{(4k+1)}(0) = 1$$
$$g^{(4k+2)}(0) = 0$$
$$g^{(4k+3)}(0) = -1.$$

Thus

$$\sin x = x - \frac{x^3}{3!} + \frac{x^5}{5!} + \dots = \sum_{k=0}^{\infty}\frac{(-1)^{2k+1}x^{2k+1}}{(2k+1)!}.$$

It can be shown that this series converges for all values of x.

1.11 Differentiation and Summation

Sometimes we may need to differentiate a function of a variable x which may be represented by a finite or infinite series. Suppose we are interested in differentiating the following function of x

$$f(x) = (a + bx)^n$$

The above function may be represented by a finite series obtained by using the binomial expansion

$$(a + bx)^n = \sum_{k=0}^{n} \binom{n}{k} a^k (bx)^{n-k} = \sum_{k=0}^{n} \binom{n}{k} a^k b^{n-k} x^{n-k}$$

We can verify that the derivative of the function and the derivative of its representation as a finite sum are equal. Differentiating both sides of the above equation we have

$$nb(a+bx)^{n-1} = \frac{d}{dx} \left[\sum_{k=0}^{n} \binom{n}{k} a^k b^{n-k} x^{n-k} \right] = \sum_{k=0}^{n} \binom{n}{k} a^k b^{n-k} \frac{d}{dx} x^{n-k}$$

$$= \sum_{k=0}^{n} (n-k) \frac{n!}{k!(n-k)!} a^k b^{n-k} x^{n-k-1} = n \sum_{k=0}^{n} \frac{(n-1)!}{k!(n-k-1)!} a^k b^{n-k} x^{n-k-1}$$

$$= nb(a+bx)^{n-1}$$

The following example demonstrates this principle for an infinite sum

Example 1.8 Let

$$f(x) = \frac{1}{1 - ax}. \tag{1.14}$$

From power series expansion (section 1.4) we know that for $|ax| < 1$

$$f(x) = \frac{1}{1 - ax} = 1 + ax + a^2 x^2 + a^3 x^3 + ... = \sum_{k=0}^{\infty} a^k x^k. \tag{1.15}$$

We once again demonstrate that the derivative of a function and the derivative of its representation as a sum are equal. Differentiating both sides of (1.14) we obtain

$$\frac{d}{dx} f(x) = \frac{a}{(1 - ax)^2}.$$

Similarly differentiating both sides of (1.15) we obtain

$$\frac{d}{dx} f(x) = \sum_{k=0}^{\infty} \frac{d}{dx} (ax)^k = \sum_{k=0}^{\infty} ka(ax)^{k-1}$$

$$= a[1 + 2ax + 3(ax)^2 + 4(ax)^3 + ...] = \frac{a}{(1-ax)^2}.$$

In general the principle holds if the function of a variable to be differentiated can be expressed as a sum of functions in the same variable and the derivative of each term is finite. That is if

$$f(x) = \sum_{i=1}^{n} f_i(x)$$

then

$$\frac{d}{dx} f(x) = \sum_{i=1}^{n} \frac{d}{dx} f_i(x).$$

We have presented several simple examples of how we can carry out the differentiation inside a summation sign. This result can be generalized to include functions of more than one variable when differentiating with respect to more than one variable as in the example below.

Example 1.9 Let

$$f(x, y, z....w) = \sum_{i=1}^{\infty} f_i(x, y, z, ...w).$$

Then

$$\frac{\partial^k}{\partial x^j \partial w^{k-j}} f(x, y, z....w) = \sum_{i=1}^{\infty} \frac{\partial^k}{\partial x^j \partial w^{k-j}} f_i(x, y, z, ...w),$$

if all derivatives are finite.

1.12 Some Properties of Integration

Unless explicitly stated, throughout this text integrable implies Riemann integrable. Riemann integrable functions have the following properties.

(1) If $f_1(x)$ and $f_2(x)$ are two integrable functions of x on $[a, b]$, and c_1 and c_2 are two finite constants. Then

$$\int_{a}^{b} \left[(c_1 f_1(x) + c_2 f_2(x)) \right] dx = c_1 \int_{a}^{b} f_1(x) dx + c_2 \int_{a}^{b} f_2(x) dx.$$

(2) If $f(x)$ is integrable on $[a, b]$ and $l_1 \le f(x) \le l_2$ for all x in $[a, b]$, then

$$l_1(b-a) \le \int_a^b f(x)dx \le l_2(b-a).$$

(3) If $f_1(x)$ and $f_2(x)$ are integrable on $[a,b]$ and if $f_1(x) \le f_2(x)$ for all x on $[a,b]$, then

$$\int_a^b f_1(x)dx \le \int_a^b f_2(x)dx.$$

(4) If $f(x)$ is integrable on $[a,b]$ and if $a < c < b$, then

$$\int_a^b f(x)dx = \int_a^c f(x)dx + \int_c^b f(x)dx.$$

(5) If $f(x)$ is integrable on $[a,b]$, then so is $|f(x)|$ and $\left| \int_a^b f(x)dx \right| \le \int_a^b |f(x)| dx$.

1.13 Integration by Parts

In this method the integrand is factored into the product of two functions, each a function of the variable that we are trying to integrate over. The concept on which this integration is based can be attributed to two nineteenth century mathematicians G.F. Riemann and T.J. Stieltjes (Khouri 1993) and is derived from the following formula for differentiation of a product

$$\frac{d}{dx}[u(x)v(x)] = u(x)\frac{d}{dx}v(x) + v(x)\frac{d}{dx}u(x)$$

or

$$u(x)\frac{d}{dx}v(x) = \frac{d}{dx}[u(x)v(x)] - v(x)\frac{d}{dx}u(x).$$

Integrating both sides we obtain

$$\int u(x)\frac{d}{dx}v(x) = \int \frac{d}{dx}[u(x)v(x)\} - \int v(x)\frac{d}{dx}u(x)$$

$$= u(x)v(x) - \int v(x)\frac{d}{dx}u(x).$$

This method makes use of the fact that $\int \frac{d}{dx}v(x) = v(x)$.

Integration by parts does not always lend itself to all types of integration, but when it does, it considerably simplifies the process of integration.

Example 1.10 Assume that we need to integrate $I = \int x^2 \sin x dx$.

Let

$$u(x) = x^2 \text{ and } \frac{d}{dx}v(x) = \sin x dx.$$

Then, since

$$\int (\sin x)dx = -\cos x$$

we find

$$I = -x^2 \cos x + \int 2x \cos x dx. \qquad (1.16)$$

For the second expression on the right hand side of (1.16) we may write

$$J = 2\int x \cos x dx = 2[x \sin x - \int \sin x dx]$$

then

$$I = -x^2 \cos x + J = -x^2 \cos x + 2x \sin x + 2 \cos x.$$

Sometimes the integration may be performed iteratively using this method as in the example below. ∎

Example 1.11 Define $I_n = \int x^n e^x dx \quad (n = 1, 2, ...)$.

Then integration by parts results in the following iterative relationship

$$I_n = e^x x^n - n\int x^{n-1} e^x dx = e^x x^n - nI_{n-1}. \qquad (1.17)$$

Integrating I_{n-1} by parts and using the relationship (1.17) above we have

$$I_n = x^n e^x - n[e^x x^{n-1} - (n-1)\int x^{n-2} e^x dx]$$
$$= x^n e^x - ne^x x^{n-1} + n(n-1)I_{n-2}.$$

If we continue to integrate we will have the following relationship

$$I_n = e^x[x^n - nx^{n-1} + n(n-1)x^{n-2} - .... + (-1)^n n!].$$

∎

1.14 Region of Feasibility

A *region of feasibility* is that geometric construct that exactly defines the function whose integral we wish to evaluate. In this section we will provide a graphical representation of a system of constraints in two variables. We will then develop a region where a set of relationships between two variables hold simultaneously. This region is termed the *feasible region*, a term borrowed from *linear program-*

ming (Dantzig G.B. 1963) where a linear function of variables is optimized subject to a set of constraints called inequalities. Feasible regions are of interest to us since events whose probability we need to find will be identifiable by means of feasible regions (Taha 1971). The concept is illustrated using a simple example.

Example 1.12 Obtain the feasible region where the following set of constraints hold simultaneously.
$x + y \geq 1, 2x + y \leq 2, x \geq 0$

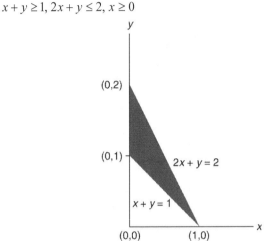

Figure 1.1 Feasible region for the constraints in Example 1.12

First the conditions $x, y \geq 0$ imply that the feasibility region must lie in the first quadrant. The shaded region in the above figure satisfies all the three constraints. If we were to integrate a function $f(x, y)$ under these constraints, then the integration has to be performed for values of x and y that lie in the feasible region. Because of the shape of the feasible region, we will divide the feasible region into two parts A (in grey) and B (in black) as seen below in Fig 1.2. The limits of integration for x in the region A can be seen to be from $x = 1 - y$ to $x = 1 - \frac{y}{2}$. In this region y will vary from $y = 0$ to $y = 1$. Similarly, in region B, x varies from $x = 0$ to $x = 1 - \frac{y}{2}$ and $y = 1$ to $y = 2$. Hence the integration of $f(x, y)$ under the given constraints is performed as follows

$$\iint_{\Omega} f(x,y)dxdy = \int_{0}^{1}\int_{1-y}^{1-\frac{y}{2}} f(x,y)dxdy + \int_{1}^{2}\int_{0}^{1-\frac{y}{2}} f(x,y)dxdy.$$

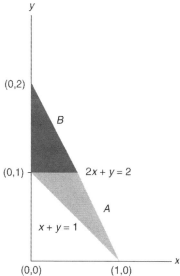

Figure 1.2 Feasible region for the constraints in Example 1.12 when divided into two parts.

Example 1.13 Let us graph the feasible region corresponding to the following constraint

$$|x - y| < 1$$

In the absence of any other constraint on x and y the above equation yields the following relationship

$$-1 < x - y < 1. \tag{1.18}$$

The left inequality in (1.18) yields

$$y < x + 1$$

and the right inequality yields

$$y > x - 1.$$

Thus the solution set consists of the points between the straight lines $y = x + 1$ and $y = x - 1$. The feasible region is the shaded area in Figure 1.3

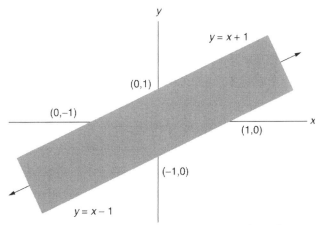

Figure 1.3 Feasible region corresponding to the constraint $|x - y| < 1$

1.15 Multiple Integration

Sometimes a function of more than one variable needs to be integrated for all feasible values of all the variables. For example, $f(x,y)$, a function of both x and y may be feasible only in certain regions as illustrated by the example below.

Example 1.14 Let $f(x,y) = e^{-(x+y)}$, $x + y \geq 1$, and $0 < x < y < \infty$.

Note that in order to integrate $f(x,y)$ we need to identify the feasible region where the integration of $f(x,y)$ can be carried out. From Figure 1.4 it can be easily seen that the feasible region consists of two parts (black and grey)

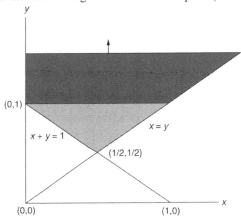

Figure 1.4 Feasible region for Example 1.14

For the first part (gray area) the integration $f(x, y)$ will be carried out for

$$x = 1 - y \text{ to } x = y \text{ and for } y = \frac{1}{2} \text{ to } y = 1.$$

and in the black region, from $x = 0$ to y, and $y = 1$ to ∞. Hence the integration of $f(x, y)$ will result in

$$\int_{\frac{1}{2}}^{1} \int_{1-y}^{y} e^{-(x+y)} dx dy + \int_{1}^{\infty} \int_{0}^{y} e^{-(x+y)} dx dy$$

$$= \int_{\frac{1}{2}}^{1} e^{-y} \left[-e^{-x} \right]_{x=1-y}^{y} dy + \int_{1}^{\infty} e^{-y} \left[-e^{-x} \right]_{x=0}^{y} dy.$$

(1.19)

After a little simplification of the right hand sides of (1.19) we obtain

$$\int_{\frac{1}{2}}^{1} e^{-y} \left[-e^{-x} \right]_{x=1-y}^{y} + \int_{1}^{\infty} e^{-y} dy \left[-e^{-x} \right]_{x=0}^{y}$$

$$= \int_{\frac{1}{2}}^{1} \left(e^{-1} - e^{-2y} \right) + \int_{1}^{\infty} \left(e^{-y} - e^{-2y} \right)$$

$$= \left[y e^{-1} + \frac{1}{2} e^{-2y} \right]_{y=\frac{1}{2}}^{1} + \left[-e^{-y} + \frac{1}{2} e^{-2y} \right]_{y=1}^{\infty} = e^{-1}$$

An alternative expression for integration of $f(x, y)$ over the feasible region would be

$$I = \int_{0}^{\frac{1}{2}} \int_{1-x}^{1} e^{-(x+y)} dy dx + \int_{\frac{1}{2}}^{1} \int_{x}^{1} e^{-(x+y)} dy dx + \int_{1}^{\infty} \int_{0}^{y} e^{-(x+y)} dx dy.$$

Integrating and summing the double integrals on the RHS above shows that $I = e^{-1}$.

∎

This notion is based on the theorem developed by the Italian mathematician Guido Fubini (1879-1943) in 1910 and is known as Fubini's Theorem. Most double integrations lend themselves to this kind of interchange. In Chapter 5 we will frequently use this tool.

Example 1.15 Let us integrate $f(x, y) = x + y$ subject to the constraints

$$x^2 + y^2 \leq 1, \ x > 0, y > 0 .$$

The feasible region is then the first quadrant of the circle below.

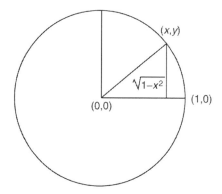

Figure 1.5 Feasible region for Example 1.15

In this case, $\iint_{x\,y} f(x,y)dxdy = \int_0^1 \int_0^{\sqrt{1-x^2}} (x+y)dydx = \int_0^1 \left[xy + \frac{y^2}{2} \right]_0^{\sqrt{1-x^2}} dx = \frac{2}{3}.$

We could change the order of integration as follows

$$\iint_{y\,x} f(x,y)dxdy = \int_0^1 \int_0^{\sqrt{1-y^2}} (x+y)dxdy = \int_0^1 \left[xy + \frac{x^2}{2} \right]_{x=0}^{\sqrt{1-y^2}} dy = \frac{2}{3}.$$

1.15.1 *Changing Order of Summation in the Discrete Case*

Suppose we want to sum a function $F(x,y)$ of x and y in such a way that $x \geq 1, y \geq 2$ and $x + y \leq r$, x, y integers Then the sum may be represented as

$$S = \sum_{x=1}^{r-2} \sum_{y=2}^{r-x} F(x,y)$$

The feasible region for this case would be the pairs of (x,y) integers in the shaded area in Figure 1.6 below.

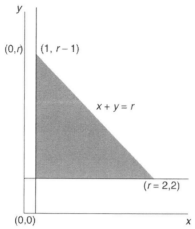

Figure 1.6 Feasible region for the constraints $x \geq 1$, $y \geq 2$ and $x + y \leq r$.

S above can as well be written in a slightly different form as $\displaystyle\sum_{y=2}^{r-1}\sum_{x=1}^{r-y} F(x,y)$.

This is possible if we look at the problem from the perspective of y instead of x as we did in the continuous case.

Example 1.16 To illustrate possibility of change of the order of summation where the variables take integer values,
let

$$S = \sum_{l=1}^{\infty}\sum_{m=l}^{\infty} f(l)p(m-l).$$

The double summation may be written as

$$S = \sum_{l=1}^{\infty}\sum_{r=0}^{\infty} f(l)p(r).$$

Similarly, $\displaystyle\sum_{l=1}^{\infty}\sum_{k=l}^{\infty} f(l)p(k) = \sum_{k=1}^{\infty}\sum_{l=1}^{k} f(l)p(k).$

The above result can be proved by summing over all integer values of k and l that satisfy the limits of summation

1.16 Differentiation under the Integration Sign (Leibnitz's Rule)

We will just give the formula here and refer the reader to a book on calculus for details.

Let $H(x) = \int\limits_{a(x)}^{b(x)} h(x, y)dy.$

Where $a(x)$ and $b(x)$ are functions of x and $h(x, y)$ is a differentiable function of x, then,

$$\frac{d}{dx}H(x) = \int\limits_{a(x)}^{b(x)} \frac{d}{dx}h(x, y)dy + \left(\frac{d}{dx}b(x)\right)h\{x, b(x)\} - \left(\frac{d}{dx}a(x)\right)h\{x, a(x)\}. \quad (1.20)$$

The above rule of integration is called Leibnitz's Rule (Courant 1955).

Example 1.17 Define

$$H(x) = \int\limits_0^\infty e^{-xy}dy = \frac{1}{x}\left[-e^{-xy}\right]_0^\infty = \frac{1}{x}$$

Also

$$\frac{d}{dx}H(x) = \int\limits_0^\infty \frac{d}{dx}(e^{-xy})dy = \int\limits_0^\infty -ye^{-xy}dy = -\frac{1}{x^2},$$

since $\frac{d}{dx}b(x) = 0$ and $\frac{d}{dx}a(x) = 0.$

Example 1.18 Let

$$H(x) = \int\limits_x^{2x}(x^2y + y^2)dy = 2x^4 - \frac{1}{2}x^4 + \frac{8}{3}x^3 - \frac{1}{3}x^3$$

We will then have

$$\frac{d}{dx}H(x) = \int\limits_x^{2x} 2xydy + 2[x^2(2x) + 4x^2] - [x^3 + x^2]$$

$$= 8x^3 - 2x^3 + 7x^2 = 6x^3 + 7x^2.$$

1.17 Jacobian

The concept of Jacobian is often used when a set of variables $x_1, \ldots, x_n$ is transformed into a new set of variables $y_1, \ldots, y_n$ in a one-to-one mapping such that

$$x_1 = f_1(y_1, \ldots, y_n)$$
$$x_2 = f_2(y_1, \ldots, y_n)$$
$$\vdots$$
$$x_n = f_n(y_1, \ldots, y_n).$$

Assume that the partial derivatives of x_i ($i = 1, 2, \ldots n$) with respect to y_j ($j = 1, 2, \ldots, n$) exist. Then the n by n determinant J whose $(i, j)^{th}$ element is $\dfrac{\partial x_i}{\partial y_j}$ $i, j = 1, 2, \ldots, n$ is called the *Jacobian* of x $x_1, \ldots, x_n$ with respect to $y_1, \ldots, y_n$.
and is represented (Swokowski 1979) by

$$J = \frac{\partial(x_1, \ldots, x_n)}{\partial(y_1, \ldots, y_n)}.$$

Alternatively,

$$J = \begin{vmatrix} \dfrac{\partial x_1}{\partial y_1} & \dfrac{\partial x_1}{\partial y_2} & \cdots & \dfrac{\partial x_1}{\partial y_n} \\ \dfrac{\partial x_2}{\partial y_1} & \dfrac{\partial x_2}{\partial y_2} & \cdots & \dfrac{\partial x_2}{\partial y_n} \\ \cdots & \cdots & \cdots & \cdots \\ \dfrac{\partial x_n}{\partial y_1} & \dfrac{\partial x_n}{\partial y_2} & \cdots & \dfrac{\partial x_n}{\partial y_n} \end{vmatrix} \qquad (1.21)$$

Furthermore under suitable restrictions

$$\int_A h(x_1, \ldots, x_n) dx_1 \ldots dx_n = \int_B h(f_1(y_1, \ldots y_n), \ldots, f_n(y_1, \ldots y_n)) |J| dy_1 \ldots dy_n$$

where the f_i's are functions of the y's and B is a proper transform of the region A into the new system. We will be using the change of variable concept in Chapters 3 and 5 when dealing with variable transformations. ∎

Example 1.19 Let the two variables y and z be transformed into r and θ as follows
$$y = r \sin \theta \text{ and } z = r \cos \theta$$
then the Jacobian for this transformation is defined as

$$J = \begin{vmatrix} \dfrac{dy}{dr} & \dfrac{dy}{d\theta} \\ \dfrac{dz}{dr} & \dfrac{dz}{d\theta} \end{vmatrix} = \begin{vmatrix} \sin\theta & r\cos\theta \\ \cos\theta & -r\sin\theta \end{vmatrix} = -r$$

The absolute value of the determinant J is r.
This result can be used to evaluate the integral

$$\int_{-\infty}^{\infty}\int_{-\infty}^{\infty} e^{-\left(y^2+z^2\right)/2}\, dy dz = \int_{0}^{2\pi}\int_{0}^{\infty} e^{-r^2/2}\, r dr d\theta = \int_{0}^{2\pi} d\theta = 2\pi \;.$$

1.18 Maxima and Minima

Several applications require maximizing or minimizing a function of some variable x. If we can find a point x_M such that $g(x_M) \geq g(x)$ for all feasible values of x then x_M is said to be the "global maximum" value of x. Similarly if x_m is such that $g(x_m) \leq g(x)$ for all   feasible values of x we say that $g(x_m)$ has a "global minimum" at x_m.

In practice, we may be interested in a point x_0 that maximizes $g(x)$ in an interval around x_0. In this case, x_0 is called a "local maximum", "local minimum" is defined similarly.

In order to obtain the local maximum or local minimum value of a continuous function $g(x)$ the *Principle of Maxima and Minima* is used. Basically the principle requires the first derivative of $g(x)$ be set equal to zero and the resulting equation be solved for x. Let x' be the solution of the equation

$$\frac{d}{dx} g(x) = 0.$$

We say that $g(x)$ is a maximum at x' if $\dfrac{d^2}{dx^2} g(x)\big|_{x=x'} > 0$, and if

$\dfrac{d^2}{dx^2} g(x)\big|_{x=x'} < 0$ $g(x)$ is a minimum at x'. For $g(x,y)$ a function of two variables x and y, the first step is to set the partial derivatives with respect to x and y equal to zero and then solve for x and y. If (x', y') is the solution to the following set of equations

$$\frac{\partial}{\partial x} g(x, y) = 0$$

$$\frac{\partial}{\partial y} g(x, y) = 0$$

and if for a particular point (x', y') the determinant

$$\Delta^* = \begin{vmatrix} \dfrac{\partial^2}{\partial x^2} g(x, y) & \dfrac{\partial^2}{\partial x \partial y} g(x, y) \\[4mm] \dfrac{\partial^2}{\partial x \partial y} g(x, y) & \dfrac{\partial^2}{\partial y^2} g(x, y) \end{vmatrix}_{\substack{x=x' \\ y=y'}} > 0 \qquad (1.22)$$

then, (x', y') is a maximum if the following two conditions hold simultaneously.

$$\frac{\partial^2}{\partial x^2} g(x, y)\bigg|_{\substack{x=x' \\ y=y'}} < 0 \text{ and } \frac{\partial^2}{\partial y^2} g(x, y)\bigg|_{\substack{x=x' \\ y=y'}} < 0$$

and minimum if

$$\frac{\partial^2}{\partial x^2} g(x, y)\bigg|_{\substack{x=x' \\ y=y'}} > 0 \text{ and } \frac{\partial^2}{\partial y^2} g(x, y)\bigg|_{\substack{x=x' \\ y=y'}} > 0 .$$

If the determinant Δ^* is less than zero neither maximum nor minimum at (x', y') exists. If $\Delta^* = 0$ the test fails.

Example 1.20 Let

$$g(x, y) = x^2 + xy + y^2 - 3x + 2$$

Then by setting

$$\frac{\partial}{\partial x} g(x, y) = 2x + y - 3 = 0$$

$$\frac{\partial}{\partial y} g(x, y) = x + 2y = 0$$

we obtain $x' = 2$ and $y' = -1$.

Note that for these values of x and y the determinant (1.22) is greater than zero as well as $\frac{\partial^2}{\partial x^2} g(x, y) > 0$ and $\frac{\partial^2}{\partial y^2} g(x, y) > 0$. Hence, $x' = 2$ and $y' = -1$

provides us with a minimum value of -1 for $g(x, y)$. Notice that in this case the local minima is also the global minima.

■

1.19 Lagrange Multipliers

The method of Lagrange multipliers is used to obtain the maxima and minima of functions of several variables subject to a set of equality constraints (Draper and Klingman 1967). The constraints may or may not be linear in the variables. In general we may need to obtain the maximum or minimum value of a function $f(x_1, x_2, ...x_k)$ of k variables subject to m equality constraints :

$$g_j(x_1, x_2, ..., x_k) = 0, \quad j = 1, 2, ..., m.$$

We proceed by creating a function F as follows

$$F = F(x_1, x_2, ...x_k, \lambda_1, \lambda_2, ...\lambda_m) = f(x_1, x_2, ...x_k) - \sum_{j=1}^{m} \lambda_j g_j(x_1, x_2, ...x_k)$$

where the quantities λ_i, $i = 1, 2, ...m$ are called Lagrange multipliers and are unknown and determined by the solution to the following two equations

$$\frac{d}{dx_i} F(x_1, x_2, ..., x_k, \lambda_1, \lambda_2, ..., \lambda_m) = 0, \quad i = 1, 2, ...k$$

$$\frac{d}{d\lambda_j} F(x_1, x_2, ..., x_k, \lambda_1, \lambda_2, ..., \lambda_m) = 0 \quad j = 1, 2, ...m.$$

Note that

$$\frac{\partial}{\partial \lambda_i} F = 0 = g_i(x_1, ..., x_k) \text{ for } i = 1, 2, ...k$$

are the constraints themselves, so F needs to be differentiated partially with respect to the x's only. Solution of the form $x_i = x_i', i = 1, 2, ...k$ from the above $m+k$ equations provides values of x_i which may be tested for maxima or minima if the following conditions are satisfied

$$\frac{d}{dx_i} F = 0 \text{ at } x_i = x_i' \text{ for } i = 1, 2, ...k$$

and for which Δ^* the determinant of the second order derivatives is defined as

$$\Delta* = \begin{vmatrix} \dfrac{\partial^2 F}{\partial x_1^2} & \dfrac{\partial^2 F}{\partial x_1 \partial x_2} & \cdots & \dfrac{\partial^2 F}{\partial x_1 \partial x_k} \\ \dfrac{\partial^2 F}{\partial x_1 \partial x_2} & \dfrac{\partial^2 F}{\partial x_2^2} & \cdots & \dfrac{\partial^2 F}{\partial x_2 \partial x_k} \\ \cdots & \cdots & \cdots & \cdots \\ \dfrac{\partial^2 F}{\partial x_1 \partial x_k} & & \cdots & \dfrac{\partial^2 F}{\partial x_k^2} \end{vmatrix}_{\{x_i\}=\{x_i'\}}$$

If $\Delta^* > 0$ then, $f(x_1, \ldots, x_n)$ is a maximum at $\{x_i\} = \{x_i'\}$ if $\dfrac{\partial^2 F}{\partial x_i^2}\bigg|_{\{x_i = x_i'\}} < 0$

and a minimum at $\{x_i\} = \{x_i'\}$ if $\dfrac{\partial^2}{\partial x_i^2} F\bigg|_{x=x_i} > 0$

The test for maxima and minima fails if $\left|\Delta^*\right| \leq 0$ and the function needs to be investigated near $\{x_i\} = \{x_i'\}$. If the accompanying constraints are not equality constraints but instead are inequalities, the conditions necessary for a maximum or minimum are known as the Kuhn-Tucker conditions (Kuhn and Tucker 1956).

Example 1.21 Suppose we are to determine the maximum and or minimum values of x_1 and x_2 in the function

$$f(x_1, x_2) = 3x_1^2 + x_2^2 - x_1 x_2$$

subject to the constraint

$$2x_1 + x_2 = 21.$$

Define $\qquad F(x_1, x_2, \lambda) = 3x_1^2 + x_2^2 - x_1 x_2 - \lambda(2x_1 + x_2 - 21)$

The following relationships are obtained by differentiation

$$\frac{\partial F}{\partial x_1} = 6x_1 - x_2 - 2\lambda = 0$$

$$\frac{\partial F}{\partial x_2} = 2x_2 - x_1 - \lambda = 0 \tag{1.23}$$

$$\frac{\partial F}{\partial \lambda} = 2x_1 + x_2 - 21 = 0$$

The corresponding second order derivatives are

$$\frac{\partial^2 F}{\partial x_1^{\,2}} = 6$$

$$\frac{\partial^2 F}{\partial x_2} = 2$$

Solving for x_1 and x_2 in (1.23) we obtain $x_1 = 35/6$, $x_2 = 56/6$, $\lambda = 77/6$
For these values of x_1, x_2 and λ it may be seen that $\Delta^* > 0$ therefore a minimum
value of $f(x_1, x_2)$ is obtained at $x_1 = 35/6$, $x_2 = 56/6$.

■

1.20 L'Hôspital's Rule

L'Hôspital's Rule is applicable in the evaluation of indeterminate forms. That is
forms of the type $\dfrac{0}{0}$ or $\dfrac{\infty}{\infty}$.

If $f_1(x)$ and $f_2(x)$ are two continuously differentiable functions of x in an interval
(a,b) that contains c such that

$$\lim_{x \to c} f_1(x) = 0 = \lim_{x \to c} f_2(x)$$

or

$$\lim_{x \to c} f_1(x) = \infty = \lim_{x \to c} f_2(x).$$

Then

$$\lim_{x \to c} \frac{f_1(x)}{f_2(x)} = \lim_{x \to c} \left[\frac{\dfrac{df_1(x)}{dx}}{\dfrac{df_2(x)}{dx}} \right]$$

provided   that

$$\frac{df_1(x)}{dx} \text{ and } \frac{df_2(x)}{dx} \text{ exist.}$$

L'Hôspital's rule may be applied successively to the ratio of the two functions until the first determinate form is arrived at. In other words if

$$\lim_{x \to c} \left[\frac{\dfrac{df_1(x)}{dx}}{\dfrac{df_2(x)}{dx}} \right]$$

is again of the indeterminate form, obtain

$$\lim_{x \to c} \left[\frac{\dfrac{d^2 f_1(x)}{dx^2}}{\dfrac{d^2 f_2(x)}{dx^2}} \right]$$

Continue differentiating the numerator and the denominator until a finite limiting value for the ratio is obtained.

Example 1.22 Let $f(x) = \dfrac{e^x - \log(x+1) - 1}{x^2}$

we need to determine $\lim_{x \to 0} f(x)$. Since $\lim_{x \to 0} f(x)$ is of the $\dfrac{0}{0}$ form we differentiate the numerator and the denominator to obtain

$$\lim_{x \to 0} \frac{\dfrac{d}{dx}[e^x - \log(x+1) - 1]}{\dfrac{d}{dx}(x^2)} = \lim_{x \to 0} \frac{e^x - \dfrac{1}{x+1}}{2x} \ .$$

This is again of the $\dfrac{0}{0}$ form. Hence we differentiate one more time to obtain

$$\lim_{x \to 0} \left[\frac{\dfrac{d}{dx}\left(e^x - \dfrac{1}{x+1}\right)}{\dfrac{d}{dx}(2x)} \right] = 1$$

Example 1.23 Let $f(x) = \dfrac{\log x}{x-1}$

and we need to determine

$$\lim_{x \to \infty} f(x) = \lim_{x \to \infty} \left[\frac{\log x}{x-1} \right].$$

Clearly the limit is of the $\dfrac{\infty}{\infty}$ form. Differentiating the numerator and the denominator with respect to x and taking the limit we obtain

$$\lim_{x \to \infty} \left[\frac{\dfrac{d}{dx}(\log x)}{\dfrac{d}{dx}(x-1)} \right] = \lim_{x \to \infty} \left[\frac{\dfrac{1}{x}}{1} \right] = 0.$$

This rule may be extended to include forms of the type $\infty \times 0, 0^0, 1^\infty, \infty^0$.

Let

(a) if $\lim\limits_{x \to c} f_1(x) = \infty$ and $\lim\limits_{x \to c} f_2(x) = 0$ then $\lim\limits_{x \to c}\left[f_1(x)f_2(x) \right]$ is of the

$\infty \times 0$ form. Hence

$$\lim_{x \to c}\{f_1(x)f_2(x)\} = \lim_{x \to c} \left[\frac{f_1(x)}{\dfrac{1}{f_2(x)}} \right]$$

which is of the $\dfrac{\infty}{\infty}$ form.

(b) if $\lim\limits_{x \to c} f_1(x) = 0 = \lim\limits_{x \to c} f_2(x)$

then

$$\lim_{x \to c}[f_1(x)]^{f_2(x)}$$

is of the 0^0 form and may be written as

$$\lim_{x \to c} \exp \left[\frac{f_2(x)}{\dfrac{1}{\log f_1(x)}} \right] = \exp \left[\lim_{x \to c} \frac{f_2(x)}{\dfrac{1}{\log f_1(x)}} \right]$$

which is $\dfrac{0}{0}$ form.

(c) If $\lim\limits_{x \to c} f_1(x) = 1$, and $\lim\limits_{x \to c} f_2(x) = \infty$ then

$$\lim_{x \to c}[f_1(x)]^{f_2(x)}$$

is of the 1^∞ form and may be written as

$$\exp\left[\lim_{x \to c} \frac{f_2(x)}{\dfrac{1}{\log f_1(x)}}\right]$$

which is of $\dfrac{\infty}{\infty}$ form

(d) If $\lim_{x \to c} f_1(x) = \infty$, and $\lim_{x \to c} f_2(x) = 0$ then

$$\lim_{x \to c}[f_1(x)]^{f_2(x)}$$

is of the ∞^0 form and may be written as

$$\exp\left[\lim_{x \to c} \frac{f_2(x)}{\dfrac{1}{\log f_1(x)}}\right]$$

which is of $\dfrac{0}{0}$ form

If $\lim_{x \to c} f_1(x) = \infty = \lim_{x \to c} f_2(x)$ then

$$\left[\frac{\dfrac{1}{1}}{f_1(x)} - \frac{\dfrac{1}{1}}{f_2(x)}\right] = \left[\frac{\dfrac{1}{f_2(x)} - \dfrac{1}{f_1(x)}}{\dfrac{1}{f_1(x)f_2(x)}}\right]$$

is of the $\dfrac{0}{0}$ form.

∎

Example 1.24 Evaluate $\lim_{x \to \infty} x(e^{\frac{1}{x}} - 1)$. This is an example where the limit will be of the $\infty \times 0$ form.

Rewriting the expression as

$$\lim_{x \to \infty} \frac{e^{\frac{1}{x}} - 1}{\dfrac{1}{x}}$$

i.e., converting it into $\dfrac{0}{0}$ form. Now differentiating the numerator and the de nominator we have

$$\lim_{x\to\infty}\frac{e^{\frac{1}{x}}-1}{\frac{1}{x}}=\lim_{x\to\infty}\frac{-\frac{1}{x^2}e^{\frac{1}{x}}}{-\frac{1}{x^2}}=\lim_{x\to\infty}e^{\frac{1}{x}}=1.$$

■

Example 1.25 Evaluate $\lim_{x\to 0}x^x$. In this example, the limiting value is of the 0^0 form

Rewriting,

$$\lim_{x\to 0}x^x=\lim_{x\to 0}e^{\log x^x}=\lim_{x\to 0}e^{\left[\frac{\log x}{\frac{1}{x}}\right]}=e^{\left[\lim_{x\to 0}\frac{\log x}{\frac{1}{x}}\right]}.$$

$$\text{Now,}\lim_{x\to 0}\frac{\log x}{\frac{1}{x}}=\lim_{x\to 0}\frac{\frac{1}{x}}{-\frac{1}{x^2}}=\lim_{x\to 0}(-x)=0.$$

Hence,

$$\lim_{x\to 0}x^x=e^0=1.$$

■

1.21 Partial Fraction Expansion

Partial fraction expansions are useful when integrating functions with denominators of degrees greater than one. Sometimes the denominator of a fraction of the form

$$\frac{1}{a_nx^n+a_{n-1}x^{n-1}+...+a_1x+a_0} \qquad (1.24)$$

may be written as

$$\frac{1}{(x-b_1)(x-b_2)....(x-b_n)}$$

where b_i's are the real roots of the denominator of (1.24) and no two b's are the same. In such cases we may be able to write

$$\frac{1}{a_nx^n+a_{n-1}x^{n-1}+...+a_1x+a_0}=\frac{A_1}{(x-b_1)}+\frac{A_2}{(x-b_2)}+...\frac{A_n}{(x-b_n)}$$

where A_i's are determined from the following identity

$$1 \equiv A_i \prod_{j=1, j \neq i}^{n} (b_i - b_j)$$

or

$$A_i = \frac{1}{(b_i - b_1)(b_i - b_2)....(b_i - b_{i-1})(b_i - b_{i+1})...(b_i - b_n)}.$$

If b_i appears k_i ($i = 1, 2, ...m$) times in the denominator such that $\sum_{i=1}^{m} k_i = n$, then

$$\frac{1}{a_n x^n + a_{n-1} x^{n-1} + a_{n-2} x^{n-2} + ...a_0} = \frac{A_{11}}{(x - b_1)} + \frac{A_{12}}{(x - b_1)^2} + ... \frac{A_{1k_1}}{(x - b_1)^{k_1}}$$

$$+ \frac{A_{21}}{(x - b_2)} + \frac{A_{22}}{(x - b_2)^2} + ... \frac{A_{2k_2}}{(x - b_2)^{k_2}}$$

$$+ .... \frac{A_{m1}}{(x - b_m)} + ... \frac{A_{mk_m}}{(x - b_m)^{k_m}}.$$

A_{ij}'s are then determined in a fashion similar to the previous case.

Often times not all roots of the denominator are available as demonstrated in the example below.

Example 1.26 The function $\dfrac{1}{(x^3 - 8)}$ has only one root $x = 2$ in the denominator

Hence this function may be written as

$$\frac{1}{(x^3 - 8)} = \frac{1}{(x - 2)(x^2 + 2x + 4)}$$

The right hand side of the above expression may be written as

$$\frac{A_1}{(x - 2)} + \frac{A_2 + A_3 x}{(x^2 + 2x + 4)}$$

$A_1, A_2,$ and A_3 can be determined from the following identity

$$1 \equiv A_1(x^2 + 2x + 4) + A_2(x - 2) + A_3 x(x - 2)$$

Set $x = 2,$ and obtain $A_1 = \dfrac{1}{12}.$

Set $x = 0,$ and obtain $A_2 = -\dfrac{1}{3}.$

Set $x=1,$ and obtain $A_3 = -\dfrac{1}{12}.$

Partial fraction expansions are useful when integrating functions with denominators of degrees greater than one.

1.22 Cauchy-Schwarz Inequality

Consider a set of real numbers $a_1, a_2, \ldots, a_n$ and another set of real numbers $b_1, b_2, \ldots, b_n$. Then it can be easily shown that

$$\left(\sum_1^n a_i^2 \right)\left(\sum_1^n b_i^2 \right) - \left(\sum_1^n a_i b_i \right)^2 = \sum_{\substack{i=1}}^{n} \sum_{\substack{j=1 \\ i \neq j}}^{n} \left(a_i b_j - a_j b_i \right)^2 \geq 0 \qquad (1.25)$$

Rearranging terms in (1.25) it can be shown that

$$\left(\sum_{i=1}^n a_i b_i \right)^2 \leq \left(\sum_{i=1}^n a_i^2 \right)\left(\sum_{i=1}^n b_i^2 \right) \qquad (1.26)$$

The inequality (1.26) is known as the Cauchy-Schwarz inequality and leads to the establishment of the following inequality.

$$\sqrt{\sum_{i=1}^n (a_i + b_i)^2} \leq \sqrt{\sum_{i=1}^n a_i^2} + \sqrt{\sum_1^n b_i^2}.$$

The above inequality is named after Herman Minkowski (1864-1909) and is known as the Minkowski's inequality.

The counterpart of the Cauchy-Schwarz inequality in integral calculus is as follows:

If $u(x)$ and $v(x)$ are such that $u^2(x)$ and $v^2(x)$ are Riemann integrable on $[a, b]$, then

$$\left[\int_a^b u(x)v(x)dx \right]^2 \leq \left[\int_a^b u^2(x)dx \right]\left[\int_a^b v^2(x)dx \right].$$

The limits of integration may be finite or infinite

The counterpart for Minkowski's inequality assumes that $|u(x)|^p$ and $|v(x)|^p$ are Riemann integrable on $[a,b]$ and for $1 \le p < \infty$ and is

$$\left[\int_a^b |u(x) + v(x)|^p \, dx\right]^{1/p} \le \left[\int_a^b |u(x)|^p \, dx\right]^{1/p} + \left[\int_a^b |v(x)|^p \, dx\right]^{1/p}$$

1.23 Generating Functions

If the power series

$$B(s) = b_0 + b_1 s + b_2 s^2 + \dots = \sum_{i=0}^{\infty} b_i s^i$$

converges in some interval $a < s < b$ then $B(s)$ is called a generating function of the sequence $\{b_i\}$. Generating functions were introduced by Laplace in 1812 (Widder, 1946) and the variable s is a working variable and does not appear in the solution hence has no significance as such. For $b_i = 1$, $i = 0, 1, \dots$

$$B(s) = 1 + s + s^2 + \dots s^n + \dots = \frac{1}{1 - s}$$

The above generating function is a geometric series. Given a generating function , the series associated with it can be obtained by inversion. That is, the coefficient of s^n will provide us with b_n. In general consider the case where if the generating function is of the form

$$G(s) = \frac{A(s)}{C(s)}.$$

$A(s)$ and $C(s)$ are polynomials of order m and n $(m < n)$ respectively with no common roots. Furthermore assume that all the roots $s_1, s_2, \dots s_n$ of $C(s)$ are distinct and $s_1 = \text{Minimum}(s_1, s_2, \dots s_n)$. It can then be shown (Feller 1957 and Chiang 1979) that the coefficients g_k in the series $G(s) = g_0 + g_1 s + g_2 s^2 + \dots$ may be obtained as

$$g_k \sim \left[\frac{-A(s_1)}{C'(s_1)}\right] \frac{1}{s_1^{k+1}}.$$

here $C'(s)$ is the first derivative of $C(s)$ at $s = s_1$. The requirement of all distinct roots in the denominator may be relaxed so long as s_1 is a simple root.

Generating functions are an important tool used in applied probability and mathematical statistics (Moye and Kapadia 2000) and will be discussed in Chapter 3. They are also useful in inverting *difference equations* and determining their exact solution.

Example 1.28 Let

$$b_k = \frac{1}{3}\left(\frac{3}{7}\right)^{k-1}, \, k = 1, 2, \ldots L.$$

Then the generating function associated with the b's is

$$B(s) = s\sum_{k=1}^{L}\frac{1}{3}\left(\frac{3s}{7}\right)^{k-1} = \frac{s}{3}\left(\frac{1-\left(\frac{3s}{7}\right)^{L}}{1-\frac{3s}{7}}\right) = \frac{7s}{3(7-3s)}\left(1-\left(\frac{3s}{7}\right)^{L}\right).$$

Example 1.29 Let

$$B(s) = \frac{1}{(6s^2 - 5s + 1)}.$$

Then b_k the coefficient of s^k may be found by inverting $B(s)$ as follows

$$B(s) = \frac{1}{(1-3s)(1-2s)} = \frac{A}{(1-3s)} + \frac{B}{(1-2s)}.$$

Using the partial fraction expansion technique we obtain $A = 3$ and $B = -2$. Hence

$$B(s) = \frac{3}{(1-3s)} - \frac{2}{(1-2s)}.$$

The coefficient of s^k is then obtained by inverting $B(s)$ and we have

$$b_k = \left(3^{k+1} - 2^{k+1}\right).$$

1.24 Difference Equations

The most general form of a difference equation is defined as

$$p_0(k)y_{k+n} + p_1(k)y_{k+n-1} + \ldots + p_n(k)y_k = R(k) \tag{1.27}$$

It consists of terms involving members of the sequence $\{y_k\}$, and coefficients $p_i(k)$ of the elements of the sequence $\{y_k\}$ in the equation. These coefficients may or may not be a function of k. When the coefficients are not functions of k, the difference equation has constant coefficients. Difference equations with coefficients that are functions of k are described as difference equations with variable coefficients. In general, difference equations with constant coefficients are easier to solve than those difference equations that have variable coefficients.

If the term $R(k)$ on the right side of (1.27) is equal to zero, then the difference equation is *homogeneous*. If the right side of the equation is not zero, then the equation becomes a *non homogeneous* difference equation. For example, the following difference equation

$$y_{k+2} = 6y_{k+1} - 3y_k, \ k = 0,1,2,... \tag{1.28}$$

is homogeneous.
The equation

$$y_{k+2} = 6y_{k+1} - 3y_k + 12, \ k = 0,1,2,... \tag{1.29}$$

is non homogeneous because of the inclusion of the term 12.

Finally, the order of a difference equation is obtained as the difference between the largest subscript of y ($k+n$ in this case) and the smallest subscript of y, that is k. Thus the order of this difference equation is $k + n - k = n$, and the equation is characterized as an nth order difference equation. If $R(k) = 0$, we describe equation as an nth order homogenous difference equation, and if $R(k) \neq 0$, the equation is an nth order non homogeneous difference equation. As an example, the family of difference equations described by

$$y_{k+2} = 6y_{k+1} - 3y_k + 12, \ k = 0,1,2,... \tag{1.30}$$

where y_0 and y_1 are known constants is a second-order non-homogeneous difference equation with constant coefficients, while

$$3y_{k+4} + (k+3)y_{k+3} + 2^k y_{k+2} + (k+1)y_{k+1} + 4ky_k = (k+1)(k+2)$$
$$\tag{1.31}$$

would be designated as a fourth-order, non homogeneous difference equation with variable coefficients.

Difference equations may be solved iteratively as in the following case

$$y_{k+1} = ay_k, \ k = 0,1,2,...$$

for all k > 0; if y_0 is known, the solution to the above equation is

$$y_k = a^k y_0, \ k = 0,1,2,...$$

Next consider

$$y_{k+1} = y_k + b, \ k = 0,1,2,...$$

Then

$$
\begin{aligned}
y_1 &= ay_0 + b \\
y_2 &= ay_1 + b = a(ay_0 + b) + b = a^2 y_0 + ab + b \\
y_3 &= ay_2 + b = a(a^2 y_0 + ab + b) + b = a^3 y_0 + a^2 b + ab + b \\
y_4 &= ay_3 + b = a^4 b + a^3 b + a^2 b + ab + b
\end{aligned}
\tag{1.32}
$$

and so on. Unfortunately, this iterative approach to the solution which is very useful for first-order difference equations becomes somewhat complicated if the order is increased. The easiest method for solving difference equations is by using the generating functions approach.

Example 1.30 Consider the following first order difference equation

$$ay_{k+1} = by_k + c, \ k = 0,1,2,... \tag{1.33}$$

assume that y_0 is a known constant. Multiply each side of equation (1.33) by s^k.

$$as^k y_{k+1} = bs^k y_k + cs^k \tag{1.34}$$

summing over all values of k and defining

$$G(s) = \sum_{k=0}^{\infty} s^k y_k \tag{1.35}$$

equation (1.34) in terms of $G(s)$ may be written as

$$as^{-1}[G(s) - y_0] = bG(s) + \frac{c}{1-s}$$

This above equation can be easily solved for $G(s)$ as follows

$$a[G(s) - y_0] = bsG(s) + \frac{cs}{1-s}$$

$$G(s)[a - bs] = \frac{cs}{1-s} + ay_0$$

leading to

$$G(s) = \frac{cs}{(1-s)(a-bs)} + \frac{ay_0}{(a-bs)}.$$

Now, y_k is the coefficient of s^k in the expression for $G(s)$. The inversion is straightforward. After inverting each term, we use the summation principle to complete the inversion of $G(s)$. Each of these two terms requires focus on the expression $\frac{1}{(a-bs)}$. Its inversion for $a > 0$ is simply

$$\frac{1}{a-bs} = \frac{1}{a\left[1 - \frac{b}{a}s\right]} = \frac{1}{a}\sum_{i=0}^{\infty}\left(\frac{b}{a}\right)^i s^i$$

Attending to the second term on the right of $G(s)$

$$\frac{ay_0}{a-bs} = y_0 \sum_{i=0}^{\infty}\left(\frac{b}{a}\right)^i s^i$$

Similarly,

$$\frac{cs}{(1-s)(a-bs)} = \frac{cs}{a}\sum_{j=0}^{\infty}\sum_{i=0}^{\infty}\left(\frac{b}{a}\right)^i s^{i+j}$$

The final solution can now be written as

$$y_k = \frac{c}{a}\sum_{i=1}^{k}\left(\frac{b}{a}\right)^{k-i} + y_0\left(\frac{b}{a}\right)^k, \ k = 1, 2, 3, \ldots$$

which is the solution to the first-order, non homogenous difference equation with constant coefficients. For solution to difference equations with variable equations refer to Moye and Kapadia (2000).

1.25 Vectors, Matrices and Determinants

Systolic blood pressure levels for an individual at n different points in time may be represented by a row or a column of observations represented in the form $(x_1, x_2, \ldots, x_n)$. Such an ordered set of observations may be represented by $\mathbf{x}$ called a row (or a column) vector. If $\mathbf{y}$ represents the systolic blood pressure scores $(y_1, y_2, \ldots, y_n)$ on a second individual, then the vector of the sum of their scores may be obtained by adding the two vectors and creating a third vector $\mathbf{z}$ whose elements are the sum of corresponding elements of $\mathbf{x}$ and $\mathbf{y}$ with elements $(x_1 + y_1, x_2 + y_2, \ldots, x_n + y_n)$ and we write $\mathbf{z} = \mathbf{x} + \mathbf{y}$.

The arithmetic operations of addition, subtraction and scalar multiplication are also defined for vectors.

1.25.1 Vector Addition

$$\mathbf{x} + \mathbf{y} = \mathbf{y} + \mathbf{x}$$
$$\mathbf{x} + (\mathbf{y} + \mathbf{z}) = (\mathbf{x} + \mathbf{y}) + \mathbf{z}.$$

In this book we will refer to a column vector as a vector and denote the row vector as the *transpose* of a column vector (see section 1.25.6 on the transpose of matrices).

1.25.2 Vector Subtraction

$$\mathbf{x} - \mathbf{y} = -(\mathbf{y} - \mathbf{x})$$
$$\mathbf{x} + (\mathbf{y} - \mathbf{z}) = (\mathbf{x} + \mathbf{y}) - \mathbf{z}.$$

1.25.3 Vector Multiplication

The multiplication of a vector $\mathbf{x}$ by a scalar quantity c leads to a vector represented by a vector $c\mathbf{x}$ where each element of $\mathbf{x}$ is multiplied by c. Also, multiplication of two vectors $\mathbf{x}$ and $\mathbf{y}$ is possible if the number of elements in both $\mathbf{x}$ and $\mathbf{y}$ is the same and one is a row vector and the other is a column vector.

1.25.4 Matrices

A collection of m row vectors each with n elements or a collection n column vectors each with m elements forms what is know as a matrix M of order $m \times n$.

1.25.5 Matrix Addition

Addition of two matrices is possible if both are of the same order. Hence if there are two $m \times n$ matrices M_1 and M_2 of the form

$$M_1 = \begin{pmatrix} a_{11} & \cdots & a_{1n} \\ \vdots & \ddots & \vdots \\ a_{m1} & \cdots & a_{mn} \end{pmatrix}, M_2 = \begin{pmatrix} b_{11} & \cdots & b_{1n} \\ \vdots & \ddots & \vdots \\ b_{m1} & \cdots & b_{mn} \end{pmatrix}$$

then their sum $M = M_1 + M_2$ is

$$M = \begin{pmatrix} a_{11} + b_{11} & \cdots & a_{1n} + b_{1n} \\ \vdots & \ddots & \vdots \\ a_{m1} + b_{m1} & \cdots & a_{mn} + b_{mn} \end{pmatrix}$$

Hence the sum of two matrices of order $m \times n$ is a matrix of order $m \times n$ the $(i, j)^{th}$ element in which is obtained by adding the $(i, j)^{th}$ element in each matrix. Subtraction of matrices is based on the same principle as addition except the $(i, j)^{th}$ element in the matrix M is obtained by element by element subtraction. For example if the matrix M_2 is subtracted from matrix M_1, then the $(i, j)^{th}$ element in the matrix $M_1 - M_2$ is $a_{ij} - b_{ij}$.

1.25.6 Product of Matrices

If two matrices A and B are to be multiplied, then their order has to be such that the number of rows in one must equal the number of columns in the other. For example if we want to determine the product AB then if A is a $m \times n$ matrix, then B must be a $n \times k$ ($k > 0$) matrix and the ij^{th} element in AB will be the product of the ith row in A by the jth column in B. For example

$$AB = \begin{pmatrix} a_{11} & a_{12} & a_{13} \\ a_{21} & a_{22} & a_{23} \\ a_{31} & a_{32} & a_{33} \end{pmatrix} \times \begin{pmatrix} b_1 \\ b_2 \\ b_3 \end{pmatrix} = \begin{pmatrix} a_{11}b_1 + a_{12}b_2 + a_{13}b_3 \\ a_{21}b_1 + a_{22}b_2 + a_{23}b \\ a_{31}b_1 + a_{32}b_2 + a_{33}b \end{pmatrix}$$

and if a matrix is premultiplied by a scalar quantity c, then all elements of the matrix are multiplied by c.

$$cA = c \begin{pmatrix} a_{11} & a_{12} & a_{13} \\ a_{21} & a_{22} & a_{23} \\ a_{31} & a_{32} & a_{33} \end{pmatrix} = \begin{pmatrix} ca_{11} & ca_{12} & ca_{13} \\ ca_{21} & ca_{22} & ca_{23} \\ ca_{31} & ca_{32} & ca_{33} \end{pmatrix}$$

When the rows and columns of a matrix M are interchanged the new matrix M' is called the transpose of the matrix M. The *inverse* of a square matrix M of order $n \times n$ is defined as a matrix whose $(i, j)^{th}$ element is obtained as the product of $(-1)^{i+j}$ and the determinant of an $(n-1)\times(n-1)$ matrix obtained by eliminating the ith row and the jth column of M. The inverse matrix of M is represented by M^{-1} and $MM^{-1}=I$.

1.25.7 Determinants

The determinant of a matrix is a scalar quantity obtained by performing specified operations on the elements of the matrix. Determinants are associated with square matrices only. For example the determinant of a square matrix A of the form

$$A = \begin{pmatrix} a_{11} & a_{12} \\ a_{21} & a_{22} \end{pmatrix}$$

is

$$|A| = a_{11}a_{22} - a_{12}a_{22}$$

In general the determinant of an $n \times n$ matrix A whose $(i, j)^{th}$

element is a_{ij} is

$$\sum_{1}^{n}(-1)^{i+j} a_{ij} |A_{ij}|$$

where $|A_{ij}|$ is the determinant of an $(n-1)\times(n-1)$ matrix obtained by deleting the ith row and j^{th} column of matrix A defined below.

$$A = \begin{pmatrix} a_{11} & ...a_{1j}.. & a_{1n} \\ a_{i1} : & \ddots.a_{ij}.. & :a_{in} \\ a_{n1} & \cdots a_{nj}.. & a_{nn} \end{pmatrix}$$

1.26 Real Numbers

Rational numbers consist of positive and negative integers and fractions and decimal fractions. Mathematically, a rational number can be expressed as the ratio of two integers m and n ($n > 0$) such that that they are *relative prime* (the numerator and the denominator have no common factor.

Irrational numbers consist of numbers that are not rational, but can be expressed as a limit of sequence of rational numbers. The *real number system* includes all rational and irrational numbers and does not include even roots of negative numbers (imaginary numbers) and complex numbers (that have both real and imaginary components).

If S is a set of real numbers, then U is an *upper bound* for S if for each $x \in S$ we have $x \leq U$. A number U' is called the *least upper bound* for S if U' is an upper bound for S and is less than or equal to any other upper bound of S. Clearly the least upper bound of a set S is unique if it exists. Every non-null set of real numbers S which has an upper bound has a least upper bound. The least upper bound of a set S of real numbers is sometimes represented as $\sup_{x \in S} x$. *Greatest lower bound* is defined similarly and occasionally represented as $\inf_{x \in S} x$.

Note that the notion of the least upper bound (the greatest lower bound) is different from the concept of the maximum (minimum) value. The latter refers to a number that is the greatest in a set of numbers. The maximum value may not always exist in a real-number set, but the least upper bound always exists in a bounded set. For example, the interval $[0,1)$ has the least upper bound 1 and has no maximum value on it. In the case that both the least upper (greatest lower) bound and the maximum (minimum) value exist, they are identical. In the above example, 0 is both the greatest lower bound and the minimum value. The concepts of the least upper bound and the greatest lower bound will be utilized in later chapters.

Problems

1.   Find the area bounded by
$$y = 4x - 4, \; y = \frac{1}{3}x^2 \text{ and } y = 6 - x, \; x, y > 0$$

2.   Evaluate the following integrals
$$\int_{-1}^{0} (2x + x^2 - x^3)dx$$

$$\int_{0}^{2a} (e^x + x^3)dx$$

$$\int_{-1}^{2} (x^2 + x)(3x + 1)dx$$

3.   Determine the values of x, y and z (if any) that maximize or minimize the function
$$f(x, y, z) = e^{4x^2 + 2y^2 + z^2 - 5xy - 4z}$$

4.   Determine the values of $x_1, x_2,$ and x_3 that maximize the function
$$f(x_1, x_2, x_3) = 6x_1x_2 + 4x_2x_3 + 6x_2 - 3x_3^2 - x_2^2$$
Subject to the constraint

$$x_1 + 2x_2 + x_3 = 75$$

5.   Integrate the following equations

$$\int \frac{\ln(x+1)}{\sqrt{x+1}} dx$$

$$\int x^n \ln x \, dx$$

$$\int x a^x \, dx$$

$$\int \frac{dx}{x^2 + 8x + 10}$$

6.   Obtain the following limits

$$\lim_{x \to 0} (1 + x^2)^{\frac{1}{x^2}}.$$

$$\lim_{x \to \infty} (x - 2) e^{-x^2}.$$

$$\lim_{x \to 0} \frac{2 - 3e^{-x} - e^{-2x}}{2x^2}.$$

$$\lim_{x \to \infty} x^2 e^{-x}.$$

7.   Obtain the generating functions associated with the following difference equations. Assume y_0 and y_1 are known.

$$-61 y_{k+1} + 55 y_k = -71$$

$$72 y_{k+2} - 93 y_{k+1} - 72 y_k = -51$$

$$39 y_{k+1} + 27 y_k = 24$$

8.   Solve the following difference equation

$$y_{k+2} = 7 y_{k+1} - 12 y_k, \; k = 0, 1, 2, \dots$$

9.   Use the partial fraction concept to invert the following expressions

$$f(x) = \frac{1}{(x-a)(x-b)(x-c)(x-d)}.$$

$$f(x) = \frac{2x + 3}{x^3 - 1}.$$

$$f(x) = \frac{2x^2 + 3x + 1}{(x-1)(x-2)(x-3)}.$$

References

Beckenbach, E. F. and Pólya, G. (1964), *Applied combinatorial mathematics,* John Wiley, New York.

Chiang, C.L. (1979), *Stochastic Processes and Their Applications,* Robert E. Krieger Publishing Company, Huntington, New York.

Courant, R. (1955), *Differential and Integral Calculus,* (Translated by McShane J.E. Vols I and II) Interscience Publishers Inc, New York.

CRC Standard Mathematical Tables (1971) 19th ed., The Chemical Rubber Company, Cleveland.

Dantzig, G.B. (1963), *Linear Programming and Extensions,* Princeton University Press, Princeton. New Jersey.

Draper, J.E. and Klingman, J.S. (1967), *Mathematical Analysis,* Harper and Row, New York.

Feller, W. (1965) *An Introduction to Probability Theory and its Applications,_Vol I.* 2nd ed., John Wiley and Sons, New York.

Goulden, I.P. and Jackson, D.M. (1983), *Combinatorial Enumeration,* John Wiley and Sons, New York.

Gray, M.W. (1972), *Calculus with Finite Mathematics for Social Sciences,* Addison-Wesley, Reading, Mass.

Khouri, A.I. (1993), *Advanced Calculus with Applications in Statistics,* John Wiley and Sons, New York.

Kuhn, H.W and Tucker, A.W. eds., (1956) *Linear Inequalities and Related Systems,* Princeton University Press, Princeton, New Jersey..

Moye, L.A. and Kapadia, A.S. (2000), *Difference Equations with Public Health Applications*, Marcel Dekker, New York.

Robbins, H.A. (1955), Remark on Stirling's formula, *American Mathematical Monthly*, Vol 62.

Royden, H.L. (1968), *Real Analysis.* 2nd ed., The Macmillan Company, New York.

Rudin, W. (1976), *Principles of Real Analysis,* McGraw-Hill, New York.

Stirling, J. (1730), *Methodus Differentialis.*

Swokowski, E.W. (1979), *Calculus with Analytical Geometry,* Prindle, Weber and Schmidt, Boston.

Taha, H.A. (1971), *Operations Research an Introduction,* The Macmillan Company, New York.

Widder, D.V. (1946), *The Laplace Transforms,* Princeton University Press, Princeton, New Jersey.

2

Probability Theory

2.1 Introduction

The development of the concepts of *probability* is both extensive and complicated. Its rudiments have been identified in ancient Far Eastern writings. The generation of modern probability laws is credited to the work of European mathematicians such as Pierre de Fermat, Pierre-Simon Laplace, Carl Fredrick Gauss, and Thomas Bayes, among others. These mathematicians lived 400 years ago and worked on experiments that we would describe as games of chance. However, the twentieth century has seen new developments of probability. The work of Einstein, the development of measure theory, and the seminal contribution of the Russian mathematician Kolmogrov have led to a new round of explosive growth in this challenging and ubiquitous field.

However, it is a unique feature of probability that despite the specialization of the field, laypeople are comfortable with its central concepts. While skilled mathematicians and probabilists provide new, extensive, and intensive mathematical computations for probabilities, non mathematicians are perfectly comfortable with the concept of probability and have no difficulty providing their own estimates of these quantities. We will begin our discussion of probability theory with a discussion of this interesting concept.

2.2 Subjective Probability, Relative Frequency and Empirical Probability

2.2.1 Subjective Probability

To the layperson the definition of probability can appear to be confusing. There are several definitions of probability (or *chance* or *likelihood*) each depending on who is defining it and the context in which it is defined. For example, a mother who does not approve of her daughter's choice of a mate may give their marriage a fifty percent chance of lasting five years. Nobody can deny that the mother's personal prejudice was embedded in her gloomy assessment. This type of probability is based solely on one person's personal beliefs, cultural opinions and attitudes and has no sound data to back it nor can it be generalized to other situations of similar nature. Such probabilities are biased and are referred to as *subjective probabilities.*

One very unique, interesting, and frustrating feature about probability is that, as it is most commonly expressed in everyday life, estimates concerning the probability of an event's occurrence appear to be a mixture of qualitative as well as quantitative components. Lay people frequently use the word "probability" when they really mean to imply "possibly". To these users probability is not a computational tool but is instead useful to convey their own personal sense or belief about whether an event will or will not occur. This is the hallmark of subjective probability. Often this amorphous mix of qualitative and quantitative measures is revealed within the context of the use of the term.

The need for subjective probability arises from an inability to always accurately answer the question of which one event will occur. We therefore combine quantitative with qualitative measures. One common example of such a mixture is predictions for the activity on New York Stock Exchange. The economic fortunes of millions of individuals are both directly and indirectly influenced by the movement of the Dow Jones Average, a composite measure of stock value. By and large, increases in stock prices are viewed much more favorably than decreases in these prices, and the investor, whose primary interest is making an investment earning, is encouraged when the price of her stock increases, since this is seen as increasing the value of her share of the market.

Investors are therefore most interested in the future movement of the stock market will it be positive or negative? Determinations about the future movement of the stock market are very valuable, and much computational effort has been invested in trying to make accurate predictions about where the stock market will be in a day, a week, a month, a year, or ten years into the future. In fact, some of the most advanced computational, probabilistic and statistical tools have been used to make these predictions, including but not limited to time series analysis, regression analysis, Brownian motion, random walk models and statistical decision analysis. These procedures are computationally intensive, and yield quantitative predictions - unfortunately, these predictions are sometimes inaccurate. For example, the Great Stock Market Crash (known as Black Friday) in October 1929 was not predicted by the probability models of the time. In the following sixty years, more advance computational and probabilistic methods were developed. However as revealed by the market crash of 1987 these newer models are also not totally reliable and the results are still not satisfactory.

The computation of an exact probability from a reliable probability model is separate and apart from the statement of whether the event will occur or not. This is perhaps the true reason for the commonly felt need to use subjective probability. It is a natural and human response for people to attempt to fill the gap between the computational/analytical result of the probability model and the precise prediction of the actual event with their own experience. We fill in the missing pieces required for the answer to the question "will the event occur or not? " with a piece of ourselves, based on not just our intuition, knowledge, experience, expertise, and training, but also with our personality, biases, point of view, hopes and fears. Attempts have been made to formally connect subjective probability to a precise mathematical statement in the form of *utility function* (often used in *statistical decision making*). Further development may be able to transform this lesser scientific notion to a rigorous science.

2.2.2 Probability as Relative Frequency
A second notion associated with the occurrence of an event and often used in probability theory is the concept of *relative frequency*. This is the ratio of the number of times an event occurs in a series of trials and the total number of trials. Hence, if an event E occurs r times in a series of n independent trials then the relative frequency of the occurrence of E is r/n. The ratio r/n clearly lies between 0 and 1. This ratio is not always the same in subsequent experiments of size n. For example if in a data set consisting of n college students the relative frequency of women is r/n, this does not imply that each and every data set of n college students will contain r women. We will see in Chapter 8 that as the size of the experiment increases, the relative frequency estimate of the probability becomes more precise. Commonly, the conduct of the experiment permits either direct deduction or a complicated calculation of the probability of the event.

2.2.3 Empirical Probabilities
Empirical probabilities are based on the occurrences of events in repeated experiments all conducted under physically uniform conditions. Such probabilities together with an experimenter's intuitive interpretation of probability and inductive reasoning collectively define the probabilities most often used in statistical theory. It is well known that an unbiased coin when tossed repeatedly will render tails about 50% of the time. This does not mean that each and every experiment where an unbiased coin is tossed 50 times, tails will show up in exactly 25 of the tosses. However, if this experiment were to be repeated several times and the total number of tails in each set of 50 tosses were to be averaged out, on an average there would be 25 tails in a set of 50 tosses.

The idea of randomness is associated with the definition of probability. Randomness plays an important part in s*ampling theory* and it implies choosing from a collection of elements (individuals or objects) such that each element has the same chance of being chosen. This also allows us to apply the calculus of probabilities *a priori*. The justification for all this is *empirical*.

Assume that we observe an experiment whose outcome cannot be foreseen with certainty. The probability of a possible event represents the likelihood of this event and is always assigned a number between 0 and 1, inclusively. Mathematically, the probability is defined to be a real-valued set function that assigns an event a number in $[0, 1]$ (The notion of events and their corresponding probabilities are defined in subsequent sections).

Recent meteorological experience provides an example of the differences between the computation of probability using empirical probability and probability calculations using different analytical tools. The Gulf Coast of the United States (Texas, Louisiana, Mississippi, Alabama, and Florida) is commonly at risk of being struck by a hurricane. In the fall, toward the end of the hurricane season, the risk of a hurricane striking Texas decreases. For example, in 100 years, only two hurricanes have ever entered Texas (called "making landfall") in October. In addition, sophisticated computer models are also used to track the position of hurricanes. These models use satellite imagery, information from reconnaissance aircraft that fly into the storm, data from the movement of low pressure and high pressure waves affecting the continental United States, as well as the physics of fluid mechanics to compute the projected path of the storm. These computations are used to compute strike probabilities, i.e. the probability that the storm will make landfall at a given location.

For example, on October 1, 2002, millions of people along the gulf coast watched with growing alarm as Hurricane Lily increased in destructive power and threatened to make landfall somewhere on the gulf coast of the United States. Using a relative frequency argument, it was estimated that the probability of Lily hitting the coast of Texas was 0.02. However, the complications from the sophisticated mathematical models produced strike probabilities that were ten times higher, causing alarm and voluntary evacuations of some areas in southeast Texas. The computations based on arguments other than relative frequencies were proving to be more persuasive then the historical record in influencing human behavior. In the end, Hurricane Lily struck not in Texas but in neighboring Louisiana, causing millions of dollars in damage.

2.3 Sample Space

Set theory is very useful in the understanding of probability theory. Central to the application of set theory to probability is the idea of sample space. The *sample space* contains all possible outcomes of an experiment. Each possible outcome or collection of outcomes of an experiment may be called an event, and sample space is, in reality, an event space. As an illustration, consider the simple experiment of tossing a coin one time. The only possible outcomes are a single head (H) or a single tail (T). If two coins are tossed (or if a coin is tossed twice) the possible outcomes are H_1H_2, H_1T_2, T_1H_2 and T_1T_2 where the subscript to the alphabet refers to the trial number. For example, the sequence H_1T_2 refers to a head in the first coin (or toss) and tail in the second coin (or toss). Hence the sample space S in the first case will consist of outcomes H and T where as in the second case it will consist of outcomes H_1H_2, H_1T_2, T_1H_2 and T_1T_2.

Next, suppose the purpose of an experiment is to observe a randomly chosen male aged 85 or older to see if he has Alzheimer's disease. There are two possible outcomes of this experiment: *Yes* and *No*. In other words the man chosen at random may (*Yes*) or may not (*No*) have Alzheimer's disease. Since 35% of male population in this age group have the Alzheimer's disease (see Pfeffer et al., 1987), a value 0.35 will be assigned to the event $\{Yes\}$ to represent its likelihood of occurrence. A mathematical representation of this probability is written as $P(Yes) = .35$.

A simple event is defined as one possible outcome as a result of an experiment. In the case of tossing two coins, the event of interest may be $T_1 T_2$. If the tossing of two coins resulted in the occurrence of $H_1 T_2$ then we say that the event of interest did not occur. Similarly a combination event may be defined as a collection of several simple events. Continuing with our example, the event of interest may be the occurrence of $H_1 H_2$ or $T_1 T_2$. Hence if $H_1 H_2$ were to occur the event of interest would have occurred.

Example 2.1 An unbiased die is rolled until the number 6 shows up. What are the elements of the sample space S of all possible outcomes associated with this experiment?

Let us consider two outcomes 6 and $\bar{6}$ corresponding to the occurrence or non occurrence of number 6 in the roll of a die. Furthermore, let us subscript these outcomes to represent the roll with which they are associated. For example, 6_3 represents the occurrence of a 6 in the 3rd roll of the die. Similarly, $\bar{6}_5$ represents the non-occurrence of number 6 in the 5th roll of the die. Since the die will continue to be rolled until a 6 appears the sample space consists of all possible sequences of outcomes associated with this experiment

$$S = \{6_1, \bar{6}_1\, 6_2, \bar{6}_1\; \bar{6}_2\; 6_3, \cdots \bar{6}_1\; \bar{6}_2\; \bar{6}_3 \cdots \bar{6}_{k-1}\; 6_k \cdots$$

∎

With this as a background, we may proceed with the following definitions.

Suppose A and B are two events (either simple events or combination events) that are contained in the sample space S.

Definition 2.1 The impossible event (denoted by the null set $\varnothing$) is an event that cannot be constructed from events or combinations of events from the sample space.

Definition 2.2 A^c (described as "A complement") is comprised of all events that do not belong to A.
Associated with events A and B are two other events that express the joint occurrence of A, B and their compliments. These events are:

Definition 2.3 The event $A \cup B$ (stated as A *union* B) consists of all points in the sample space that belong to either A or B.

Definition 2.4 The event $A \cap B$ (stated as A *intersection* B) consists of all points that are common to both A and B (see Figure 2.1). This definition also implies the simultaneous realization of A and of B. If A and B do not have any common points then they do not overlap (i.e., $A \cap B = \varnothing$) and the events A and B are said to be *disjoint.*
We easily see that

$$A \cup A^c = S.$$

and

$$A \cap A^c = \varnothing.$$

Definition 2.5 If every $x \in B$ implies $x \in A$ then $A \supset B$ (stated as A *contains* B) or $B \subset A$ (stated as B is contained in A). If $B \subset A$ and $A \subset B$ then $A = B$.

Definition 2.6. $B - A$ denotes the event that B but not A occurs. This event can also be written as $B \cap A^c$ or $A^c \cap B$ (see Figure 2.2).
           We may apply these definitions to our first example of tossing a coin. Let A be the event consisting of $H_1 T_2, T_1 T_2, H_1 H_2$ in the sample space of all possible outcomes resulting from tossing two coins. If we define B as the event $\{H_1 T_2, H_1 H_2\}$ then clearly B is contained in A and all points in B also lie in A.
Note that
A^c is the event $\{T_1 H_2\}$ and B^c *is* the event consisting of $T_1 T_2$ and $T_1 H_2$.

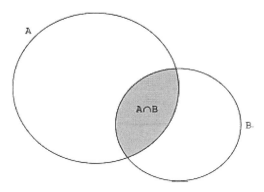

Figure 2.1 Intersection of events A and B

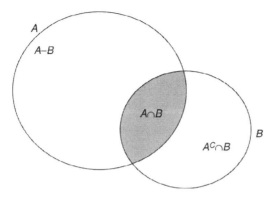

Figure 2. 2 Complementary event $A^c \cap B$

There are well known relationships between sets which are historically known as laws.

Commutative Law

$$A \cup B = B \cup A$$

and

$$A \cap B = B \cap A$$

Now consider three events A, B and C on a sample space S. Then the following laws of set theory hold (see Figure 2.3)

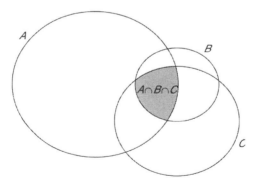

Figure 2.3 Intersection of three events (A, B, and C)

Associative Law

$$(A \cup B) \cup C = A \cup (B \cup C) \text{ and } (A \cap B) \cap C = A \cap (B \cap C)$$

Distributive Law

$(A \cup B) \cap C = (A \cap C) \cup (B \cap C)$
and
$(A \cap B) \cup C = (A \cup C) \cap (B \cup C).$

DeMorgan's Law
$(A \cup B)^{C} = A^{C} \cap B^{C}$
and
$(A \cap B)^{C} = A^{C} \cup B^{C}.$

The proof of the Commutative and Associative Laws are left as an exercise for the reader. We provide a proof for the first of the Distributive Laws and the second of these laws can be proven analogously.

Assume that the sample space S contains events A, B and C, i.e. $A, B, C \subset S$. Consider a set of points x. Then for any $x \in (A \cup B) \cap C$, it is implied that $x \in (A \cup B)$ and $x \in C$. Since $x \in (A \cup B)$ it implies that $x \in A$ or $x \in B$ and $x \in C$. Therefore, $x \in A \cap C$ or $x \in B \cap C$ that is equivalent to the statement $x \in (A \cap C) \cup (B \cap C)$..

Hence (by Definition 5) $(A \cup B) \cap C \subset (A \cap C) \cup (B \cap C).$

Assume now that $x \in (A \cap C) \cup (B \cap C).$

Then $x \in (A \cap C)$ or $x \in (B \cap C).$

If $x \in (A \cap C)$, then $x \in A$ and $x \in C$. Since $x \in A$ implies $x \in (A \cup B)$, we have $x \in (A \cap C)$ implies $x \in [(A \cup B) \cap C].$

Similarly, if $x \in (B \cap C)$, then $x \in [(A \cup B) \cap C].$

Hence, in either case $x \in [(A \cup B) \cap C].$

We have therefore shown that $(A \cap C) \cup (B \cap C) \subset (A \cup B) \cap C$. This establishes the equivalency of $(A \cup B) \cap C$ and $(A \cap C) \cup (B \cap C).$ ∎

To prove the first of the two relationships in the DeMorgan's Law, we proceed as follows (the second relationship may be derived similarly).

Since $x \in (A \cup B)^{c}$, we have $x \notin A \cup B$, that implies $x \notin A$ and $x \notin B$, or equivalently, $x \in A^{c}$ and $x \in B^{c}$. These imply $x \in A^{c} \cap B^{c}$. We can conclude that $(A \cup B)^{c} \subset A^{c} \cap B^{c}.$

Conversely, $x \in A^{c} \cap B^{c}$ implies $x \notin A$ and $x \notin B$, that is $x \notin A \cup B$. Hence $A^{c} \cap B^{c} \subset (A \cup B)^{c}.$

Therefore $(A \cup B)^c = A^c \cap B^c$.

2.4 Decomposition of Union of Events: Disjoint Events

For two events A_1 and A_2, it is obvious that

$$A_1 \cap (A_2 - A_1 \cap A_2) = A_1 \cap (A_2 \cap A_1^c) = \varnothing.$$

Hence the two events

A_1 and $A_2 \cap A_1^c$ are disjoint .

Therefore,

$$A_1 \cup A_2 = A_1 \cup (A_2 \cap A_1^c).$$

Similarly, for three independent events $A_1, A_2,$ and A_3, we may observe that

$A_1, A_2 \cap A_1^c$ and $A_3 \cap (A_1 \cup A_2)^c$

are disjoint and

$$A_1 \cup A_2 \cup A_3 = A_1 \cup (A_2 \cap A_1^c) \cup \{(A_3 \cap (A_1 \cup A_2)^c\}.$$

In general for n events, $A_1, A_2, \dots$ and A_n, it may be observed that

$A_1, A_2 \cap A_1^c, \{A_3 \cap (A_1 \cup A_2)^c\}, \dots, \{A_n \cap (A_1 \cup A_2 \cup \dots \cup A_{n-1})^c\}$

are pair wise disjoint and

$$A_1 \cup A_2 \cup A_3 \dots \cup A_n = A_1 \cup (A_2 \cap A_1^c) \cup \{A_3 \cap (A_1 \cup A_2)^c\}$$
$$\dots \dots \cup \{A_n \cap (A_1 \cup A_2 \dots A_{n-1})^c\}. \tag{2.1}$$

Note that a collection of events is called pairwise disjoint if all pairs of events are disjoint.

2.5 Sigma Algebra and Probability Space

Among several events under consideration, a new event generated from these events may also interest the experimenter. In probability theory, the event generated by (i) the sample space itself, (ii) the complement of another event, or (iii) the countable unions of events, is of interest. The collection of events that are closed under the formation of these three operations is called a *sigma algebra* (σ - *algebra)*.

Mathematically, a collection of events Ω from a sample space S form a sigma algebra if

(i) $S \in \Omega$;

(ii) $A \in \Omega$ implies $A^c \in \Omega$.

(iii) $A_n \in \Omega$ for $n = 1,2,\ldots$ implies $\bigcup_{n=1}^{\infty} A_n \in \Omega$.

Note that, by DeMorgan's law, (i) can be replaced by (i') $\varnothing \in \Omega$ and (iii) can be replaced by (iii') $A_n \in \Omega$, for $n = 1,2,\ldots$, implies $\bigcap_{n=1}^{\infty} A_n \in \Omega$, in the presence of (ii).

A collection of events is called an algebra if (iii) is replaced by the finite unions.

There are two simple examples of σ-algebra: one is the collection of all subsets of S and the other is the collection of the sample space and the null set. All other examples of σ-algebra are difficult to visualize and are usually generated by a collection of visible events. For example, consider the sample space consisting of all positive numbers less than or equal to 1, i.e. $S = (0,1]$. Let Ω_0 be the collection of all finite disjoint unions of subintervals,

$A = (a_1, b_1] \cup \cdots \cup (a_n, b_n]$ where $a_1 \leq a_2 \leq \ldots \leq a_n$.

Clearly,

$S = (0, 0.5] \cup (0.5, 1] \in \Omega_0.$,

$A^c = (0, a_1] \cup (b_1, a_2] \cup \cdots \cup (b_{n-1}, a_n] \cup (b_n, 1] \in \Omega_0$ and all finite unions of A's belongs to Ω_0. Hence, Ω_0 is an algebra. If we construct another collection of events Ω which contains Ω_0 and allows countable unions of A's to belong to it, then Ω is a σ-algebra. However, not all elements in Ω can be expressed in an explicit form. This is just like when we learn irrational numbers: we understand their existence, but we cannot write down all of them. In section 2.12, an element in Ω will be revisited and called a *Borel set*.

Note that, if $\{A_i\}_{i=1}^{\infty}$ is a sequence of monotonically increasing events, defined as $A_1 \subset A_2 \subset \cdots \subset A_n \subset A_{n+1} \subset \cdots$, then

$$\lim_{i \to \infty} P(A_i) = P\left(\bigcup_{i=1}^{\infty} A_i\right).$$

If $\{A_i\}_{i=1}^{\infty}$ is a sequence of monotone decreasing events, defined as $A_1 \supset A_2 \supset \cdots \supset A_n \supset A_{n+1} \supset \cdots$, then

$$\lim_{i \to \infty} P(A_i) = P\left(\bigcap_{i=1}^{\infty} A_i\right).$$

In probability theory, the triplet (S, Ω, P) is called the probability space, where S is the sample space, Ω is the σ-algebra containing the events from S, and P is the set function defined on Ω satisfying the following;

(*i*) $P(A) \geq 0$ when $A \in \Omega$.

(*ii*) $P(S) = 1$.

(iii) $P(\bigcup_{i=1}^{\infty} A_i) = \sum_{i=1}^{\infty} P(A_i)$ whenever $A_i \in \Omega, 1 \le i < \infty,$ and $A_i \cap A_j = \varnothing, i \ne j.$

2.6 Rules and Axioms of Probability Theory

A probability function maps a set to a number. There are some rules that a probability function must follow. It must assign a non-negative number to the occurrence of an event. The probability of the entire sample space S must equal 1. If two events cannot occur simultaneously, the probability of their union must be the same as the sum of the probabilities of these two events. The rules in mathematical notation are:

(1) $P(A) \ge 0$, for any event A.

(2) $P(S) = 1$.

(3) $P(A \cup B) = P(A) + P(B)$ if $A \cap B = \varnothing$.

These three properties are referred to as the three axioms of probability (Feller 1957).

Example 2.2 In a 1999 study (Kapadia et al), the course of stay in a Pediatric Intensive Care Unit (PICU) as the patients move back and forth between severity of illness states in a children's hospital in the Texas Medical Center in Houston was studied. The main objective of the study was to investigate the dynamics of changing severity of illness as measured by the Pediatric Risk of Severity (PRISM) scores, in the pediatric intensive care unit. The PRISM score comprises of 7 physiological and 7 laboratory variables, each reflecting an organ system dysfunction and ranges in value from 1 to 20. Data on the PRISM scores were recorded daily on 325 patients admitted and released between November 1994 and February 1995. PRISM scores were then classified into three categories. High (PRISM score >10), Medium ($4 < $ PRISM score ≤ 10) and Low (PRISM score ≤ 4). Initial probabilities associated with these scores are 0.176, 0.315, and 0.509 respectively. (i) What is the probability that when a patient arrives at the PICU his/her PRISM score is not classified as High? (ii) What is the probability that the initial PRISM score is classified as either Low or High? (iii) What is the probability that initial PRISM score is classified as at least Medium?

     If the score is not going to be classified as High it must be classified as Low or Medium. The probability associated with this event is derived as follows: P(Low $\cup$ Medium)=P(Low)+P(Medium) =0.509+0.315=0.824, since the events Low and Medium cannot occur simultaneously,
P(Low $\cup$ High)=0.509+0.176=0.685 and
P(at least Medium)=P(Medium)+P(High)=0.315+0.176=0.491. ∎

     Events that cannot occur simultaneously are called *mutually exclusive*. In this case the occurrence of one precludes the occurrence of the other. If an experiment must result in the occurrence of exactly one of the events $A_1, A_2, \cdots, A_n$, then the events A_i ($i = 1, ..., n$) are called *mutually exclusive and exhaustive*, that is $A_1 \cup A_2 \cup \cdots \cup A_n = S$ and $A_i \cap A_j = \varnothing$, for $i \neq j$. In the example above, the events are mutually exclusive because no patient may be classified into more than one category. For example, if a patient is classified into the Low PRISM score category he/she cannot be classified into the Medium or the High category i.e., the occurrence of one event precludes the occurrence of any other, therefore the events are mutually exclusive. Also, since the sum of the probabilities associated with each of the categories is one, the set of events is exhaustive.

Alternatively, if the experiment may result in the simultaneous occurrence of both A_i and A_j, then A_i and A_j may or may not be independent of one another and

$P(A_i \cap A_j) = P(A_i)P(A_j)$ if A_i and A_j are independent.

$\neq P(A_i)P(A_j)$ if A_i and A_j are dependent.

For example, let A_i be the event that number i shows up when an unbiased die is rolled. Then $P(A_i) = \dfrac{1}{6}$ for $i = 1, ..., 6$.

Suppose an experiment is conducted where a die is rolled just once. In such an experiment A_i and A_j ($i, j = 1, 2, \cdots 6$ *and* $i \neq j$) cannot occur simultaneously. Therefore they are mutually exclusive and $P(A_i \cap A_j) = 0$.

Now assume that A_i represents the event that number i shows up in the first roll of the die and A_j is the event that number j shows up in the second roll ($i, j = 1, 2, \cdots 6$). Then

$$P(A_i \cap A_j) = \frac{1}{6} \cdot \frac{1}{6} = \frac{1}{36}.$$

In the above example, what shows up in the second roll is independent of what shows up in the first roll and therefore A_i and A_j are independent events.

Next let A_i be the same as above but now define A_j as the event that occurs on the second roll but is different from what occurred on the first roll. In this case $P(A_i \cap A_j) = \dfrac{1}{6} \cdot \dfrac{5}{6} = \dfrac{5}{36}$ and A_i and A_j are dependent events. In general, if $A_1, A_2, \cdots, A_n$ are independent events, then

$$P(A_1 \cap A_2 \cap \cdots \cap A_n) = P(A_1)P(A_2) \cdots P(A_n).$$

The above relationship does not hold when A_i's are dependent events (that is if at least two of the A's are dependent). Note that events that are independent are not necessarily mutually exclusive. For example, A_1 may be the event that the card chosen at random from a well shuffled deck of cards is a queen and

A_2 the event that it is a heart. Then $P(A_1) = \dfrac{4}{52} = \dfrac{1}{13}$, $P(A_2) = \dfrac{13}{52} = \dfrac{1}{4}$, and

$P(A_1 \cap A_2) = P(\textit{the queen of hearts}) = \dfrac{1}{52}$, which is equal to $P(A_1)P(A_2)$. Obviously, A_1 and A_2 are not mutually exclusive.

Mutually exclusive non-null events are necessarily dependent since the occurrence of one precludes the occurrence of the other. Hence if A_1 and A_2 are independent events and $P(A_1)$, $P(A_2) > 0$, then they cannot be mutually exclusive.

The four laws of set theory described in section 2.3 are translated into probability theory as follows:

(i) $P(A_1 \cup A_2) = P(A_2 \cup A_1)$ and $P(A_1 \cap A_2) = P(A_2 \cap A_1)$.

(ii) $P[(A_1 \cup A_2) \cup A_3] = P[A_1 \cup (A_2 \cup A_3)]$ and
$P[(A_1 \cap A_2) \cap A_3] = P[A_1 \cap (A_2 \cap A_3)]$,

(iii) $P[(A_1 \cup A_2) \cap A_3)] = P[(A_1 \cap A_3) \cup (A_2 \cap A_3)]$ and
$P[(A_1 \cap A_2) \cup A_3)] = P[(A_1 \cup A_3) \cap (A_2 \cup A_3)]$,

(iv) $P((A_1 \cup A_2)^c) = P(A_1^c \cap A_2^c)$ and $P((A_1 \cap A_2)^c) = P(A_1^c \cup A_2^c)$.

Note if $A_1 \subset A_2$ then $P(A_1) \le P(A_2)$ and $P(A_1 \cup A_2) = P(A_2) \le 1$.

And that A_i^c is defined as the complement of A_i and hence satisfies the following equations

$$P(A_i) + P(A_i^c) = 1 \text{ and } P(A_i \cap A_i^c) = P(\emptyset) = 0.$$

The following relationships hold if A_1 and A_2 are independent

(1) $P(A_1 \cap A_2^c) = P(A_1)P(A_2^c)$

(2) $P(A_1^c \cap A_2) = P(A_1^c)P(A_2)$

(3) $P(A_1^c \cap A_2^c) = P(A_1^c)P(A_2^c)$.

Proof. (1) Observe that $A_1 = (A_1 \cap A_2) \cup (A_1 \cap A_2^c)$. Since $A_1 \cap A_2$ and $A_1 \cap A_2^c$ are mutually exclusive, we have

$$P(A_1) = P(A_1 \cap A_2) + P(A_1 \cap A_2^c).$$

Hence

$$P(A_1 \cap A_2^c) = P(A_1) - P(A_1 \cap A_2)$$
$$= P(A_1) - P(A_1)P(A_2) = P(A_1)[1 - P(A_2)] = P(A_1)P(A_2^c).$$

Note that the assumption of independence was used in the above derivation.

The relationship (2) may be obtained by interchanging A_1 and A_2 in (1). The proof of the third relationship is derived based on (2) above as well as the following fact.

$$A_1^c = (A_1^c \cap A_2) \cup (A_1^c \cap A_2^c).$$

Therefore,

$$P(A_1^c) = P[(A_1^c \cap A_2) \cup (A_1^c \cap A_2^c)] = P(A_1^c \cap A_2) + P(A_1^c \cap A_2^c)$$

$$P(A_1^c \cap A_2^c) = P(A_1^c) - P(A_1^c \cap A_2) = P(A_1^c) - P(A_1^c)P(A_2)$$

$$= P(A_1^c)[1 - P(A_2)] = P(A_1^c)P(A_2^c).$$

For two arbitrary events A_1 and A_2, $P(A_1 \cup A_2)$ represents the probability that either A_1 occurs or A_2 occurs or both occur. In other words it is the probability that at least one among A_1 and A_2 occurs. Hence, we are going to prove that

$$P(A_1 \cup A_2) = P(A_1) + P(A_2) - P(A_1 \cap A_2). \tag{2.2}$$

Now,

$$A_1 \cup A_2 = A_1 \cup (A_2 \cap A_1^c)$$

implies

$$P(A_1 \cup A_2) = P(A_1) + P(A_2 \cap A_1^c),$$

since A_1 and $(A_2 \cap A_1^c)$ are disjoint.

Note that

$$A_2 = (A_2 \cap A_1) \cup (A_2 \cap A_1^c)$$

$$P(A_2) = P(A_2 \cap A_1) + P(A_2 \cap A_1^c)$$

$$P(A_2 \cap A_1^c) = P(A_2) - P(A_2 \cap A_1)$$

Hence by substituting for $P(A_2 \cap A_1^c)$ we obtain (2.2).

(Note that the subtraction of $P(A_1 \cap A_2)$ in (2.2) comes from the fact that $A_1 \cap A_2$ is contained in set A_1 as well as in the set A_2 and must be counted only once). Rewriting (2.2) we obtain

$$P(A_1 \cap A_2) = P(A_1) + P(A_2) - P(A_1 \cup A_2)$$

i.e.

$$P(A_1 \cap A_2) \le P(A_1) + P(A_2)$$

and

$$P(A_1 \cap A_2) \ge P(A_1) + P(A_2) - 1.$$

The above relationship is referred to as the Bonferroni's inequality. It can be applied in developing a conservative testing strategy to avoid propagating the Type 1 error in statistical hypotheses testing (Chapter 9).

Boole's Inequality Let $A, A_1, A_2,...A_n...$ be events such that

$$A \subset \bigcup_1^\infty A_i = (A_1 \cup A_2 \cup .....A_n \cup .....), \text{ then } P(A) \le \sum_1^\infty P(A_i).$$

Proof: Since $A \subset \bigcup_1^\infty A_i$,

we have

$$A = A \cap \bigcup_1^\infty A_i. \tag{2.3}$$

Now,

$$\bigcup_{i=1}^n A_i = A_1 \cup A_2 \cup A_3 ... \cup A_n = A_1 \cup (A_2 \cap A_1^c) \cup \{A_3 \cap (A_1 \cup A_2)^c\}$$

$$........ \cup \{A_n \cap (A_1 \cup A_2 ....A_{n-1})^c\}.$$

Since all the events on the RHS of the above relationship are pairwise disjoint and together with a generalized version of the Distributive Law, equation (2.3) may be expressed as

$$P(A) = P(A \cap \bigcup_{i=1}^\infty A_i)$$

$$= P[A \cap [A_1 \cup (A_2 \cap A_1^c) \cup \{A_3 \cap (A_1 \cup A_2)^c\}$$

$$........ \cup \{A_n \cap (A_1 \cup A_2 ....A_{n-1})^c\} \cup \cdots]]$$

Since

$$\{A \cap A_1\}, \{A \cap (A_2 \cap A_1^c)\}, [A \cap \{A_3 \cap (A_1 \cup A_2)^c\}],...$$

$$..[A \cap \{A_n \cap (A_1 \cup A_2 \cup ...A_{n-1})^c\}]...$$

are all disjoint we have

$$P(A) = P(A \cap A_1) + P\{A \cap (A_2 \cap A_1^c)\} + \cdots + P[A \cap \{A_n \cap (A_1 \cup A_2 \cup ...A_{n-1})^c\}] \cdots$$

But,

$$(A \cap A_1) \subset A_1, (A \cap (A_2 \cap A_1^c)) \subset A_2, ..., (A \cap \{A_n \cap (A_1 \cup A_2 ... \cup A_{n-1})^c\}) \subset A_n, ...$$

Therefore

$$P(A) \le P(A_1) + P(A_2)....$$

$$= \sum_{1}^{\infty} P(A_i).$$

Example 2.3 In an elementary school, there are 400 students. Among them, ten did not make the honor roll in any of the three subjects: Reading, Mathematics and Science. Thirty did not make the honor roll in Reading or Science, 20 did not make the honor roll in Reading or Mathematics and 30 did not make the honor roll in Mathematics or Science. There were also 24, 36 and 42 students who made the honor roll in two subjects but not in Reading, Math and Science, respectively.

Let A_1, A_2 and A_3 be the event that a randomly chosen student from this elementary school, makes the honor roll in Reading, Math and Science respectively. Assume that A is the event that this randomly chosen student makes the honor roll in at least one of the three subjects, then we know that $A = A_1 \cup A_2 \cup A_3$ and we are going to prove that

$$P(A) \le P(A_1) + P(A_2) + P(A_3).$$

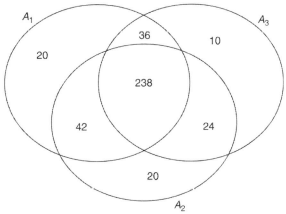

Figure 2.4 Venn diagram representing the students by honor roll category

Since 30 students did not make the honor roll in Math or Science and 10 did not make the honor roll in any subject, there were 20 students who made the honor roll in Reading only. There were another 42 students who made the honor roll in Reading and Math only, and 36 students who made the honor roll in Reading and Science. Using similar argument, we can calculate that 20 students made the honor roll in Math only and 10 students made the honor roll in Science only.

From all this information, we can summarize that 10 students did not make the honor roll in any subject, 20+20+10 students made the honor roll in only one subject and 36+42+24 students made the honor roll in exactly two subjects. Subtracting all these numbers from 400, the total number of students that made the honor roll in all three subjects would be 238 . Therefore, the probability that a randomly chosen student made the honor roll in Reading is

$$P(A_1) = \frac{20+42+238+36}{400} = \frac{336}{400} = 0.84$$

Similarly , the probability that a randomly chosen student made the honor roll in Math is

$$P(A_2) = \frac{20+24+238+42}{400} = 0.81 \text{ and } P(A_3) = \frac{308}{400} = 0.77.$$

Now,
$A \equiv$ randomly chosen student makes the honor roll in at least one of the three subjects.

$$P(A) = 1 - P \text{ (honor roll in none of the subjects)}$$

$$= 1 - \frac{10}{400} = 0.975.$$

In this case note that $P(A) \le P(A_1) + P(A_2) + P(A_3)$.

2.7 Conditional Probability

Consider two outcomes A_1 and A_2 of an experiment. Then the conditional probability of the occurrence of A_1 given A_2 has occurred is defined as

$$P(A_1 | A_2) = \frac{P(A_1 \cap A_2)}{P(A_2)} \text{ if } P(A_2) > 0.$$

The above relationship can be written as $P(A_1 \cap A_2) = P(A_1 | A_2)P(A_2)$

and be extended to $P(A_1 \cap A_2) = P(A_2 | A_1)P(A_1)$. If A_1 and A_2 are independent events then $P(A_1 | A_2) = P(A_1)$ and $P(A_2 | A_1) = P(A_2)$. However for the case when the A_i's are dependent the relationship can be generalized to n events as follows:

$$P(A_1 \cap A_2 \cap \cdots \cap A_n)$$
$$= P(A_1)P(A_2 | A_1)..P(A_{n-1} | A_1 \cap \cdots \cap A_{n-2})P(A_n | A_1 \cap \cdots \cap A_{n-1}).$$

2.8 Law of Total Probability

For a set of mutually exclusive and exhaustive events $D_1, D_2, \cdots, D_k$ in S, define an event B which can occur in conjunction with any of the D's, then

$$B \cap D_i \text{ for } i = 1 \cdots k$$

are also mutually exclusive. The total or marginal probability of the occurrence of B is defined as

$$P(B) = P(B \cap D_1) + P(B \cap D_2) + \cdots + P(B \cap D_k).$$

Example 2.4 In a study about aircraft noise and schools around London's Heathrow Airport, schools were classified into three categories: Government (G), Church (CH) and Grant Maintained (GM), and the noise level around were low(L), moderate (M) and high (H). It was reported that 60% of the low-noise schools, 70% of moderate-noise schools and 83% of high-noise schools are government owned. It was also reported that, among all schools, 47.2% were ranked as low-noise, 38.2% were ranked as moderate-noise and 14.6% were ranked high-noise. What is the probability that a school randomly chosen from among those schools is Government? (Haines, Stansfeld and Head 2002).

    Let L, M and H denote the events that the school is ranked as low, moderate and high noise respectively

    Then $P(G|L) = .6$, $P(G|M) = .7$ and $P(G|H) = .83$, using the Law of Total probability, the probability of event G may be obtained as follows:

$$\begin{aligned} P(G) &= P(G \cap L) + P(G \cap M) + P(G \cap H) \\ &= P(L)P(G|L) + P(M)P(G|M) + P(H)P(G|H) \quad . \\ &= (0.472)(0.6) + (0.382)(0.7) + (0.146)(0.83) = 0.672 \end{aligned}$$

2.9 Bayes Theorem

Public health problems often require classifying diseases by International Classification of Disease(ICD) categories such as $D_1, D_2 ... D_k$ which are mutually exclusive and exhaustive. Assume that the prevalence associated with each of these diseases is known as $P(D_i)$, $i=1,2,..,k$. From past research (or by physician's intuition before any tests are performed on the patient) the conditional probabilities of a set of symptoms B given each of theses diseases are known as $P(B|D_i)$. In such a case it may be of interest to obtain the conditional probability of a particular disease after a patient has undergone clinical tests relevant to the set of symptoms. For example B may consist of symptoms such as chest pain, cough, phlegm, breathlessness, and loss of weight. The set of mutually exclusive and exhaustive possible medical diagnosis $\{D_i\}$ $i=1,2...k$ may be any one of asthma, heart disease or lung cancer etc. Clearly,

$$\sum_{i=1}^{k} P(D_i) = 1$$

Now assume that the patient undergoes a battery of tests such as chest X-ray, blood work, pulmonary function test, EKG etc. All this information is put together in order to make the most appropriate diagnosis.

The probabilities of the form $P(D_i)$ are termed as *prior probabilities* and represent the prevalence of D_i in the population. In other words they provide the clinician with possibilities of the various diseases before any clinical test results on the patient are available. Probabilities of the form $P(D_i|B)$ are called *posterior probabilities* and are associated with the occurrence of the different diseases conditional on the set of symptoms after ("post") test results are available. The relationship of the posterior to the prior probabilities is described by what has come to be known as Bayes's Rule named after the Rev. Thomas Bayes (1702-1761). Using the definition of conditional probabilities,

$$P(D_i | B) = \frac{P(D_i \cap B)}{P(B)}. \qquad (2.4)$$

Writing the numerator of the above equation as

$$P(D_i \cap B) = P(D_i)P(B|D_i) \quad i = 1,2...k$$

and using the law of total probability, we have

$$P(B) = \sum_{i=1}^{k} P(D_i \cap B) = \sum_{i=1}^{k} P(D_i)P(B|D_i). \qquad (2.5)$$

Combining equations (2.4) and (2.5), we obtain

$$P(D_i | B) = \frac{P(D_i \cap B)}{P(B)} = \frac{P(D_i \cap B)}{\sum_{i=1}^{k} P(D_i)P(B|D_i)} .$$

The above probabilities are referred to as posterior probabilities.

Example 2.5 In two studies (Lairson, et al.1992, and Pugh, et al.1993) undertaken by the Veterans Administration Hospital in San Antonio, Texas, a group of 351 diabetic patients were screened for the presence of diabetic retinopathy D in at least one eye. Each patient was examined both by the usual ophthalmologic examination performed by a physician as well as by photographs of the eye taken by a wide angle (45 degrees) Cannon CR3 model retinal camera. The objective of the study was to see whether the wide angle retinal camera used either non-mydriatically (without pharmacological dilation) or mydriatically (after pharmacological dilation) was comparable to the ophthalmologist's examination in detecting diabetic retinopathy. The results by the three methods of examination were compared to a reference standard of stereoscopic fundus photographs interpreted by experts at the University of Wisconsin Fundus Photograph Reading Center. Effectiveness was measured by the number of true positives detected by each method. The Table below gives the conditional probabilities of test results for the three screening modalities.

In Table 2.1 below, the superscripts + and − above a D or T represents whether the event is positive or negative. For example D^+ represents the presence of the diseases as determined by the University of Wisconsin Fundus Photograph Reading Center (gold standard) and T^+ represents the test result being positive as determined by a screening modality. Similarly, D^- represents the disease being not present and T^- represents the test result being negative. The sensitivity and specificity of a screening modality are defined by $P(D^+|T^+)$ and $P(D^-|T^-)$. We are going to obtain the posterior probabilities $P(D^+|T^+)$ and $P(D^+|T^-)$ for the three screening modalities.

Table 2.1 Screening Modalities

Probability	Ophthalmologic Examination	Undilated 45° Photographs	Dilated 45° Photographs
N	347	351	351
$P(D^+)$	0.21	0.21	0.21
$P(D^-)$	0.79	0.79	0.79
$P(T^+|D^+)$	0.33	0.61	0.81
$P(T^-|D^+)$	0.67	0.39	0.19
$P(T^+|D^-)$	0.01	0.15	0.03
$P(T^-|D^-)$	0.99	0.85	0.97

In this example the conditional probabilities of the test results (positive or negative) for the three screening modalities are known. Also known is the prevalence of the disease. The required posterior probabilities for the screening modality Ophthalmologic Examination are derived below and for other screening modalities may be obtained similarly.

$$P(D^+|T^+) = \frac{P(D^+ \cap T^+)}{P(T^+)}.$$

Now, $P(D^+ \cap T^+) = P(D^+)P(T^+|D^+)$ and the total probability $P(T^+)$ is

$$P(T^+) = P(D^+ \cap T^+) + P(D^- \cap T^+)$$
$$= P(D^+)P(T^+ \mid D^+) + P(D^-)P(T^+ \mid D^-)$$
$$= 0.21(0.33) + 0.79(0.01) = 0.0772.$$

Therefore using results summarized in Table 2.1
$$P(D^+ \mid T^+) = 0.0693 / 0.0772 = 0.897$$

Similarly,

$$P(D^+ \mid T^-) = \frac{P(D^+ \cap T^-)}{P(T^-)}.$$

Now, $P(D^+ \cap T^-) = P(D^+)P(T^- \mid D^+)$ and the total probability $P(T^-)$ is

$$P(T^-) = P(D^+ \cap T^-) + P(D^- \cap T^-)$$
$$= P(D^+)P(T^- \mid D^+) + P(D^-)P(T^- \mid D^-)$$
$$= 0.21(0.67) + 0.79(0.99) = 0.923.$$

Hence,
$$P(D^+ \mid T^-) = 0.141 / 0.923 = 0.153.$$

2.10 Sampling with and without Replacement

Consider a room containing m men and n women where each trial consists of choosing a person at random from the room. If W_i (M_i) represents the event that the person chosen in the ith trial is female (male), $i = 1, 2, \cdots$, then $P(W_1) = \dfrac{n}{m+n}$

and $P(M_1) = \dfrac{m}{m+n}$. Now suppose that the person chosen in each trial is sent back to the room before the next trial is conducted. Then the outcome from the second trial will not be affected by the outcome from the first trial. In this case, the sequence of trials is called *sampling with replacement* and we can conclude that

$$P(W_2 \mid W_1) = P(W_2) = \frac{n}{m+n}.$$

Therefore, the trials are independent and $P(W_i) = \dfrac{n}{m+n}$, for all i.

If, before the second trial, the person chosen in the first trial is not allowed to return, then,

$$P(W_2) = P(W_2 \mid W_1)P(W_1) + P(W_2 \mid M_1)P(M_1)$$
$$= \frac{n-1}{m+n-1}\frac{n}{m+n} + \frac{n}{m-1+n}\frac{m}{m+n} = \frac{n(n-1)+nm}{(m+n)(m+n-1)}.$$

In the case of *sampling with replacement*, we have $P(W_2) = P(W_1)$. In the second case the sequence of trials is called *sampling without replacement* and $P(W_2) \neq P(W_1)$.

Lemma 1 The number of discernible ways in which r balls may be drawn from among n balls under conditions of sampling with replacement assuming ordering is not important and the balls are distinguishable is (Bose-Einstein assumption, Feller 1957)

$$\binom{n+r-1}{r}.$$

Proof. This problem is the same as the distribution of r indistinguishable balls in n cells such that r_i balls are placed in the ith cell where $r_i \geq 0$ in such a way that $\sum_{i=1}^{n} r_i = r$.

Represent the n cells by n-1 bars. For example, to build 3 cells one may require 2 bars since the 1st cell does not have to begin with a bar and the last cell does not have to end with a bar. For $n = 3$ $r = 4$, the 3 cells will have the appearance $||$. If we are required to place 4 balls in these 3 cells there are many ways of doing this. We can for instance place all the balls in the 1st cell and nothing in the two remaining cells. In this case our configuration of bars and balls will look like ****| | (here a * represents a ball and nothing in a cell implies an empty cell). Or we can place all 4 balls in cell two and nothing in cells 1 and 3. Or we can place 2 balls in cell 3 and one each in cells 1 and 2. In the later case the configuration will look like *|*|**. We need to keep track of the number of ways of arranging the 2 bars and the 4 stars (a total of 6 things occupying 6 places) in an arbitrary order and we are required to calculate the number of ways of choosing 4 places out of them. Since we have n-1 bars representing the n cells and r balls, a total of n-1+r places which can appear in an arbitrary order and we need to choose r places from among these $n-1+r$ places. Hence the result is proved. ∎

This problem is different from that of sampling without replacement in two different aspects: (1) r may not be less than or equal to n and (2) the same ball can not be drawn more than once.

Example 2.6 Consider 10 indistinguishable balls to be distributed among 6 urns. What is the probability that each urn has at least one ball? Using Bose-Einstein model which assumes that each of the distinguishable arrangements is equally likely.

According to Lemma 1, there are $\binom{6+10-1}{10}$ distinguishable ways of placing these balls into the 6 urns. If each urn is to have at least one ball the remaining 4 balls may be distributed among the 6 urns in $\binom{6+4-1}{4}$ ways. Therefore the required probability, using the Bose-Einstein assumption is

$$\frac{\binom{6+4-1}{4}}{\binom{6+10-1}{10}}.$$

■

Example 2.7 There are openings for Health Services Research Ph.Ds at 7 pharmaceutical firms (one opening in each firm). Assume that there are 5 eligible candidates and that each interviewed for all 7 positions. What is the probability that each candidate will receive at least one offer? Assume that the positions are not identical.

Assume that the 5 candidates are equally likely to receive each of the 7 offers. There are 5^7 equally likely ways in which the 7 offers may be distributed among the 5 candidates. There are two possible scenarios under which each candidate will receive at least one offer: (1) one candidate receives three offers and the rest of the candidates receive one offer each (2) two candidates receive two offers each and the other three receive one offer each.

For scenario (1) there are $\binom{7}{3}$ different ways in which a particular candidate receives 3 offers and all the others receive one each. Since there are 5 candidates, there are $\binom{5}{1}$ ways in which any one of them could receive 3 offers while the others receive only one each in $4!$ ways. Hence there are $\binom{7}{3}\binom{5}{1}4!$ possible combinations to fit into this scenario. Similarly, if two candidates are to receive two offers each and the rest only one offer each, there are $\binom{5}{2}\binom{7}{2}\binom{5}{2}3!$ possible combinations for this scenario. Hence the probability that at least one offer will be received by each candidate is

$$\frac{\binom{5}{1}\binom{7}{3}4!+\binom{5}{2}\binom{7}{2}\binom{5}{2}3!}{5^7}=.21504 .$$

■

Note that this derivation is based on Maxwell-Boltzmann model (Read, 1983).

Example 2.8 Consider 50 Fortune 500 companies each with a Chief Executive Officer (CEO) and a President. A committee is being formed of 50 members to investigate corporate excesses. Assuming that the committee members will be chosen at random from among these 100 corporate officers (CEO's and Presidents), what is the probability that there will be one member from each of these 50 companies?

First of all there are $\binom{100}{50}$ ways of selecting 50 committee members from a pool of 100 individuals. Next, there are two ways of choosing a corporate officer from each company. Since there are 50 companies there are a total of 2^{50} ways of choosing one officer from each company. Hence the probability of choosing one officer from each company in the committee is

$$\frac{2^{50}}{\binom{100}{50}} .$$

To obtain the probability that a particular company is represented, it is best to first obtain the probability that a particular company is not represented and then obtain its complement. The probability that a particular company is not represented in the committee will be

$$p=\frac{\binom{98}{50}}{\binom{100}{50}} .$$

The required probability is $1-p$.

■

2.11 Probability and Simulation

We have seen that probability can be computed in several ways, including the use of the axioms of probability, the relative frequency method, logical reasoning, and geometric shape. With the development of computer technology, the probability of an event can also be approximated by computer simulation. Using simulation to compute an event's probability is especially convenient when the event is generated by 1) a sequence of complicated procedures or 2) is dependent on a compound probability model. For illustration purposes, consider a box containing four red balls and one white ball. A ball is randomly selected (i.e. we assume that all balls are equally likely to appear in a draw). The probability that the ball drawn is white can be approximated by the following procedures:

Divide the unit interval [0, 1) into five subintervals of equal length: [0, 0.2), [0.2, 0.4), [0.4, 0.6), [0.6, 0.8) and [0.8, 1.0). (Each subinterval is associated with a particular ball). Pick a random number from the computer. If the random number falls in [0, 0.2), we assume the white ball is drawn otherwise a red ball is drawn. Repeat the last procedure a large number of times (say 10,000 times);

Calculate the ratio of the number of times the selected random numbers fall in [0, 0.2) and the total number of drawings (in this case 10,000)

This ratio is approximately equal to the probability of choosing a white ball in a draw in which each ball has the same chance of being chosen. A SAS program simulating this ball drawing scheme for 10,000 times, has found that the approximate probability of getting a white ball in a draw is 0.1958.

The theoretical concepts of simulation as well as the notion of random number will be discussed in Chapter 11. The issue that the probability using simulation is different from the one calculated directly, will also be discussed.

Example 2.9 If the calculation of the probability in Example 2.7 is to be obtained by simulation, one would simulate the offer of each of the 7 positions to any of 5 candidates using the method similar to that of randomly choosing one ball from a box containing 5 balls as discussed above. After 7 positions are offered, the condition that at least one offer has been made to each candidate, will be checked. If this condition is met, then this iteration will be counted as a "success". At the end of the iterations, the ratio of the total number of "successes" to the total number of iterations should be close to the probability computed in Example 2.7. The SAS simulation computed the probability of this event as 0.2134 compared to 0.2150 when the combinatorial method was applied.

2.12 Borel Sets

A fine example of the application of set theory is its usefulness in describing collections of numbers on the real number line. The real number line is itself composed of numbers so tightly packed that we cannot even describe two adjacent numbers without leaving out infinitely many numbers between them. For example, if we are challenged with two rational numbers a and b, we may easily set aside the claim that no rational number (i.e. a number that is the quotient of two integers that are relatively prime numbers) lies between them by providing the counter example $(a+b)/2$. Of course the real numbers are more tightly packed than the rational numbers and we say that the real numbers are *dense*. That means any real number can be a limit of a sequence of rational numbers. Thus collections or sequences of real numbers cannot be used to completely enumerate the real line.

However, sequences of sets can completely contain any segment of the real line that we would like. We will, for illustrative purposes, confine ourselves to the segment of the real line between 0 and 1. Since we want to enclose the ends of the line (i.e. the point "0" and the point "1"), we will say that this interval is "closed"; the interval contains all of the real numbers between zero and one including the points "0" and "1". We describe an interval as *open* if it does not contain the endpoints of the interval.

We will define a Borel set as an interval of real numbers or a set produced by countably many unions and/or countably many intersections. Borel sets may be open or closed. For example [⅓, ⅔], is a Borel set that contains an infinite number of real numbers between ⅓ and ⅔ including the endpoints. There are uncountable Borel sets that are contained in the [0, 1] interval.

In mathematics, it is common to find non-overlapping sets that contain another set. This is a tool of study that is often applied to real line. For example, we might consider how we would do this for the interval (0, 1] (i.e,. the interval that contains all of the numbers that are greater than zero up to and including 1). An example of non-overlapping sets which contain the (0, 1] real numbers is a collection of Borel sets that are right semi-closed. For example, the sets, (0, ⅓], (⅓, ⅔], (⅔, 1], is a collection of three such Borel sets. They are "non-overlapping" which, using set theory terminology, means that the intersection of any of these sets is the null set. Their union however is equal to (0,1].

2.13 Measure Theory in Probability

Recall from the introduction to this chapter that one of the important developments of the application of mathematical analysis to probability and statistics was the development of measure theory. The focus of this section is a discussion of the application of measure theory to probability. In probability, measure theory will be implemented to accumulate probability over related but quite dissimilar events.

Measure theory has improved the reach of the theory of probability by greatly expanding the scope of events for which a probability can be computed. For example if we would randomly choose a number from a unit interval, what is the probability that this number is a rational number. It has accomplished this by extending the ability of probabilists to compute probability. Measure theory focuses on the process of accumulation. The major goal of probability is the computation of the probability of an event. Since this effort may be redefined as accumulating probability over a collection of outcomes, we should not be surprised that measure theory can be useful in this accumulation process. The direct application of measure theory is its capacity to dramatically increase the resilience of probabilists in computing the probability of a collection of outcomes which appear to be in very different categories or classes.

What appears complicated about measure theory is that the process of accumulation is commonly not just the simple application of one rule of measurement, but the use of different procedures to measure a quantity. However, what appears at first glance to be a complicated myriad of arbitrary rules turns out upon further examination to be precisely the combination of procedures required to accurately accumulate the required quantity. Some of the results of the application of measure theory to probability are spectacular and a little counterintuitive. However, we will start with some very elementary illustrations first.

2.14 Probability Accumulation

Students of probability might begin to appreciate the role of measure theory by considering exactly what they do when they must compute the probability of an event. The solution of this problem requires that students recognize the event for which they want to compute the probability, the counting being governed by the nature of the experiment. However, there are natural settings in which the process of accumulation must be altered.

Rather than being an unusual or an unnatural procedure, flexibility in the process of quantification is the exact requirement for the precise measurement of the desired quantity. As a simple example, consider the task of measuring earned wealth accumulation for a typical 5-year-old US boy over the course of his life. At the beginning of the process, everything that this five year old possesses (e.g. clothes) is purchased by his parents. The wealth that he has truly earned comes solely from his own weekly allowance, an allowance that can be measured by simply counting up the value of the coins that he is either paid or that he occasionally finds. Since gathering coins is the only way by which he accumulates his own independently earned worth, we are content to only count the value of coins.

However, as the boy grows he starts to accumulate wealth in other ways. One natural development is to earn not just coinage but paper money. This change poses a dilemma for our measurement of his wealth accumulation. If we continued to base our measure of his wealth solely on the value of coins, we would miss an important new source (perhaps the greatest source) of his earned wealth, and consequently, the estimate of this earned wealth would be in error. We therefore very naturally alter our wealth counting mechanism by now including a new counting mechanism – the accumulation of the value of paper money. Note here that the tools used to count money have changed (from coin value to a combination of coin value and paper money value), but not the goal of measuring his accumulated wealth. Our counting mechanism had to flexibly adapt to the new economic situation if it was to remain accurate. Since accuracy is the key, we change the manner in which we count but we remain true to the process of assessing wealth accumulation.

Further changes in how we accumulate wealth are required as our subject prepares to go to college. How should the process adopt to the mechanism of the boy (now a young man) who uses his own money to buy a car? Simply continuing to merely count coin and paper money value as a measure of his independently acquired wealth would possibly produce inaccuracies. Our accumulation process, already having adapted once to the inclusion of paper money, must adapt again to include a new form of wealth. Since self purchased material items must be included in the consideration of the young man's wealth, our counting rules must now count coin value, paper money value, and the extent to which that money is exchanged for material things. Again our rules of accumulation have had to adapt to the changing, increasingly complex reality of the circumstances[1].

Just as the rule of wealth counting had to adapt to the increasing complexity of wealth accumulation, so our methods of accumulating probability will have to flexibly respond to the measure of probability in increasingly complex experiments.

Example 2.10 Attracting Objects

As an application of this process to probability, consider the following scenario. An object in the atmosphere is falling to the ground. Based on its trajectory, its speed and the time, this object has been computed to land within a one mile by one mile square of land. Our task is to compute the probability that this object will land at any arbitrary location within this square. As a first approximation, it is natural to assume all locations in the square are equally likely as the point of impact. From this assumption, a useful tool to use to compute probability is area. This notion is in complete accordance with the rules for computing (or accumulating) probability. The total area of this grid is 1 (square mile). The probability that the object lands in a particular region depends on the area of that region. For example, the probability that the object will land within a rectangle that is completely contained in this square and is 0.25 miles across and 0.5 miles long is $(0.25)(0.50) = 0.125$. Using area as probability is what in measure theory is called *Lebesque measure*, as described in Chapter 1. We can write the probability that the object falls in region A wholly contained in the square is simply the area of A. Using this rule, the computation of probability is simply the measurement (or accumulation of area).

The implication of this rule is that the probability that the object lands on any particular singleton point (e.g. the location (0.34, 0.71) is zero. This is true in general for any particular point, a natural consequence of the underlying assumption that accumulating probability is synonymous with accumulating area, and that points by definition do not have area or can be viewed by shrinking from a series of circles centered at this point and with radius shrinking to 0. While the recognition that points do not have area is as undeniable as the recognition that paper money is not coin money, this observation may not reflect the reality of this situation.

Consider the following scenario. Two lines equally divided this unit square are drawn and the middle third is excluded. For the remaining two rectangles, two similar lines each are drawn and the middle strips are removed. If we continue this process, what is the remaining area? In other words, how can we proceed to calculate the probability that the meteor lands in this area? Can the measurement of this area be carried out by a few simple procedures such as that in

---

[1] The consideration of depreciation of these material assets over time is yet one more easily added addition to our increasingly complex rules in estimating this individual's wealth.

elementary geometry? In this scenario, only measure theory can help calculate the remaining area. Note that this area is called the Cantor set in this unit square.

From the measure theory perspective consider the following complication. At a certain number of particular locations within the one mile square there is a combination of metals that act as attractors for the inbound object. Define a location X, Y on the surface as (X, Y). At these clearly identifiable points, the probability that the object lands at these points is not proportional to the area at all. For example let there be three of these "attractors", with coordinates, (0.25, 0.75), (0.50, 0.50), and (0.75, 0.90), respectively. Let the probability that the objects lands at these points be P(0.25,0.75) = 0.30, P(0.50,0.50) = 0.20, and P(0.75, 0.90) = 0.15. How can the probability of the object landing anywhere in the square be computed now?

The application of measure theory in this matter is clear. This theory tells us that we are free to construct a function which measures probability not as area alone and not as point mass alone (so called counting measure), but as a combination, as long as how we accumulate the probability is consistent with the axioms enunciated earlier in this chapter. We can therefore proceed as follows. The probability associated with the three attractor points is $0.30 + 0.20 + 0.15 = 0.65$. The probability that the object lands anyplace else in the grid is 0.35. So, we can accumulate measure of the region A as 0.35 times the area of A plus the sum of the probabilities associated with each of the three attractors within any of the regions of A. This may be written as

$$P(A) = 0.35 Area(A) + 0.35 I_A((0.25, 0.75))$$
$$+ 0.20 I_A(0.50, 0.50) + 0.15 I_A(0.75, 0.90)$$

where $I_A(x)$ is the indicator function, equal to 1 if $x \in A$ and 0 if $x \notin A$. We can observe how this works. Let A be the circle of radius 0.25 centered at (0.5, 0.5). Then the area of this circle is $.0625\pi$. This circle contains the point (0.50, 0.50). Thus, $P(A) = 0.35(0.0625\pi) + 0.20 = .0219\pi + 0.20$ or approximately 0.269.

2.15 Application of Probability Theory: Decision Analysis

Decision analysis is a technique for making sound decisions under conditions of uncertainty. The theory was originally developed for business applications by Raiffa and Schlaifer (1961) who have pioneered work in the field mostly from the managerial perspective. More recently, this tool has found its way into medicine and public health as well as cost effectiveness analysis. Weinstein and Fineberg (1980) in their ground breaking book *Clinical Decision Analysis* give several examples of application of this technique in clinical problems.

In public health, as in medicine, the decisions that need to be made regarding a problem or patient under consideration have to be stated and the final outcomes associated with the various decisions are clearly defined. The problem in hand is then divided into a string of smaller decisions and choices each with its set of possible outcomes. The realization of an outcome may lead to yet another string of outcomes. In other words, the starting point may branch itself out into a whole lot of branches each corresponding to a possible outcome. A graph of this process appears like a tree with its branches. Hence the term "decision tree" is often associated with this type of decision making process.

By convention the decision tree is built from left to right. The tree consists of several nodes; these are the points at which either a choice (or decision) is made or it is a point at which several outcomes beyond the control of the decision maker may occur. The later points are called chance nodes. The branches stemming from choice nodes do not have probabilities associated with them. Only the branches stemming from chance nodes have probabilities associated with them. In Figure 2.4 the symbol □ will represent a choice node and **O** will represent a chance node. The sum of all probabilities associated with possible outcomes from a chance node must add to one. In order to make an optimal decision at a choice node, the decision maker must fold back by selecting the branch that is associated with the highest average probability in favor of the outcome of interest or the most preferred outcome. At chance nodes the decision maker averages out the probabilities on all the branches that emanate from that node and makes the decision based on the path associated with the maximum weighted average probability of the most desirable outcome (which may be a monetary gain, number of days of hospitalization averted or five year survival probability). In general, the probability of an event at any point in time is the conditional probability of that event given all the events preceding it have occurred.

It must be repeatedly emphasized that clinical decision analysis is not a substitute for sound medical judgment of the clinician but only provides the decision maker with an additional perspective. It also provides a point of view based on probabilities of various outcomes such as five year mortality, duration of stay in the hospital, cancer or no cancer etc. which are estimated using national data bases. In other words, if the patient were to fit a certain "average" profile and if thus and such procedures were performed on him or her, the probabilities associated with the possible outcome of the procedure represent the likelihood of certain outcomes associated with the procedure. It can be argued that this process of decision making simplifies reality and most physicians are not trained to think in probability terms. Furthermore an argument made against its use is that it is difficult to justify when the patient has multiple health problems. However, the proponents of this theory believe that since the probabilities associated with possible outcomes are based on extensive impartial evidence, the decision maker arrives at a determination consistent with the evidence.

Example 2.11 A Case Report
Mrs. Smith is 65 years old and is admitted into the hospital with severe weakness, jaundice, fatigue, and abdominal pain. On a physical examination together with some laboratory work the physician observes an enlarged liver, and elevated liver

enzymes. Because the symptoms are somewhat troublesome (she had been talking to a friend who has had liver cancer) Mrs Smith wants to know from her attending physician if she is a candidate too.

Based on the information, the physician may wait and watch to see if the symptoms will disappear, or she may recommend a liver biopsy. The cost of liver biopsy is around $7000 including a one day stay in the hospital. Assume that liver biopsy is considered the best diagnostic test available in detecting cancer of the liver. And if one is detected, then the next step is to perform surgery to remove the tumor and surrounding tissue if feasible. Mrs. Smith is quite concerned about undergoing surgery since her mother underwent surgery for breast cancer and died on the operating table from an anesthesia related medical error. Hence Mrs. Smith has serious apprehensions about undergoing a biopsy and subsequent surgery if needed. Her physician is an expert in Decision Sciences and was able to visualize the sequence of events that may occur subsequent to a liver biopsy and was aware of the probabilities associated with those events. Since data on liver cancer survival are available for only five years, the physician was interested in survival at five years as the outcome of interest. Figure 2.4 below provides us with a simplistic decision tree for the pathway liver biopsy based on several assumptions:

(1) The liver biopsy will definitely detect cancer if there is one (perfect information)
(2) If a biopsy on an individual with these symptoms indicates no cancer then we assume it to imply no severe liver disease
(3) The probabilities associated with the different branches stemming out of chance nodes are hypothetical.

The reasoning presented may be generalized to the case where at each chance node and each choice node there are more than two possibilities.

Assigning a value 1.0 to being alive at five years and a value 0 to being dead by five years, we can calculate the value of each chance node in the tree by multiplying the probability of the event with the value of the outcome as follows:

At chance node D the weighted probability of survival is : $0.5(1) + 0.5(0) = 0.5$

At chance node E it is: $0.2(1) + 0.8(0) = 0.2$
At chance node F it is: $.01(1) + 0.99(0) = .01$
At chance node C it is: $0.7(.50) + 0.3(0) = 0.35$

At choice node 3, the choice is between five year survival with probability 0.2 with medical management versus five year survival probability of 0.01 under no medical management. Clearly at choice node 3, the decision maker will make a decision in favor of the medical management. Folding back to choice node 2, the choice is between surgery and no surgery with medical management. The values associated with these choices are 0.35 and 0.2 respectively. Here again the choice is clear and the decision maker will clearly choose the surgery route since it offers a probability of 0.35 for five years of survival. The value associated with chance node A is:

$$0.35(0.8)(1) + 0.95(0.2)(1) = 0.47.$$

Next let us look at the no biopsy route. The value associated with chance node G and H is 0.3 and 0.01 respectively, clearly the choice at choice node 4 is, if no biopsy is performed to go via medical management route with an associated probability of five year survival of 0.30.

Now the decision maker at choice node 1 is faced with the choice of whether or not to perform biopsy. When Mrs Smith is informed about the choices with associated probabilities for five year survival, she compares the probability 0.47 of five year survival if she goes for biopsy and subsequent surgery (if required) versus 0.30 probability of five year survival if she goes the no biopsy route. The choice is clear. However in our example since Mrs. Smith is very afraid of anesthesia, she may take a chance and decide not to have a biopsy. In this case her weighted probability of survival to five years is only 0.30.

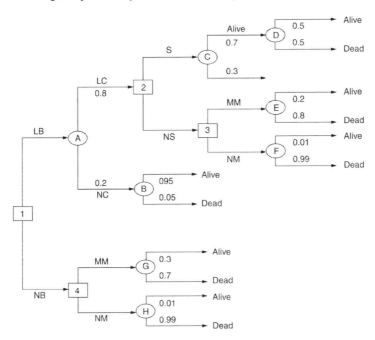

LB=Liver Biopsy , NB= No Liver Biopsy, LC=Liver Cancer,
NC=No Liver Cancer, S=Surgery, NS=No Surgery
MM= Medical Management of the Disease, NM= No Medical Management of the Disease

Figure 2.5 Decision Tree for the Pathway Liver Biopsy

Problems

1. An unbiased coin is tossed k times and exactly k_1 heads showed up. Prove that the probability that the number of heads exceeded the number of tails throughout the tossing is $\dfrac{2k_1 - k}{k}$.

2. Two names are picked at random from a list of k different names. Describe the sample space generated by this process and state the total number of points in the sample space

3. Generalize Problem 2 to the case where k_1 names are drawn at random from a list of k name

4. Four friends hang their identical looking (except for their wallets in the inside pocket) raincoats when they arrive at a friend's home for dinner. Assume they all leave together (two hours later) and each picks up a raincoat at random. Find the probability that each picks up the wrong raincoat.

5. A crate contains M toasters out of which m are defective (assume that the toasters are randomly arranged within the crate). A person tests them one at a time until a defective toaster is found. Describe the sample space associated with this operation. Find the probability that in the first k tests exactly 3 defective toasters are found.

6. On Sundays the neighborhood grocery store opens at noon and closes at 5pm. A woman walks into the store at time t_1 and is out at time t_2 (both t_1 and t_2 are measured in hours on the time axis with 12 pm as the origin). Describe the sample space of (t_1, t_2).

7. Events $B_1, B_2, \cdots B_M$ are such that the probability of the occurrence of any specified m of them is p_m, $m = 1, 2, \cdots M$.
   a) Show that the probability of the occurrence of one or more of them is

   $$\binom{M}{1} p_1 - \binom{M}{2} p_2 + \cdots + (-1)^{m-1} \binom{M}{M} p_M.$$

   b) The probability of the occurrence of m or more of the events $B_1, B_2, \cdots B_M$ is

$$\binom{m-1}{m-1}\binom{M}{m}P_m - \binom{m}{m-1}\binom{M}{m+1}P_{m+1} + \cdots + (-1)^{M-m}\binom{M-1}{m-1}\binom{M}{M}P_M.$$

c) The probability of the occurrence of exactly m of the events $B_1, B_2, \cdots B_M$ is

$$\binom{m}{m}\binom{M}{m}P_m - \binom{m+1}{m}\binom{M}{m+1}P_{m+1} + \cdots + (-1)^{M-m}\binom{M}{m}\binom{M}{M}P_M.$$

8.   Show that $2^n - n - 1$ conditions must be satisfied for n events to be independent.

9.   Using the definition of conditional probability show that for two events A and B, $P(A \mid B) + P(A^c \mid B) = 1$, provided that $P(B) \neq 0$.

10.   Mrs. Gray is 75 years old and has enrolled in a karate class. The probability that she will receive some federal grant for enrolling in the karate class is 0.30. If she gets the grant, then the probability she will get her "black belt" in karate is 0.85 and if she does not get the federal grant the probability that she will complete the course and get her "black belt" is only 0.45. What is the probability that she will get her "black belt"?

11.   The airline employs only three baggage checkers Adam, Brian and Carl. Adam, who checks 38% of all bags fails to put a security clearance tag on the bag 2% of the time, Brian, who checks 22% of the bags fails to tag the bags 8% of the time and Carl, who checks 40% of the bags fails to tag the bags 5% of the time. Find the probability that a bag cleared by these three men fails to get tagged.

12.   In an elegant restaurant in Manhattan the manager knows that the probability is 0.5 that a customer will have a glass of wine before dinner. If she does have a glass of wine, the probability is 0.66 that she will not order beef for main course and if she does not have a glass of wine before dinner the probability that she will not order a beef dish is 0.33. If she has a glass of wine and orders beef, then the probability of not ordering a dessert is 10%. On the other hand if she does have a glass of wine but does not order beef the probability that she will not order a dessert is 0.22. If she will not have a glass of wine and will not order beef then the probability of her not ordering a dessert is 0.50. The probability of ordering a dessert given that she did not have any wine and ordered beef is 0.18. What is the probability that she will order a dessert?

13. A box contains 90 good and ten defective light bulbs. If ten bulbs are chosen one at a time (without replacement) what is the probability that none of them are defective?

14. What is the probability that two throws with three dice each will show the same configuration a) the dice are distinguishable b) they are not?

15. Mrs. Jones's closet contains N sets of matching pants and jackets. If $2n$ pieces of clothing are removed from her closet, then what is the probability that there will be at least one complete set (i.e., matching pant and jacket)?

16. A population of N individuals contains Np women and Nq men $(p+q=1$, and Np and Nq are integers). A random sample of size r is chosen with replacement. Show that the probability of choosing exactly k women is

$$\binom{r}{k} p^k q^{r-k}.$$

17. The following problem refers to the classical occupancy problem (Maxwell-Boltzmann Statistic)
N balls are distributed among n cells. Show that the probability that exactly m cells remain empty is

$$\frac{\binom{n}{m}\binom{N-1}{N-n+m}}{\binom{N+n-1}{N}}.$$

18. A biased die when rolled has a probability of 0.5 of a six showing up all other numbers have a probability of 0.1. An individual is chosen to keep rolling the dice until number six or five shows up in four consecutive rolls. What is the probability that at most 10 rolls will be required?

19. A computer company is planning on setting up their new software headquarters anywhere in the US. They need to decide between 10 locales $L_1, L_2, \cdots L_{10}$. Let A_1 be the event that they will chose between locales L_1 or L_2, A_2 the event that they will choose between L_3 or L_4, A_3 the event that they will choose between L_5 and L_6, A_4 the event that they will choose between L_7 or L_8 and A_5 the event that they will choose between L_9 or L_{10}. List the elements in each of the following sets
    (a) A_2^c (b) A_4^c (c) $A_3 \cup A_1$ (d) $A_5 \cup A_3$ (e) $(A_1 \cup A_3)^c$.

20. Prove that if $A_1 \cap A_2 = A_1$ then $A_1 \cap A_2^c = \varnothing$.

21. The probabilities associated with grades A, B, C, D or F for a student in a Mathematical Statistics course are 0 .2, 0.3, 0.2, 0.2, and 0.1 respectively. What is the probability that a student will get (a) at least a C (b) neither an A nor an F (c) at most a B?

22. For roommates A and B living in a university town, the probability that A will vote for Mr. Brown in the upcoming primaries is 0.20, the probability that B will vote for Mr. Brown is 0.41 and the probability that they will both vote for the same candidate (Mr. Brown) is 0.13. What is the probability that neither of them will vote for Mr. Brown?

23 In a study about aircraft noise and schools around London's Heathrow Airport, schools were classified into three categories: government, church, and grant maintained. The noise levels were ranked as low, moderate and high. It was reported that 60% of the low-noise schools, 70% of moderate-noise schools and 83% of high-noise schools were government owned. It was also reported that, among all schools, 47.2% were ranked as low-noise, 38.2% were ranked as moderate-noise and 14.6% were ranked high-noise. What is the probability that a school randomly chosen from government-owned schools will be ranked as low noise? (*J. of Epidemiology & Community Health* (2002) 56:139-144)

24. For three events A_1, A_2 and A_3 show that
   (i) $(A_1 \cap A_2)^c = A_1^c \cup A_2^c$ (ii).

25. An urn contains 6 red chips numbered 1, 2, 3, 4, 5 and 6 and 4 blue chips numbered 1, 2, 3 and 4. Two chips are drawn at random one after the other and without replacement. What is the probability that their sum will be a) 10 b) at least 6 c) at most 8 d) an even number?

26. Let B_1, B_2 and B_3 be three boxes containing 3, 4 and 5 numbered compartments respectively. There are three red balls marked 1, 2 and 3, there are 4 blue balls marked 1, 2, 3 and 4 and there are 5 white balls marked 1, 2, 3, 4 and 5. Red balls are placed in B_1 at random, blue balls in B_2 at random and the white balls in B_3 at random. The balls are placed at random in the boxes in such a way that no compartment receives more than one ball. What is the probability that only one of the boxes will show no matches?

27. Show that A_1 and A_2 are independent if A_1 and A_2^c are independent.

28. Let A, B and C be three independent events. Show that A^c, B^c and C^c are all independent.

29. Mr. Fields receives 10 emails per week. What is the probability that he gets at least one email each day?

30. Mrs. Fisher sends 6 e-mails, one each to Drs. Smith, Brown, Jones, Green, Chong and Patel. There are no identifying names on the e-mail addresses. Assume that Mrs. Fisher sends these e-mails at random to the 6 doctors. What is the probability that a) exactly one doctor will receive the correct e-mail b) at least 3 will receive the correct e- mail.

31. A student teaching assistant needs to return the graded homework problems to the professor's office. She is given k computerized access cards to the faculty member's office. She tries the cards successively one at a time (sampling without replacement). What is the probability that she will succeed on the kth trial?

32. If k balls are placed at random in k urns, what is the probability that exactly two urns remain empty?

33. The Dean of a medical school requires the formation of a committee for faculty affairs. The committee is to consist of 4 clinical faculty members and four faculty members in basic sciences. The school has 100 clinical faculty of which 10 are women and 60 basic sciences faculty of which 20 are women. What is the probability that the committee will have at least two women?

34. Two players roll a fair die 2 times each. Each player receives a score based on the sum of the numbers that show up in the rolls. What is the probability that after four rolls their total score is the same?

35. It has been observed that 15% of children exposed to the West Nile virus acquire the disease. In a math class there are 40 children all of whom have been exposed to the virus. What is the probability that there will be at least one ill and at least one healthy child in the class?

36. A researcher develops a screening test for diabetic retinopathy among diabetics. She uses the screening test on known diabetic retinopathy patients and known diabetics without retinopathy. She finds that the test has a true negative rate of 95% (i.e., the screening test gives a negative result to those

who are known negatives) and a true positive rate of 80% (i.e., the screening test gives a positive result to those who are known positives). Satisfied with her development, she is now going to apply this test to a population of diabetics where it is known that the prevalence of the disease is 2.1%. Using Bayes Rule determine the probability that someone with negative test actually does not have the disease.

37. The Department of Biostatistics in a major university has 10 faculty members $A_1, A_2, \cdots, A_{10}$. The faculty is required to select a chairperson and a vice chairperson. Assume that one faculty member cannot occupy both the positions. Write down the sample space associated with these selections. What is the probability that A_1 is either a chair or a vice chair?

38. Let A_1 and A_2 be two events such that $P(A_1) = p_1 > 0$ and $P(A_2) = p_2 > 0$ and $p_1 + p_2 > 1$. Show that $P(A_2 \mid A_1) \geq 1 - \left[\dfrac{1-p_2}{p_1} \right]$

39. Let $A_1, A_2, \cdots A_n$ be a set of possible events. Show that (The Principle of Inclusion-Exclusion)

$$P(\bigcup_{i=1}^{n} A_i) = \sum_{i=1}^{n} P(A_i) - \sum_{i_1 < i_2}^{n} P(A_{i_1} \cap A_{i_2})$$

$$+ \sum_{i_1 < i_2 < i_3}^{n} P(A_{i_1} \cap A_{i_2} \cap A_{i_3}) + \cdots + (-1)^{n-1} P(\bigcap_{i=1}^{n} A_i).$$

40. Generalize Bonferroni's Inequality to n events $A_1, A_2, \cdots A_n$ by showing

$$\sum_{1}^{n} P(A_i) - \sum_{i<j}^{n} P(A_i \cap A_j) \leq P(\bigcup_{i=1}^{n} A_i) \leq \sum_{i=1}^{n} P(A_i)$$

References

Bartoszynski, R. and Niewiadomska-Bugaj, M. (1996), *Probability and Statistical Inference,* John Wiley and Sons, New York.
Bayes, T. (1763), An essay towards solving a problem in the doctrine of chances, *Philosophical Transactions,* 53.
Berger, J.O. (1985), *Statistical Decision Theory and Bayesian Analysis,* 2nd ed., Springer-Verlag, New York.

Casella, G. and Berger R.L. (2000), *Statistical Inference,* 2nd ed., Duxbury Press, Belmont, Mass.

Chung, K.L. (1974), *A Course in Probability Theory,* Academic Press, New York.

Colton, T. (1974), *Statistics in Medicine,* Little, Brown and Company, Boston.

Cox, R. and Hinkley, D.V. (1974), *Theoretical Statistics*, Chapman and Hall, London.

Dudewicz, E.J. and Mishra, S.N. (1988), *Modern Mathematical Statistics*, John Wiley and Sons, New York.

Feller, W. (1965), *An Introduction to Probability Theory and Its Applications,* Vol I, 2nd ed., John Wiley and Sons, New York.

Feller, W. (1971), *An Introduction to Probability Theory and Its Applications*, Vol II, John Wiley and Sons, New York.

Freund, J.F. (1971), *Mathematical Statistics*, 2nd ed., Prentice Hall, Englewood Cliffs, New Jersey.

Haines, M., Stansfeld, S., Head, J. and Job, R. (2002), Multilevel modeling of air craft noise on performance tests in schools around Heathrow *Epidemiology & Community Health* 56:139-144. airport London, *J. of*

Hogg, R.V. and Craig, A.T. (1978), *Introduction to Mathematical Statistics*, 4th ed., Macmillan Publishing Company, New York.

Kapadia, A.S., Chan, W., Sachdeva, R., Moye, L.A. and Jefferson, L.S. (1998), Predicting duration of stay in a pediatric care unit: A Markovian approach, *European Journal of Operations Research.*

Kaufmann, A., Grouchko, D. M. and Cruon, R.M (1977), *Mathématiques pour l'étude de la fiabilité des systèmes,* vol 124, Academic Press, New York.

Kendall, M.G. (1952), *The Advanced Theory of Statistics,* Vol I, 5th ed., Charles Griffin and Company, London.

Lairson, D.R., Pugh, J.A., Kapadia, A.S. et al., (1992), Cost effectiveness of alternative methods for diabetic retinopathy screening, *Diabetics Care*, Vol 15.

Lindley, D. V. (c1972), *Bayesian Statistics*: *A Review*, Society for Industrial and Applied Mathematics, Philadelphia.

Mood, A.M., Graybill, F.A., and Boes D.C. (1974), *Introduction to the Theory of Statistics,* 3rd ed., McGraw-Hill, New York.

Mosteller, F. (1965), *Fifty Challenging Problems in Probability with Solutions*, Addison-Wesley, Reading, Mass.

Pfeffer, R.I., Afifi, A. A. and Chance, J. M. (1997), Prevalence of Alzheimer's disease in a retirement community, *American Journal of Epidemiology*, 123(3).

Pugh, J.A., Jacobsen, J., VanHeuven, W.A.J., Watters, A., Tuley, M., Lairson, D.R., Lorimor, R.J., Kapadia, A.S. and Valez R. (1993), Screening for diabetic retinopathy in a primary care setting: The wide angle retinal camera , *Diabetic Care*, Vol 16.

Raiffa, H. and Schlaifer, R. (1968), *Applied Statistical Decision Theory*, 2nd ed., MIT Press, Cambridge, Mass.

Rao, C.R. (1952), *Advanced Statistical Methods in Biometric Research*, John Wiley and Sons, New York.

Rao, C.R. (1977), *Linear Statistical Inference and Its Applications*, 2nd ed., John Wiley and Sons, New York.

Read, C.B. (1983), Maxwell-Boltzmann, and Bose-Einstein Statistics, *Encyclopedia of Statistical Sciences*, Vol 3.

Rice, J.A. (1995), *Mathematical Statistics and Data Analysis,* 2nd ed.,Duxbury Press , Belmont, Mass.

Rohatgi, V.K. (1975), A*n Introduction to Probability Theory and Mathematical Statistics*, John Wiley and Sons, New York.

Rosner, B. (2000), *Fundamentals of Biostatistics,* 5th ed., Duxbury Press, New York.

Satake, I. and Koh S. (1962), *An Introduction to Probability and Mathematical Statistics,* Academic Press, New York.

Stigler S.M. (1986), *The History of Statistics: The Measurement of Uncertainty before 1900*, Harvard University Press, Cambridge, Mass.

Weinstein, M.C. and Fineberg, H.V. (1980), *Clinical Decision Analysis*, Saunders, Philadelphia.

Wilks, S.S. (1962), *Mathematical Statistics*, John Wiley and Sons, New York.

3

Random Variables

3.1 Introduction

We saw in Chapter 2 that probability is a function that maps an event in sample space to a number between 0 and 1. The random variable is the tool we need to describe the set on which the probability function operates. The values that a random variable takes are associated with the outcomes of an experiment. For example when a fair die is rolled, then any of the numbers 1, 2,…6 may appear and the random variable X may be defined as the number of 5's that may show up in 10 rolls of the die. In this case the sample space S will consist of all possible outcomes in the rolling of a die 10 times. There will in fact be a total of 6^{10} elements in S since each roll of a die may result in 1 or 2 or….6. The outcome of interest in this experiment may be whether there were zero 5's or one 5 or two 5's…or ten 5's. Hence we may define X as a random variable which takes values $0,1,…,10$ with probabilities $p_0, p_1…., p_{10}$ where p_i is the probability that in 10 rolls of a die there are exactly i 5's. A function X defined on a sample space S together with a probability function P on S is called a *random variable*. Since the random variables are based on the elements of the sample space S and the probability functions are based on the subsets of the sample space the probability function may be perceived to operate on X. Therefore an intelligent selection of a random variable will lead to the identification of probability for events of interest in S if the values that X takes is a subset s of S such that $s \in S$. In this case the function X transforms points of S into points on the x axis and we say that the probability function operates on X.

3.2 Discrete Random Variables

A random variable X is *discrete* if it takes a finite or countable number of values. Figure 3.1 below is a graphical representation of a discrete random variable that takes integer values in the interval (0, 5). We denote by p_i the probability that the random variable takes the value i.

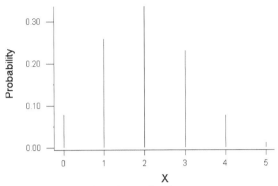

Figure 3.1 A graph of p_i

Example 3.1 An unbiased die is rolled 10 times and X is the random variable denoting the number of times 5 appeared. The number of 5's is between zero and ten. Define X as the random variable that represents the number of fives that occur in 10 rolls of a die. X is then a function that maps each point in the sample space to a number from 0 to 10. For example, a sample point that contains no 5 will be mapped to the number 0. We may represent the possible values X may take together with the corresponding probabilities in the form represented in Table 3.1 below.

Table 3.1 Probability distribution of X

x	0	1	2	3	4	5	6	7	8	9	10
$P(X = x)$	p_0	p_1	p_2	p_3	p_4	p_5	p_6	p_7	p_8	p_9	p_{10}

In this case $\sum_{i=0}^{10} p_i = 1.$ (3.1)

In the above example, the values of p_i can be obtained by actual enumeration.

For example, $P(X = 0)$ represents the probability that 5 appears zero times in 10 rolls of the die. The event $\{w, X(w) = 0\}$ denotes the event that all trials (rolls) result in an outcome other than 5. Let $T_{i,5}$ represent the event that the ith trial results in a 5. Similarly, define $T_{j,\bar{5}}$ as the event that on the jth trial something other than 5 appears. In order for the number of fives to be zero, the sequence of events in 10 trials must be of the form $T_{1,\bar{5}}, T_{2,\bar{5}}, \ldots T_{10,\bar{5}}$. Each of the events in this sequence of events occurs with probability $p_0 = \left(\dfrac{5}{6}\right)^{10}$. Similarly, in order for exactly one 5 to appear any of the following sequence of events must occur

$$T_{1,5}, T_{2,\bar{5}}, ... T_{10.\bar{5}}$$

$$T_{1,\bar{5}}, T_{2,5}, ... T_{10.\bar{5}}$$

....

$$T_{1,5}, T_{2,5}, ... T_{10.5}$$

By reasoning that we developed in Chapter 2, each of these sequences occurs with probability $\dfrac{1}{6}\left(\dfrac{5}{6}\right)^9$. Since there are 10 such sequences, the probability that is associated with the occurrence of exactly one 5 in 10 trials is

$$p_1 = 10 \left(\frac{5}{6}\right)^9 \left(\frac{1}{6}\right).$$

Similarly, by actual enumeration, the probability of two fives in ten trials can be calculated by listing all the different locations in the sequence of events which may result in the appearance of a 5. Since there are 10 such locations two 5's can appear in exactly $\dbinom{10}{2}$ ways. The corresponding probability is

$$p_2 = \binom{10}{2}\left(\frac{5}{6}\right)^8 \left(\frac{1}{6}\right)^2$$

In the same way the probability that the number of fives appear j times for $0 \le j \le 10$, may be obtained as $p_j = \dbinom{10}{j}\left(\dfrac{5}{6}\right)^{10-j} \left(\dfrac{1}{6}\right)^j$

Note that

$$p_0 + p_1 + ...p_{10} = 1$$

Recall that a discrete random variable X assumes only a finite and countable number of values. The probability mass function of a discrete random variable X that assumes values is $x_1, ..., x_n$ is $p(x_1), ..., p(x_n)$ where $p(x_j) = P(X = x_j)$ $j = 1, 2, ...$. In the above example, the quantity $P(X = x_i)$ represents the probability mass function associated with X. The probability mass function of X together with the values that X takes is the *probability distribution* of X. Note that $\displaystyle\sum_{\substack{\text{all possible}\\\text{values of } i}} p(x_i)$ 1. If one were to consider a subset A of the values $x_1, x_2, ...$ then the probability that the random variable takes a value in A is defined as

$P(x \in A) = \sum_{x \in A} p(x)$ where the summation is performed over all values of X that are contained in A. In the die example if A corresponds to the event that in 10 rolls of the die number 5 appears only even number of times, then

$$P(x \in A) = \sum_{j=0,2,4,6,8,10} \binom{10}{j}\left(\frac{5}{6}\right)^{10-j}\left(\frac{1}{6}\right)^{j}$$

As another example of a discrete random variable consider the situation where a general needs to formulate the strategy of moving his army around the enemy. It could move either around the enemy's left flank or around the right flank. Let $Y=0$ be the event that the army moves around the left flank. Similarly let $Y=1$ be the event that the army moves around the right flank. Although the general could make a decision based on predictable military strategies he decides instead to make it randomly. He chooses random numbers between 1 and 8 and $Y = 0$ when $X = 1, 2, 3, 4$ and $Y = 1$ when $X = 5$ to 8. In this example Y is a discrete event that is based on the values of X.

Chapter 4 will introduce the readers to several frequently used discrete distributions along with their properties. We will also consider functions of random variables. It is important to note that if X is a discrete random variable, then any function $h(X)$ of X is also a discrete random variable and it assumes values $h(x_i)$ with probabilities $p(x_i)$.

3.3 Cumulative Distribution Function

Associated with each random variable X (discrete or continuous) is a function defined by $P(X \le x) = F_X(x)$. This function represents the probability that the random variable X takes values smaller than or equal to x. The function $F_X(x)$ is also known as the *cumulative probability* or the *cumulative distribution function* (cdf) or simply the *distribution function* of the random variable.

Note that a cumulative density function is different from a probability density function and a probability mass function. In the case of discrete variables the probability mass function defines the probability that a discrete variable takes a specific value. These are also called *point probabilities*. The cumulative probability or $P(X \le x)$ is the sum of these point probabilities up to a certain value x.

Example 3.2 In section 3.2 we derived the probability distribution of X the number of times number 5 appeared in 10 rolls of a fair die. If we wish to obtain the probability $P(X \le 5)$, then we may write in terms of the cumulative density function notation as

$$F_X(5) = P(X \le 5) = \sum_{x=0}^{5} \binom{10}{x}\left(\frac{1}{6}\right)^x \left(\frac{5}{6}\right)^{10-x}.$$

Note that, for any random variable that takes values in the interval (a, b),

$$F_X(b) = 1 \text{ and } F_X(a) = 0, \text{ and}$$
$$P(a_1 \le X \le a_2) = F_X(a_2) - F_X(a_1) \text{ for } a_1 < a_2$$

The graph below illustrates the above probabilities. The graph below represents The probability $F_X(x)$.

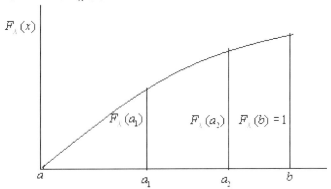

Figure 3.2 Graph of $F_X(x)$

A function $F_X(x)$ is a cdf if and only if

(a)   It is a non decreasing function of x,
(b) $\lim_{x \to -\infty} F_X(x) = 0$ and $\lim_{x \to \infty} F_X(x) = 1$
(c) $\lim_{\varepsilon \downarrow 0} F_X(x + \varepsilon) = F_X(x)$ i.e., $F_X(x)$ is right continuous.

3.4 Continuous Random Variables

If $F_X(x)$ is a continuous function of x, then the associated random variable X is called a continuous random variable. For a continuous random variable X, the range of values that X may take consist of uncountable many points. In this case, $P(X \le x) = F_X(x)$ is the same as $P(X < x) = \lim_{\varepsilon \to 0} F_X(x - \varepsilon)$. Therefore, the probability that the random variable X takes a particular value x is equal to zero, i.e.

$$P(X = x) = P(X \le x) - P(X < x) = 0.$$

In the continuous case, if the cdf is continuously differentiable, then there exists a function $f_X(x)$ such that $F_X(x) = \int_{-\infty}^{x} f_X(t)dt$. Here $f_X(x)$ is continuous except for countably many points and is referred to as the *probability density function* (or simply the density function) of X, and sometimes it is said that "X is distributed as $f_X(x)$" and $F_X(x)$ is referred to as absolutely continuous cdf. For the rest of the book, the cdf of continuous random variables will be assumed to be absolutely continuous.

In this case, the probability that X lies in the interval $(x, x + \Delta x)$ is approximately equal to $f_X(x)\Delta x$ where Δx is the length of an infinitesimally small interval. The continuous analog to summing probabilities is the formulation of an integral where the integration is performed over an interval. For $a < b$

$$P(X \in (a,b)) = \int_a^b f_X(x)dx.$$

For a density function $f_X(x)$, the total area under the *f(x)* curve is unity. Note that in the case X takes value in the finite interval (c, d)

$$\int_c^d f_X(x)dx = 1 \tag{3.2}$$

implying $f_X(x) = 0$ for $x \notin (c, d)$. By the Fundamental Theorem in Calculus it can be shown that

$$\frac{d}{dx}F_X(x) = f_X(x) \tag{3.3}$$

provided the derivative of $F_X(x)$ exists.

In Chapter 5 we will introduce several *continuous density functions* and their properties.

■

Example 3.3 Let X have a density function defined as

$$f_X(x) = 3x^2, 0 < x < 1$$

Then we may find the cumulative distribution function for X such that for

$$F_X(x) = \begin{cases} 0 & x \le 0 \\ x^3 & 0 < x < 1 \\ 1 & x \ge 1 \end{cases}$$

■

The best known and most commonly used continuous distribution in statistics is the *normal distribution*. The shape of its density function $f_X(x)$ is symmetric about its average value, social scientists commonly refer to it as the "bell" curve. U.S Population age data when graphed by age intervals roughly displays this property of symmetry about the average age. This is also true for income data.

The continuous analog of the probability mass function is the probability density function. As we said earlier, when X is a continuous random variable, the probability that X (a one dimensional or *univariate* random variable) takes a particular value is zero and instead $f_X(x)\,dx$ defines the probability that X lies between x and x+dx. In this case, the summation sign is replaced by an integral, and $f_X(x)$ is referred to as the probability density function or simply, density function. In other words the probability density function defines the weight or measure that is placed on an interval on the real line. For e.g., the probability density function associated with a normally distributed random variable with parameters μ and σ^2 is

$$f_X(x) = \frac{1}{\sigma\sqrt{2\pi}} e^{-\frac{1}{2\sigma^2}(x-\mu)^2} \qquad -\infty < x < \infty$$

For mathematical convenience, we sometimes use the indicator function $I_{(-\infty,\infty)}$ to write the pdf as $f_X(x) = \dfrac{1}{\sigma\sqrt{2\pi}} e^{-\frac{1}{2\sigma^2}(x-\mu)^2} I_{(-\infty,\infty)}$. This is the notation we will be using in later chapters.

3.5 Joint Distributions

By definition a *joint* probability function involves more than one random variable. If the probability function involves exactly two random variables then we may also refer to such probability functions as *bivariate*. The joint (bivariate) distribution function of random variables X_1 and X_2 is defined as follows

$$F_{X_1,X_2}(x_1,x_2) = P(X_1 \le x_1, X_2 \le x_2)$$

If X_1 and X_2 are discrete and assume discrete values in the intervals (a,b) and (c,d) respectively then the distribution function may be written as

$$F_{X_1,X_2}(x_1,x_2) = \sum_{i=a}^{x_1}\sum_{j=c}^{x_2} P(X_1 = i, X_2 = j)$$

The probability mass function for the discrete case is defined as

$$P(X_1 = x_1, X_2 = x_2) = p_{x_1,x_2}$$

such that

$$\sum_{x_1}\sum_{x_2} P(X_1 = x_1, X_2 = x_2) = 1$$

Similarly if X_1 and X_2 are continuous variables in the intervals (a,b) and (c,d) respectively, then the joint distribution (bivariate) function of X_1 and X_2 can be written as

$$F_{X_1,X_2}(x_1,x_2) = \int_a^{x_1} \int_c^{x_2} f_{X_1,X_2}(t_1,t_2)\,dt_2\,dt_1,$$

where $f_{X_1,X_2}(x_1,x_2)$ is the joint (bivariate) density function of X_1 and X_2 such that

$$\int_a^b \int_c^d f_{X_1X_2}(x_1,x_2)\,dx_2\,dx_1 = 1$$

The *marginal* density of X_1 may be obtained by integrating (or summing) over all values of X_2 subject to the constraints on their relationship.

Example 3.4 Consider the following joint distribution of two continuous random variables X_1 and X_2

$$f_{X_1,X_2}(x_1,x_2) = x_1 + x_2, \quad 0 < x_1, x_2 < 1$$

Then

$$F_{x_1,x_2}(x_1,x_2) = \int_0^{x_1} \int_0^{x_2} (t_1 + t_2)\,dt_2\,dt_1 = \int_0^{x_1} \left(t_1 x_2 + \frac{x_2^2}{2} \right) dt_1$$

$$= \left(\frac{x_1^2 x_2}{2} + \frac{x_2^2 x_1}{2} \right) = \frac{x_1 x_2}{2}(x_1 + x_2), \quad 0 < x_1, x_2 < 1$$

3.6 Independent Random Variables

Two random variables X_1 and X_2 are *independent* if for all possible subsets of real numbers A and B

$$P(X_1 \in A, X_2 \in B) = P(X_1 \in A)P(X_2 \in B)$$

Furthermore, if two random variables X_1 and X_2 are identically distributed then

$$F_{X_1}(t) = F_{X_2}(t) \quad \text{for all } t$$

Definition 3.1 Variables X_1 and X_2 are independent if and only if
$P(X_1 = x_1, X_2 = x_2) = P(X_1 = x_1)P(X_2 = x_2)$ for the discrete case and
$f_{X_1,X_2}(x_1,x_2) = f_{X_1}(x_1)f_{X_2}(x_2)$ for all x_1 and x_2 for the continuous case.
We will prove the assertion for the discrete case

Proof. Assume that X_1 and X_2 are both discrete and independent random variables and let A be the event $X_1 = x_1$ and B be the event $X_2 = x_2$. Then

$$P(X_1 = x_1, X_2 = x_2) = P(A \cap B)$$
$$= P(A)P(B)$$
$$= P(X_1 = x_1)P(X_2 = x_2)$$

Now to prove the converse let us assume that

$$P(X_1 = x_1, X_2 = x_2) = P(X_1 = x_1)P(X_2 = x_2) \text{ for all } x_1 \text{ and } x_2.$$

We will show that X_1 and X_2 are independent. Let A and B be two sets of real numbers, then

$$P(X_1 \in A, X_2 \in B) = \sum_{x_1 \in A_1} \sum_{x_2 \in A_2} P(X_1 = x_1, X_2 = x_2)$$
$$= \sum_{x_1 \in A_1} \sum_{x_2 \in A_2} P(X_1 = x_1)P(X_2 = x_2)$$
$$= \left\{ \sum_{x_1 \in A_1} P(X_1 = x_1) \right\} \left\{ \sum_{x_2 \in A_2} P(X_2 = x_2) \right\}$$
$$= P(X_1 \in A)P(X_2 \in B)$$

This proves our assertion. The proof for the continuous case is left for the reader to derive.

∎

3.7   Distribution of the Sum of Two Independent Random Variables

Let X_1 and X_2 be two independent discrete random variables and we are interested in the distribution of $Z = X_1 + X_2$, we may write

$$P(Z = z) = P(X_1 + X_2)$$
$$= \sum_r P(X_1 = r)P(X_2 = z - r)$$

The above summation is over all possible values of r. The distribution of Z is the *convolution* of the distributions of X_1 and X_2 for only integer values of X such that $P(X_1 = i) = p_i$ and $P(X_2 = j) = q_j$ then,

$$P(Z = z) = \sum_{i=0}^{z} P(X_1 = i)P(X_2 = z - i)$$

$$= \sum_{i=0}^{z} p_i q_{z-i}.$$

Example 3.5 Let X and Y be two independently distributed random variables with the following probability mass functions

$$P_X(X = x) = \frac{e^{-\theta_1} \theta_1^{x}}{x!} \quad 0 \le x < \infty$$

and

$$P_Y(Y = y) = \frac{e^{-\theta_2} \theta_2^{y}}{y!} \quad 0 \le y < \infty.$$

Define $Z = X + Y$

Then,

$$P(Z = z) = \sum_{x=0}^{z} \frac{e^{-\theta_1} \theta_1^{x}}{x!} \frac{e^{-\theta_2} \theta_2^{z-x}}{(z-x)!}$$

$$= e^{-(\theta_1 + \theta_2)} \sum_{x=0}^{z} \frac{\theta_1^{x}}{x!} \frac{\theta_2^{z-x}}{(z-x)!} \frac{z!}{z!}$$

$$= \frac{e^{-(\theta_1 + \theta_2)}}{z!} \sum_{x=0}^{z} \frac{z!}{x!(z-x)!} \theta_1^{x} \theta_2^{z-x}$$

$$= \frac{e^{-(\theta_1 + \theta_2)}(\theta_1 + \theta_2)^{z}}{z!}, \quad z = 0, 1, 2, \cdots$$

∎

The last equality is obtained using the binomial theorem from Chapter 1. The probability distribution assumed for X and Y is the *Poisson* distribution with parameters θ_1 and θ_2. Note that the above result shows that the sum of two independent Poisson variables X and Y also has a Poisson distribution with a parameter which is the sum of the parameters of X and Y. The Poisson distribution will be discussed in Chapter 4.

The continuous analog of the distribution function of Z is as follows. Let Z be the random variable which is the sum of two independent continuous random variables, then

$$P(Z \le z) = P(X_1 + X_2 \le z)$$

$$F_Z(z) = \int \int_{x_1 + x_2 \le z} f_{X_1, X_2}(x_1, x_2) dx_1 dx_2$$

$$= \int_{-\infty}^{\infty} \left(\int_{-\infty}^{z-x_1} f_{X_2}(x_2)dx_2 \right) f_{X_1}(x_1)dx_1$$

$$= \int_{-\infty}^{\infty} f_{X_1}(x_1) F_{X_2}(z-x_1)dx_1$$

The density function of Z may be obtained by differentiating both sides with respect to z and using the rule of differentiation under the integration sign (see Chapter 1) as

$$f_Z(z) = \int_{-\infty}^{\infty} f_{X_1}(x_1) f_{X_2}(z-x_1)dx_1 \tag{3.4}$$

We say that the density function $f_Z(z)$ is the *convolution* of the density functions $f_{X_1}(x_1)$ and $f_{X_2}(x_2)$ and may be expressed as

$$\{f_Z(z)\} = \{f_{X_1}(z)\} * \{f_{X_2}(z)\}$$

Example 3.6   Let X and Y be independently distributed according to the following density functions

$$f_X(x) = \theta_1 e^{-\theta_1 x} \quad 0 < x < \infty \text{ and}$$

$$f_Y(y) = \theta_2 e^{-\theta_2 y} \quad 0 < y < \infty$$

Let $Z=X+Y$. Then, using (3.4) above, the density function of Z is given by

$$f_Z(z) = \int_{x=0}^{z} f_X(x) f_Y(z-x)dx$$

$$= \int_{x=0}^{z} \theta_1 e^{-\theta_1 x} \theta_2 e^{-\theta_2(z-x)} dx$$

$$= \frac{\theta_1 \theta_2}{(\theta_1 - \theta_2)} \left[e^{-\theta_2 z} - e^{-\theta_1 z} \right] \quad 0 < z < \infty, \theta_1 > \theta_2$$

Notice that for $\theta_1 = \theta_2$, the last equality becomes

$$f_Z(z) = \theta_2^2 z e^{-\theta_2 z} \quad 0 < z < \infty$$

In Chapter 5, the density function of X and Y known as the *exponential density* function is discussed in greater detail. We will also show that the sum of k identical and independent exponential variables will be a useful distribution in its own right.

3.8 Moments, Expected Values and Variance

If X is a discrete random variable which takes values $x_1, x_2 .... x_N$ with probabilities $p_1, p_2, ... p_N$ such that $\sum_{i=1}^{N} p_i = 1$, then the *mean* of X is defined as

$$\mu = \sum_{i=1}^{N} x_i p_i .$$

The *mathematical expectation* of a discrete random variable is given by

$$E(X) = \sum_{i=1}^{N} x_i p_i$$

and is the same as the mean of the distribution of the random variable. For the continuous case, if X lies in the interval (a, b) and the density function of X is defined as $f_X(x)$ then the mean or expectation of X is defined as

$$\mu = \int_a^b x f_X(x) dx, \quad -\infty < a < b < \infty \tag{3.5}$$

In the discrete case N may be replaced by ∞ if the variable takes all integer values in $(1, \infty)$. Similarly a and b may be replaced by $-\infty$ and ∞. The mean value of a random variable is also called the first *moment* of the distribution of the random variable. The rth moment about the origin of the distributions is defined as

$$\mu'_r = \sum_{i=1}^{N} x_i^r p_i \text{ and }$$

$$\mu'_r = \int_a^b x^r f(x) dx$$

for discrete and continuous distributions respectively. The moments about the mean are similarly defined as

$$\mu_r = \sum_{i=1}^{N} (x_i - \mu)^r p_i \text{ and }$$

$$\mu_r = \int_a^b (x - \mu)^r f(x) dx \tag{3.6}$$

The *variance* is the second moment about the mean and is the sum of squares of the weighted deviations from the mean. It is a measure of dispersion (scatter) around the mean. In the above relationships, μ_2 is the variance in each case. The variance is very often represented by the symbol σ^2.

If X and Y have a joint density function of the form $f_{X,Y}(x, y), \quad a < x < b, c < y < d$.

Then the expectation of their product is defined as

$$E(XY) = \int_a^b \int_c^d xy f_{X,Y}(x,y)dydx$$

Furthermore it can be easily seen that if X and Y are independent of one another

$$E(XY) = E(X)E(Y)$$

The above result also holds for X and Y discrete.

Note that the above quantities are defined as expected values only if they are absolutely convergent.

Example 3.7 If X represents a random variable which takes a value 1 with probability p and a value 0 with probability $1-p$. This variable is one of the simplest of all random variables. In this case we say that X follows a *Bernoulli distribution*. Its expected value may be obtained as

$$E(X) = 1.p + 0.(1-p) = p$$

This random variable will be developed in greater detail in Chapter 4.

∎

Example 3.8 If X represents the number that shows up when an unbiased dice is rolled, then X takes values 1, 2, 3, 4, 5, and 6 with probabilities $\dfrac{1}{6}$ and its expected value may be obtained as

$$E(X) = 1.\frac{1}{6} + 2.\frac{1}{6} + 3.\frac{1}{6} + 4.\frac{1}{6} + 5.\frac{1}{6} + 6.\frac{1}{6} = \frac{7}{2}.$$

∎

We will obtain the expected values of more complicated distributions in Chapters 4 and 5. For continuous distributions the expected values are defined as

$$E(X^r) = \int_a^b x^r f(x)dx, \ r = 1, 2, \ldots,$$

For a discrete random variable

$$E(X^r) = \sum_x x^r P(X = x).$$

The *variance* of a random variable is defined as

$$E\{X - E(X)\}^2 = E(X^2) - [E(X)]^2.$$

Example 3.9 Let a continuous random variable X have the following density function

$$f_X(x) = 2x, \quad 0 < x < 1 \text{ then}$$

$$E(X) = \int_0^1 x\,(2x)dx = 2\int_0^1 x^2 dx = \frac{2}{3}$$

and

$$E(X^2) = \int_0^1 x^2(2x)dx = \frac{1}{2}.$$

The variance of X is given by

$$Var(X) = E(X^2) - [E(X)]^2$$

$$= \frac{1}{2} - \frac{4}{9} = \frac{1}{18}.$$

If $g(X)$ is a function of a random variable it is also a random variable. Therefore,

$$E[g(x)] = \sum_x g(x)p_x$$

for the discrete case and

$$E[g(x)] = \int_x g(x)f_X(x)dx$$

for the continuous case.
In Example 3.9 assume that

$$g(X) = (3X^2 - 1)$$

then

$$E[g(X)] = \int_0^1 g(x)f_X(x)dx = \int_0^1 (3x^2 - 1)2xdx = \int_0^1 (6x^3 - 2x)dx$$

$$= \frac{3}{2} - 1 = \frac{1}{2}$$

and

$$E[g(X)^2] = \int_0^1 [(3x^2 - 1)^2 2xdx = \int_0^1 (18x^5 - 12x^3 + 2x)dx$$

$$= 3 - 3 + 1 = 1$$

$$Var[g(X)] = 1 - \frac{1}{4} = \frac{3}{4}.$$

Lemmas 3.2-3.5

(i) $g(x) = a$, then $E\{g(X)\} = a$

(ii) If $g(x) = ax + b$, then $E\{g(X)\} = aE(X) + b$ and if
$g(x) = a_0 + a_1 x + a_2 x^2 + \quad a_k x^k$ then
$E\{g(X)\} = a_0 + a_1 E(X) + a_2 E(X^2) + ...a_k E(X^k)$

(iii) $Var[g(X)] = E[g(X) - E\{g(X)\}]^2$,
$g(x) = aX + b$, then $Var(aX + b) = a^2 Var(X)$
(iv) $E[X - E(X)] = E(X) - E(X) = 0$
Proofs of Lemmas 3.2-3.5 are obvious enough for the readers to derive on their own.
Lemma 3.6 $E(X - b)^2$ is minimized when $b = E(X)$
Proof.
$$\begin{aligned} E(X - b)^2 &= E[X - E(X) + E(X) - b]^2 \\ &= E[X - E(X)]^2 + E[E(X) - b]^2 + 2E\left[\{X - E(X)\}\{E(X) - b\}\right] \\ &= E[X - E(X)]^2 + E[E(X) - b]^2 + 2[E(X) - b]E[X - E(X)] \\ &= E[X - E(X)]^2 + E[E(X) - b]^2 \end{aligned}$$
The RHS in the above Lemma is the sum of two positive quantities and the sum can be minimized when $b = E(X)$ It is believed that a small variance implies that large deviations from the mean are less likely. The following inequality known as the Chebychev's inequality presents this point more precisely and will be discussed in detail in Chapter 7.

Inequality 3.1 Let X be a random variable with mean μ and variance σ^2. Then for any $t > 0$ the following holds
$$P[|X - \mu| \geq t] \leq \frac{\sigma^2}{t^2}$$
In order to prove the above inequality we are assuming that the variable is continuous with density function $f_X(x)$. The proof for the discrete case can be obtained similarly.

Proof.

$$\sigma^2 = \int_{\Omega_X} (x - \mu)^2 f_X(x) dx \geq \int_{|x - \mu| \geq t} (x - \mu)^2 f_X(x) dx$$

$$\sigma^2 = \int_{\Omega_X} (x-\mu)^2 f_X(x)dx \geq \int_{|x-\mu| \geq t} (x-\mu)^2 f_X(x)dx$$

$$\geq t^2 \int_{|x-\mu| \geq t} f_X(x)dx = t^2 P[|X-\mu| \geq t]$$

The first inequality above is due to the fact that a partial integration of a non-negative integrand is always less than or equal to the full integration of the same integrand. The second inequality is just a comparison of two integrals.

A slight variation of the above inequality is the result presented below. Let $h(X)$ be a non negative function of X. Then we know that for $t>0$

$$E[h(X)] = \int_{-\infty}^{\infty} h(x)f_X(x)dx \geq \int_{h(x) \geq t} h(x)f_X(x)dx$$

$$\geq t \int_{h(x) \geq t} f_X(x)dx = tP[h(X) \geq t]$$

or

$$P[h(X) \geq t] \leq \frac{E[h(X)]}{t}. \qquad \blacksquare$$

Example 3.10 If X is the number that appears in rolling an unbiased die, then as we obtained in Example 3.8

$E(X) = \mu = \dfrac{7}{2}$ and the variance may be calculated as $E(X^2)-[E(X)]^2 = \sigma^2 = \dfrac{35}{12}$.

In this case according to the Chebychev's Inequality

$$P\left[\left|X - \frac{7}{2}\right| \geq 2.5\right] = \frac{1}{3} \leq \left[\frac{35/12}{(2.5)^2}\right] = 0.47$$

In reality in this example the probability of an absolute deviation greater than 2.5 is zero. The Chebychev's inequality merely states that the probability is less than or equal to 0.47. $\qquad \blacksquare$

The direct application of Chebychev's inequality reveals that for any random variable X that has a mean μ and variance σ^2 the following relationship

holds $\qquad\qquad P[|X - \mu| \geq 2\sigma] \leq \dfrac{\sigma^2}{(2\sigma)^2} = 0.25$.

This is an important result in statistics reinforcing the fact that at least 75% of the observations in a population lie within two standard deviations of the mean regardless of the distribution from which the random sample is drawn.

Lemma 3.7 If X is a non negative continuous random variable such that the integrand and the expectation both exist then,

$$E(X) = \int_0^\infty [1 - F_X(x)]dx.$$

Proof.

$$\int_0^\infty \{1 - F_X(x)\}dx = \int_{x=0}^\infty \int_{y=x}^\infty f_X(y)dy dx = \int_{y=0}^\infty \int_{x=0}^y f_X(y)dx dy$$

$$= \int_{y=0}^\infty y f_X(y)dy = E(X)$$

Please refer to Chapter 1 for more on changing the order of integration.

It can be easily shown that the discrete analog of the above relationship

$$E(X) = \sum_{x=0}^\infty [1 - F_X(x)] \text{ also holds}$$

3.8.1 Sampling Distribution of Sample Mean and Variance

Let $X_1,\ldots,X_n$ be a sample of values from a population with mean μ and variance σ^2. We will assume that these values are drawn using the method of *simple random sampling* which requires that each and every sample of size n has exactly the same probability of being selected. Then $X_1,\ldots,X_n$ can be considered as a set of n identically distributed random variables such that $E(X) = \mu$ and $Var(X) = \sigma^2$.

If $\bar{X}$ is the sample mean, then $\bar{X} = \dfrac{1}{n}\sum_{i=1}^n X_i$

Since the X's are all independently and identically (iid) distributed variables

$$E(\bar{X}) = E\left(\frac{1}{n}\sum_{i=1}^n X_i\right) = \frac{1}{n}\sum_{i=1}^n E(X_i) = \frac{1}{n}(n\mu) = \mu$$

Similarly,

$$Var(\bar{X}) = E\left(\bar{X} - \mu\right)^2 = E\left[\frac{1}{n}\sum_{i=1}^n X_i - \mu\right]^2$$

$$= \frac{1}{n^2}E\left[\sum_{i=1}^n (X_i - \mu)\right]^2 = \frac{1}{n^2}\sum_{i=1}^n \sigma^2 = \frac{\sigma^2}{n}$$

Notice that in the above derivation since the X's are independent $E[(X_i - \mu)(X_j - \mu)] = E(X_i - \mu)E(X_j - \mu) = 0$ for $i \neq j$.

Define the sample variance as

$$S^2 = \frac{\sum_{i=1}^{n}(X_i - \bar{X})^2}{n-1}$$

The right hand side of above expression may also be written as

$$S^2 = \frac{1}{n-1}\left[\sum_{i=1}^{n}(X_i - \mu + \mu - \bar{X})^2\right]$$

$$= \frac{1}{n-1}\left[\sum_{i=1}^{n}(X_i - \mu)^2 + n(\mu - \bar{X})^2 + 2(\mu - \bar{X})\sum_{i=1}^{n}(X_i - \mu)\right]$$

$$= \frac{1}{n-1}\left[\sum_{i=1}^{n}(X_i - \mu)^2 + n(\mu - \bar{X})^2 - 2n(\mu - \bar{X})^2\right]$$

$$= \frac{1}{n-1}\left[\sum_{i=1}^{n}(X_i - \mu)^2 - n(\mu - \bar{X})^2\right]$$

$$= \frac{1}{n-1}\left[\sum_{i=1}^{n}(X_i - \mu)^2 - \frac{1}{n}\left(\sum_{i=1}^{n}(X_i - \mu)\right)^2\right]$$

(3.7)

After a little manipulation the above relationship can be written as

$$S^2 = \left[\frac{1}{n}\sum_{1}^{n}(X_i - \mu)^2 - \frac{1}{n(n-1)}\sum_{1}^{n}\sum_{\substack{1 \\ i \neq j}}^{n}(X_i - \mu)(X_j - \mu)\right]$$

(3.8)

We will now obtain the mean and variance of S^2

$$E(S^2) = \frac{1}{n-1}E\left(\sum_{i=1}^{n}(X_i - \bar{X})^2\right) = \frac{1}{n-1}\sum_{1}^{n}\left[E(X_i^2) + E(\bar{X}^2) - 2E(X_i\bar{X})\right]$$

We know that, $E(\bar{X}^2) = \frac{\sigma^2}{n} + \mu^2$, $E(X_i^2) = \sigma^2 + \mu^2$ and

$$E(X_i\bar{X}) = E\left[X_i\left(\frac{1}{n}\sum_{i=1}^{n}X_i\right)\right] = \frac{1}{n}E(X_iX_1 + X_iX_2 + ... + X_i^2 + ... + X_iX_n)$$

$$= \frac{1}{n}(n\mu^2 + \sigma^2)$$ here we have used the fact that the X's are independent

and that the expectation of their product is the product of the expectations i.e.,

$$E(X_iX_j) = E(X_i)E(X_j) = \mu^2$$

Hence,

$$E(S^2) = \frac{1}{n-1}\left[n(\sigma^2 + \mu^2) + n\left\{\frac{\sigma^2}{n} + \mu^2\right\} - 2n\frac{1}{n}(n\mu^2 + \sigma^2)\right]$$

$$= \frac{1}{n-1}(n-1)\sigma^2 = \sigma^2$$

The above relationship shows that S^2 is an unbiased estimate of σ^2. We will learn more about unbiased estimators in Chapter 8.

When we square both sides of S^2 in (3.8) and take the expected value of S^4, the expected values associated with terms of the form

$$(X_i - \mu)(X_j - \mu), (X_i - \mu)^2(X_j - \mu), (X_i - \mu)^2(X_j - \mu)(X_k - \mu)$$

will be zero for $i \neq j \neq k$ since the X's are all independent.. Hence

$$E(S^4) = \frac{1}{n^2}\left(n\mu_4 + n(n-1)\sigma^4\right) + \frac{2n(n-1)}{n^2(n-1)^2}\sigma^4 = \frac{\mu_4}{n} + \frac{n^2 - 2n + 3}{n(n-1)}\sigma^4$$

where $\mu_4 = E(X_i - \mu)^4$ (i.e., the fourth moment about the mean).

Hence $Var(S^2) = E(S^4) - (\sigma^2)^2 = \frac{1}{n}\left(\mu_4 - \frac{n-3}{n-1}\sigma^4\right)$

3.8.2 Chebychev's Inequality

Let $\bar{X}$ denote the mean of a random sample of n observations from a population with mean μ and variance σ^2. Then the distribution of $\bar{X}$ has a mean μ and variance $\frac{\sigma^2}{n}$. Hence for any $\varepsilon > 0$ using Chebychev's inequality (Inequality 3.1)

$$P\left[\left|\bar{X} - \mu\right| \geq \varepsilon\right] \leq \frac{\sigma^2}{n\varepsilon^2}$$

$$\lim_{n\to\infty} P\left[\left|\bar{X} - \mu\right| \geq \varepsilon\right] \leq \lim_{n\to\infty} \frac{\sigma^2}{n\varepsilon^2} = 0$$

Hence for large n, $\bar{X}$ converges in probability to μ for $\sigma > 0$

Similarly it is shown below that S^2 converges in probability to σ^2 (see Chapter 7) According to the Chebychev's inequality

$$P\left[\left|S^2 - \sigma^2\right| \geq \varepsilon\right] \leq \frac{Var(S^2)}{\varepsilon^2} = \frac{\mu_4 - \frac{n-3}{n-2}\sigma^4}{\varepsilon^2 n}.$$

Taking limits of both sides in the above relationship we have

$$\lim_{n\to\infty} P\left[\left|S^2 - \sigma^2\right| \geq \varepsilon\right] \leq \lim_{n\to\infty} \frac{\mu_4 - \dfrac{n-3}{n-2}\sigma^4}{\varepsilon^2 n} = 0$$

∎

Lemma 3.8 Let X be a random variable and define $g(x)$ as a non degenerate function such that $g(x) > 0$. Then

$$E\left[\frac{1}{g(X)}\right] > \frac{1}{E[g(X)]}$$

This is a special case of the well known Jensen's inequality
Proof. Begin with

$$g(X) = E[g(X)]\left[1 + \frac{g(X) - E[g(X)]}{E[g(X)]}\right]$$

$$E\left[\frac{g(X) - E[g(X)]}{E[g(X)]}\right] = 0$$

Therefore,

$$E\left[\frac{1}{g(X)}\right] = E\left[\frac{1}{E[g(X)]\left[1 + \dfrac{g(X) - E[g(X)]}{E[g(X)]}\right]}\right]$$

$$= \frac{1}{E[g(X)]} E\left[\frac{1}{1 + \dfrac{g(X) - E[g(X)]}{E[g(X)]}}\right]$$

Now, using long division,

$$\frac{1}{1 + \dfrac{g(X) - E[g(X)]}{E[g(X)]}} = 1 - \frac{g(X) - E[g(X)]}{E[g(X)]} + \frac{\left[\dfrac{g(X) - E[g(X)]}{E[g(X)]}\right]^2}{1 + \dfrac{g(X) - E[g(X)]}{E[g(X)]}}$$

Hence

$$E\left[\frac{1}{g(X)}\right] = \frac{1}{E[g(X)]} E\left[1 - \frac{g(X) - E[g(X)]}{E[g(X)]} + \frac{\left[\dfrac{g(X) - E[g(X)]}{E[g(X)]}\right]^2}{1 + \dfrac{g(X) - E[g(X)]}{E[g(X)]}}\right]$$

$$= \frac{1}{E[g(X)]}\left|1 + E\left|\frac{\left[\dfrac{g(X) - E[g(X)]}{E[g(X)]}\right]^2}{1 + \dfrac{g(X) - E[g(X)]}{E[g(X)]}}\right|\right| = \frac{1}{E[g(X)]}[1 + A], A > 0$$

Now,

$$1 + \frac{g(X) - E[g(X)]}{E(g(X))} = \frac{g(X)}{E[g(X)]} > 0$$

we have

$$E\left[\frac{1}{g(X)}\right] \geq \frac{1}{E[g(X)]}.$$

3.9 Covariance and Correlation

If there are two random variables X and Y with mean values μ_X and μ_Y then the *covariance* between X and Y is defined as $E\{(X - \mu_X)(Y - \mu_Y)\}$ and may be written as

$$Cov(X,Y) = E(XY) - E(X)E(Y)$$

Covariance provides us with an idea of how the values of Y vary with the values of X. If for example as X increases, Y tends to increase then in this case large values of Y will be observed when large values of X are observed. Furthermore, if $X > E(X)$ implies $Y > E(Y)$ then Covariance (X,Y) will be a positive quantity.

If as X increases Y tends to decrease or vice a versa, then on an average they move in opposite directions and the covariance between them is negative. In this case $X > E(X)$ is associated with $Y < E(Y)$. If however as X increases, Y sometimes tends to increase and at other times Y tends to decrease then there is no pattern to their relationship and one cannot with any certainty determine the direction of the relationship between X and Y. In this case X and Y are independent of each other and the covariance of X and Y is 0. However, covariance value of zero does not imply that X and Y are necessarily independent as in the case when X takes values say -1 and 1 (with probability $1/2$) and $Y = X^2$ takes only the value 1, then the covariance between X and Y is zero even though Y is a function of X.

Example 3.11 In order to study the functional decline in patients with Parkinson's disease, the Unified Parkinson's Disease Rating Scale (UPDRS) was utilized (Jankovic and Kapadia 2000) to study the relationship between years of observation (X) and the change in the total UPDRS score (Y). the following data are a sample of ten patients from the longitudinal follow-up data

Table 3.2 X and Y values

X	3.84	10.75	6.52	6.97	6.34	11.56	8.32	5.63	5.74	7.51
Y	34	35	10	49	22	−6	−25	23	7	17

From the above table the following sample statistics may be computed

$$\bar{X} = 7.318, \bar{Y} = 16.60$$

$$\sum_{i=1}^{10} X_i Y_i = 1073$$

$$\frac{1}{10} \sum_{i=1}^{10} (X_i - \bar{X})(Y_i - \bar{Y}) = -14.18$$

∎

Let X be a random variable and $f_1(X)$ and $f_2(X)$ be any two functions of X such that $E[f_1(X)]$ and $E[f_2(X)]$ and $E[f_1(X)f_2(X)]$ exist. Then the following inequalities hold.

Inequality 3.2 If $f_1(X)$ is nondecreasing function of X and $f_2(X)$ is a non increasing function of X, then

$$E[f_1(X)f_2(X)] \le E[f_1(X)]E[f_2(X)]$$

Proof. Since both $f_1(X)$ and $f_2(X)$ vary in the opposite directions, by definition the covariance will be negative. Hence

$$Cov[f_1(X), f_2(X)] = E[f_1(X)f_2(X)] - E[f_1(X)]E[f_2(X)] \le 0$$

and the inequality is established.   ∎

Inequality 3.3 If $f_1(X)$ and $f_2(X)$ are both a non-decreasing function of X or if both are non-increasing functions of X, then
$$E[f_1(X)f_2(X)] \ge E[f_1(X)]E[f_2(X)]$$

Proof. Since both $f_1(X)$ and $f_2(X)$ vary in the same direction their covariance by definition will be positive. Hence

$Cov[f_1(X), f_2(X)] = E[f_1(X)f_2(X)] - E[f_1(X)]E[f_2(X)] \geq 0$

and the inequality is established. The above two inequalities are sometimes referred to as the Covariance Inequalities. ∎

We will now obtain the covariance between linear functions of two sets of random variables Z and W

Let

$Z = a_1X_1 + a_2X_2 + ... + a_mX_m$ and $W = b_1Y_1 + b_2Y_2 + ...b_nY_n$ be two linear functions of random variables $X_1, X_2, ..., X_m$ and $Y_1, Y_2, ...Y_n$, where

a_i and b_j $i = 1, 2, ...m$, $j = 1, 2, ...n$ are constants then,

$Cov(Z, W) = E(ZW) - (E(Z)E(W)$

$$= E\left\{\left(\sum_{i=1}^{m} a_iX_i\right)\left(\sum_{j=1}^{n} b_jY_j\right)\right\} - E\left(\sum_{i=1}^{m} a_iX_i\right)E\left(\sum_{j=1}^{n} b_jY_j\right)$$

$$= E\left\{\sum_{i=1}^{m}\sum_{j=1}^{n} a_ib_jX_iY_j\right\} - \left\{\sum_{i=1}^{m}\sum_{j=1}^{n} a_ib_jE(X_i)E(Y_j)\right\}$$

$$= \sum_{i=1}^{m}\sum_{j=1}^{n} a_ib_jCov(X_i, Y_j).$$

Lemma 3.9

If X_1 and X_2 are assumed to be independent random variables then

$Cov(X_1, X_2) = 0$

Proof. Since X_1, X_2 are independent .

$E(X_1, X_2) = E(X_1)E(X_2)$.

Therefore

$Cov(X_1, X_2) = E(X_1)E(X_2) - E(X_1)E(X_2) = 0$. ∎

The above results can now be combined to show that for two random variables X_1 and X_2

Lemma 3.10

$Var(a_1X_1 \pm a_2X_2) = a_1^2Var(X_1) + a_2^2Var(X_2) \pm 2a_1a_2Cov(X_1, X_2)$

Proof. Using the definition of variance

$Var(a_1X_1 \pm a_2X_2) = E[a_1X_1 \pm a_2X_2]^2 - [E(a_1X_1 \pm a_2X_2)]^2$

Write

$$E[a_1 X_1 \pm a_2 X_2]^2 = E[a_1^2 X_1^2 \pm 2a_1 a_2 X_1 X_2 + a_2^2 X_2^2]$$
$$= a_1^2 E(X_1^2) \pm 2a_1 a_2 E(X_1 X_2) + a_2^2 E(X_2^2)$$

and

$$[E(a_1 X_1 \pm a_2 X_2)]^2 = [a_1 E(X_1) \pm a_2 E(X_2)]^2$$
$$= a_1^2 [E(X_1)]^2 + a_2^2 [E(X_2)]^2 \pm 2a_1 a_2 E(X_1) E(X_2)$$

Subtracting the second expression from the first and rearranging terms we have proved the lemma. For X_1 and X_2 independent and $a_1 = a_2 = 1$,

$$Var(X_1 + X_2) = Var(X_1) + Var(X_2).$$

This result may be generalized to prove that if $X_1, X_2, ...X_n$ are n independent random variables

$$Var(X_1 + X_2 + ...X_n) = \sum_{i=1}^{n} Var(X_i).$$

Furthermore if the X's are independent and identically distributed with variance σ^2

$$Var(X_1 + X_2 + ...X_n) = nVarX_1 = n\sigma^2.$$

The *coefficient of correlation* between two random variables X and Y is defined as follows

$$Corr(X,Y) = \frac{Cov(X,Y)}{\sqrt{Var(X)Var(Y)}}.$$

The above quantity is usually denoted by ρ if it represents the correlation of (X,Y) values in the population and is often represented by r where

$$r = \frac{\sum_{i=1}^{n}(x_i - \bar{x})(y_i - \bar{y})}{\sqrt{\sum_{i=1}^{n}(x_i - \bar{x})^2 \sum_{i=1}^{n}(y_i - \bar{y})^2}}$$

when it represents the correlation between X and Y in a sample of n (X,Y) values. The later is also referred to as the *sample correlation coefficient*. In either case the coefficient of correlation lies between -1 and $+1$. The correlation of coefficient is not defined if one (or both) term in the denominator is zero. Using the above formula, the coefficient of correlation for data in Table 3.2 is -0.312. ∎

3.10 Distribution of a Function of a Random Variable

As mentioned in Example 3.8 any function $h(X)$ of a random variable X is also a random variable. Hence it will have a probability mass function or a probability density function depending on whether X is discrete or continuous. The distribution of $h(X)$ may itself be of interest and may be obtained by inverting the distribution function of X. For example, let the variable X have the density function defined below.

$$f_X(x) = 1 \text{ for } 0 < x < 1$$

This is a special case of a probability density function known as the *uniform density*. If we let

$$h(x) = x^2.$$

Then

$$P(h(X) \le y) = P(X^2 \le y) = P(X \le \sqrt{y}) - P(X \le -\sqrt{y}) = P(X \le \sqrt{y})$$

Note that $P\left(X \le -\sqrt{y}\right) = 0$ since the variable X does not take negative values. Hence

$$F_{x^2}(y) = F_X(\sqrt{y})$$

or

$$f_{x^2}(y) = \frac{d}{dy} F_X\left(\sqrt{y}\right) = \frac{1}{2\sqrt{y}} f_X\left(\sqrt{y}\right).$$

In general, if X is a continuous variable and if $h(X)$ is a one to one function then it is possible to easily obtain the density function of $h(X)$ by the above procedure. We are assuming that $h(X)$ is differentiable. There are two possible scenarios to consider here

(i) $h(X)$ is an increasing function of X

(ii) $h(X)$ is a decreasing function of X

For the first scenario, notice the shaded area in Figure 3.3 below represents $P(h(X) \le y)$. The corresponding point on the X-axis is $h^{-1}(y)$. It can then be easily seen that

$$P[h(X) \le y] = P[X \le h^{-1}(y)]$$

or equivalently by

$$F_{h(X)}(y) = F_X[h^{-1}(y)]$$

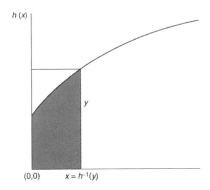

Figure 3.3 $h(x)$ an increasing function of x

Differentiating both sides with respect to y we obtain

$$f_{h(X)}(y) = f_X[h^{-1}(y)]\left[\frac{d}{dy}h^{-1}(y)\right] \tag{3.9}$$

In the second scenario, notice in Figure 3.4 that because of the shape of the $h(X)$ curve $P(h(X) \le y)$ is represented by the area to the right of the point $h^{-1}(y)$ on the X axis. Hence in this case

$$P[h(X) \le y] = P[X > h^{-1}(y)]$$
$$F_{h(X)}(y) = 1 - F_X[h^{-1}(y)]$$

Differentiating both sides with respect to y we obtain

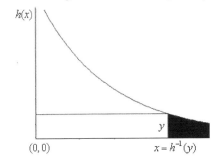

Figure 3.4 $h(x)$ a decreasing function of x

$$f_{h(X)}(y) = -f_X[h^{-1}(y)]\left[\frac{d}{dy}h^{-1}(y)\right]$$ (3.10)

Results (3.9) and (3.10) above can be more succinctly represented by

$$f_{h(X)}(y) = f_X[h^{-1}(y)]\left|\frac{d}{dy}h^{-1}(y)\right|$$

$$= 0 \text{ otherwise}$$

The absolute value in the above relationship is needed to guarantee that the density is non-negative. Since if $h(x)$ is a decreasing function of x, $\frac{d}{dy}h^{-1}(y)$ will be negative. We will now obtain the probability density function of some more complicated functions of X by using the results obtained above.

■

Example 3.12 Let X be a continuous random variable with probability density function

$$f_X(x) = \frac{1}{\sigma^2}xe^{-\frac{1}{2}\left[\frac{x}{\sigma}\right]^2} \quad 0 < x < \infty.$$

Define $h(x) = e^x$

$$P[h(X) \le y] = P(e^x \le y) = P(x \le \log y)$$
$$F_{h(X)}(y) = F_X(\log y)$$

or

$$f_{h(X)}(y) = f_X(\log y)\left[\frac{d}{dy}(\log y)\right]$$

$$f_{h(X)}(y) = \frac{\log y}{y\sigma^2}e^{-\frac{(\log y)^2}{2\sigma^2}} \quad 1 \le y < \infty$$

■

Example 3.13
Define a continuous random variable X with pdf

$$f_X(x) = \frac{1}{\sqrt{2\pi}}\exp\left(-\frac{x^2}{2}\right), \quad -\infty < x < \infty$$

Define

$$h(X) = |X|.$$

Note that for $0 < x < \infty$, $0 < h(x) < \infty$ and for $-\infty < x < 0, 0 < h(X) < \infty$.

Hence $0 < y < \infty$

$$P[h(X) \le y] = P(|X| \le y) = P(-y \le X \le y)$$
$$= P(X \le y) - P(X \le -y)$$
$$= P(X \le y) - P(X \le -y)$$
$$= F_X(y) - F_X(-y).$$

Differentiating both sides of the above relationship with respect to y we obtain

$$f_{h(X)}(y) = f_X(y) + f_X(-y) = \sqrt{\frac{2}{\pi}} \exp\left(-\frac{y^2}{2}\right), 0 < y < \infty$$

It can be easily shown that

$$E[h(X)] = \sqrt{\frac{2}{\pi}} \text{ and the } Var[h(X)] = 1 - \frac{2}{\pi}.$$

■

A very interesting density function is the density function of a distribution function as illustrated in the following example.

Example 3.14 Let the continuous random variable X have the following cumulative distribution function $F_X(x) = P(X \le x)$.
Now assume that the inverse of the distribution function defined as
$F_X^{-1}(y) = \inf\left\{x \mid F_X(x) \ge y\right\}$ for $0 < y < 1$. Then
for $h(X) = F_X(X)$

$$P[h(X) \le y] = P[F_X(X) \le y] = P[X \le F_X^{-1}(y)]$$
$$= F_X[F_X^{-1}(y)] = y$$

■

In Chapter 5 we will learn that $h(X)$ follows a uniform distribution on $(0,1)$ and the usefulness of this transformation will be displayed in greater detail in Chapters 6 and 11.

.

3.11 Multivariate Distributions and Marginal Densities

Sometimes a data point may not be characterized by its value on the X axis alone but instead it may have a Y component also. For example, we may be interested in looking at the ages as well as heights of a set of school children. In this situation we deal with joint values of two variables X and Y. The *joint probability mass function* of the two variables may be written as

$$P(X = x_i, Y = y_j) = p_{ij}$$

If the variable X takes integer values in the range $[m_1, n_1]$ and Y takes integer values in the range $[m_2, n_2]$ then if we assume that that each combination of X and Y values is equally likely , the joint probability density of X and Y is

$$P(X = i, Y = j) = \frac{1}{(n_1 - m_1 + 1)} \frac{1}{(n_2 - m_2 + 1)}$$

where

$$\sum_{i=m_1}^{n_1} \sum_{j=m_2}^{n_2} p_{ij} = 1.$$

Example 3.15 Consider two random variables X and Y such that

$$P(X = x, Y = y) = \binom{10}{x}\left(\frac{1}{3}\right)^x \left(\frac{2}{3}\right)^y$$

and assume that both take integer values in the interval $[0,10]$ subject to the constraint $X + Y = 10$. The above joint density function of X and Y is defined only for integer values of X and Y in the feasible region represented by the line $X+Y=10$ in Figure 3.5 below.

If we were to sum the joint density function of X and Y above for all possible values of Y then the resulting function of X i.e., $P(X = x)$ is known as the marginal density function of X.

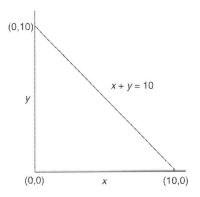

Figure 3.5 Feasible region for Example 3.14

$$P(X = x) = \left[\binom{10}{x} \left(\frac{1}{3} \right)^x \left(\frac{2}{3} \right)^y \right]_{y=10-x}$$

$$= \binom{10}{x} \left(\frac{1}{3} \right)^x \left(\frac{2}{3} \right)^{10-x}$$

This is because in the above example, if X were to take a value x, the only value that Y could take is 10-x. This is an example of a *degenerate bivariate* distribution when X and Y are dependent on one another in such a way that for each value of X there is only one value that Y can assume.

∎

Let us consider another example where the dependence between X and Y is more pronounced.

Example 3.16 Consider a set of 10 independent trials each of which can result in the occurrence of events E_1, E_2, and E_3 with probabilities p_1, p_2, and p_3 respectively. Assume that each trial must result in only one of these events. Representing the number of times the event E_i $(i = 1,2,3)$ occurs by the random variable $X_i (i = 1,2,3)$, the joint probability of $X_i = x_i (i = 1,2,3)$ where

$x_1 + x_2 + x_3 = 10$ (or $x_3 = 10 - x_1 - x_2$) is given by the following joint probability

$$P\left(X_1 = x_1, X_2 = x_2, X_3 = 10 - x_1 - x_2 \right)$$

$$= \frac{10!}{x_1! x_2! (10 - x_1 - x_2)!} p_1^{x_1} p_2^{x_2} \left(1 - p_1 - p_2 \right)^{10 - x_1 - x_2}, \quad 0 \le x_1 + x_2 \le 10$$

The *marginal probability* of X_1 is obtained by summing the above joint probability for all values of x_2 from $x_2 = 0$ to $10 - x_1$. That is

$$P(X_1 = x_1) = \sum_{x_2=0}^{10-x_1} \frac{10!}{x_1! x_2! (10 - x_1 - x_2)!} p_1^{x_1} p_2^{x_2} (1 - p_1 - p_2)^{10 - x_1 - x_2} \quad 0$$

$$= \frac{10!}{x_1!} p_1^{x_1} \sum_{x_2=0}^{10-x_1} \frac{1}{x_2! (10 - x_1 - x_2)!} p_2^{x_2} (1 - p_1 - p_2)^{10 - x_1 - x_2}$$

$$= \frac{10!}{x_1! (10 - x_1)!} p_1^{x_1} \sum_{x_2=0}^{10-x_1} \frac{(10 - x_1)!}{x_2! (10 - x_1 - x_2)!} p_2^{x_2} (1 - p_1 - p_2)^{z - x_2}$$

$$= \frac{10!}{x_1! (10 - x_1)!} p_1^{x_1} [p_2 + (1 - p_1 - p_2)]^{10 - x_1},$$

$$= \frac{10!}{x_1! (10 - x_1)!} p_1^{x_1} [(1 - p_1)]^{10 - x_1}, \quad 0 \le x_1 \le 10$$

Similarly the marginal distribution of X_2 may be obtained by interchanging x_1 and x_2 in the above equation. In Chapter 4 we will discuss the above distribution also known as the *binomial distribution* as well as the *multinomial* distribution in detail. ∎

In general the m- dimensional random variable $(X_1, X_2, ..., X_m)$ is a m dimensional discrete random variable if it can assume values of the form $(x_1, x_2, ...x_m)$ and provided the vector of values is finite and denumerable. The joint probability of these variables is defined by

$$P(X_1 = x_1, X_2 = x_2, ..., X_m = x_m) = p(x_1, x_2, ...x_m) \qquad (3.11)$$

If A is any subset of the set of values $X_1, X_2, ..., X_m$ can assume, then

$$P[(X_1, X_2, ..., X_m) \in A] = \sum_{(x_1, ..., x_n) \in A} P(X_1 = x_1, X_2 = x_2, ...X_m = x_m).$$

That is, we sum the above equation for all the $X's$ in such a way that the constraints imposed on the $X's$ are observed.

The marginal distribution $m_{X_1, X_2, ...X_k}(x_1, x_2, ..., x_k)$ of a subset of the X_i 's, for example $(X_1, X_2, ..., X_k) \, k < m,$ may be obtained by summing the right hand side of (3.11) for all the other $m - k$ variables.

The conditional distribution of $(X_{k+1}, X_{k+2}, ..., X_m)$ given $X_1 = x_1, X_2 = x_2, ...X_k = x_k$ may be obtained as

$$h_{X_{k+1}, X_{k+2}, ...X_m | X_1, ..., X_k} (x_{k+1}, x_{k+2}, ..., x_m | x_1, x_2, ..., x_k)$$

$$= \frac{p(x_1, x_2, ...x_m)}{m_{X_1, X_2, ..., X_k}(x_1, x_2, ...x_k)}.$$

If variables X and Y under consideration are continuous, then their joint density function $f_{X,Y}(x, y)$ is defined as the function such that

$$\int_{-\infty}^{s} \int_{-\infty}^{t} f_{X,Y}(x, y) \, dxdy = P(X \le s, Y \le t)$$

The marginal density of X is obtained by integrating $f_{X,Y}(x, y)$ over all values of y subject to the constraints on their relationship.

Example 3.17 Let X and Y have the following joint distribution

$$f_{X,Y}(x,y) = \frac{(n-1)(n-2)}{(1+x+y)^n}, \quad x > 0, y > 0.$$

The marginal density function of x is obtained as follows

$$m_X(x) = \int_0^\infty \frac{(n-1)(n-2)}{(1+x+y)^n}dy$$

$$= \frac{(n-2)}{(1+x)^{n-1}}, \quad x > 0$$

Similarly, the marginal density of Y is $w_Y(y) = \frac{(n-2)}{(1+y)^{n-1}}, \quad y > 0$

Example 3.18 Let the random variables X and Y have the joint density function

$$f_{X,Y}(x,y) = \frac{1}{(1-3e^{-2})}e^{-x}e^{-y}$$

subject to the constraints $x+y < 2, x > 0, y > 0$.
In this case the feasible region where both X and Y are valid is the area bounded by the line $x+y = 2$ and by the x and y axes. The marginal density of X is obtained by integrating over all possible values of y such that $x+y < 2$.
The feasible region is the shaded region in Figure 3.6. Hence,

$$g(x) = \frac{1}{(1-3e^{-2})}\int_{y=0}^{2-x} e^{-x}e^{-y}dy = \frac{1}{(1-3e^{-2})}(e^{-x}-e^{-2})$$

Notice that

$$\int_{x=0}^{2} g_X(x)dx = 1$$

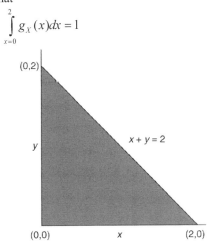

Figure 3.6 The feasible region for Example 3.17

As in the discrete case the conditional distribution of Y given X is

$$w_{Y|X}(y|x) = \frac{f_{X,Y}(x,y)}{m_X(x)}, \quad m_X(x) > 0$$

The multivariate distribution of a set of variables $Y_1, Y_2, ..., X_m$ is defined in a way similar to the discrete case.

3.12 Conditional Expectations

Let the joint mass function of X and Y be $p_{X,Y}(x,y)$. Then the *conditional expectation* of Y given $X = x_j$ is defined as

$$E(Y|X=x_j) = \sum_i y_i \frac{p_{ji}}{p_j}, p_j > 0, \text{ where } p_{ji} = P(X=x_j, Y=y_i)$$

for the discrete case. Similarly, for the continuous case

$$E(Y|X) = \int_y yh(y|x)dy, \text{ where } h(y|x) \text{ is the}$$

conditional density function of Y given X and may also be written as

$$h_{Y|X}(y|x) = \frac{f(x,y)}{g(x)}, g(x) > 0.$$

Similarly

$$E[w(Y)|X] = \int_y w(y)h(y|x)dy.$$

The *conditional variance* of Y for a given X is similarly defined as

$$E[Y - E(Y|X=x_j)]^2 \text{ in the discrete case and,}$$

$$E[Y - E(Y|X)]^2 \text{ in the continuous case.}$$

Example 3.19
Let

$$f_{X,Y}(x,y) = \begin{cases} e^{-y} & 0 < x < y < \infty \\ 0 & \text{otherwise} \end{cases}$$

Then, $g_X(x) = \int_x^\infty f_{X,Y}(x,y)dy = \int_x^\infty e^{-y}dy = e^{-x}$

$$h_{X,Y}(y|x) = \begin{cases} \dfrac{f_{X,Y}(x,y)}{g_X(x)} = \dfrac{e^{-y}}{e^{-x}} = e^{-(y-x)}, & y > x \\ 0 & \text{otherwise} \end{cases}$$

$$E(Y|X) = \int_x^\infty y e^{-(y-x)} dy = x + 1$$

and

$$Var(Y|X) = \int_x^\infty (y - x - 1)^2 e^{-(y-x)} dy = 1.$$

∎

Lemma 3.11-3.12

For any two random variables X and Y, the following relationships hold

3.11 $E[(X + Y)|Y] = E(X|Y) + Y$

3.12 $E(XY|Y) = YE(X|Y)$

Lemmas 3.11 and 3.12 are quite obvious and based on the fact that the random variable on which the expectation is conditioned is treated as a constant.

Lemmas 3.13-3.16 assume that the expectations exist and are interchangeable

Lemmas 3.13-3.14

3.13 $E[E(Y|X)] = E(Y)$

3.14 $E(XY) = E[XE(Y|X)]$

For Lemmas 3.13-3.14 the outer expectation is with respect to X

Proof of Lemma 3.13 for the continuous case.

$$E[E(Y|X)] = \int_x \left[\int_y y \frac{f(x,y)}{g(x)} dy \right] g(x) dx, \; g(x) \neq 0$$

$$= \int_y \left[\int_x y \frac{f(x,y)}{g(x)} g(x) dx \right] dy \text{ (by changing the order of integration)}$$

$$= \int_y y \left[\int_x f(x,y) dx \right] dy$$

$$= \int_y y f_Y(y) dy = E(Y).$$

∎

Proof of Lemma 3.14

$$E[XE(Y|X)] = \int_x x\{\int_y yh(y|x)dy\}g(x)\,dx$$

$$= \int_x\int_y xyf(x,y)dxdy \text{ (by changing the order of integration)}$$

$$= E(XY)$$

3.13 Conditional Variance and Covariance

Conditional variance and covariance are defined as follows

$$(i)\, Var(X|Z) = E\left[\{X - E(X|Z)\}^2\,|Z\right]$$

$$(ii)\, Cov[(X,Y)|Z] = E[[\{X - E(X|Z)\}\{Y - E(Y|Z)\}|Z]]$$

Lemmas 3.15-3.16

For any random variable X, Y and Z

$$Var(X) = E\left[Var(X|Y)\right] + Var\left[E(X|Y)\right]$$

$$Cov(X,Y) = E[Cov\{(X,Y)|Z\}] + Cov[E(X|Z), E(Y|Z)].$$

Proof of Lemma 3.15

$$Var(X) = E[X - E(X)]^2$$

$$= E[X - E(X|Y) + E(X|Y) - E(X)]^2$$

$$= E[X - E(X|Y)]^2 + E[E(X|Y) - E(X)]^2 +$$

$$2E[\{X - E(X|Y)\}\{E(X|Y) - E(X)\}].$$

In the derivation of Lemma 3.15 we will use Lemmas 3.11-3.13 as well as the definitions of conditional variance and covariance. In the above equation the first term on the right hand side of the last equality may be written as

$$E[X - E(X|Y)]^2 = E[E\{X - E(X|Y)\}^2\,|Y] = E[Var\,(X|Y)]$$

Similarly it can be shown that

$$E[E(X|Y) - E(X)]^2 = E[E(X|Y) - E\{E(X|Y)\}]^2 = Var[E(X|Y)]$$

$$2E[\{X - E(X|Y)\}\{E(X|Y) - E(X)\}] = 2E[E\{[X - E(X|Y)][E(X|Y) - E(X)]|Y\}]$$

The right hand side of the above equation may be written as

$$2E[[E(X|Y) - E(X)]E\{[X - E(X|Y)]|Y\}] = 0$$

Hence

$$Var(X) = E[Var(X|Y)] + Var[E(X|Y)]$$

∎

Proof of Lemma 3.16

$$E[Cov\{(X,Y|Z)\}] = E\Big[E[\{X - E(X|Z)\}\{Y - E(Y|Z)\}|Z] \Big]$$

$$= E\Big[E[\{XY - XE(Y|Z) - YE(X|Z) + E(X|Z)E(Y|Z)\}|Z] \Big]$$

$$= E\Big[E\{XY|Z\} \Big] - E[E(X|Z)E(Y|Z)]$$

$$- E[E(Y|Z)E(X|Z)] + E[E(X|Z)E(Y|Z)]$$

$$= E(XY) - E[E(X|Z)E(Y|Z)] \tag{3.12}$$

Next

$$Cov\Big[E(X|Z), E(Y|Z) \Big] = E\Big[E(X|Z)E(Y|Z) \Big]$$

$$- \Big[E\{E(X|Z)\} \Big]\Big[E\{E(Y|Z)\} \Big]$$

$$= E\Big[E(X|Z)E(Y|Z) \Big] - E(X)E(Y) \tag{3.13}$$

Adding (3.12) and (3.13) above the proof of Lemma 3.16 is completed.

∎

3.14 Moment Generating Functions

As the name suggests, *moment generating functions* (mgf) generate moments of a distribution. The moments of a distribution are important for determining its probability mass and its density function. The mgf of a distribution is defined as

$$M_X(t) = E(e^{tX})$$

if it exists for every value of t in the interval $|t| \le t_0$.

For a random variable X.

$$E(e^{tX}) = E\left(1 + \frac{tX}{1!} + \frac{t^2 X^2}{2!} + \dots \frac{t^n X^n}{n!} + \dots \right)$$

$$= 1 + tE(X) + \frac{t^2}{2!}E(X^2) + \dots \frac{t^n}{n!}E(X^n) + \dots$$

If we were to differentiate $E(e^{tX})$ n times with respect to t and set $t = 0$ we will get the nth moment of the distribution of X if it exists.

This technique of obtaining the moments is illustrated for the following discrete distribution

$$P(X = x) = \frac{e^{-\lambda}\lambda^x}{x!} \qquad x = 0,1,2,....$$

The form of the above distribution is the well known *Poisson distribution*. This distribution plays a very important role in the field of stochastic processes particularly in queuing theory. The mgf for the Poisson distribution is

$$M_X(t) = \sum_{i=0}^{\infty} \frac{e^{xt}e^{-\lambda}\lambda^x}{x!} = e^{-\lambda}\sum_{i=0}^{\infty} \frac{(e^t\lambda)^x}{x!}$$

$$= e^{-\lambda}e^{e^t\lambda} = e^{-\lambda(1-e^t)}$$

The rth moment can now be obtained by differentiating the above function r times and setting $t = 0$. For illustration purposes, set $r = 1$ and obtain

$$E(X) = \frac{d}{dx}M_X(t)\Big|_{t=0} = e^{-\lambda(1-e^t)}\lambda e^t\Big|_{t=0} = \lambda$$

For the sum $Z = \sum_{i=1}^{n} X_i$ of a set of independently distributed random variables, the moment generating function of Z is defined as

$$E\left[\exp\left\{t\sum_{i=1}^{n} X_i\right\}\right] = \prod_{i=1}^{n} M_{X_i}(t).$$

If $X_1,...,X_n$ are a set of n jointly distributed random variables, then the moment generating function of $X_1,...,X_n$ is defined as $E\left[\exp\left\{\sum_{i=1}^{n} t_i X_i\right\}\right]$.

3.15 Characteristic Functions

For some random variables, the moment generating function may not exist. Hence it may not be possible for us to obtain

$$\sum_x e^{tx} P(X = x) \text{ or } \int_x e^{tx} f_X(x)dx.$$

In such cases a function called the *characteristic function* of the form $C(t) = E[\exp(itX)]$ may for real t may be a lot more useful. A density function of the form $f_X(x) = \dfrac{A}{(1+x^2)^k}$ is a case in point.

The characteristic function like the moment generating function also generates moments. The rth moment about the origin may be obtained by using the following relationship

$$\mu_r' = (-i)^r \frac{d^r}{dt^r} C(t)\big|_{t=0}$$

Example 3.20 Consider the following density function (the Cauchy distribution)

$$f_X(x) = \frac{1}{\pi(1+x^2)} \quad -\infty < x < \infty$$

The corresponding characteristic function will be

$$C_X(t) = \int_0^\infty \frac{2e^{itx}}{\pi(1+x^2)} dx$$

$$= \int_0^\infty \frac{2(\cos tx + i \sin tx)}{\pi(1+x^2)} dx$$

which, since *sintx* is an odd function (ie., *sin(-tx) = -sintx*), reduces to

$$\int_0^\infty \frac{2\cos tx}{\pi(1+x^2)} dx$$

$$= e^{-|t|}$$

We will learn in Chapter 5, section 5.12, that for this distribution only $E[|X|^k]$ exists . These quantities are also known as the absolute moments.

3.16 Probability Generating Functions

The skillful use of *generating functions* for discrete random variables defined in Chapter 1, section 1.23 is invaluable in solving families of difference equations. However, generating functions have developed in certain specific circumstances. Consider the circumstance where the sequence $\{b_k\}$ are probabilities. Then, using our definition of $G(s)$ we see that

$$P(s) - \sum_{k=0}^\infty P(x=k)s^k \text{ where } \sum_{k=0}^\infty p_k = 1.$$

In probability it is often useful to describe the relative frequency of an event in terms of the probability of an outcome of an experiment. Among the most useful of these experiments are those for which the random variable takes only non-negative values and for which the outcomes are discrete. These discrete models have many applications, and it is often helpful to recognize the generating function associated with the models. The generating function associated with such a model is described as the *probability generating function* (pgf). We will continue to refer to these probability generating functions using the notation P(s). This section will describe the use of probability generating functions and derive the probability generating function for the commonly used discrete distributions. Note that in this context of probability, the generating function can be considered as an expectation, i.e., $P(s) = E(s^X)$ and $P(1) = 1$.

Consider one of the simplest of experimental probability models, the Bernoulli trial. Consider an experiment which can result in only two possible outcomes E_1 and E_2. Let X be a random variable such that if the experiment results in the occurrence of E_1 then $X = 1$ and if the experiment results in the occurrence of E_2 then $X = 0$. Furthermore assume that E_1 occurs with probability p and E_2 occurs with probability $1 - p$. The probability generating function for this distribution of X is

$$P(s) = \sum_{k=0}^{1} p_k s^k = qs^0 + ps^1 = q + ps, \quad \text{where } q = 1 - p.$$

Note that in this computation we can collapse the range over which the sum is expressed from the initial range of 0 to ∞ to the range of 0 to 1 since $p_k = 0$ for $k = 2, 3, 4, \ldots$

If the probability generating function can be expressed as an infinite series in s using the inversion methods described in (Moyé and Kapadia 2000) then p_k can be obtained as the coefficient of s^k. If the probability generating function for a random variable is more complicated and can be expressed as a ratio of the form

$$P(S) = \left[\frac{A(s)}{C(s)} \right].$$

then an approximation to p_k is given by the following formula

$$p_k = \left[\frac{-A(s_1)}{C'(s_1)} \right] \frac{1}{s_1^k}$$

where s_1 is the smallest root of $C(s)$ and $C'(s)$ is the derivative of $C(s)$.

$$\frac{d}{ds} P_X(s) \Big|_{s=1} = E(X)$$

$$\frac{d^2}{ds^2} P(s) \Big|_{s=1} = E[X(X-1)]$$

$$Var(X) = \frac{d^2}{ds^2} P(s) \Big|_{s=1} + \frac{d}{ds} P_X(s) \Big|_{s=1} - \left[\frac{d}{ds} P_X(s) \Big|_{s=1} \right]^2.$$

In general

$$E[X(X-1)(X-2)....(X-k)] = \frac{d^{k+1}}{ds^{k+1}} P(s)\Big|_{s=1}.$$

Such moments are often referred to as *factorial moments*. Note that the moment generating function for a discrete random variable is a special case of the probability generating function since $M_X(t) = P(e^t)$.

Example 3.21 Let X be a discrete random variable having the following probability mass function (negative binomial distribution)

$$P(X = k) = \binom{k-1}{r-1} p^r q^{k-r}, \quad p+q=1 \text{ and } k = r, r+1,...$$

The probability generating function can be derived as follows

$$P(s) = \sum_{k=r}^{\infty} s^k \binom{k-1}{r-1} p^r q^{k-r} = \sum_{k=r}^{\infty} \binom{k-1}{r-1} s^{k-r} p^r q^{k-r}$$

$$= \sum_{k=r}^{\infty} \binom{k-1}{r-1} (qs)^{k-r} p^r = p^r \sum_{k=r}^{\infty} \binom{k-1}{r-1} (qs)^{k-r}$$

Continuing, the quantity on the right hand side of the above relationship can be written as

$$p^r \sum_{k=r}^{\infty} \binom{k-1}{k-r} (qs)^{k-r} = p^r \sum_{n=0}^{\infty} \binom{r+n-1}{n} (qs)^n = \left[\frac{p}{1-qs}\right]^r$$

Hence,

$$E(X) = \frac{d}{ds} P(s)\Big|_{s=1} = r \frac{(1-p)}{p}$$

$$Var(X) = \frac{d^2}{ds^2} P(s)\Big|_{s=1} + \frac{d}{ds} P_X(s)\Big|_{s=1} - \left[\frac{d}{ds} P_X(s)\Big|_{s=1}\right]^2$$

$$= r \frac{(1-p)^2(1+r)}{p^2} + r \frac{(1-p)}{p} - \left[r \frac{(1-p)}{p}\right]^2 = r \frac{(1-p)}{p^2}.$$

Assume that $X_1, X_2, ..., X_n$ are integer valued random variables, with joint probability mass function

$$P(X_1 = k_1, X_2 = k_2, ..., X_n = k_n) = p_{k_1,...,k_n}$$

Then

$$\sum_{k_1}\sum_{k_2}...\sum_{k_n} p_{k_1,...,k_n} = 1.$$

The joint probability generating function is defined for the multivariate case as

$$P_{Y_1...Y}(s_1, s_2, ...s_n) = \sum_{k_1}\sum_{k_2}...\sum_{k_n} p_{t_1,...,t_n} s_1^{k_1} s_2^{k_2} ... s_n^{k_n}$$

$$= E(s_1^{x_1} s_2^{x_2} ...s_n^{x_n}).$$

In this case we have

$$E[X_i(X_i - 1)....(X_i - k)] = \frac{d^{k+1}}{ds_i^{k+1}} P(s_1, s_2, ..s_i, ...s_n)\Big|_{s_j=1, j=1,2,...}$$

3.16.1 Probability Generating Function of a Random Number of Random Variables

Probability generating function of a sum $Z = \sum_{i=1}^{n} X_i$ of n random variables $X_1, ..., X_n$ is defined as

$$P_Z(s, s, ..., s) = E(s^Z) = E(s^{X_1} s^{X_2} ....s^{X_n}).$$

Furthermore, if the X's are independent variables, then

$$P_Z(s, s, ..., s) = \prod_{i=1}^{n} E(s^{X_i}).$$

Probability generating function of a random sum of random variables are known as *compound distributions* and can be obtained as follows.

Let $Z = \sum_{i=1}^{N} X_i$ where the X's are independent and identically distributed each with a probability generating function $G(s)$ and N is a random variable with probability mass function and a probability generating function $Q(s)$. Then the probability generating function of Z is derived as follows. Let

$$P(s) = E(s^{Z_N}) = E[E(s^{Z_n} | N = n)]$$

Now,

$$E(s^{Z_n} | N = n) = [G(s)]^n$$

Therefore

$$P(s) = E[\{G(s)\}^N] = Q[G(s)]$$

Problems

1. Let X be a random variable with probability mass function

   $$P(X = x) = \frac{16}{15}\left(\frac{1}{2}\right)^x \quad x = 1,2,3,4...$$

   Find $P(X \le 3)$

2. A coin is tossed until a tail appears. What is the probability mass function of the number of tosses? What is the expected number of tosses?

3. Six cards are drawn without replacement from an ordinary deck of cards. Find the joint probability distribution of the number of nines and the number of eights.

4. Find the mean and variance of X when X has the density function

   $$f_X(x) = \frac{4}{(x+1)^5}, \quad 0 < x < \infty$$

5. Let the random variable X have the moment generating function

   $$M_X(t) = e^{\frac{t^2}{2}}.$$ For c a constant, find the mgf of (1) $cX + b$ (2) $2cX$

6. Let the joint density function of X and Y be $f_{X,Y}(x,y)$ and let $g(x)$ and $h(y)$ be functions of X and Y, respectively. Show that

   $$E[g(X)h(Y)|X = x] = g(X)E[h(Y)|X = x].$$

7. Could the function $E\left[\dfrac{1}{1+tX}\right]$ be used to generate the moments of a variable X?

8. If X has the density $f_X(x) - \dfrac{x}{8}, \quad 0 < x < 4$ find the density function of X^2

9. A random variable X has density function $f_X(x) = xe^{-\frac{1}{2}x^2}, \quad x > 0$. Find $P(x < 4)$.

10. Consider a circle with radius X where X has the density function $f_X(x) = \mu e^{-\mu x}, \quad \mu > 0, \ 0 < x < \infty$, find the density function of the area of the circle.

11. If X has the density function $f_X(x) = \dfrac{1}{\pi(1+x^2)}, \quad -\infty < x < \infty$ find the cumulative distribution function $F_X(x)$.

12. Let X be a random variable whose cumulative density function is given by

$$F_X(x) = \begin{cases} 0 & for\ x < 0 \\ \dfrac{x}{3} & for\ 0 \le x < 1 \\ \dfrac{x}{2} & for\ 1 \le x < 2 \\ 1 & for\ x \ge 2 \end{cases}$$

Find $P\left(\dfrac{1}{2} \le X \le \dfrac{3}{2}\right)$

13. Consider the discrete random variable Y such that

$$P(Y = y) = \frac{1}{y(y+1)}, \quad y = 1, 2, 3, \ldots$$

Find $P(Y \le y)$ and show that $E(Y)$ does not exist.

14. Random variable X has the density function (Pareto density)

$$f_X(x) = \frac{(1+\lambda)}{x^{2\lambda+1}} \quad if\ 1 < x < \infty, \lambda > 0$$

Find the expected value and variance of X

15. Let U have the density function $f_U(u) = 1$, $0 < u < 1$. Find the density function of $X = U^3$. Find the expected value and the variance of X.

16. Prove Lemmas 3.11 and 3.12 for the discrete as well as the continuous cases

17. The random variable X is such that

$$P[|X - 1| = 2] = 0$$

What is $P(X = 3)$ and $P(X = -1)$?

18. The continuous random variables X and Y have the joint density function $f_{X,Y}(x, y) = a(2x + y)$, $0 \le y \le x \le 1$. Find (1) the marginal density of Y (2) the conditional density of X given Y and (3) the density function of $Z-X+Y$,

19. If X is an integer valued random variable, show that

$$P(X = k) = F_X(k) - F_X(k-1), \quad for\ all\ k$$

20.   If the probability mass function of a random variable X is
$$P(X = k) = (1-p)^{k-1} p, \quad k = 1, 2, 3 \dots,$$
show that
$$P(X > n+k-1 | X > n-1) = P(X > k).$$

21.   The cumulative density function of a random variable X is
$$F_X(x) = 1 - e^{-\lambda x^\mu} \quad x \ge 0, \ \lambda, \mu > 0. \text{ Find the density function of } X.$$

22.   A median M of a distribution is a value such that
$$\int_{x=-\infty}^{M} f_X(x)dx = \int_{x=M}^{\infty} f_X(x)dx.$$
Find the median of the following distributions
(1) $f_X(x) = 2x, \ 0<x<1$ and (2) $f_X(x) = \dfrac{1}{\pi(1+x^2)}, \quad -\infty < x < \infty$

23.   A teaching assistant in a statistics course needs to open the professor's off-ice door in order to place the graded examination papers on the professors desk. She has exactly six keys in her pocket all of which look alike. Only one of them will open the door. She picks a key at random (without re-placement) from the bunch of keys at her disposal. Find the expected num-ber of keys that she will have to try before she can actually open the door.

24.   Let X_1, X_2 and X_3 be three random variables with the following joint prob ability
$$P(X_1 = x_1, X_2 = x_2, X_3 = x_3) = \frac{x_1 x_2 x_3}{72}$$
$$x_1 = 1, 2; x_2 = 1, 2, 3; x_3 = 1, 3$$
Find the marginal probability function of X_1

25.   In problem 24 find the marginal probability distribution of X_2 and of $X_1, X_3 | X_2$. Obtain the corresponding probability generating functions.

26.   A biased die when rolled has a probability 0.5 of a six appearing. All other numbers have a probability 0.1 of appearing. An individual is chosen to keep rolling the die until a five or a six shows up. Find the expected number of rolls required.

27.   An urn contains 6 red balls numbered 1, 2,..., 6 and 4 blue balls numbered 1, 2, 3, 4. Two balls are drawn at random one after the other (sampling

without replacement). What is the expected sum of the numbers on the two balls.

28. From the generating function $P(s) = \dfrac{4}{1+3s}$ find the coefficient of s^k.

29. From the generating function $P(s) = \dfrac{2}{6s^2 - 5s + 1}$ find the coefficient of s^k.

30. Let X be the number of successes in n independent trials with a constant probability of success p. Denote the probability of $X = k$ by
$$P_{k,n}, \quad k = 1, 2, ..., n.$$
Verify the following equation
$$P_{k,n} = pP_{k-1,n-1} + (1-p)P_{k,n-1}.$$
Derive the expression for the p.g.f. of X.

31. Prove the following inequality for a positive random variable Y using the Cauchy-Schwarz inequality $E\left(\dfrac{1}{Y}\right) > \dfrac{1}{E(Y)}$

32. Find the covariance and the coefficient of correlation between X and Y if
$$Y = a + bX.$$

33. If X and Y have the joint density function $f_{X,Y}(x,y) = x + y, \quad 0 < x, y < 1$. Find the variance of $X^2 Y^2$.

References

Bartoszynski, R. and Niewiadomska-Bugaj, M. (1996), *Probability and Statistical Inference,* John Wiley and Sons, New York.

Casella, G. and Berger, R.L. (2000), *Statistical Inference,* 2nd ed., Duxbury Press, Belmont, Mass

Chung, K.L. (1974), *A Course in Probability Theory*, Academic Press, New York.

Colton, T. (1974), *Statistics in Medicine*, Little, Brown and Company, Boston.

Cox, R. and Hinkley, D.V. (1974), *Theoretical Statistics*, Chapman and Hall, London.

Dudewicz, E.J. and Mishra, S.N. (1988), *Modern Mathematical Statistics*, John Wiley and Sons, New York.

Feller, W. (1965), *An Introduction to Probability Theory and Its Applications*, Vol 1, 2nd ed., John Wiley and Sons, New York.

Feller, W. (1971), *An Introduction to Probability Theory and Its Applications,* Vol II, John Wiley and Sons, New York.

Freund, J.F. (1971), *Mathematical Statistics*, 2nd ed., Prentice Hall , Englewood Cliffs, New Jersey.

Hogg, R.V. and Craig, A.T. (1978), *Introduction to Mathematical* Statistics, 4th ed., Macmillan Publishing Company, New York.

Jankovic, J. and Kapadia, A.S. (2001), A Functional Decline in Parkinson's Disease, *Archives of Neurology,* 58:1611-1615

Kendall, M.G. (1952), *The Advanced Theory of Statistics*, Vol I, 5th ed.,Charles Griffin and Company, London.

Mood, A.M. Graybill, F.A. and Boes D.C. (1974), *Introduction to the Theory of Statistics*, 3rd ed., McGraw-Hill, New York.

Moye, L.A. and Kapadia, A.S. (2000), *Difference Equations with Public Health Applications*, Marcel Dekker, New York.

Rao, C.R. (1952), *Advanced Statistical Methods in Biometric Research*, John Wiley and Sons, New York.

Rao, C.R. (1977), *Linear Statistical Inference and Its Applications,* 2nd ed., John Wiley and Sons, New York .

Rice, J.A. (1995), *Mathematical Statistics and Data Analysis,* 2nd ed., Duxbury Press, Belmont, Mass.

Rohatgi, V.K. (1975), *An Introduction to Probability Theory and Mathematical Statistics,* John Wiley and Sons, New York.

Wilks, S.S. (1962), *Mathematical Statistics*, John Wiley and Sons, New York.

4

Discrete Distributions

4.1 Introduction

In this chapter we will introduce several discrete distributions that will be useful at various points in the development of the remaining chapters in this book. While making important contributions to probability theory, each of these distributions is valuable in its own way. In addition the occurrence of several real life phenomena are governed by their probabilities. We will notice that associated with these distributions will be quantities known as the parameters of the distribution and each distribution may have more than one parameter. The shape of the distribution will vary with changes in the values of the parameters. Hence, we will be presenting families of distributions rather than a specific distribution. In Chapters 8 and 10 we will present methodologies for estimating these unknown parameters.

4.2 Bernoulli Distribution

This distribution is based on the idea of conducting an *experiment* consisting of several *independent trials* in such a way that each trial has exactly two possible outcomes (this concept was introduced in Chapter 3). These outcomes are usually referred to as "*success*" or "*failure*". For example a research effort that is designed to study the prevalence of coronary heart disease (CHD) in a group of unrelated adults may recruit participants at a construction site. In this case the response of each construction worker may correspond to a trial and the trial may be classified as a "success" if the worker has CHD or if he does not as a "failure". In this example the response for n workers corresponds to n trials. Similarly, tossing of a coin may be classified as a "success" if the toss results in a head and a failure if the toss results in a tail. If we were to toss the coin n times we would be involved in an experiment consisting of n trials. Such trials are referred to as " Bernoulli" trials and are named after James Bernoulli (1713) who is recognized as one of the earlier probabilists.

      Putting this information in a probability framework, a random variable X follows a Bernoulli distribution if

$$X = 1 \text{ with probability } p$$
$$X = 0 \text{ with probability } 1 - p.$$

Here p is the *parameter* associated with this distribution. Of course p can take any value in the interval $[\,0,1]$. Note that in the CHD example above, p is the probability that a construction worker has CHD and $1 - p$ is the probability that he has no CHD. The expected value (mean) and variance of X (defined in Chapter 3) may be obtained as follows.

$$E(X) = (1)p + (0)(1 - p)$$

$$Var(X) = E[X - E(X)]^2 = E(X^2) - [E(X)]^2$$
$$= p - p^2 = p(1 - p).$$

The moment generating function (mgf) of X will then be

$$M_X(t) = E(e^{tx}) = pe^t + (1 - p)e^0 = 1 - p(1 - e^t)$$
$$= (pe^t + q) \text{ where } q = 1 - p \qquad (4.1)$$

The kth moment $E(X^k)$ of this distribution may be obtained by differentiating $M_X(t)$ in (4.1) k times and setting $t = 0$ as follows

$$E(X^k) = \frac{d^k}{dt^k} M_X(t)\Big|_{t=0} = pe^t\Big|_{t=0} = p$$

Similarly the probability generating function $P(s)$ corresponding to this distribution will be

$$P(s) = E(s^X) = ps + (1 - p) = ps + q \qquad (4.2)$$

Note that differentiating $P(s)$ in (4.2) above more than once and setting $s = 1$ will yield zero values for the probabilities reflecting the fact that the variable takes only values 0 and 1 and that any other value has 0 probability associated with it.

We may expand on the Bernoulli distribution by considering two random variables X and Y with parameters p_1 and p_2 then the distribution of $Z = X+Y$ may be obtained as follows

$Z = 0$ with probability $(1-p_1)(1-p_2)$
$Z = 1$ with probability $p_1(1-p_2) + p_2(1-p_1) = p_1 + p_2 - 2p_1p_2$
$Z = 2$ with probability p_1p_2

Moments of the distribution of $Z = X+Y$ may be obtained using the concept of moment generating function of two independent variables as described in Chapter 3, section 3.7.

$$M_Z(t) = E(e^{tZ}) = M_X(t)M_Y(t) = [(1 - p_1(1 - e^t)][1 - p_2(1 - e^t)]$$
$$= 1 - p_1 - p_2 + p_1 p_2 + (p_1 + p_2 - 2p_1 p_2)e^t + p_1 p_2 e^{2t}$$

(4.3)

Using the mgf (4.3) above $E(Z^k)$, $k > 0$ may be calculated as

$$\frac{d^k}{dt^k} M_Z(t)\Big|_{t=0}$$

$$E(Z^k) \quad = (p_1 + p_2 - 2p_1 p_2)e^t + 2^k p_1 p_2 e^{2t}\Big|_{t=0}$$
$$= p_1 + p_2 - 2p_1 p_2 + 2^k p_1 p_2$$

For $p_1 = p_2 = p$.
$$M_Z(t) = q^2 + 2pqe^t + p^2 e^{2t} = (q + pe^t)^2, q = 1 - p$$

4.3 Binomial Distribution

The *binomial distribution* is the distribution of the number of successes in n independent trials where the probability of success in each trial is p and the probability of failure in each trial is q where $q = 1 - p$. This distribution is often represented by B(n,p). Hence if we were to perform a series of Bernoulli trials in such a way that associated with the ith trial is the random variable X_i which takes a value one with probability p (if the trial results in a success) and takes a value zero (if the trial results in a failure) with probability $q(= 1 - p)$ and if we define the random variable X as follows

$$X = X_1 + X_2 + \dots + X_n$$

then, X is the sum of n Bernoulli variables. The probability distribution of X is obtained using the following logic. The variable X takes the value 1, if only one of the X_i's takes a value 1 and all others take a value 0. This can happen in $\binom{n}{1}$ ways. For $X = 2$ to occur, exactly two of the X_i's may take a value 1 each with probability p and all others must be zeros each with probability q. This can occur in $\binom{n}{2}$ ways. In general, for X to take a value r, exactly r of the X_i's must take a value 1 and all others must take a value 0. This event occurs in $\binom{n}{r}$ ways and each of these is associated with a string of p's of length r and a string of q's of length $(n-r)$. The probability associated with X taking the value r therefore is

$$P(X = r) = \binom{n}{r} p^r q^{n-r}, \ p+q = 1, 0 \le p \le 1 \tag{4.4}$$

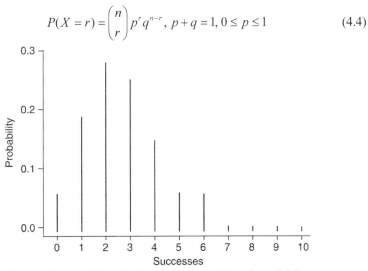

Figure 4.1 Binomial probability distribution for $n = 10$ and $p = 0.25$

From (4.4) we can obtain a recurrence relationship among the binomial probabilities as follows

$$P(X = r+1) = P(X = r)\frac{(n-r)p}{(r+1)q}, \ r = 0,1,...n-1$$

This relationship shows that $P(X = r)$ increases with r so long as $r \le np$ and decreases with r if $r > np$.

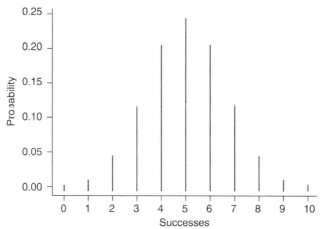

Figure 4.2 Binomial probably distribution for $n = 10$ and $p = 0.5$

Any variable that has a probability density function of the form (4.4) is said to follow the binomial distribution with parameters p and n and is often represented by $B(n, p)$.. Note that $P(X = r)$ is the rth term in the expansion of $(p + q)^n$.
(see Chapter 1, section 1.4)

The moment generating function of the binomial distribution will be obtained using two different methods.

Method 1

$$M_X(t) = E\left(e^{tX}\right) = \sum_{r=0}^{n} \binom{n}{r} p^r q^{n-r} e^{tr} = \sum_{r=0}^{n} \binom{n}{r} \left(pe^t\right)^r q^{n-r}$$

$$= \left(q + pe^t\right)^n. \tag{4.5}$$

The corresponding probability generating function may be obtained by replacing e^t by s in (4.5) and is
$$P(s) = (q + ps)^n$$

Method 2
Since $X = X_1 + X_2 + \ldots + X_n$
and each of the X's in the above relationship are independent Bernoulli variables, with mgf

$$\left(pe^t + q\right) \tag{4.6}$$

the mgf of X will be the product of n such terms. Hence

$$M_X(t) = (q + pe^t)^n. \tag{4.7}$$

The probability generating function can be similarly found by replacing

$$P(s) = (q + ps)^n. \tag{4.8}$$

4.3.1 Moments of the Binomial Distribution
It will now be illustrated that the moments for the binomial distribution may be derived using three different methods.

Method 1
We know from Chapter 3 that since the X_i's are independent random variables

$$E(X) = E(X_1 + X_2 + ... X_n) = \sum_{i=1}^{n} E(X_i) = \sum_{i=1}^{n} p = np$$

$$Var(X) = \sum_{i=1}^{n} Var(X_i) = \sum_{i=1}^{n} p(1-p) \tag{4.9}$$

$$= np(1-p) = npq$$

Method 2
Using the mgf (4.7)

$$E(X) = \frac{d}{dt}(pe^t + q)^n \Big|_{t=0} = n(pe^t + q)^{n-1} pe^t \Big|_{t=0} = np$$

and

$$Var(X) = E(X^2) - [E(X)]^2$$

$$E(X^2) = \frac{d^2}{dt^2}(pe^t + q)^n \Big|_{t=0} = np[1 + (n-1)p]$$

$$Var(X) = np(1-p) = npq$$

Method 3
In this method we use the binomial distribution (4.4) directly to obtain the first two moments and the variance of X

$$E(X) = \sum_{r=0}^{n} r \binom{n}{r} p^r q^{n-r} = \sum_{r=0}^{n} r \frac{n!}{r!(n-r)!} p^r q^{n-r}$$

$$= \sum_{r=1}^{n} \frac{n(n-1)!}{(r-1)!(n-r)!} p^r q^{n-r} = np \sum_{r=1}^{n} \binom{n-1}{r-1} p^{r-1} q^{n-r}$$

$$np \sum_{m=0}^{n-1} \binom{n-1}{m} p^m q^{n-1-m} = np(p+q)^{n-1} = np$$

To obtain the variance we first obtain $E(X^2)$ as follows

$$E(X^2) = \sum_{r=0}^{n} r^2 \binom{n}{r} p^r q^{n-r} = \sum_{r=0}^{n} r(r-1+1) \frac{n!}{r!(n-r)!} p^r q^{n-r}$$

$$= \sum_{r=2}^{n} \frac{n(n-1)(n-2)!}{(r-2)!(n-r)!} p^r q^{n-r} + E(X) = n(n-1) p^2 \sum_{r=2}^{n} \binom{n-2}{r-2} p^{r-2} q^{n-r}$$

$$+ np = n(n-1) p^2 \sum_{m=0}^{n-2} \binom{n-2}{m} p^m q^{n-2-m} + np = n(n-1) p^2 (p+q)^{n-2}$$

$$+ np = n(n-1) p^2 + np = n(n-1) p^2 + np$$

The variance is then obtained as

$$Var(X) = n(n-1) p^2 + np - (np)^2 = npq \tag{4.10}$$

Example 4.1 Suppose that the probability that a lung transplant patient who receives a blood transfusion develops Human Immunodeficiency Virus (HIV) infection is 0.001. An individual receives blood transfusion ten times during the course of her treatment. We are interested in determining the probability that she is not infected by the HIV virus.

To obtain this probability we will treat each encounter with the transfusion as an independent trial which has a probability of success (contracts HIV virus) as 0.001 and failure (does not contract HIV virus) with probability 0.999. In order for the patient not to get infected by the HIV virus in all these trials, we require that none of the trials result in a "success". Hence we need to obtain the probability of no successes in 10 trials when the probability of success in each trial is 0.001. Using the binomial formula (4.4) with $n = 10$, $p = 0.001$ and $r = 0$, we obtain the required probability as $(0.999)^{10} = 0.990$ a very large probability of not contacting HIV via blood transfusion. This is indeed good news for transplant patients.

Example 4.2 Let X have the $B(20, 0.2)$ distribution. Suppose we are required to obtain $P(X < 5)$. Then

$$P(X < 5) = \sum_{i=0}^{4} P(X = i) = \sum_{i=0}^{4} \binom{20}{i} (0.2)^i (0.8)^{20-i}$$
$$= 0.6296$$

Now suppose we are required to find a value of n such that $P(X > 0) = 0.97$
i.e., we need to first determine $P(X > 0)$ as follows

$$P(X > 0) = 1 - P(X = 0) = 1 - (0.8)^n$$

Next we solve the equation $1 - (0.8)^n = 0.97$
For n and obtain

$$(0.8)^n = 0.03 \tag{4.11}$$

The solution to (4.11) above may be obtained either using logarithms or from the Binomial tables and is obtained as $n=16$

Example 4.3 Suppose the probability of a male child being born (ignoring multiple births) is 0.5. We are required to find the probability that a family of 4 children will have (1) all girls (2) at most one boy (3) number of girls in the family is between 2 and 3, inclusively.

$$P(4 \ girls) = \left(\frac{1}{2}\right)^4 = 0.0625$$

$$P(\text{at most one boy}) = P(0 \text{ boys}) + P(\text{exactly 1 boy})$$

$$= 0.0625 + \binom{4}{1}\left(\frac{1}{2}\right)^4 = 0.0625 + 0.2500 = 0.3125$$

$$P(\text{between 2 and 3 girls}) = P(2 \text{ girls}) + P(3 \text{ girls})$$
$$= 0.375 + 0.250 = 0.625$$

Lemma 1* Let $q_{r,n}$ represent the probability that in a series of n Bernoulli trials there are exactly r failures. We will verify the following difference equation

$$q_{n-k,n} = qq_{n-k-1,n-1} + pq_{n-k,n-1} \tag{4.12}$$

In this lemma

$$q_{n-k,n} = \binom{n}{n-k} q^{n-k} p^{k}, \quad q_{n-k-1,n-1} = \binom{n-1}{n-k-1} q^{n-k-1} p^{k}$$

$$q_{n-k,n-1} = \binom{n-1}{n-k} q^{n-k} p^{k-1}$$

Substituting for $q_{n-k,n}$, $q_{n-k-1,n-1}$, and $q_{n-k,n-1}$ in (4.12) the difference equation is easily verified. ■

Lemma 2* Let $F_X(x:n,p)$ be the cumulative distribution function of the binomial random variable X representing the number of successes in n trials where the probability of success is p.

We will show that

$$F_X(x:n,p) = 1 - F_X(n-x-1:n,q) \tag{4.13}$$

Proof

$$F_X(x:n,p) = \sum_{r=0}^{x} \binom{n}{r} p^r q^{n-r} = \binom{n}{0} p^0 q^{n-0} + \binom{n}{1} p^1 q^{n-1} + \ldots + \binom{n}{x} p^x q^{n-x} \tag{4.14}$$

and

$$F_X(n-x-1:n,q) = \sum_{r=0}^{n-x-1} \binom{n}{r} q^r p^{n-r} \tag{4.15}$$

$$= \binom{n}{0} q^0 p^n + \binom{n}{1} q^1 p^{n-1} + \ldots + \binom{n}{n-x-1} q^{n-x-1} p^{n-(n-x-1)}$$

Adding (4.14) and (4.15) and using the facts $\binom{n}{r} = \binom{n}{n-r}$ and $(p+q)^n = 1$ relationship(4.13) is established ■

4.3.2 Renewal Processes*

In *renewal processes* we consider a series of independent trials and observe the occurrence of an event E of interest. Each time the event E occurs the process of observation starts from scratch. The following two probabilities are essential to the explanation of the occurrence of E.

$f_n = P(E$ occurs for the first time on the nth trial)

$u_n \equiv P(E$ occurs on the nth trial not neccessarily for the first time)

Then defining $u_0 = 1$ and $f_0 = 0$ a relationship between u_n and f_n can be derived based on the following reasoning.

For the event to occur on the nth trial not necessarily for the first time, the event could have occurred for the first time on the nth trial and then occurred in zero trials. The joint probability of this event is $f_n u_0$ (2). The event could have occurred for the first time on the $(n-1)^{th}$ trial and then in one trial. The joint probability of this event is $f_{n-1} u_1$ and so on. In general the event could have occurred for the first time on the ith trial and then again on the nth [i.e., in $(n-i)$ trials $(i = 1, 2, ..., n)$] the joint probability of this event is $f_i u_{n-i}$. Hence the probability of occurrence of the event on the nth trial can be written as

$$u_n = f_n u_0 + f_{n-1} u_1 + f_{n-2} u_2 + ... + f_1 u_{n-1}$$
$$= \sum_{i=1}^{n} f_i u_{n-i}, \quad n = 1, 2, ... \tag{4.16}$$

Here,
$$f_r u_{n-r} \equiv P(E \text{ occurs for the first time on the } r^{th} \text{ trial and then again on}$$
$$\text{the } (n-r)^{th} \text{ trial}).$$

Define
$$U(s) = \sum_{n=0}^{\infty} u_n s^n, \quad F(s) = \sum_{n=0}^{\infty} f_n s^n.$$

Multiplying both sides of (4.16) by s^n and summing over all n from $n = 1$ to $n = \infty$ we have

$$U(s) - 1 = \sum_{n=1}^{\infty} \sum_{i=1}^{n} f_i u_{n-i} s^n = \sum_{i=1}^{\infty} \sum_{n=i}^{\infty} f_i s^i u_{n-i} s^{n-i} = \sum_{i=1}^{\infty} f_i s^i \sum_{j=0}^{\infty} u_j s^j$$

$$U(s) - 1 = U(s)F(s)$$

$$U(s) = \frac{1}{1 - F(s)}$$

4.3.3 Success Run of Length r*

In a series of Bernoulli trials if r consecutive trials result in the occurrence of a success then we say that a success run of length r has occurred. Failure runs of length r (Moye and Kapadia 2000) are similarly defined. Consider a series of n independent trials with a constant probabil**Error! Bookmark not defined.**ity of success p. If "success" occurs r times in succession, we say that the event E has

occurred (or a run of length r has occurred). We need to determine the probability that at least one run of r successes occurs in n trials (Uspensky 1937).
Let

$f_n \equiv P$ (a success run of length r occurs for the first time on the nth trial)

$u_n \equiv P$ (a success run of length r occurs on the nth trial, not necessarily for the first time)

Then the probability that at least one run of r successes occurs in n trials is given by

$$f_r + f_{r+1} + ... f_n$$

where f_i is the coefficient of s^i in $F(s)$ and $F(s) = 1 - \dfrac{1}{U(s)}$.

The probability that the last r trials n-r+1, n-r+2,..., n-1, n result in successes is p^r. In these trials E could occurs at the trial n or n-1 or n-2,...or n-r+1. Hence

$$p^r = u_n + u_{n-1}p + u_{n-2}p^2 + ... + u_{n-r+1}p^{r-1}, n \geq r$$
$$u_1 = u_2 = ... = u_{r-1} = 0, u_0 = 1$$
(4.17)

Multiplying (4.17) by s^n and summing over all n from $r, r+1, r+2,...$

$$\sum_{n=r}^{\infty}\sum_{i=0}^{r-1}u_{n-i}p^i s^n = \sum_{n=r}^{\infty}p^r s^n = \frac{(ps)^r}{(1-s)}$$
(4.18)

The left hand side of (4.18) can be written as

$$\sum_{n=r}^{\infty}u_n s^n + ps\sum_{n=r}^{\infty}u_{n-1}s^{n-1} + (ps)^2\sum_{n=r}^{\infty}u_{n-2}s^{n-2} + ... + (ps)^{r-1}\sum_{n=r}^{\infty}u_{n-r+1}s^{n-(r-1)}$$

$$= [U(s)-1] + (ps)[U(s)-1] + (ps)^2[U(s)-1] + ... + (ps)^{r-1}[U(s)-1]$$

$$= [U(s)-1]\frac{1-(ps)^r}{1-ps}$$

Equating both sides of (4.18) we have

$$[U(s)-1]\frac{1-(ps)^r}{1-ps} = \frac{(ps)^r}{(1-s)}$$

or

$$U(s) = \frac{1-s+(1-p)p^r s^{r+1}}{(1-s)(1-p^r s^r)}.$$

Therefore f_n the probability of the first occurrence of a success run of length r can be obtained as the coefficient of s^n in $F(s)$.

4.3.4 Sum of Two Binomial Variables

Consider two independent binomial variables X and Y. Assume that X has the binomial $B(n_1, p_1)$ distribution and Y has the $B(n_2, p_2)$ distribution. Then define $Z = X+Y$. Since X and Y are independent random variables, the probability generating function of Z will be

$$P_Z(s) = E\left(s^Z\right) = E(s^X s^Y) = P_X(s)P_Y(s) = (q_1 + p_1 s)^{n_1} (q_2 + p_2 s)^{n_2} \qquad (4.19)$$

Hence, for $n_1 + n_2 \geq k$

$$P(Z = k) = \text{Coefficient of } s^k \text{ in } (q_1 + p_1 s)^{n_1} (q_2 + p_2 s)^{n_2}$$

$$= \sum_{i=\max(0, k-n_2)}^{\min(n_1, k)} \binom{n_1}{i} \binom{n_2}{k-i} p_1^i p_2^{k-i} q_1^{n_1-i} q_2^{n_2-(k-i)}$$

Alternatively, for $p_1 = p_2 = p$ the probability generating function of the sum of two $B(n_1, p)$ and $B(n_2, p)$ variables is the same as the pgf of a $B(n_1 + n_2, p)$ variable. The probability mass function for this case may be derived for the following three cases:

(1) For $0 \leq k \leq n_1 < n_2$

$$P(Z = k) = p^k (1-p)^{n_1+n_2-k} \sum_{i=0}^{k} \binom{n_1}{i} \binom{n_2}{k-i} = \binom{n_1+n_2}{k} p^k (1-p)^{n_1+n_2-k}$$

(2) For $n_1 < k \leq n_2$

$$P(Z = k) = p^k (1-p)^{n_1+n_2-k} \sum_{i=0}^{n_1} \binom{n_1}{i} \binom{n_2}{k-i} = \binom{n_1+n_2}{k} p^k (1-p)^{n_1+n_2-k}$$

(3) For $n_1 < n_2 < k \leq n_1 + n_2$

$$P(Z = k) = p^k (1-p)^{n_1+n_2-k} \sum_{i=k-n_2}^{n_1} \binom{n_1}{i} \binom{n_2}{k-i} = \binom{n_1+n_2}{k} p^k (1-p)^{n_1+n_2-k}$$

The above results are obtained using the Vendermonde convolution (Chapter 1, section 1.4) which states that

$$\sum_{i=0}^{n} \binom{x}{i} \binom{y}{n-i} = \binom{x+y}{n}$$

4.3.5 Relative Frequency

Let X be a $B(n.p)$ variable and let us define a new variable $Y = \dfrac{X}{n}$. Then Y is called the relative frequency of the occurrence of a success in a series of n .
Berrnoulli trials.
Clearly

$$E(Y) = p \text{ and } Var(Y) = \frac{p(1-p)}{n} = \frac{pq}{n}$$

Using Chebychev's Inequality (Inequality 3.1, Chapter 3, section 3.8) for a fixed t (>0) we can write the following inequality

$$P\Big[|Y - p| \geq t\Big] \leq \frac{Var(Y)}{t^2} = \frac{p(1-p)}{nt^2} \qquad (4.20)$$

Taking the limits of both sides of (4.20) as $n \to \infty$ we have
$\lim_{n \to \infty} P\Big[|Y - p| \geq t\Big] = 0$, $P\Big[|Y - p| \geq t\Big] = 0$ implying that the relative frequency of success for large n is close to the actual probability of success. This will be described in detail in Chapter 7.

4.4 Multinomial Distribution

The multinomial distribution is a generalization of the binomial distribution to the case when $m+1$ mutually exclusive outcomes $E_0, E_1, E_2, ..., E_m$ are associated with each trial. The n trials are independent and the probability of the occurrence of E_i at any trial is assumed to be p_i. The event E_0 occurs with probability $p_0 (= 1 - \sum_{i=1}^{m} p_i)$.

If we associate random variables $X_0, X_1, X_2, ..., X_m$ with events $E_0, E_1, E_2, ..., E_m$ such that in a series of n trials X_i represents the number of times event E_i occurs then the probability

$$P(X_1 = n_1, ..., X_m = n_m) \qquad (4.21)$$

$$= \frac{n!}{n_1! n_2! ... n_m! (n-k)!} p_1^{n_1} p_2^{n_2} ... p_m^{n_m} p_0^{n-k}, \quad \sum_{1}^{m} n_i = k$$

The above combinatorial expression was described in Chapter 1 section 1.2.2 and represents the number of ways of choosing $n_i (i = 1, 2, ..., m)$ objects of the ith type from among a total of n objects. We will assume that p_0 is like q in the binomial case and is associated with the occurrence of E_0 which is assumed to occur if none of the $E_i (i = 1, 2, ..., m)$ occur.

The distribution (4.21) is called *multinomial* because the right hand term is the general term in the expansion of $(p_0 + p_1 + p_2 + ... p_m)^n$. Its main application is in sampling with replacement when the items being sampled are categorized into more than two categories. Summing over all values of n_m in (4.21) we obtain

$$P(X_1 = n_1, ..., X_{m-1} = n_{m-1}) =$$

$$\frac{n!}{n_1! n_2! ... n_{m-1}! (n - \sum_{i=1}^{m-1} n_i)!} p_1^{n_1} p_2^{n_2} ... p_{m-1}^{n_{m-1}} p_0^{n - \sum_{i=1}^{m-1} n_i} \sum_{n_m=0}^{n - \sum_{i=1}^{m-1} n_i} \frac{(n - \sum_{i=1}^{m-1} n_i)!}{n_m! (n - \sum_{i=1}^{m} n_i)! p_0^{n_m}} p_m^{n_m}$$

$$= \frac{n!}{n_1! n_2! ... n_{m-1}! (n - \sum_{i=1}^{m-1} n_i)!} p_1^{n_1} p_2^{n_2} ... p_{m-1}^{n_{m-1}} p_0^{n - \sum_{i=1}^{m-1} n_i} \sum_{n_m=0}^{n - \sum_{i=1}^{m-1} n_i} \binom{n - \sum_{i=1}^{m-1} n_i}{n_m} \left(\frac{p_m}{p_0} \right)^{n_m}$$

$$= \frac{n!}{n_1! n_2! ... n_{m-1}! (n - \sum_{i=1}^{m-1} n_i)!} p_1^{n_1} p_2^{n_2} ... p_{m-1}^{n_{m-1}} p_0^{n - \sum_{i=1}^{m-1} n_i} \left[\frac{p_0 + p_m}{p_0} \right]^{n - \sum_{i=1}^{m-1} n_i}$$

$$= \frac{n!}{n_1! n_2! ... n_{m-1}! (n - \sum_{i=1}^{m-1} n_i)!} p_1^{n_1} p_2^{n_2} ... p_{m-1}^{n_{m-1}} \left[1 - \sum_{i=1}^{m-1} p_i \right]^{n - \sum_{i=1}^{m-1} n_i}$$

Hence the marginal probability distribution function of the joint distribution of $X_1, ..., X_{m-1}$ also follows a multinomial distribution. Proceeding as above the marginal probability distribution of X_i will have the form

$P(X_i = n_i) = \dfrac{n!}{n_i! (n - n_i)!} p_i^{n_i} (1 - p_i)^{n - n_i}$ which is a binomial distribution with pa-

rameters (n, p_i). That is X_i is a binomial variable with probability of "success" p_i and probability of "failure" $\sum_{\substack{j=1 \\ j \neq i}}^{k} p_j = 1 - p_i$. Hence,

$E(X_i) = np_i$ and $Var(X_i) = np_i(1 - p_i)$

        Similarly the marginal distribution of X_i and X_j has the trinomial distribution

$$P(X_i = n_i, X_j = n_j) = \frac{n!}{n_i! n_j! (n - n_i - n_j)!} p_i^{n_i} p_j^{n_j} (1 - p_i - p_j)^{n - n_i - n_j}$$

and

$$E(X_i X_j) = \sum_{n_i} \sum_{n_j} \frac{n_i n_j n(n-1)(n-2)!}{n_i (n_i - 1)! n_j (n_j - 1)! (n - n_i - n_j)!} p_i p_i^{n_i - 1} p_j p_j^{n_j - 1}$$

$$\times (1 - p_i - p_j)^{n - 2 - (n_i - 1) - (n_j - 1)} \quad (4.22)$$

$$= n(n-1) p_i p_j$$

Since X_i and X_j are not independent their covariance may be obtained as follows

$$Cov(X_i, X_j) = E(X_i X_j) - E(X_i)E(X_j)$$

$$= n(n-1) p_i p_j - (n p_i)(n p_j) \quad (4.23)$$

$$= -n p_i p_j, \quad i \neq j$$

Example 4.4 Consider a deck of playing cards containing 4 suits and 13 denominations from which cards are chosen at random one at a time and after each draw are put back in the deck (sampling with replacement). Hence the probability of choosing any particular card from the deck stays constant. Assume that we choose 10 cards. What is the probability of drawing exactly 2 hearts, 3 spades, and 1 club? Notice that this is a problem requiring the use of the multinomial distribution.

Probability (heart) = $13/52 = 0.25$
Probability (spade) = $13/52 = 0.25$
Probability (club) = $13/52 = 0.25$

Since we have specified the number of hearts, spades and clubs it is implied that $10 - 2 - 3 - 1 = 4$ diamonds will appear in this sample of 10 cards. The probability of the appearance of a diamond at each draw will be $1 - 0.25 - 0.25 - 0.25 = 0.25$. Hence the required probability is

$$\binom{10}{2, 3, 1, 4} (0.25)^{10} = \frac{10!}{2! 3! 1! 4!} (0.25)^{10} = 0.0120$$

4.4.1 Conditional Probability Distributions

Consider the probability distribution of X_j conditional on X_i. That is, we may be interested in the distribution of $X_j | X_i = n_i$

$$P(X_j = n_j | X_i = n_i) = P(X_j = n_j, X_i = n_i) / P(X_i = n_i).$$

Now $P(X_j = n_j, X_i = n_i) = \begin{pmatrix} n \\ n_i, n_j, (n - n_i - n_j) \end{pmatrix} p_i^{n_i} p_j^{n_j} (1 - p_i - p_j)^{n - n_i - n_j}$

and

$$P(X_i = n_i) = \begin{pmatrix} n \\ n_i, n - n_i \end{pmatrix} p_i^{n_i} (1 - p_i)^{n - n_i} .$$

Hence

$$P(X_j = n_j | X_i = n_i) = \begin{pmatrix} n - n_i \\ n_j \end{pmatrix} \left(\frac{p_j}{1 - p_j} \right)^{n_j} \left(1 - \frac{p_j}{1 - p_j} \right)^{n - n_i - n_j} .$$

The above conditional probability is binomial. To obtain $E(X_j | X_i = n_i)$ we proceed as follows

$$E(X_j = n_j | X_i = n_i) = \sum_{n_j=0}^{n-n_i} n_j \begin{pmatrix} n - n_i \\ n_j \end{pmatrix} \left(\frac{p_j}{1 - p_i} \right)^{n_j} \left(1 - \frac{p_j}{1 - p_i} \right)^{n - n_i - n_j}$$

$$= (n - n_i) \left(\frac{p_j}{1 - p_i} \right) \left(1 - \frac{p_j}{1 - p_i} \right)^{n - n_i - 1} \sum_{n_j=1}^{n-n_i} \begin{pmatrix} n - n_i - 1 \\ n_j - 1 \end{pmatrix} \left(\frac{1 - p_i}{1 - p_i - p_j} \right)^{n_j - 1} \left(\frac{p}{1 -} \right)$$

$$= (n - n_i) \left(\frac{p_j}{1 - p_i} \right) \left(1 - \frac{p_j}{1 - p_i} \right)^{n - n_i - 1} \left(1 + \frac{p_j}{1 - p_i - p_j} \right)^{n - n_i - 1}$$

$$= \frac{(n - n_i) p_j}{(1 - p_i)}$$

$Var(X_j | X_i = n_i)$ can be similarly obtained as

$$Var(X_j | X_i = n_i) = E(X_j^2 | X_i = n_i) - [E(X_j | X_i = n_i)]^2$$

$$E(X_j^2 | X_i = n_i) = (n - n_i)(n - n_i - 1) \frac{p_j^2}{(1 - p_i)^2} + (n - n_i) \frac{p_j}{(1 - p_i)}$$

$$Var(X_j | X_i = n_i) = (n - n_i) \frac{(1 - p_i - p_j) p_j}{(1 - p_i)^2}$$

4.5 Hypergeometric Distribution

The hypergeometric distribution is often used in the field of statistical quality control where products are constantly checked to see if they meet specifications laid down by the manufacturer. More generally, assume there is a lot containing N units of a product. It is known that M of these items are defective and $N-M$ are non defective. A sample of size n is drawn at random and without replacement and the manufacturer is interested in determining the probability distribution of the variable X representing the number of defective items that show up in a sample of size n. The basic sample space consists of $\binom{N}{n}$ sample points. Here each sample point is a set of n items from a lot containing N items. Furthermore, since it is required that exactly X of these be from among the M defective items, there are $\binom{M}{X}$ ways of choosing X items from the M defective ones. Similarly there are $\binom{N-M}{n-X}$ ways of choosing $n-X$ items from the $N-M$ non defective ones. Hence there are $\binom{M}{X}\binom{N-M}{n-X}$ ways of choosing exactly X defective and exactly $(n-X)$ non defective items. Therefore the probability associated with this combination of defective and non defective items in a sample of size n is given by

$$P(X = x) = \frac{\binom{M}{x}\binom{N-M}{n-x}}{\binom{N}{n}}, \quad Max[M+n-N,0] \le x \le Min(n,M) \quad (4.24)$$

In the derivation below it is assumed that $0 \le x \le n$ since n is presumed to be small compared to both N and M.

$$\sum_{x=0}^{n} \frac{\binom{M}{x}\binom{N-M}{n-x}}{\binom{N}{n}} = \frac{1}{\binom{N}{n}} \sum_{x=0}^{n} \binom{M}{x}\binom{N-M}{n-x} = 1$$

The above result is obtained using the Vendermonde Equality (Chapter 1, Section Section 1.4) The hypergeometric distribution will be used as the basis of a statistical hypotheses test known as the Fisher's Exact Test in Chapter 9.

Example 4.5 Consider a lot containing m defective items and $N-m$ acceptable items. The units are tested one at a time without replacement. Let X represent the number of defective items found before the first acceptable one. We need to derive the probability distribution of X and obtain its mean. Define

$$P(X = i) = P_i = P(i \text{ defectives items found before a good item})$$

$$= \frac{m(m-1)(m-2)...(m-i+1)(N-m)}{N(N-1)(N-2)...(N-i+1)(N-i)}$$

$$= \frac{m!i!(N-i-1)!(N-m)}{i!(m-i)!N!}$$

$$= \frac{N-m}{N} \frac{\binom{m}{i}}{\binom{N-1}{i}} = \frac{N-m}{N} \binom{m}{i} \binom{N-1}{i}^{-1} \qquad i = 0,1,...m$$

Before we can show that $\sum_{1}^{m} P_i = 1$ the following relationship in combinatorics (Frisch 1926) is assumed

For $x > z, x$ and z integers

$$\sum_{k=j}^{n} \binom{z}{k} \binom{x}{k}^{-1} = \frac{x+1}{(x-z+1)} \left[\binom{z}{j} \binom{x+1}{j}^{-1} - \binom{z}{n+1} \binom{x+1}{n+1}^{-1} \right],$$

Using the above result

$$\sum_{i=0}^{m} P_i = \frac{N-m}{N} \sum_{i=0}^{m} \binom{m}{i} \binom{N-1}{i}^{-1}$$

$$= \frac{N-m}{N} \frac{N-1+1}{N-1-m+1} \left[\binom{m}{0} \binom{N}{0}^{-1} - \binom{m}{m+1} \binom{N}{m+1}^{-1} \right] = 1$$

The above derivation is based on the fact that $\binom{m}{j} = 0$ for $j > m$. The mean of X is obtained as follows

$$E(X) = \sum_{i=1}^{m} iP_i = \frac{N-m}{N} \sum_{i=1}^{m} \sum_{j=i}^{m} \binom{m}{j} \binom{N-1}{j}^{-1}$$

$$= \frac{N-m}{N} \sum_{i=1}^{m} \frac{N-1+1}{N-1-m+1} \binom{m}{i} \binom{N}{i}^{-1}$$

$$= \sum_{i=1}^{m} \binom{m}{i} \binom{N}{i}^{-1} = \frac{N+1}{N-m+1} \frac{m}{N+1} = \frac{m}{N-m+1}$$

Notice that in the above derivation Frisch's formula was used twice.

4.5.1 Approximation to Binomial Distribution

In this section we will show that if N and $M \to \infty$ in such a way that

$\dfrac{M}{N} \to p$ the hypergeometric probability mass function (1.21) approaches the

probability mass function of the binomial distribution (4.4). First Stirling's formula from Chapter 1, section 1.6 is applied to all the factorials in (4.24) containing N and M. For example for large M and N

$$M! \approx M^{M+\frac{1}{2}} \sqrt{2\pi} e^{-M}$$

$$N! \approx N^{N+\frac{1}{2}} \sqrt{2\pi} e^{-N}$$

$$(M-x)! \approx (M-x)^{M-x+\frac{1}{2}} \sqrt{2\pi} e^{-(M-x)}$$

$$(N-n)! \approx (N-n)^{N-n+\frac{1}{2}} \sqrt{2\pi} e^{-(N-n)}$$

$$(N-M)! = (N-M)(N-M-1)...(N-M-(n-x-1)(N-m-(n-x))!$$

Using the fact that

$$\lim_{M \to \infty} \left(\frac{M}{M-x}\right)^M = e^x$$

and

$$\lim_{N \to \infty} \left(\frac{N}{N-n}\right)^N = e^n$$

After a little algebraic manipulation (4.24) simplifies to

$$\binom{n}{x} p^x (1-p)^{n-x}$$

4.5.2 Moments of the Hypergeometric Distribution

For the derivation of the moments of the hypergeometic distribution assume $0 \le x \le n$

$$E(X) = \frac{1}{\binom{N}{n}} \sum_{x=1}^{n} \frac{M!}{(x-1)!(M-x)!} \binom{N-M}{n-x}$$

$$= \frac{M}{\binom{N}{n}} \sum_{x=1}^{n} \binom{M-1}{x-1}\binom{N-1-(M-1)}{n-1-(x-1)}.$$

Let $x - 1 = y$, then, the RHS of the above expression may be written as

$$= \frac{M}{\binom{N}{n}} \sum_{y=0}^{n-1} \binom{M-1}{y}\binom{N-1-(M-1)}{n-1-y}$$

$$= \frac{M}{\binom{N}{n}} \binom{N-1}{n-1} = \frac{Mn}{N}.$$

and

$$E(X^2) = \frac{M}{\binom{N}{n}} \sum_{x=1}^{n} (x-1+1)\frac{(M-1)!}{(x-1)!(M-x)!}\binom{N-M}{n-x}$$

$$= \frac{M}{\binom{N}{n}} \left[(M-1)\sum_{x=2}^{n} \binom{M-2}{x-2}\binom{N-2-(M-2)}{n-2-(x-2)} + \binom{N-1}{n-1} \right].$$

The second term in the above equation is obtained using the Vendermonde convolution (see Chapter 1 section 1.4)

$$E(X^2) = \frac{M}{\binom{N}{n}} \left[(M-1)\binom{N-2}{n-2} + \binom{N-1}{n-1} \right]$$

$$= \frac{M(M-1)}{\binom{N}{n}}\binom{N-2}{n-2} + \frac{M}{\binom{N}{n}}\binom{N-1}{n-1}$$

After a little simplification we obtain

$$E(X^2) = \frac{M(M-1)(n-1)n}{N(N-1)} + \frac{Mn}{N}$$

and

$$Var(X) = \frac{Mn(n-N)(M-N)}{N^2(N-1)}$$

Example 4.6 Suppose that a lot of 100 oranges has been delivered to a New York City restaurant. The restaurant owner is going to randomly select a sample from the lot of oranges which (unknown to the restaurant owner) is supposed to have six rotten ones. He will accept the lot if in his sample of size n no rotten orange shows up. We need to determine n such that the probability of accepting the lot is less than 0.1.

Let X be the random variable representing the number of rotten oranges that show up in a random sample of n oranges. We need to obtain a value of n such that $P(X = 0) \leq 0.1$ after substituting $N = 100$, $M = 6$ and $x = 0$ in (4.24).

$$P(X = 0 | N = 100, M = 6, x = 0, n) = \frac{\binom{6}{0}\binom{94}{n}}{\binom{100}{n}}$$

Computation of the above expression reveals that for $n = 31$, $P(X = 0) = 0.10056$ and for $n = 32$, $P(X = 0) = 0.09182$. Hence we conclude that $n \geq 32$.

4.6 k-Variate Hypergeometric Distribution*

Assume that we have a total of N items classified into $k+1$ categories (or classes). Suppose that N_i $(i = 1, 2, .., k+1)$ items among these N belong to the ith class and we want to draw a sample of n objects from among the N elements such that x_i $(i = 1, 2, ..., k+1)$ of these in the sample belong to the ith class.

Here again there are $\binom{N}{n}$ possible ways of choosing a sample of n elements out of which there are exactly

$$\binom{N_1}{x_1}\binom{N_2}{x_2}...\binom{N_{k+1}}{x_{k+1}}$$

many possibilities such that x_i elements from the ith class ($i = 1, 2, ... k+1$) are chosen. Hence the probability that the sample of size n drawn from among the N items has exactly x_i items from the ith class is given by

$$P(x_1, x_2..., x_k) = \frac{\binom{N_1}{x_1}\binom{N_2}{x_2}...\binom{N_{k+1}}{x_{k+1}}}{\binom{N}{n}}$$

where $\sum_{i=1}^{k+1} x_i = n$ and $\sum_{i=1}^{k+1} N_i = N$

4.7 Geometric Distribution

Consider an infinite sequence of Bernoulli trials where the probability of "success" is p and the probability of "failure" is $q(=1-p)$. Let X be a random variable representing the number of failures before the first success. The variable X is said to have a *geometric distribution.* Sometimes X is referred to as the *Pascal variable.* Here,

$$P(X = k) = q^k p, \quad k = 0,1,\ldots$$

The moment generating function of X is obtained as follows

$$M_X(t) = E(e^{tX}) = p \sum_{k=0}^{\infty} (qe^t)^k = \frac{p}{(1-qe^t)}.$$

The expected value and the variance of X are easily obtained from the mgf as

$$E(X) = \frac{d}{dt} M_X(t)\Big|_{t=0} = \frac{q}{p} \text{ and}$$

$$E(X^2) = \frac{d^2}{dt^2} M_X(t)\Big|_{t=0} = \frac{q(1+q)}{p^2},$$

$$Var(X) = \frac{q}{p^2}.$$

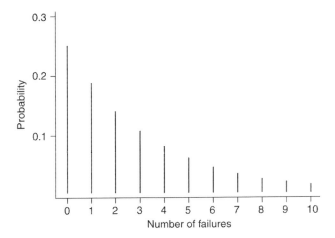

Figure 4.3 Geometric distribution for $p = 0.25$

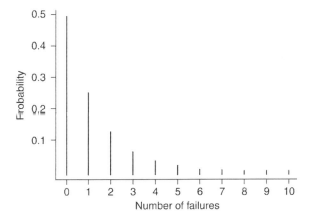

Figure 4.4 Geometric distribution for $p = 0.5$

The geometric distribution is often referred to as the *failure time distribution*. Its continuous analog is the *exponential distribution* which we will study in great depth in Chapter 5, section 5.5.

Example 4.7 Let X be a random variable following a geometric distribution. To compute $P(X > n)$

$$P(X > n) = q^{n+1}p + q^{n+2}p + q^{n+3}p + ...$$
$$= pq^{n+1}(1 + q + q^2 + ...) = q^{n+1}$$

Similarly

$$P(X \leq n) = p + qp + q^2p + q^3p + .. + q^n p$$
$$= p(1 + q + q^2 + ... + q^n) = (1 - q^{n+1})$$

Example 4.8 Let X and Y be two independent random variables each distributed according to a geometric distribution with parameter p, such that

$$P(X = x) = q^x p \text{ and } P(Y = y) = q^y p, \quad x, y = 0,1,2,...$$

We are interested in determining the distribution of $U = Min(X,Y)$

Case (1)
Let $U = Min(X,Y)$

and

let $V = X - Y$.

If $X = Min(X,Y)$ then $X < Y$ and V takes negative values only.

The joint distribution of U and V is then

$$P(U=u, V=v) = p^2 q^{2u-v} = (pq^{2u})(pq^{-v})$$

$$P(U=u) = p^2 q^{2u} \sum_{-\infty}^{-1} q^{-v} = p^2 q^{2u} q(1+q+q^2+...)$$

$$= p^2 q^{2u+1} \frac{1}{1-q} = pq^{2u+1}, \ u = 1, 2, ...$$

Case (2)

Let $U = Y = Min(X,Y)$ then $V = X - Y = X - U$ or $X = V + U$

$$P(U=u, V=v) = p^2 q^{2u+v}, \ u, v = 1, 2, ...$$

$$P(U=u) = p^2 q^{2u} \sum_{v=1}^{\infty} q^v = pq^{2u+1}$$

Case (3)

$X = Y, U = X = Y$ and $V = 0 = X - Y$

$$P(U=u, V=v) = p^2 q^{2u} = P(U=u)$$

■

The geometric distribution has the unique property that at any point in time the number of failures until the next occurrence of a "success" is independent of the number of failures since the last occurrence of a "success". This is equivalent to saying

For $i > j$

$$P(X > i \mid X > j) = P(X > i \text{ and } X > j) / P(X > j)$$
$$= P(X > i - j) \tag{4.25}$$

This property associated with the geometric distribution is called the *memoryless property* of the geometric distribution.

4.8 Negative Binomial Distribution

The distribution of the number of failures before the first success generalizes to what has come to be known as the *negative binomial distribution.* Here we are interested in the distribution of the number of failures before the rth success.

Let the random variable Y represent the number of failures before the rth success.

Then $Y + r - 1$ trials will have occurred before the rth success and $(Y+r)^{th}$ trial is a success. The probability of this event is

$$P(Y=y) = \binom{y+r-1}{y} q^y p^r \ \ y = 0, 1, 2, ..., q = 1 - p \tag{4.26}$$

The variable Y is said to have a negative binomial distribution with parameters r and p. The distribution described in (4.26) is also called the *binomial waiting time distribution.* The probability mass function indicates that one must wait through $y + r - 1$ trials before r successes can occur. The moment generating function of Y can be obtained as follows

$$M_Y(t) = E(e^{tX}) = \sum_{y=0}^{\infty} \binom{y+r-1}{y}(qe^t)^y p^r$$

$$= p^r \sum_{y=0}^{\infty} \binom{r}{y}(-qe^t)^y = \frac{p^r}{(1-qe^t)^r}. \qquad (4.27)$$

The relationship in (4.27) is derived by using equation (1.6) in Chapter 1 and the Newton's binomial formula discussed in Chapter 1, section 1.4. which states

$$\binom{-t}{k} = (-1)^k \binom{t+k-1}{k}.$$

4.8.1 Moments of the Negative Binomial Distribution

$$E(Y) = \frac{d}{dt} M_X(t)\Big|_{t=0} = \frac{d}{dt}\left[\frac{p}{1-qe^t}\right]^r\Big|_{t=0}$$

$$= \frac{rqp^r e^t}{(1-qe^t)^{r+1}}\Big|_{t=0} = \frac{rq}{p}$$

$$E(Y^2) = \frac{d^2}{dt^2} M_X(t)\Big|_{t=0} = r(r+1)p^r q^2 e^{2t}(1-qe^t)^{-r-2}$$

$$+ rp^r qe^t (1-qe^t)^{-r-1}\Big|_{t=0}$$

$$= p^r qr \left[\frac{1}{p^{r+1}} + \frac{q(r+1)}{p^{r+2}}\right]$$

$$= \frac{qr(1+qr)}{p^2}$$

$$Var(Y) = E(Y^2) - [E(Y)]^2 = \frac{qr}{p^2}.$$

4.8.2 Relationship to the Geometric Distribution

Consider a sequence of Bernoulli trials with probability of success p and let Y_i ($i = 1, 2, ...n$) represent the number of failures between the $(i\text{-}1)^{th}$ and i^{th} success. Then Y_i has the geometric distribution with parameter p. Define

$$W - Y_1 + Y_2 + ... + Y_n$$

Then W is the random variable representing the number of failures before the nth success. Hence W must have the negative binomial distribution.

Since the variables are independent and $P(Y_i = k) = q^k p$, $(q = 1 - p)$ the moment generating function associated with each Y_i is

$$M_{Y_i}(t) = \frac{p}{1 - qe^t} \tag{4.28}$$

and the mgf of W is obtained by raising the mgf in (4.28) to the nth power. Hence,

$$M_W(t) = \left(\frac{p}{1 - qe^t} \right)^n$$

which is clearly the mgf of a negative binomial variable with parameters p and n.

4.8.3 Truncated Distributions*

Definition: A 0-*truncated* discrete random variable X_T has a probability mass function

$$P(X_T = x) = \frac{P(X = x)}{P(X > 0)}, \quad x = 1, 2, \ldots$$

We are going to show that starting with the 0- truncated negative binomial distribution, if we let $r \to 0$ we get a distribution called the *logarithmic series* distribution or *the logarithmic* distribution with the following probability mass function

$$P(X = x) = \frac{-(1-p)^x}{x \log p}, \quad x = 1, 2, \ldots, \quad 0 < p < 1$$

The 0-truncated negative binomial distribution has the following form

$$P(X_T = x) = \frac{\binom{x + r - 1}{r - 1} p^r q^x}{1 - p^r}, \quad x = 1, 2, \ldots$$

The moment generating function of the above distribution is

$$M_{X_T}(t) = \frac{p^r}{(1 - qe^t)^r (1 - p^r)} - \frac{p^r}{(1 - p^r)}$$

Letting $r \to 0$ in the above mgf and using L'Hôspital's rule the moment generating function $M_{X_T}(t)$ simplifies to

$$\lim_{r \to 0} M_{X_T}(t) = \frac{\log(1 - e^t + pe^t)}{\log p}$$

which is the mgf of the logarithmic distribution defined above.

4.8.4 Sum of k Independent Negative Binomial Variables*

Consider a sequence of k independent variables $X_1, X_2, ... X_k$ each having a negative binomial distribution with parameters p and l_i $(i = 1, 2, .., k)$. Then the mgf of X_i is

$$M_{X_i}(t) = \left(\frac{p}{1 - qe^t} \right)^{l_i}, q = 1 - p.$$

Since the $X's$ are independent, the mgf of their sum $V = \sum_{i=1}^{k} X_i$ will be the product of the mgf of each of the $X's$. Hence

$$M_V(t) = \prod_{i=1}^{k} M_{X_i}(t) = \left(\frac{p}{1 - qe^t} \right)^{l_1 + l_2 + ... l_k}.$$

The form of the mgf of V indicates that it also has a negative binomial distribution with parameters p and $l_1 + l_2 + ... + l_k$

Lemma 4.1 Suppose X has a binomial distribution with parameters n and p and Y is a negative binomial variable with parameters r and p, then it can be shown that

$$F_X(r - 1) = 1 - F_Y(n - r) \qquad (4.29)$$

$F_X(r - 1) = P(\text{number of successes in } n \text{ trials is } \leq r - 1)$

$\qquad = P(r^{th} \text{ success occurs on the } (n+1)^{th} \text{ or } (n+2)^{th} \text{ or } ...\text{trial})$

$\qquad = P(\text{at least } \{n - (r - 1)\} \text{ failures occur before the } r^{th} \text{ success})$

$\qquad = P(Y \geq n - r + 1) = 1 - P(Y \leq n - r)$

The mathematical form of (4.29) can be written as

$$\sum_{x=0}^{r-1} \binom{n}{x} p^x q^{n-x} = 1 - \sum_{y=0}^{n-r} \binom{y + r - 1}{y} q^y p^r.$$

Lemma 4.2 Let Y have a negative binomial distribution with parameters r and p. Then it can be shown that for $U = 2pY$

$$\lim_{p \to 0} M_U(t) = \frac{1}{(1 - 2t)^r}, \; |t| < \frac{1}{2}.$$

Proof. The moment generating function of Y is

$$M_Y(t) = E(e^{tY}) = \left(\frac{p}{1-qe^t}\right)^r, \; q = 1 - p.$$

The moment generating function of U is

$$M_U(t) = E(e^{tU}) = E(e^{(2pt)Y}) = \left(\frac{p}{1-qe^{2pt}}\right)^r. \tag{4.30}$$

Taking limits as $p \to 0$ of both sides of (4.30) and using the L'Hôspital's rule we have

$$\lim_{p \to 0} M_U(t) = \lim_{p \to 0}\left(\frac{p}{1-qe^{2pt}}\right)^r = \left(\frac{1}{1-2t}\right)^r.$$

We will learn in Chapter 5, section 5.9 that this is the moment generating function of a χ^2 variable with $2r$ degrees of freedom. Note that in this case, a random variable which originally had a discrete distribution now, in the limit, follows a continuous distribution. This limiting process is discussed in Chapter 7.

Example 4.9 Let X be the number of failures preceding the rth success in an infinite series of independent trials with a constant probability of success p. Denote the probability $P(X = k) = p_{k,r}$, $k = 0,1,2...$. Let us first obtain the difference equation for $p_{k,r}$. The probability generating function using the difference equation approach will be derived next.

There are two mutually exclusive ways in which k failures can precede the rth success. (1) the first trial results in a success and k failures take place before the $(r-1)^{th}$ success with probability $p_{k,r-1}$ (2) the first trial results in a failure and $(k$-1) failures occur before the rth success with probability $p_{k-1,r}$. Hence the relationship expressed as a difference equation is

$$p_{k,r} = pp_{k,r-1} + qp_{k-1,r}. \tag{4.31}$$

In the above difference equation we define $p_{0.0} = 1$ and $G_r(s)$ as the probability generating function

$$\sum_{k=0}^{\infty} p_{k,r}s^k = G_r(s), \; G_0(s) = 1.$$

Multiplying both sides of (4.31) by s^k and summing over all k we obtain

$$G_r(s) = qsG_r(s) + pG_{r-1}(s)$$

$$G_r(s) = \frac{p}{(1-qs)}G_{r-1}(s) = \left(\frac{p}{1-qs}\right)^r \quad G_0(s) = \left(\frac{p}{1-qs}\right)^r, \quad r = 0,1,2,\ldots$$

which is the probability generating function of a negative binomial variable.

4.9 Negative Multinomial Distribution[^]

In the *negative multinomial distribution* we consider a sequence of independent trials such that each trial may result in either a success with probability p or one of k different types of failures $E_1, E_2, \ldots E_k$ with probabilities $q_1, q_2, \ldots, q_k$ in such a way that $p + \sum_{i=1}^{k} q_i = 1$ Then if Y_i, $i = 1, 2, \ldots k$ are the random variables representing the number of occurrences of E_i (failure of type i each with probability q_i) before the rth success, then

$$P(Y_1 = y_1, Y_2 = y_2, \ldots Y_k = y_k) = \binom{y_1 + y_2 + \ldots + y_k + r - 1}{y_1, y_2, \ldots y_k, r-1} q_1^{y_1} q_2^{y_2} \ldots q_k^{y_k} p^r . \quad (4.32)$$

The result in (4.32) is based on the fact that before the rth success occurs $y_1 + y_2 + \ldots y_k$ failures of different types will have occurred along with $(r-1)$ successes and then the $(y_1 + y_2 + \ldots y_k + r)^{th}$ trial results in a success.

4.9.1 Conditional Distribution of $Y_2 | Y_1$

Assume that $k = 2$ in (4.32). Then,

$$P(Y_1 = y_1, Y_2 = y_2) = \binom{y_1 + y_2 + r - 1}{y_1, y_2, r - 1} q_1^{y_1} q_2^{y_2} (1 - q_1 - q_2)^r \quad (4.33)$$

and

$$P(Y_2 = y_2 | Y_1 = y_1) = \frac{P(Y_1 = y_1, Y_2 = y_2)}{P(Y_1 = y_1)}. \quad (4.34)$$

To obtain $P(Y_1 = y_1)$ proceed by summing over all values of y_2 in (4.33).

$$P(Y_1 = y_1) = \frac{q_1^{y_1}(1 - q_1 - q_2)^r}{y_1!(r-1)!} \sum_{y_2=0}^{\infty} \frac{(y_2 + y_1 + r - 1)!}{y_2!} q_2^{y_2}$$

$$= \frac{q_1^{y_1}(1 - q_1 - q_2)^r (y_1 + r - 1)!}{y_1!(r-1)!} \sum_{y_2=0}^{\infty} \frac{(y_2 + y_1 + r - 1)!}{y_2!(y_1 + r - 1)!} q_2^{y_2}$$

$$= q_1^{y_1}(1 - q_1 - q_2)^r \binom{y_1 + r - 1}{r - 1} \sum_{y_2=0}^{\infty} \binom{y_2 + y_1 + r - 1}{y_1 + r - 1} q_2^{y_2}$$

$$= q_1^{y_1}(1 - q_1 - q_2)^r \binom{y_1 + r - 1}{r - 1} \sum_{y_2=0}^{\infty} \binom{-y_1 - r}{y_2}(-q_2)^{y_2}$$

$$= q_1^{y_1}(1 - q_1 - q_2)^r \binom{y_1 + r - 1}{r - 1}\left(\frac{1}{1 - q_2}\right)^{y_1 + r}.$$

Substituting the above result along with the right hand side of (4.33) into (4.34), after a little simplification we obtain the negative binomial distribution

$$P(Y_2 = y_2 | Y_1 = y_1) = \binom{y_1 + y_2 + r - 1}{y_1 + r - 1} q_2^{y_2}(1 - q_2)^{y_1 + r}. \qquad (4.35)$$

Next we derive the mean and variance of the above conditional distribution

$$E(Y_2 | Y_1 = y_1) = \sum_{y_2=0}^{\infty} y_2 \frac{(y_1 + y_2 + r - 1)!}{y_2!(y_1 + r - 1)!} q_2^{y_2}(1 - q_2)^{y_1 + r}$$

$$= (y_1 + r)q_2(1 - q_2)^{y_1 + r} \sum_{y_2=1}^{\infty} \frac{(y_1 + y_2 + r - 1)!}{(y_2 - 1)!(y_1 + r)!} q_2^{y_2 - 1} = \frac{(y_1 + r)q_2}{(1 - q_2)}$$

$$E(Y_2^2 | Y_1 = y_1) = (y_1 + r)q_2(1 - q_2)^{y_1 + r} \sum_{y_2=1}^{\infty} \frac{(y_2 - 1 + 1)(y_1 + y_2 + r - 1)!}{(y_2 - 1)!(y_1 + r)!} q_2^{y_2 - 1}$$

$$= (y_1 + r)(y_1 + r + 1)q_2^2(1 - q_2)^{y_1 + r} \sum_{l=0}^{\infty} \frac{(y_1 + l + r + 1)!}{y_2!(y_1 + r + 1)!} q_2^l + \frac{(y_1 + r)q_2}{(1 - q_2)}$$

$$= \frac{(y_1 + r)(y_1 + r + 1)q_2^2}{(1 - q_2)^2} + \frac{(y_1 + r)q_2}{(1 - q_2)}$$

$$Var(Y_2 | Y_1 = y_1) = \frac{(y_1 + r)q_2}{(1 - q_2)^2}$$

4.10 Poisson Distribution

Another important discrete distribution is the *Poisson distribution*. A variable X is said to follow a Poisson distribution with parameter λ if

$$P(X = k) = \frac{e^{-\lambda}\lambda^k}{k!}, \quad k = 0,1,2,...$$

This distribution is usually assocociated with applications that require dealing with counts. In particular most queuing theory (waiting line) applications require the distribution of the number of patrons that arrive for service at a counter or the number of items produced by a factory in a certain amount of time given the production rate etc. It is also used when the distribution of elements in an area is required given the density of elements per unit area. The Poisson distribution has been applied to estimate the number of telephone calls received during peak time, at a switching station or the count of bacterial colonies per petri plate in microbiology experiments.

4.10.1 Moments of the Poisson Distribution
The moments for this distribution are easily obtained as follows

$$E(X) = \sum_{k=0}^{\infty} k \frac{e^{-\lambda}\lambda^k}{k!} = \sum_{k=1}^{\infty} \frac{e^{-\lambda}\lambda^k}{(k-1)!} = \lambda$$

and the variance

$$Var(X) = E(X^2) - [E(X)]^2 = \sum_{k=0}^{\infty} k^2 \frac{e^{-\lambda}\lambda^k}{k!} - \lambda^2 = \sum_{k=1}^{\infty} k(k-1+1)\frac{e^{-\lambda}\lambda^k}{k!} - \lambda^2$$

$$= \sum_{k=1}^{\infty} k(k-1)\frac{e^{-\lambda}\lambda^k}{k!} + E(X) - \lambda^2$$

$$= \lambda^2 + \lambda - \lambda^2 = \lambda.$$

The probability generating function for X is

$$P(s) = E(s^X) = e^{-\lambda} \sum_{k=0}^{\infty} \frac{(\lambda s)^k}{k!} = e^{-\lambda(1-s)}.$$

Graphs that depict the Poisson distribution for different values of λ are shown below (Figures 4.5 and 4.6)

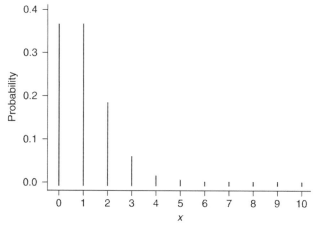

Figure 4.5 Poisson distribution for $\lambda = 1$

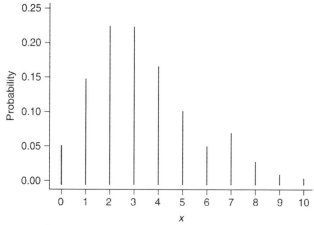

Figure 4.6 Poisson distribution for $\lambda = 5$

Example 4.10

Let the number of houses in different neighborhoods in a city have a Poisson distribution. For what value of the parameter of this distribution will the probability that a randomly chosen neighborhood has at least two houses be greater than 0.99.

In this problem λ, the parameter of the Poisson distribution is not known. We need to determine λ from the following relationship (Let X be the Poisson variable with parameter λ)

Then,
$$P(X \geq 2) = 1 - P(X = 0) - P(X = 1) \geq 0.99$$
or
$$e^{-\lambda} + \lambda e^{-\lambda} \leq 0.01$$
$$\lambda = 6.64$$
The above solution can be easily computed

Definition If the variable X_T has the following probability distribution

$$P(X_T = k) = \frac{P(X = k)}{P(X > 0)}, \quad k = 1, 2, \ldots$$

where X follows a Poisson distribution, then X_T is said to follow a *truncated Poisson distribution* with parameter λ (it is truncated from below).

Example 4.11 Consider a truncated Poisson random variable X_T with parameter λ we need to determine the mean and variance of the truncated variable X_T

Then,

$$E(X_T) = \sum_{k=1}^{\infty} kP(X_T = k) = \frac{E(X)}{P(X > 0)} = \frac{\lambda}{1 - e^{-\lambda}}$$

$$Var(X_T) = \frac{E(X^2)}{P(X > 0)} - \left[\frac{E(X)}{P(X > 0)} \right]^2$$

$$= \frac{1}{P(X > 0)} \left[\lambda + \lambda^2 - \frac{\lambda^2}{1 - e^{-\lambda}} \right]$$

$$= \frac{\lambda}{(1 - e^{-\lambda})^2} \left(1 - e^{-\lambda} - \lambda e^{-\lambda} \right)$$

4.10.2 Sum of Two Poisson Variables

Let X and Y be two independent Poisson random variables with parameters λ_1 and λ_2 respectively. Then their probability generating functions are $P(s) = e^{-\lambda_1(1-s)}$ and $Q(s) = e^{-\lambda_2(1-s)}$. Hence the probability generating function of $Z = X + Y$ is $e^{-(\lambda_1 + \lambda_2)(1-s)}$ which is the probability generating function of a Poisson variable with parameter $(\lambda_1 + \lambda_2)$. In general the sum of k Poisson variables each with a parameter λ_i $i = 1, 2, \ldots, k$ is a Poisson variable with parameter $\sum_{i=1}^{k} \lambda_i$.

Example 4.12 Let $X_1, \ldots, X_n$ be independent and identically distributed (iid) Poisson variables with parameter λ. Our goal is to determine the conditional distribution of X_i given $\sum_{i=1}^{n} X_i$.

In the derivation below we will use the known fact that sum of m independent Poisson variables each with parameter λ has a Poisson distribution with parameter $m\lambda$.

$$P\left(X_i = k \left| \sum_{i=1}^{n} X_i = l \right. \right) = \frac{P(X_i = k)P\left(\sum_{\substack{j=1 \\ j \neq i}}^{n} X_j = l - k \right)}{P\left(\sum_{i=1}^{n} X_i = l \right)}$$

$$= \frac{e^{-\lambda}\lambda^k e^{-(n-1)\lambda}[(n-1)\lambda]^{l-k} l!}{(l-k)! k! e^{-n\lambda}(n\lambda)^l}$$

$$= \binom{l}{k}\left(\frac{1}{n}\right)^k \left(1 - \frac{1}{n}\right)^{l-k} .$$

The above derivation uses the fact that the sum of $(n$-$1)$ independent and identically distributed (iid) Poisson variables each with parameter λ follows a Poisson distribution with parameter $(n-1)\lambda$. The above result shows that the required conditional probability follows a binomial distribution with parameters $\left(l, \frac{1}{n}\right)$

■

4.10.3 Limiting Distributions

(1) Poisson probabilities can be obtained as limits of the binomial distribution by using the pgf of a binomial variable in such a way that $\lim_{n \to \infty} np = \lambda$.

$$\lim_{n \to \infty}(q + ps)^n = \lim_{n \to \infty}[1 - p(1-s)]^n = \lim_{n \to \infty}\left[1 - \frac{np(1-s)}{n}\right]^{\frac{-n}{np(1-s)}[-np(1-s)]}$$

$$= \lim_{n \to \infty}\left[1 - \frac{\lambda(1-s)}{n}\right]^{-\frac{n}{\lambda(1-s)}[-\lambda(1-s)]}$$

$$= e^{-\lambda(1-s)}$$

The expression on the left hand side in the above relationship is the probability generating function (pgf) of the binomial distribution and the right hand side is the pgf of the Poisson distribution. This observation will be expanded in Chapter 7.

(2) Show that when $N, n, M \to \infty$, in such a way that, $\dfrac{Mn}{N} \to \lambda$ the hypergeometric distribution approaches the Poisson distribution.

Using Stirling's formula (Chapter 1, section 1.6) and taking appropriate limits, after some simplification we have

$$\lim_{n,N,M\to\infty} \frac{\binom{M}{x}\binom{N-M}{n-x}}{\binom{N}{n}} = \lim \frac{1}{x!}\left(\frac{nM}{N}\right)^x \frac{(N-M)^{n-x}}{N^{n-x}}$$

$$= \frac{1}{x!}\lambda^x e^{-\lambda}$$

(3) We will now show that if $r \to \infty$, and $q \to 0$ in such a way that $rq \to \lambda$, the negative binomial distribution approximates to the Poisson distribution as follows. The pgf of the negative binomial distribution with parameters p and r is

$$\left(\frac{p}{1-qs}\right)^r$$

Taking the limits of the above pgf as $r \to \infty$ and using the results obtained in Chapter 1, we have

$$\lim_{r\to\infty}\left(\frac{p}{1-qs}\right)^r = \lim_{r\to\infty}\left(\frac{(1-q)^r}{(1-qs)^r}\right)$$

$$= \lim_{r\to\infty}\frac{\left(1-\dfrac{rq}{r}\right)^{\frac{-r}{rq}(-rq)}}{\left(1-\dfrac{rqs}{r}\right)^{\frac{-r}{rqs}(-rqs)}} = \frac{e^{-\lambda}}{e^{-\lambda s}} = e^{-\lambda(1-s)}$$

which is the pgf of the Poisson distribution.

It can also be demonstrated that the multinomial distribution in (4.21) for $n \to \infty$ and $np_i \to \lambda_i$ $(i = 1, 2, ..., k)$ can be approximated by the following distribution

$$e^{-(\lambda_1+\lambda_2+...+\lambda_m)}\frac{\lambda_1^{n_1}\lambda_2^{n_2}...\lambda_m^{n_m}}{n_1!n_2!...n_m!} \qquad (4.36)$$

The distribution (4.36) is the *multinomial Poisson distribution*

Using the fact that $\lim_{x \to \infty} \left(1 + \dfrac{a}{x}\right)^x = e^a$ and rearranging terms on the right hand side

of (4.21) and setting $\sum_{1}^{m} n_i = k$ in the multinomial distribution (4.21)

$P(X_1 = n_1, ..., X_m = n_m)$

$= \lim_{n \to \infty} \dfrac{n^{n-k}}{n_1! n_{2!}..n_m!(n-k)!} (np_1)^{n_1} (np_2)^{n_2} ...(np_m)^{n_m} (1 - p_1 - p_2 - ..p_m)^{n-k}$ \hfill (4.37)

Now

$$\dfrac{n^{n-k}}{(n-k)!} = \dfrac{1}{\left(1 - \dfrac{k}{n}\right)\left(1 - \dfrac{(k+1)}{n}\right)...\left(1 - \dfrac{n-1}{n}\right)}$$

The fractions in the denominator of the above relationship approach 0 as $n \to \infty$. Also as $n \to \infty$ each of the np_i $(i = 1, 2, .., m)$ terms in (4.37) can be replaced by λ_i . Therefore,

$$\lim_{n \to \infty}(1 - p_1 - p_2 - ... - p_m)^{n-k} = \lim_{n \to \infty} \left[\left(1 - \dfrac{n\sum_{i=1}^{m} p_i}{n}\right)^{\frac{n}{n\sum_{i=1}^{m} p_i}} \right]^{\left(-n\sum_{i=1}^{m} p_i\right)\left(1 - \frac{k_1}{n} - \frac{k_2}{n} - ... - \frac{k_m}{n}\right)}$$

$$= e^{-(\lambda_1 + \lambda_2 + ... + \lambda_m)}.$$

In the above derivation $\lim_{n \to \infty} \left(1 - \dfrac{k_1}{n} - \dfrac{k_2}{n} - ...\right) = 1$.

Hence the limiting form of (4.37) is

$$e^{-(\lambda_1 + \lambda_2 + ... \lambda_m)} \dfrac{\lambda_1^{n_1} \lambda_2^{n_2} ... \lambda_m^{n_m}}{n_1! n_2! ... n_m!}.$$

which is the product of the distribution of m independent Poisson variables $X_i (i = 1, 2, ..., n)$ where X_i follows the Poisson distribution with parameter λ_i .

4.11 Discrete Uniform Distribution

The *discrete uniform distribution* is also sometimes referred to as the *discree rectangular distribution*. The probability mass function for this distribution is

$$P(X = i) = \dfrac{1}{n+1}, \quad i = 0, 1, .., n.$$

The moment generating function for this distribution is

$$M_X(t) = E(e^{tX}) = \sum_{x=0}^{n} \frac{e^{tx}}{n+1} = \frac{1}{n+1} \frac{1-e^{(n+1)t}}{1-e^t}$$

The corresponding probability generating function is obtained by replacing e^t by s in the above mgf. Hence

$$P(s) = \left(\frac{1-s^{n+1}}{(n+1)(1-s)} \right)$$

$$\frac{d}{ds}P(s) = \frac{1}{(n+1)} \frac{(n+1)s^{n+1} +1-s^{n+1} -(n+1)s^n}{(1-s)^2}$$

Since the limit of the derivative of $P'(s)$ as $s = 1$ is of the $\frac{0}{0}$ form we use the L'Hôspital's rule to evaluate $P'(s)$ at $s = 1$ and obtain.

$$E(X) = \frac{d}{ds}P(s)\bigg|_{s=1} = \frac{n}{2}$$

and the variance is similarly obtained as

$$Var(X) = \frac{d^2}{ds^2}P(s) + E(X) - [E(X)]^2 = \frac{n(n+2)}{12}$$

This distribution is often used in census surveys where it is assumed that the individuals are uniformly distributed within the sampling frame.

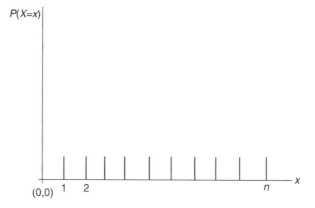

Figure 4.7 Probability mass function for the discrete uniform variable

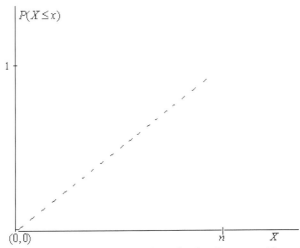

Figure 4.8 Cumulative distribution functions for the discrete uniform distribution

4.12 Lesser Known Distributions*

4.12.1 Logarithmic Distribution
A random variable X follows the *logarithmic distribution* if

$$P(X = k) = \frac{b\lambda^k}{k}, \ k = 1, 2, \dots, \ \text{where } b = \frac{-1}{\log(1-\lambda)}, \ 0 < \lambda < 1$$

The mean and variance for this distribution are

$$E(X) = \frac{b\lambda}{1-\lambda}$$

$$Var(X) = \frac{b\lambda}{(1-\lambda)^2}(1-b\lambda)$$

4.12.2 Negative Hypergeometric Distribution*
This distribution is a variation on the hypergeometric distribution in that sampling is with replacement. Using the same parameters as in Section 4.5, the probability of choosing X defective items for

$$Max[0, M - (N - n)] \leq x \leq Min[n, M)$$

is

$$P(X = x) = \frac{\binom{N - M + n - x - 1}{n - x}\binom{M + x - 1}{x}}{\binom{N + n - 1}{n}}$$

$$= \frac{\binom{-N + M}{n - x}\binom{-M}{x}}{\binom{-N}{n}}$$

4.12.3 Weighted Binomial Distribution*

Rao (1965) introduced the concept of weighted distributions. This was developed for cases when an event that occurs is not necessarily recorded in the sample. Basically, if X is a random variable with probability mass function $P_X(x)$ and suppose that when the event $X = x$ occurs, the probability of recording it is $w(x)$ then the probability mass function of the recorded distribution is defined as

$$P(X = x) = \frac{w(x)P_X(x)}{\sum_x w(x)P_X(x)}$$

For a binomial distribution with parameters n and p, the weighted probability mass function was identified by Kocherlakota and Kocherlakota (1990) as

$$P(X = x) = \frac{w(x)\binom{n}{x}p^x(1 - p)^{n-x}}{\sum_{x=0}^{n} w(x)\binom{n}{x}p^x(1 - p)^{n-x}}$$

4.13 Joint Distributions

Even though the concept of joint distributions was discussed in Chapter 3 section 3.11, we will present examples of some such distributions here. By definition, joint probability function is implied if more than one variable is involved. In this section we will consider the joint distributions of two or more variables. In general a function is called a joint probability mass function of X and Y if

$$\sum_x \sum_y P(X = x, Y - y) - 1$$

If X and Y are independent of one another, then

$P(X = x, Y = y) = P(X = x)P(Y = y)$.

The marginal density function of X may be obtained by summing over all values of Y within the (X, Y) feasible region.

Example 4.13 Four hundred adult males with angina pectoris are classified by age and weight as follows (Remington and Schork 1970)

Table 4.1 Angina Pectoris Data by Age and Weight

	Weight			
	130-149 (1)	150-169 (2)	170-189 (3)	≥ 190 (4)
Age (yrs)				
30-39 (1)	10	20	20	40
40-49 (2)	10	15	50	70
50-59 (3)	5	15	50	40
60-69 (4)	5	10	15	25

Let X represent the discrete variable age of the men, and Y their weights. Since we have categorical data with 4 age categories and 4 weight categories, we will associate numbers 1-4 with the age and weight categories. For example instead of writing $P[X \in (30,39)]$ we will simply write $P(X = 1)$ etc. Using this notation we have constructed a table of (X, Y) values and the probabilities at the intersection of the ith row and j^{th} column (i,j=1,2,3,4) represent the joint probability

$P(X = i, Y = j)$

Table 4.2 Probabilities Associated with the Angina Pectoris Data

$X \downarrow Y \rightarrow$	1	2	3	4
1	0.025	0.050	0.050	0.10
2	0.025	0.0375	0.125	0.175
3	0.0125	0.0375	0.125	0.10
4	0.0125	0.025	0.0375	0.0625

In the table above

(1) $P(X = 2, Y = 4) = 0.175$

(2) $P(X = 1) = P(X = 1, Y = 1) + P(X = 1, Y = 2)$

$\quad\quad + P(X = 1, Y = 3) + P(X = 1, Y = 4) = 0.225$

(3) $P(X = 2 | Y = 3) = \dfrac{P(X = 2, Y = 3)}{P(Y = 3)} = \dfrac{0.125}{0.3375} = 0.3704$

In the above derivations, (1) represents the joint distribution of X and Y, (2) represents the marginal distribution of X (obtained by summing over all possible values of Y) and (3) represents the conditional distribution of X for a specified value of Y. We can also use the above table to compute $P(X=Y)$

$$P(X = Y) = \sum_{i=1}^{4} P(X = Y = i) = 0.025 + 0.0375 + 0.125 + 0.0625 = 0.25$$

Example 4.14 Consider rolling two dice and let X represent the number that shows up on the first die and let Y represent the number that shows up on the second die. Then we know that

$$P(X = i, Y = j) = P(X = i)P(Y = j) = \frac{1}{36}, \ i, j = 1, 2, ..., 6$$

$$E(XY) = \frac{1}{36}\sum_{i=1}^{6}\sum_{j=1}^{6} ij = \frac{1}{36}6(7)\frac{1}{2}6(7)\frac{1}{2} = 12.25$$

Let $Z = X + Y$
We know that Z takes values $2, 3, ... 12$

Let us develop the joint distribution of Z and X. We will develop a table similar to Table 4.2 above to examine the (Z,X) values and the corresponding probabilities.

We can use the information contained in Table 4.3 below to obtain the joint probability $P(X = 3, Z > 6)$. This is obtained by summing over all values of $Z > 6$ in the row for which $X=3$. Likewise, $P(X \le 4, Z > 6)$ can be obtained by adding the probabilities at the intersection of rows 1,2,3,and 4 and columns 7,8,...,12. and we get

$$P(X \le 4, Z > 6) = \frac{10}{36}$$

Conditional probabilities can be similarly found. From Table 4.3 we can calculate the probabilities below.

$$P(Z = 10 | X > 4) = \frac{P(Z = 10, X > 4)}{P(X > 4)}$$

Now $P(X > 4) = \frac{12}{36}$ and

$$P(Z = 10, X > 4) = \frac{2}{36}$$

Table **4.3** Joint Probability Distribution of X and Z

$X \downarrow Z \rightarrow$	2	3	4	5	6	7	8	9	10	11	12
1	$\frac{1}{36}$	$\frac{1}{36}$	$\frac{1}{36}$	$\frac{1}{36}$	$\frac{1}{36}$	$\frac{1}{36}$	0	0	0	0	0
2	0	$\frac{1}{36}$	$\frac{1}{36}$	$\frac{1}{36}$	$\frac{1}{36}$	$\frac{1}{36}$	$\frac{1}{36}$	0	0	0	0
3	0	0	$\frac{1}{36}$	$\frac{1}{36}$	$\frac{1}{36}$	$\frac{1}{36}$	$\frac{1}{36}$	$\frac{1}{36}$	0	0	0
4	0	0	0	$\frac{1}{36}$	$\frac{1}{36}$	$\frac{1}{36}$	$\frac{1}{36}$	$\frac{1}{36}$	$\frac{1}{36}$	0	0
5	0	0	0	0	$\frac{1}{36}$	$\frac{1}{36}$	$\frac{1}{36}$	$\frac{1}{36}$	$\frac{1}{36}$	$\frac{1}{36}$	0
6	0	0	0	0	0	$\frac{1}{36}$	$\frac{1}{36}$	$\frac{1}{36}$	$\frac{1}{36}$	$\frac{1}{36}$	$\frac{1}{36}$

Therefore,

$$P(Z = 10, X > 4) = \frac{1}{6}$$

We will next calculate $E(XZ)$

$$E(XZ) = \sum_{i=1}^{6} \sum_{j=i+1}^{i+6} ij \frac{1}{36} = \frac{1}{36} \sum_{i=1}^{6} (6i + 21)i = 27.42$$

Marginal probability

$$P(Z > 3) = P(X = 1, Z > 3) + P(X = 2, Z > 3) + P(X = 3, Z > 3)$$
$$+ P(X = 4, Z > 3) + P(X = 5, Z > 3) + P(X = 6, Z > 3)$$
$$= \frac{33}{36}$$

∎

Example 4.15 Consider the classic urn problem with two urns each containing red and blue balls. Assume that the proportion of red and blue balls in each urn is p_i and $q_i (= 1 - p_i)$, $i = 1, 2$. A predetermined number n_1 of balls are drawn with replacement from the first urn and let X_1 be the number of red balls that show up. Then X_1 balls are drawn with replacement from the second urn and let us assume that X_2 red balls show up. We need to determine the joint distribution of X_1 and X_2.

$$P(X_1 = x_1, X_2 = x_2) = P(X_1 = x_1)P(X_2 = x_2 | X_1 = x_1)$$

$$= \binom{n_1}{x_1} p_1^{x_1} (1 - p_1)^{n_1 - x_1} \binom{x_1}{x_2} p_2^{x_2} (1 - p_2)^{x_1 - x_2}$$

$$= \frac{n_1 !(1 - p_1)^{n_1}}{(n_1 - x_1)! x_2 !(x_1 - x_2)!} \left(\frac{p_1(1 - p_2)}{1 - p_1} \right)^{x_1} \left(\frac{p_2}{1 - p_2} \right)^{x_2}, \ 0 \le x_2 \le x_1 \le n_1$$

The marginal distribution of x_2 may be obtained by summing over all values of x_1 from x_2 to n_1 in the above joint distribution as follows

$$P(X_2 = x_2) = \sum_{x_1 = x_2}^{n_1} P(X_1 = x_1, X_2 = x_2)$$

$$= \frac{n_1 !(1 - p_1)^{n_1}}{x_2 !} \left(\frac{p_2}{1 - p_2} \right)^{x_2} \sum_{x_1 = x_2}^{n_1} \left(\frac{p_1(1 - p_2)}{1 - p_1} \right)^{x_1} \frac{1}{(x_1 - x_2)!(n_1 - x_1)!}$$

$$= \frac{n_1 !(1 - p_1)^{n_1}}{x_2 !} \left(\frac{p_2}{1 - p_2} \right)^{x_2} \frac{1}{(n_1 - x_2)!} \sum_{x_1 = x_2}^{n_1} \left(\frac{p_1(1 - p_2)}{1 - p_1} \right)^{x_1} \binom{n_1 - x_2}{x_1 - x_2}$$

$$= \binom{n_1}{x_2} (1 - p_1)^{n_1} \left(\frac{p_2}{1 - p_2} \right)^{x_2} \left(1 + \frac{p_1(1 - p_2)}{1 - p_1} \right)^{n_1 - x_2} \left(\frac{p_1(1 - p_2)}{1 - p_1} \right)^{x_2}$$

$$= \binom{n_1}{x_2} (p_1 p_2)^{x_2} (1 - p_1 p_2)^{n_1 - x_2} \ 0 \le x_2 \le n_1$$

In other words the marginal distribution of X_2 is binomial with parameters n_1 and $p_1 p_2$ ∎

Example 4.16 Let X and Y be two independent random variables following the Poisson distribution with parameters λ and μ respectively. We need to find the joint distribution of $X + Y$ and Y

Let $U = X + Y$ and $V = Y$

Since X and Y are independent variables, their joint distribution is

$$f_{X,Y}(x, y) = e^{-\lambda} \frac{\lambda^x}{x!} e^{-\mu} \frac{\mu^y}{y!}, \ x, y = 0, 1, 2, \dots.$$

Hence the joint probability mass function of U and V is

$$P(U = u \text{ and } V = v) = P(X = u - v \text{ and } Y = v) = \frac{e^{-(\lambda+\mu)} \lambda^{u-v} \mu^{v}}{(u-v)!v!},$$

$$u = 0, 1, 2, ..., v = 0, 1, 2, ... \ u \geq v.$$

To obtain the conditional distribution $P(V = v | U = u)$, we use the following formula for conditional distributions

$$P(V = v | U = u) = \frac{P(U = u \text{ and } V = v)}{P(U = u)}$$

Now we know from section 4.10 that the sum of two independent Poisson variables is distributed as a Poisson variable with a parameter which is the sum of the two parameters. Therefore

$$P(U = u) = e^{(\lambda+\mu)} \frac{(\lambda+\mu)^{u}}{u!}, \ u = 0.1.2....$$

Hence

$$P(V = v | U = u) = \frac{\dfrac{e^{-(\lambda+\mu)} \lambda^{u-v} \mu^{v}}{(u-v)!v!}}{(\lambda+\mu)^{u} \dfrac{e^{-(\lambda+\mu)}}{u!}} = \binom{u}{v} \left(\frac{\mu}{\lambda+\mu}\right)^{v} \left(\frac{\lambda}{\lambda+\mu}\right)^{u-v}, v = 0, 1, 2, ... u \quad (4.38)$$

The distribution on the right hand side of (4.38) is the binomial distribution with

parameters u and $\dfrac{\mu}{\lambda+\mu}$. ∎

4.13.1 Moment and Probability Generating Functions of Joint Distributions

If $X_1, X_2, ..., X_k$ are discrete and jointly distributed random variables, then their moment generating function is defined as

$$M_{X_1,...,X_k}(t_1,...,t_k) = E(e^{t_1 X_1} e^{t_2 X_2} ... e^{t_k X_k})$$

If the random variables are independent,

$$M_{X_1,...,X_k}(t_1,...,t_k) = E(e^{t_1 X_1}) E(e^{t_2 X_2}) ... E(e^{t_k X_k})$$

Likewise the joint probability generating functions are defined as

$$P_{X_1, X_2, ... X_k}(s_1, s_2, ... s_k) = E(s_1^{X_1} s_2^{X_2} ... s_k^{X_k}) \quad (4.39)$$

To obtain the probability generating function of the marginal distribution of X_i

$$P_{X_i}(s_i) = P_{X_1,...,X_k}(1, 1, .. s_i, 1, 1, ..., 1)$$

and

$$E(X_i) = \frac{\partial}{\partial s_i} P_{X_1,...,X_k}(1, 1, .. s_i, 1, 1, ..., 1)\Big|_{s_i=1}$$

The variance of X_i is obtained from the relationship

$$Var(X_i) = E[X_i(X_i - 1)] + E(X_i) - [E(X_i)]^2$$

where $E[X_i(X_i - 1)] = \dfrac{\partial^2}{\partial s_i^2} P_{X_1,...,X_k}(1,1,..s_i,1.1....,1)\Big|_{s_i=1}$

4.14 Convolutions

We introduced the concept of convolutions in Chapter 3, section 3.7. In this section we will present an example of the use of the convolution idea in deriving the distribution of a sum of three independently distributed discrete random variables.

Example 4.17

Let us consider three independently distributed binomial variables X_1, X_2 and X_3 with parameters $(n_1, p), (n_2, p)$ and (n_3, p) respectively.

Let $Z = Y + X_3$ where $Y = X_1 + X_2$. Then using the convolution argument and the arguments presented in section 4.3 in the derivation of the probability mass function of two independent binomial variables we proceed as follows.

$$P(Z = k) = \sum_0^k P(Y = i)P(X_3 = k - i)$$

$$= p^k(1-p)^{n_1+n_2+n_3-k} \sum_{i=\max[0,k-n_3]}^{\min[n_1+n_2,k]} \binom{n_1+n_2}{i}\binom{n_3}{k-i}, \; 0 \le k \le n_1 + n_2 + n_3.$$

We will now consider four different scenarios

    (1) $0 \le k \le n_1 < n_2 < n_3$

$$P(Z = k) = p^k(1-p)^{n_1+n_2+n_3-k} \sum_{i=0}^k \binom{n_1}{i} \sum_{j=0}^{k-i} \binom{n_2}{j}\binom{n_3}{k-i-j}$$

$$= p^k(1-p)^{n_1+n_2+n_3-k} \sum_{i=0}^k \binom{n_1}{i}\binom{n_2+n_3}{k-i}$$

$$= \binom{n_1+n_2+n_3}{k} p^k(1-p)^{n_1+n_2+n_3-k}$$

    (2) $n_1 < k \le n_2 < n_3$

$$P(Z = k) = p^k(1-p)^{n_1+n_2+n_3-k} \sum_{i=0}^{n_1} \sum_{j=0}^{k-i} \binom{n_1}{i}\binom{n_2}{j}\binom{n_3}{k-i-j}$$

$$= \binom{n_1+n_2+n_3}{k} p^k(1-p)^{n_1+n_2+n_3-k}$$

(3) $n_1 < n_2 < k \le n_3$

$$P(Z = k) = p^k (1-p)^{n_1 + n_2 + n_3 - k} \sum_{i=0}^{n_1} \sum_{j=0}^{n_2 - i} \binom{n_1}{i}\binom{n_2}{j}\binom{n_3}{k-i-j}$$

$$= \binom{n_1 + n_2 + n_3}{k} p^k (1-p)^{n_1 + n_2 + n_3 - k}$$

(4) $n_1 < n_2 < n_3 < k \le n_1 + n_2 + n_3$

(a) $k \le n_1 + n_2$

$$P(Z = k) = p^k (1-p)^{n_1 + n_2 + n_3 - k} \sum_{i=0}^{n_1} \sum_{j=Max[0,k-n_3-i]}^{Min[k-i,n_2]} \binom{n_1}{i}\binom{n_2}{j}\binom{n_3}{k-i-j}$$

$$= p^k (1-p)^{n_1 + n_2 + n_3 - k} \sum_{i=0}^{n_1} \binom{n_1}{i} \sum_{j=Max[0,k-n_3-i]}^{Min[k-i,n_2]} \binom{n_2}{j}\binom{n_3}{k-i-j}$$

$$= \binom{n_1 + n_2 + n_3}{k} p^k (1-p)^{n_1 + n_2 + n_3 - k}$$

(b) $n_1 + n_2 < k \le n_2 + n_3$

$$P(Z = k) = p^k (1-p)^{n_1 + n_2 + n_3 - k} \sum_{i=0}^{n_1} \sum_{j=Max[0,k-n_3-i]}^{n_2} \binom{n_1}{i}\binom{n_2}{j}\binom{n_3}{k-i-j}$$

$$= \binom{n_1 + n_2 + n_3}{k} p^k (1-p)^{n_1 + n_2 + n_3 - k}$$

(c) $n_2 + n_3 < k \le n_1 + n_2 + n_3$

$$P(Z = k) = p^k (1-p)^{n_1 + n_2 + n_3 - k} \sum_{i=k-n_2-n_3}^{n_1} \sum_{j=k-n_3-i\}}^{n_2} \binom{n_1}{i}\binom{n_2}{j}\binom{n_3}{k-i-j}$$

$$= \binom{n_1 + n_2 + n_3}{k} p^k (1-p)^{n_1 + n_2 + n_3 - k}$$

4.15 Compound Distributions*

In Chapter 3, section 3.16, we had defined the concept of compound distributions. The distributions considered in this section result from the combination of two distributions. Douglas (1971,1980) refers to this type of distributions as "stopped—sum distributions" because the summation of observations from one distribution is stopped by the value of an observation from a second distribution. Let us illustrate this concept by means of an example

Example 4.18 Let the variable Y_i have the logarithmic distribution defined below

$$P(Y_i = k) = -\frac{p^k}{k \log(1-p)}, \quad k = 1, 2, \dots \text{ and } 0 < p < 1$$

Define the probability generating function of Y_i as

$$g_{Y_i}(s) = \sum_{k=1}^{\infty} s^k P(Y_i = k) = -\sum_{k=1}^{\infty} \frac{(ps)^k}{k \log(1-p)} = \frac{1}{\log(1-p)} \log(1-ps)$$

Consider the distribution of $Z = \sum_{i=1}^{N} Y_i$ where the Y_i are independent and identically logarithmically distributed variables. Assume that N itself is a random variable distributed according to the Poisson probability function with parameter λ. Then

$$P(N = n) = e^{-\lambda} \frac{\lambda^n}{n!} \quad n = 0, 1, \dots$$

with probability generating function

$$h_N(s) = e^{-\lambda(1-s)}$$

Using the result obtained in Chapter 3, section 3.16, the probability generating function of $Z = \sum_{i=1}^{N} Y_i$ is obtained as

$$P(s) = E[\{g(s)\}^N] = h[g(s)]$$

$$= e^{-\lambda} \exp\left[1 - \frac{\log(1-ps)}{\log(1-p)}\right]$$

$$= e^{-\lambda}(1-ps)^{\frac{\lambda}{\log(1-p)}}.$$

Set

$$l = -\frac{\lambda}{\log(1-p)} \quad \text{implying } e^{-\lambda} = e^{l \log(1-p)}.$$

Then, $P(s)$ may be simplified to

$$P(s) = \left[\frac{1-p}{1-ps}\right]^l$$

which is the pgf of a negative binomial distribution. ∎

Example 4.19 Consider a random variable Y_i that follows a geometric distribution with parameter p. Then

$$P(Y_i = y) = q^y p, \quad q = 1 - p, \; y = 0, 1, \dots$$

The probability generating function of Y_i is $\dfrac{p}{1-ps}$.

Let $Z = X_1 + Y_2 + \ldots + Y_N$

Assume that N follows a Poisson distribution with parameter λ

Then,

$$E(Z) = e^{-\lambda\left(1 - \frac{ps}{1-qs}\right)} = e^{-\lambda\left(\frac{1-s}{1-qs}\right)}.$$

This is the pgf of the Polya-Aeppli distribution (Feller 1957) which is often used when it can be assumed that objects that are to be counted occur in clusters having a Poisson distribution while the number of objects per cluster has the geometric distribution. The coefficient of s^l is then $P(Z = l)$.

The following example is due to Feller (1957)   ∎

Example 4.20 Suppose that the number of hits by lightning during time t follows a Poisson distribution with parameter λt. Assume that damage caused by each hit of lightning follows a distribution $\{h_n\}$ and has the probability generating function $h(s)$. Then the total damage caused by lightning hits has the following pgf.

$$p(s,t) = e^{-\lambda t + \lambda t h(s)} \tag{4.40}$$

The probability distribution associated with the pgf in (4.40) is *called the compound Poisson distribution.*   ∎

In ecology it is assumed that the number of animal litters in a plot has a Poisson distribution with mean proportional to the area of the plot. If $\{h_n\}$ is the distribution of the number of animals in a litter, then (4.40) is the generating function for the total number of animals in the plot.

Among integral valued discrete variables only the compound Poisson distribution has the property

$$p(s, t_1 + t_2) = p(s, t_1)p(s, t_2) \tag{4.41}$$

For the negative binomial distribution discussed in section 4.8

$p(s,t) = \left(\dfrac{p}{1-qs}\right)^t$ does have the property (4.41) for

$$\lambda = \log\frac{1}{p}, h(s) = \frac{1}{\lambda}\log\frac{1}{1-qs}, h_n = \frac{\lambda q^n}{n}$$

Hence the negative binomial distribution is a compound Poisson distribution. The distribution $\{h_n\}$ is the *logarithmic distribution.*

4.19 Branching Processes*

We will now consider an application of compound distributions (first introduced in Chapter 3, section 3.16) in branching processes first developed by Watson and Galton (1874) and the process may be visualized as follows. An individual (zero generation) may produce X offspring according to the following distribution

$$P(X = i) = p_i, i = 0,1,2,... \qquad (4.42)$$

Let us assume that the distribution of the number of offspring of all individuals of subsequent generations follow the probability mass function given in (4.42) and that the number of offspring produced by any individual in a particular generation is independent of the number of offspring produced by other members in the same generation. Denote by $P(s)$ the probability generating function associated with (4.42). If the individual in the zero generation produced S_1 offspring, then the population in the first generation would be S_1 and it has the same probability distribution as (4.42). Let X_i be the number of offspring produced by the ith individual in the first generation. The random variable representing the total number of individuals in the second generation is

$$S_2 = X_1 + X_2 + ... + X_{S_1} \qquad (4.43)$$

where S_1 and X_i are all random variables with probability generating function $P(s)$. The pgf for S_2 in (4.43) would be $P_2(s) = P(P(s))$. This is based on the concept of compound distributions introduced in Chapter 3. Proceeding in this way we find that the pgf associated with the $(n+1)^{st}$ generation would be $P_{n+1}(s) = P(P_n(s))$. The coefficient of s^r in this pgf would represent the probability of r individuals in the nth generation. For more on branching processes readers may refer to book on *stochastic processes* for example Chiang (1979).

4.20 Hierarchical Distributions

This approach takes into account any prior information/knowledge of the parameter (or parameters) in the distribution of a variable. The marginal probability mass function obtained after accounting for the prior distribution of the parameter is then used to obtain the moments of the variable of interest. Such methods have become commonplace in clinical decision making when uncertainty exists between clinical information on the patient and the presence of disease. No matter how accurate data on a patient's history, physical examination or laboratory tests results may be, there is always uncertainty associated with each piece of information on the patient. Hence the need to associate a probability distribution (prior distribution) to the uncertain parameters of a patient. Sometimes the posterior distribution of the population parameter is obtained using the Bayes Theorem. Decisions based on the prior and posterior analysis have come to be known as *Bayesian Analysis* (Berger 1985) after the famous probabilist Thomas Bayes (1730). We will apply Bayesian techniques in Chapters 8-10. For now, marginal probability mass functions after adjusting for the prior of the parameter will be referred to as hierarchical distributions. We will revisit this type of distributions when considering continuous variables in Chapter 5.

Example 4.21 Let X follow a binomial distribution with parameters n and p. . Then

$$P(X = r \mid n) = \binom{n}{r} p^r q^{n-r} \quad q = 1 - p, r = 0, 1, ..., n$$

Assume that the parameter n has a Poisson distribution with parameter λ such that

$$P(n = k) = e^{-\lambda} \frac{\lambda^k}{k!}, \ k = 0, 1, ...$$

The distribution of X is obtained by using the relationship

$$P(X = r) = \sum_{k=r}^{\infty} P(X = r \mid n) P(n = k)$$

The marginal distribution of X is therefore

$$P(X = r) = \sum_{k=r}^{\infty} \binom{k}{r} p^r q^{k-r} e^{-\lambda} \frac{\lambda^k}{k!}$$

$$= \frac{(\lambda p)^r e^{-\lambda}}{r!} \sum_{k=r}^{\infty} \frac{(q\lambda)^{k-r}}{(k-r)!}$$

$$= \frac{(\lambda p)^r}{r!} e^{-\lambda p}$$

Hence the marginal distribution of X is a Poisson distribution with parameter λp and

$$E(X) = Var(X) = \lambda p$$

Example 4.22 Let us now consider a binomial variable X with parameters (n, p_1), where n itself has a negative binomial distribution with parameter (p_2, r). Then

$$P(X = x) = \sum_k P(X = x, n = k)$$

$$= \sum_{k=x}^{\infty} \binom{k}{x} p_1^x q_1^{k-x} \binom{k+r-1}{r-1} p_2^r q_2^k$$

$$= \frac{p_1^x p_2^r (x+r-1)!}{x!(r-1)!} \sum_{l=0}^{\infty} \frac{(l+x+r-1)!}{l!(x+r-1)!} q_1^l q_2^{x+l}$$

$$= (p_1 q_2)^x p_2^r \binom{x+r-1}{x} \sum_{l=0}^{\infty} \binom{l+x+r-1}{l} (q_1 q_2)^l (1-q_1 q_2)^{x+r} \frac{1}{(1-q_1 q_2)^{x+r}}$$

$$= \binom{x+r-1}{x} \left(\frac{p_1 q_2}{1-q_1 q_2}\right)^x \left(\frac{p_2}{1-q_1 q_2}\right)^r$$

Hence the unconditional distribution of X has a negative binomial form with parameters $\left(r, \frac{p_2}{1-q_1 q_2}\right)$, $p_1 + q_1 = 1$, $p_2 + q_2 = 1$.

■

Problems

1. Verify the mean and variance of the logarithmic distribution defined in section 4.12.

2. If X has the geometric distribution such that
$$P(X = k) = \frac{1}{3}\left(\frac{2}{3}\right)^k, \quad k=0,1,2,\dots.$$
Determine the probability mass function of $Z = \frac{X}{X+1}$.

3. Find the moment generating function of the sum of two negative binomial variables X_1 and X_2 with parameters (p, r_1) and (p, r_2) respectively.

4. A fair die is cast until a 5 appears. What is the probability that it must be ast less than 5 times.

5. Let X be a binomial random variable with parameters n and p. Let $Y = n - X$. Find the covariance between X and Y.

6. For the multiple Poisson distribution compute the conditional distribution of X_1 given X_2.

7. In the negative multinomial distribution of section 4.9, let $k = 3$. Obtain

$P(X_1 = k_1 | X_2 = k_2, X_3 = k_3)$.

8.   Derive the moment generating function of the negative multinomial distribution.

9.   Let $\mu_r = E[X - E(X)]^r$ and let D denote the operation of raising the order of a moment by unity, i.e. $D\mu_r = \mu_{r+1}$. Show that
      (a) for the binomial distribution $\{(1+D)^r - D^r\}(npq\mu_0 - p\mu_1) = \mu_{r+1}$
      (b) for the Poisson distribution $\{(1+D)^r - D^r\}\lambda\mu_0 = \mu_{r+1}$.

10.   An unbiased coin is tossed n times and it is known that exactly m showed heads. Show that the probability that the number of heads exceeded the number of tails throughout the tossing is $\dfrac{2m-n}{n}$.

11.   In each of a packet of cigarettes there is one of a set of cards numbered from 1 to n. If a number N of packets is bought at random, the population of packets is large and the numbers are equally frequent, show that the probability of getting a complete set of cards is

$$1 - \binom{n}{1}\left(\frac{n-1}{n}\right)^N + \binom{n}{2}\left(\frac{n-2}{n}\right)^N + \dots + (-1)^{n-1}\binom{n}{n-1}\left(\frac{1}{n}\right)^N.$$

12.   Let X be the number of cigarette cartons that are sold per day in a particular Stop and Go store. Assume that X follows the Poisson distribution with parameter $\lambda = 10$. The profit associated with the sale of each pack is $1. If at the beginning of the week 10 cigarette cartons are in stock, the profit Y from sale of cigarette cartons during the week is $Y = \min(X, 10)$. Find the probability distribution of Y.

13.   Suppose that a woman, in her lifetime is equally likely to have 1 or 2 female offspring, and suppose that these second generation women are in turn each equally likely to have 1 or 2 females and so on in subsequent generations. What is the probability generating function of the number of 4th generation women?

14.   If X is a random variable with mean μ and variance σ^2 and has cumulative distribution function F(x). Show that (Cramer 1946)

$$F(x) \le \frac{1}{1+\left(\dfrac{x-\mu}{\sigma}\right)^2}, \ \text{if} \ x<\mu$$

$$F(x) \ge \frac{1}{1+\left(\dfrac{x-\mu}{\sigma}\right)^{-2}}, \ \text{if} \ x > \mu$$

15. A jar contains $m+n$ chips numbered $1,2,\dots,m+n$. A set of n chips are drawn at random from the jar. Show that the probability is

$$\frac{\dbinom{m+n-x-1}{m-1}}{\dbinom{m+n}{m}}$$

that x of the chips drawn have numbers exceeding all numbers on the chips remaining in the jar (Wilks 1962).

16. In the multinomial distribution of section 4.4 show that the distribution of the conditional random variable $X_m | X_1 = n_1,\dots, X_{m-1} = n_{m-1}$ is the binomial distribution with parameters $\left(n - \sum_{i=1}^{k-1} n_i ; \dfrac{p_k}{p_k + p_0} \right)$

17. In the negative binomial distribution suppose X is the number of trials required in order to obtain k successive successes. Show that the moment generating function of X is

$$(pe^t)^k (1- pe^t)(1-e^t + p^k qe^{(k+1)t})^{-1}$$

18. Show that for problem 4.17 $E(X) = \dfrac{1-p^k}{p^k q}$

19. A random variable Y is defined by $Z = \log Y$, where $E(Z) = 0$. Is $E(Y)$ greater than, less than or equal to 1?

20. Let the joint distribution of X and Y be $f_{X,Y}(x,y)$, and let $u(X)$ and $v(Y)$ be functions of X and Y, respectively. Show that

$$E[u(X)v(Y)|x] = u(x)E[v(Y)|x]$$

21. Define

$$f_X(x) = \frac{6}{\pi^2 x^2}, \quad x = 1, 2, 3, \ldots$$
$$= 0 \quad \text{otherwise}$$

Assume that $\sum_{i=1}^{\infty} \frac{1}{i^2}$ converges to $\frac{\pi^2}{6}$

Obtain the moment generating function of X.

22. Show that

$$\sum_{k=0}^{x} \binom{n}{x} p^k (1-p)^{n-k} = (n-x) \binom{n}{x} \int_{t=0}^{1-p} t^{n-x-1} (1-t)^x \, dt.$$

23. Find the mean and variance of the logarithmic (series) distribution.
24. Derive the following relations
    (1) For $n_1 < k \leq n_2$

$$P(Z = k) = p^k (1-p)^{n_1+n_2-k} \sum_{i=0}^{n_1} \binom{n_1}{i} \binom{n_2}{k-i} = \binom{n_1+n_2}{k} p^k (1-p)^{n_1+n_2-k}.$$

    (2) For $n_1 < n_2 < k \leq n_1 + n_2$

$$P(Z = k) = p^k (1-p)^{n_1+n_2-k} \sum_{i=k-n_2}^{n_1} \binom{n_1}{i} \binom{n_2}{k-i} = \binom{n_1+n_2}{k} p^k (1-p)^{n_1+n_2-k}.$$

References

Bartoszynski, R. and Niewiadomska-Bugaj, M. (1996), *Probability and Statistical Inference*, John Wiley and Sons, New York.

Bayes, T. (1763), An essay towards solving a problem in the doctrine of chances, *Philosophical Transactions*, 53.

Bernoulli, J. (1713), *Ars conjectandi* .

Berger, J.O. (1985), *Statistical Decision Theory and Bayesian Analysis*, Springer-Verlag, New.York.

Chiang, C.L. (1979), *Stochastic Proceses and Their Applications,* Robert E. Krieger Publishing Company, Huntington, New York.

Chung, K.L. (1974), *A Course in Probability Theory, Academic Press,* New York.

Casella, G. and Berger, R.L. (2000), *Statistical Inference*, 2nd ed., Duxbury Press, Belmont, Mass.

Colton, T. (1974), *Statistics in Medicine*, Little, Brown and Company, Boston.

Cox, D.R. and Hinkley, D.V. (1974), *Theoretical Statistics*, Chapman and Hall, London..

Douglas, J.B. (1971), *Stirling Numbers in Discrete Distributions: Statistical Ecology* I, Pennsylvania State University Press, University Park.

Douglas, J.B. (1980), *Analysis with Standard Contagious Distributions,* Interna tional Co-operative Publishing House, Burtonsville, MD.

Dudewicz, E.J. and Mishra, S.N. (1988), *Modern Mathematical Statistics*, John Wiley and Sons, New York.

Evans, M., Nicholas, H. and Peacock, B. (2000), *Statistical Distributions*, 3rd ed., Wiley Series in Probability and Statistics, New York.

Erlang, A.K. (1909), *Probability and Telephone Calls,* Nyt Tidsskr. Mat. Series b, Vol 20.

Feller, W. (1965), *An Introduction to Probability Theory and Its Applications*, Vol I, 2nd ed., John Wiley and Sons, New York.

Feller, W. (1971), *An Introduction to Probability Theory and Its Applications*, Vol II, John Wiley and Sons, New York.

Freund, J.F. (1971), *Mathematical Statistics*, 2nd ed., Prentice Hall, Englewoods, New Jersey.

Frisch, R. (1926), *Sur les semi-invariants*, Videnskaps-Akademiets Skrifter,Vol II, No 3, Oslo.

Hogg, R.V. and Craig, A.T. (1978), *Introduction to Mathematical Statistics*, 4th ed., Macmillan Publishing Company, New York.

Johnson, N.L. and Kotz, S.(1970), *Continuous Univariate Distributions-1*, The Houghton Mifflin Series in Statistics, New York.

Johnson, N.L.and Kotz, S. (1970), *Continuous Univariate Distributions-2*, The Houghton Mifflin Series in Statistics, New York.

Kendall, M.G. (1952), *The Advanced Theory of Statistics*, Vol I. 5th ed., Charles Griffin and Company, London.

Kocherlakota, S. and Kocherlakota, K. (1990), Tests of hypotheses for the weighted binomial distribution*, Biometrika*, 46.

Mood, A.M.,Graybill, F.A. and Boes, D.C. (1974), *Introduction to the Theory of Statistics*, 3rd ed., McGraw-Hill, New York.

Rao C.R. (1952), *Advanced Statistical Methods in Biometric Research*, John Wiley and Sons, New York.

Rao, C.R. (1977), *Linear Statistical Inference and Its Applications*, 2nd ed., John Wiley and Sons, New York.

Remington, R.D. and Schork, M.A. (1970), *Statistics with Applications to Bio logical and Health Sciences*, Prentice Hall Inc, Englewood Cliffs, New Jersey.

Rice, J.A. (1995), *Mathematical Statistics and Data Analysis*, 2nd ed., Duxbury Press, Belmont, Mass.

Rohatgi, V.K. (1975), *An Introduction to Probability Theory and Mathematical Statistics,* John Wiley and Sons, New York.

Uspensky, J.L. (1937), *Introduction to Mathematical Probability*, McGraw-Hill, New York.

Watson, H.W. and Galton, F. (1874), On the probability of extinction of families, *Journal of the Anthropological Institute*, 4.

Wilks S.S. (1962), *Mathematical Statistics*, John Wiley and Sons, New York.

5

Continuous Random Variables

In this chapter we will present continuous distributions often used in statistical inference. Their identity will become clearer in subsequent chapters.

5.1 Location and Scale Parameters

If $f_X(x)$ is the probability density function (pdf) of a random variable X, then $g_X(x) = f_X(x-a)$ is also a pdf except that each value that X may take is now reduced by a quantity a. In essence the density function for X has now shifted by a units to the right.. The parameter a is called the *location parameter*. Similarly,

$$g_X(x) = \frac{1}{b} f_X\left(\frac{x}{b}\right)$$

is also the density function of X except that all values of X are now divided by a quantity b (>0). If $b \geq 1$, the probability distribution of X is scaled "down" by an amount b. Similarly if $b < 1$ X is scaled "up". Therefore b is called a *scale parameter*. For a quantity to be a scale parameter it does not necessarily have to divide all values of X by b. Instead we could just as well multiply all values of X by b. Sometimes a distribution may contain a location and a scale parameter. You will see in section 5.4 that the normal distribution with mean μ and variance σ^2 has a location parameter μ and a scale parameter σ.

5.2 Distribution of Functions of Random Variables

There are three methods that are generally used for obtaining the distribution of a function of random variables. These methods are
1. The change of variable technique
2. The cumulative density function method
3. The moment generating function method
Several other techniques have been used by some authors but in this chapter we will use these three only.

The preference on which method to use depends on the distribution of the original variables and if they lend themselves easily to the method under consideration. We will illustrate this by means of a simple example.
Consider a continuous random variable X with the following pdf

$$f_X(x) = 3x^2, 0 < x < 1$$

Define a new variable $Y = 2X+1$

(a) Using the change of variables technique

$$X = \frac{Y-1}{2}$$

$$dx = \frac{1}{2}dy$$

To obtain the pdf of Y we substitute $\dfrac{Y-1}{2}$ for X in the pdf of X and multiply

by $\dfrac{1}{2}$ (this is the Jacobian corresponding to the transformation) and obtain

$$f_Y(y) = f_X\left(\frac{y-1}{2}\right)\frac{1}{2} = 3\left(\frac{y-1}{2}\right)^2\frac{1}{2}, 1 < y < 3$$

(b) Using the cumulative density function method

$$F_Y(y) = P(Y \le y) = P(2X + 1 \le y) = P\left(X \le \frac{y-1}{2}\right)$$

Differentiating both the sides and using the methodolgy presented
in Chapter 3, section 3.10 we have

$$f_Y(y) = f_X\left(\frac{y-1}{2}\right)\frac{1}{2} = 3\left(\frac{y-1}{2}\right)^2\frac{1}{2}, 1 < y < 3$$

(c) Moment generating function technique
The mgf of X is

$$M_X(t) = E\left(e^{tX}\right)$$

and the mgf of Y is

$$M_Y(t) = E\left(e^{tY}\right) = E\left(e^{t(2X+1)}\right) = e^t E\left(e^{2tX}\right)$$

The above derivation of the mgf of Y indicates that the mgf of Y may be ob
tained by replacing t by $2t$ in the mgf of X and multiplying by e^t.

■

In Chapter 3 section 3.10 the cumulative density function approach was
presented and in Chapter 3, section 3.14 moment generating functions were intro-

duced. The notion of change of variables was introduced in Chapter 1 section 1.17 in connection with the concept of Jacobian. Examples using these three techniques appear throught this chapter but in no particular order. We will extend the change of variables technique to the case where several variables are jointly distributed.

5.2.1 Change of Variables Technique

We will now present another useful technique that we alluded to in Chapter 1 Section 1.16 . This technique is used when we want to transform a set of continuous random variables $X_1,\ldots,X_n$ with joint density function $f_{X_1,\ldots,X_n}(x_1,\ldots,x_n)$ into a new set of variables $Y_1,\ldots,Y_n$ such that

$$Y_1 = f_1(X_1,\ldots,X_n)$$
$$Y_2 = f_2(X_1,\ldots,X_n)$$
$$\cdots \cdots \cdots$$
$$Y_n = f_n(X_1,\ldots,X_n)$$

are single valued and have a continuous first order partial derivative at all $x_1,\ldots,x_n$. Assuming that the Y_i have a unique inverse such that the $X's$ may be expressed as

$$X_1 = g_1(Y_1,\ldots,Y_n)$$
$$X_2 = g_2(Y_1,\ldots,Y_n)$$
$$\cdots \cdots \cdots \cdots$$
$$X_n = g_n(Y_1,\ldots,Y_n)$$

Then the Jacobian of the transformation defined by the determinant

$$J = \begin{vmatrix} \dfrac{\partial X_1}{\partial Y_1} & \dfrac{\partial X_1}{\partial Y_2} & \cdots & \dfrac{\partial X_1}{\partial Y_n} \\ \dfrac{\partial X_2}{\partial Y_1} & \dfrac{\partial X_2}{\partial Y_2} & \cdots & \dfrac{\partial X_2}{\partial Y_n} \\ \cdots\cdots\cdots\cdots\cdots \\ \dfrac{\partial X_n}{\partial Y_1} & \dfrac{\partial X_n}{\partial Y_2} & \cdots & \dfrac{\partial X_n}{\partial Y_n} \end{vmatrix}$$

is assumed to have a non-zero value. The joint probability density function of $Y_1,\ldots,Y_n$ is given by

$$f_{Y_1,\ldots,Y_n}(y_1,\ldots,y_n) = f_{X_1,\ldots,X_n}(g_1,\ldots,g_n)|J|$$

Where $|J|$ is the absolute value of the above determinant and $Y_1,\ldots,Y_n$ are absolutely continuous.

Examples using this type of transformation are given throughout this chapter.

5.2.2 Distribution of a Ratio

Suppose that we are interested in determining the distribution of the ratio of two independent random variables X and Y. Let

$$V = \frac{X}{Y}$$

We will derive the distribution of V for two cases.

Case I, $Y > 0$.

The feasible region corresponding to the constraint $P(V \le v) = P(X \le vY)$ is marked as A and A' in Figure 5.1a below.

$$F_V(v) = P(V \le v) = P(X \le vy)$$

$$= \int_0^\infty \int_{x=0}^{vy} f_X(x) f_Y(y) \, dx \, dy + \int_0^\infty \int_{-\infty}^{0} f_X(x) f_Y(y) \, dx \, dy \qquad (5.1)$$

$$= \int_0^\infty \int_{-\infty}^{vy} f_X(x) f_Y(y) \, dx \, dy = \int_0^\infty F_X(vy) f_Y(y) \, dy$$

Assuming that the left hand side of (5.1) is differentiable, we have

$$f_V(v) = \int_0^\infty y f_X(vy) f_Y(y) \, dy, \quad -\infty < v < \infty$$

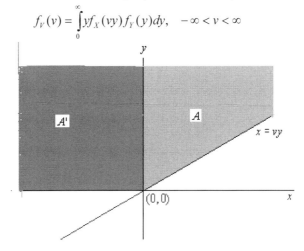

Figure 5.1a Feasible region for the case $Y > 0$

Case II, $Y < 0$

In this case the feasible region is the region marked by B and B' in Figure 5.1b. Hence

$$P(V \leq v) = P(X \leq vY) = \int_{-\infty}^{0} \int_{vy}^{\infty} f_X(x) f_Y(y) dx dy$$

$$F_V(v) = -\int_{-\infty}^{0} \overline{F}_X(vy) f_Y(y) dy + F_Y(0)$$

Differentiating both sides of the above relationship we have,

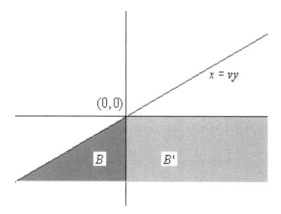

Figure 5.1b Feasible region for the case when Y *is* negative

$$f_V(v) = -\int_{-\infty}^{0} y f_X(vy) f_Y(y) dy$$

Combining the two cases ($Y \leq 0$ and $Y>0$) we have

$$f_V(v) = \int_{-\infty}^{\infty} |y| f_X(vy) f_Y(y) dy \qquad -\infty < v < \infty$$

5.2.3 *Distribution of a Product*

Suppose we have two independent random variables X and Y with probability density functions $f_X(x)$ and $f_Y(y)$ respectively. Assume that both X and Y take values in the interval $(-\infty, \infty)$. We need to determine the probability density function of XY.

Let $Z = XY$ for $-\infty < X, Y < \infty$. Then,

$$P(Z \le z) = P(XY \le z) = P\left(X \le \frac{z}{Y} \right)$$

Here again we will consider two cases

Case I $Y \ge 0$

The feasible regions corresponding to this case (Figure 5.2 regions marked by A and A') satisfies the constraint $X \le \frac{z}{Y}$ and we have

$$P(Z \le z) = \int_{0}^{\infty} \int_{x=-\infty}^{z/y} f_X(x)f_Y(y)dxdy = \int_{0}^{\infty} F_X(z/y)f_Y(y)dy$$

Differentiating both sides w.r.t z we have

$$f_Z(z) = \int_{0}^{\infty} \frac{1}{y} f_X(z/y)f_Y(y)dy, \quad -\infty < z < \infty$$

Case II $Y < 0$.

The region where the constraint $X \le \frac{z}{Y}$ is valid in this case contains the feasible region marked as B and B' in Figure 5.2. Hence

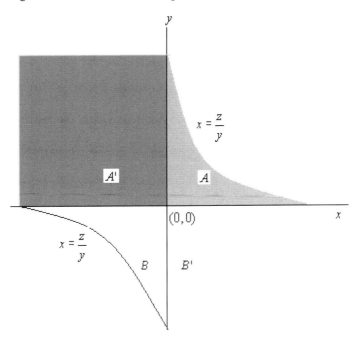

Figure 5.2 Feasible region for Case I ($Y > 0$)and Case II ($Y<0$)

$$P(Z \leq z) = \int\limits_{-\infty}^{0} \int\limits_{z/y}^{\infty} f_X(x)f_Y(y)dxdy$$

$$= F_Y(0) - \int\limits_{-\infty}^{0} F_X(z/y)f_Y(y)dy$$

Differentiating both sides w.r.t z we have

$$f_Z(z) = -\int\limits_{-\infty}^{0}\frac{1}{y}f_X(z/y)f_Y(y)dy. \text{Combining the two results (} Y \geq 0 \text{ and } Y{<}0)$$

$$= \int\limits_{-\infty}^{\infty}\left|\frac{1}{y}\right|f_X(z/y)f_Y(y)dy, \quad -\infty < z < \infty$$

If we are interested in obtaining the first two moments of the product XY ($=Z$) we may obtain $E(XY)$ and $Var(XY)$ by the following two methods.

(1) $E(XY) = E(Z) = \int\limits_{-\infty}^{\infty}zf_Z(z)dz, \ Var(XY) = \int\limits_{-\infty}^{\infty}z^2 f_Z(z)dz - [E(Z)]^2$

(2) Using the fact that X and Y are independent

$$E(XY) = E(X)E(Y)$$
$$Var(XY) = E\left[(XY)^2\right] - \left[E(XY)\right]^2$$
$$= E(X^2)E(Y^2) - \left[E(X)E(Y)\right]^2$$
$$= E(X^2)E(Y^2) - [E(X)]^2[E(Y)]^2$$

5.2.4 General Considerations Involving Multivariate Distributions

Most of the work presented in this chapter will consider elements of a population according to a single variate, and therefore their probability distribution may well be called univariate. In real life situations however, we need to consider more than one variable at a time. For example when studying IQ scores we may be interested in important predictor variables like age, education level attained, socioeconomic status, sex, etc. of the individual as well as the parent's education, whether the individual is mentally challenged or not. All these forces work jointly to determine a person's IQ score. Hence when we discuss multivariate distributions we consider elements of a population bearing in mind that each element is multidimensional and that measurement on the ith element represented by x_i is a vector of observations of the form $x_{i1}, x_{i2}, \ldots, x_{in}$.

Analogous to the definitions in Chapter 3, section 3.6, the joint distribution function involving n different variables in a population may be written as

$$F_{X_1,\ldots,X_n}(x_1,\ldots,x_n) = \int_{-\infty}^{x_1}\int_{-\infty}^{x_2}\ldots\int_{-\infty}^{x_n} f_{X_1,\ldots,X_n}(x_1,\ldots,x_n)dx_1\ldots dx_n$$

where $f_{X_1,\ldots,X_n}(x_1,\ldots,x_n)$ is the corresponding probability density such that

$$\int_{x_1}\int_{x_2}\ldots\int_{x_n} f_{X_1,\ldots,X_n}(x_1,\ldots,x_n)dx_1 dx_2\ldots dx_n = 1, \quad -\infty < x_i < \infty, i = 1,2,\ldots,n$$

The marginal distribution of X_i is obtained by integrating over $X_1,\ldots,X_{i-1},X_{i+1},\ldots X_n$ that is all the variables except X_i.

If all the variables are independent of one another then their joint density function factors into n components as follows

$$f_{X_1,\ldots,X_n}(x_1,\ldots,x_n) = f_{X_1}(x_1)f_{X_2}(x_2)\ldots f_{X_n}(x_n), -\infty < x_i < \infty \ (i = 1,2,\ldots,n)$$

In this case the distribution of X_i for any fixed X_j is the same whatever the fixed value of X_j may be. The converse is also true, that is, if the pdf's factor as above we say that the variables are independent

Having first discussed these useful concepts, we now begin a discussion of specific continuous distributions.

5.3 Uniform Distribution

A continuous random variable X is said to have a *uniform distribution* if its probability density function is of the form

$$f_X(x) = \frac{1}{b-a} \quad a < x < b$$
$$= 0 \qquad \text{otherwise}$$

This implies that the probability that the random variable will fall in any small interval of length Δx in the interval (a,b) is $\dfrac{1}{b-a}\Delta x$

This distribution is the continuous analog of the discrete uniform distribution and is of particular interest because of the mathematical ease in handling it. In this text we will be referring to a uniform distribution in the interval (a,b) by $U(a,b)$. Note that since X is a continuous random variable we may also define a uniform distribution to be on $[a,b],[a,b)$ or $(a,b]$ depending on mathematical convenience.

The cumulative distribution function for $U(a,b)$ is

$$F_X(x) = \frac{1}{b-a}\int_a^x dx = \frac{x-a}{b-a}, \quad a \le x \le b$$

Hence

$$F_x(x) = \begin{cases} 0 & x \le a \\ \dfrac{x-a}{b-a}, & a < x < b \\ 1 & x \ge b \end{cases}$$

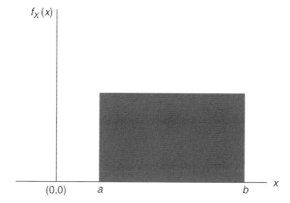

Figure 5.3 Graphical representation of the uniform distribution in the interval (a,b).

5.3.1 Moment Generating Function and Moments of the Uniform Distribution

The moment generating function for this distribution is

$$M_X(t) = \frac{1}{b-a} \int_a^b e^{tx} dx = \frac{1}{t(b-a)} \left[e^{bt} - e^{at} \right]$$

$$E(X) = \frac{b+a}{2}$$

$$Var(X) = \frac{b^2 + a^2 - 2ab}{12} = \frac{(b-a)^2}{12}$$

For $a = 0, b = 1$

$$E(X) = \frac{1}{2} \text{ and } Var(X) = \frac{1}{12}$$

Example 5.1 Let X be a $U(0,1)$ variable. We are going to derive the distribution of Z, where

$Z = -2\log X$

$$F_z(z) = P(Z \le z) = P(-2\log X \le z) = P\left(\log X > -\frac{z}{2}\right) \tag{5.2}$$

$$= P\left(X > e^{-\frac{z}{2}}\right) = 1 - P\left(X \le e^{-\frac{z}{2}}\right)$$

Hence differentiating both sides of (5.2) above , we have

$$f_z(z) = \frac{1}{2}e^{-\frac{1}{2}z} \quad 0 < z < \infty$$

which is the probability density function of an exponential random variable with parameter ½. The exponential distribution is defined in section 5.5.

5.3.2 Sum of Uniform Random Variables

Consider a set of n independent random variables $X_1, \ldots, X_n$ each with $U(a,b)$, $a < b$.

Let $Z = \sum_{i=1}^{n} X_i$

Then the moment generating function of Z is

$$M_Z(t) = \left(\frac{1}{t(b-a)}\right)^n \left(e^{bt} - e^{at}\right)^n \tag{5.3}$$

The moments may be obtained by differentiating (5.3) with respect to t and setting $t = 0$ as follows

$$E(Z) = \frac{n(b+a)}{2}$$

$$Var(Z) = \frac{n}{3}(b^2 + a^2 + ab) - \frac{n}{4}(b^2 + a^2 + 2ab)$$

$$= \frac{n}{12}(b^2 + a^2 - 2ab) = \frac{n(b-a)^2}{12}.$$

Since the $X's$ are independent variables the above mean and variance could also have been obtained by multiplying the mean and variance in section 5.3.1 by n. For $a = 0, b = 1$

$$M_Z(t) = \frac{1}{t^n}\left(e^t - 1\right)^n = \left[\frac{e^t}{t} - \frac{1}{t}\right]^n$$

$$E(Z) = \frac{n}{2}, \; Var(Z) = \frac{n}{12}$$ ■

5.3.3 Probability Density Function of the Sum of Uniformly Distributed Random Variables

Consider two $U(0,1)$ variables X_1 and X_2 and let $Z = X_1 + X_2$. In this section we will obtain the pdf of Z using convolution properties.
The joint pdf for X_1 and X_2 is

$$f_{X_1, X_2}(x_1, x_2) = 1, \; 0 < x_1, x_2 < 1$$

Since the feasible region corresponding to $0 < Z \le 1$ and $1 < Z \le 2$ are different we will derive the pdf of Z in two intervals,
(1) $0 < Z \le 1$

In this case the feasible region is bounded by the X_1 and X_2 axes and the line $X_1 + X_2 = z, 0 < z < 1$.

$$F_z(z) = P(Z \le z) = P(X_1 + X_2 \le z) = \int_0^z \int_0^{z-x_1} dx_2 dx_1 = \frac{1}{2}z^2 \text{ for } 0 < z \le 1$$

$$f_Z(z) = z, \; 0 < z < 1$$

(2) $1 < z < 2$

Note that for this case the integration is over the region below the line $X_1 + X_2 = z \; (1 < z \le 2)$ in Figure 5.4b. Hence,

$$P(Z \le z) = F_Z(z) = \int_0^1 \int_0^1 dx_1 dx_2 - \int_{z-1}^1 \int_{z-x_1}^1 dx_2 dx_1 = 1 - \int_{z-1}^1 (1 - z + x_1) dx_1$$

$$= 1 - \frac{1}{2}(2 - z)^2$$

or

$$f_Z(z) = 2 - z, \; 1 < z \le 2$$

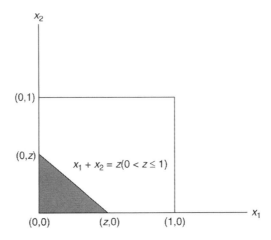

Figure 5.4a Feasible region (shaded area) for $0 \leq z \leq 1$

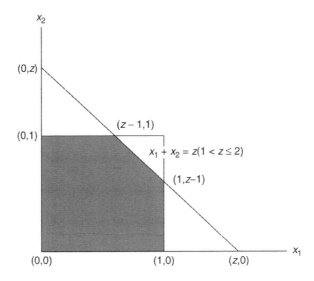

Figure 5.4b Feasible region for $1 \leq z \leq 2$

To summarize the above results

$$f_z(z) = \begin{cases} z & 0 < z \leq 1 \\ (2-z) & 1 < z \leq 2 \end{cases}$$

Example 5.2 Let us now consider three $U(0,1)$ variables X_1, X_2, and X_3. We are going to derive the pdf of $W = X_1 + X_2 + X_3 = Z + X_3$

In this case we will have to obtain the pdf for each of the following three regions, $0 < w \leq 1, 1 < w \leq 2$, and $2 < w \leq 3$

(1) For $0 < w \leq 1$ the feasible region is the shaded area in Figure 5.5a

$$P(W \leq w) = P(Z + X_3 \leq w)$$

$$= \int_0^w \int_0^{w-z} z\,dx_3\,dz = \frac{1}{6}w^3$$

Hence $f_W(w) = \frac{1}{2}w^2, \ 0 < w \leq 1$

(2) For $1 < w \leq 2$ the feasible region is the shaded area in Figure 5.5b

$$F_W(w) = \frac{1}{2}\int_0^1\int_0^1 z\,dx_3\,dz - \int_{w-1}^1 \int_{w-z}^1 z\,dx_3\,dz + \int_1^w \int_0^{w-z}(2-z)\,dx_3\,dz$$

$$= -\frac{1}{3}w^3 + \frac{3}{2}w^2 - \frac{3}{2}w$$

$$f_W(w) = \frac{[3 - (2w-3)^2]}{4} \quad 1 < w \leq 2.$$

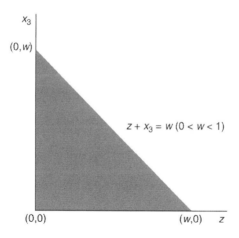

Figure 5.5a Feasible region for the case $0 < w \leq 1$

(3) $2 < w \le 3$ the only feasible region is the shaded area in Figure 5.5c. Hence,

$$P(W \le w) = P(Z + X_3 \le w) = \int_0^1 \int_0^1 z\, dx_3\, dz + \int_1^2 \int_0^1 (2 - z)\, dx_3\, dz - \int_{w-1}^2 \int_{w-z}^1 (2 - z)\, dx_3\, dz$$

$$= \frac{1}{2} + \frac{1}{6} w^3 - \frac{3}{2} w^2 + \frac{9}{2} w - \frac{9}{2}$$

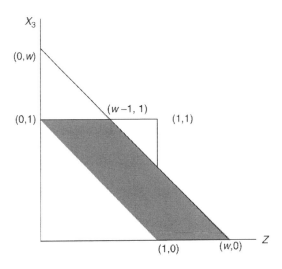

Figure 5.5b Feasible region for the case $1 < w \le 2$

(3) $2 < w \le 3$ the only feasible region is the shaded area in Figure 5.5c. Hence,

$$P(W \le w) = P(Z + X_3 \le w) = \int_0^1 \int_0^1 z\, dx_3\, dz + \int_1^2 \int_0^1 (2 - z)\, dx_3\, dz - \int_{w-1}^2 \int_{w-z}^1 (2 - z)\, dx_3\, dz$$

$$= \frac{1}{2} + \frac{1}{6} w^3 - \frac{3}{2} w^2 + \frac{9}{2} w - \frac{9}{2}$$

$$f_W(w) = \frac{(3 - w)^2}{2} \quad 2 < w \le 3$$

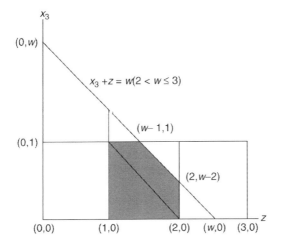

Figure 5.5c Feasible region for the case $2 < w \le 3$

The above results may be summarized as

$$f_W(w) = \begin{cases} \dfrac{1}{2}w^2 & 0 \le w < 1 \\[2mm] \dfrac{[3-(2w-3)^2]}{4} & 1 < w \le 2 \\[2mm] \dfrac{(3-w)^2}{2} & 2 < w \le 3 \end{cases}$$

Example 5.4 Let Y be a $U(0,1)$ variable. To find the pdf of $X = Y^3$

$$F_X(x) = P(X \le x) = P(Y^3 \le x) = P\left(Y \le x^{\frac{1}{3}}\right) = x^{\frac{1}{3}} \quad 0 < x < 1 \tag{5.4}$$

Differentiating both sides of (5.4) above we have

$$f_X(x) = \frac{1}{3}x^{-\frac{2}{3}}, \ 0 < x < 1$$

$$E(X) = \int_0^1 \frac{x}{3} x^{-\frac{2}{3}} dx = \frac{1}{3}\int_0^1 x^{\frac{1}{3}} dx = \frac{1}{3}\frac{3}{4} = \frac{1}{4}$$

$$E(X^2) = \frac{1}{3}\int_0^1 x^{\frac{4}{3}} dx == \frac{1}{3}\frac{3}{7} = \frac{1}{7}$$

Hence
$$E(X) = \frac{1}{4}, \quad Var(X) = \frac{9}{112}$$

Example 5.5 Let X_1 and X_2 be two $U(0,1)$ random variables. Define
$Y = \dfrac{X_1}{X_2}$, then $X_1 = YX_2$

In order to obtain the density function of Y we will consider two scenarios here
(1) $0 < y \leq 1$ (bounded region **a** in Figure 5.6a)

$$P(Y \leq y) = F_Y(y) = P\left(\frac{X_1}{X_2} \leq y\right) = P\left(X_1 \leq yX_2\right)$$

$$= \int_0^1 F(X_1 \leq yx_2 | X_2) f_{X_2}(x_2) dx_2$$

$$= \int_0^1 \int_0^{yx_2} dx_1 dx_2 = \int_0^1 yx_2 dx_2 = \frac{y}{2}.$$

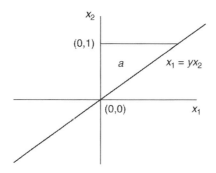

Figure 5.6a Feasible region for the case when $0 < y \leq 1$

(2) $1 < y < \infty$

$F_Y(y) = P(Y \le y) = 1 - P(Y > y) = 1 - P(X_1 > yX_2),$

(since X_1 is a $U(0,1)$ variable, in order for $P(X_1 > yX_2)$ to be non-zero yX_2 must be ≤ 1. hence the feasible region is the bounded region A in Figure 5.6b)

$$= 1 - \int_0^{\frac{1}{y}} \int_{yx_2}^{1} dx_1 dx_2 = 1 - \int_0^{\frac{1}{y}} (1 - yx_2) dx_2 = 1 - \frac{1}{2y}$$

$\vdots$

Hence the pdf of Y is

$$f_Y(y) = \begin{cases} 0, & y \le 0 \\ \dfrac{1}{2}, & 0 < y \le 1 \\ \dfrac{1}{2y^2}, & 1 < y < \infty \end{cases}$$

Example 5.6 Let X, Y and Z be independent $U(0,1)$ variables. We are interested in the pdf of

$$\frac{XY}{Z}$$

The joint density function of X, Y and Z is

$$f_{X,Y,Z}(x, y, z) = 1, \quad 0 < x, y, z < 1$$

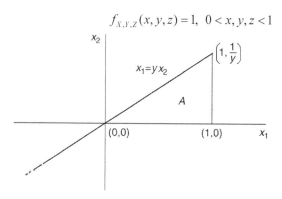

Figure 5.6b Feasible region when $1 < y < \infty$

Define

$$T = \frac{XY}{Z}, V = X \text{ and } W = Y$$

Note that since Z is a $U(0,1)$ variable the denominator in the above ratio will be less than 1.

Using the change of variables technique, the Jacobian of the transformation is

$$J = -\frac{vw}{t^2}$$

And the joint density function of

(T, W, V)

is

$$f_{T,V,W}(t,v,w) = \frac{vw}{t^2}, \quad 0 < v, w < 1, \quad vw < t < \infty$$

We will consider two cases

Case I: $0 < t \le 1$ (represented by bounded region A and bounded region A' in Figure 5.7a

$$f_T(t) = \int_0^t \int_0^1 \frac{vw}{t^2} dv dw + \int_t^1 \int_0^{\frac{t}{v}} \frac{vw}{t^2} dw dv = \frac{1}{4} - \frac{1}{2} \log t$$

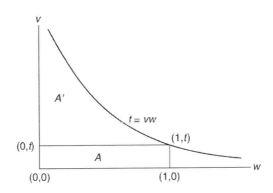

Figure 5.7a Feasible region for the case $0 < t < 1$

Case II: $t > 1$ (represented by bounded region A in Figure 5.7b)

$$f_T(t) = \int_0^1 \int_0^1 \frac{vw}{t^2} dv dw = \frac{1}{4t^2}$$

The cumulative distribution functions is

$$F_T(t) = \begin{cases} \int_0^t \left(\frac{1}{4} - \frac{1}{2}\log x \right) dx = \frac{3}{4}t - \frac{1}{2}t\log t, \ 0 < t \le 1 \\[3mm] \frac{3}{4} + \int_1^t \frac{1}{4x^2}dx = \frac{3}{4} + \frac{1}{4} - \frac{1}{4t}, \ t > 1 \end{cases}$$

Note that this cumulative density function is continuous at $t = 1$. This can be obtained by substituting t = 1 into both formulas separately and obtaining

$$F_T(1) = \lim_{t \to 1} F_T(t) = \frac{3}{4}$$

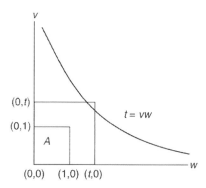

Figure 5.7b Feasible region for the case $t > 1$

Example 5.7 Let X and Y be two $U(-1,1)$ random variables. We are to determine the probability $P(2X - Y > 0)$

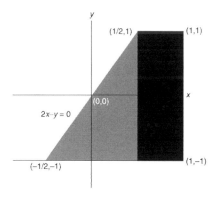

Figure 5.8 Feasible region for Example 5.7

The shaded region in Figure 5.8 represents the area where the constraint $2X - Y > 0$ is valid. To obtain the required probability we proceed as follows.

$$f_{X,Y}(x,y) = \frac{1}{4}, \quad -1 < x, y < 1$$

$$P(2X - Y > 0) = \frac{1}{4}\int\limits_{-1}^{0}\int\limits_{\frac{y}{2}}^{0}dxdy + \frac{1}{4}\int\limits_{-1}^{0}\int\limits_{0}^{\frac{1}{2}}dxdy + \frac{1}{4}\int\limits_{0}^{1}\int\limits_{\frac{y}{2}}^{\frac{1}{2}}dxdy + \frac{1}{4}\int\limits_{-1}^{1}\int\limits_{\frac{1}{2}}^{1}dxdy = \frac{1}{2}.$$

5.4 Normal Distribution

A random variable X is said to have a *normal distribution* if its probability density function is of the following form

$$f_X(x) = \frac{1}{\sigma\sqrt{2\pi}}e^{-\frac{1}{2\sigma^2}(x-\mu)^2} \quad -\infty < x < \infty \tag{5.5}$$

In this case X has a mean μ (location parameter) and variance σ^2 (where σ is the scale parameter). We often represent a normal distribution (5.5) with mean μ and variance σ^2 by $N(\mu, \sigma^2)$.

The shape of the $N(\mu, \sigma^2)$ curve is symmetrical about the mean μ and X takes values in the interval $(-\infty, \infty)$. There are points of inflection (points at which the second derivative of $f_X(x)$ is zero) at a distance of σ on either side of the mean. This distribution has had an interesting and notable background. It was first developed by De Moivre in 1753 as the limiting form of the binomial distribution and later rediscovered in the context of the theory of probability and distribution of measurements in particular when dealing with data on the heights of individuals. Later in the 19th century its importance was reestablished when it was discovered that several distributions encountered in the theory of sampling are either normal or near normal for large sample sizes.

A normal distribution with mean 0 and variance 1 is called the *standard normal distribution* and will be donated by N(0,1). The density function for a standard normal variable is

$$f_X(x) = \frac{1}{\sqrt{2\pi}}e^{-\frac{1}{2}x^2}, \quad -\infty < x < \infty$$

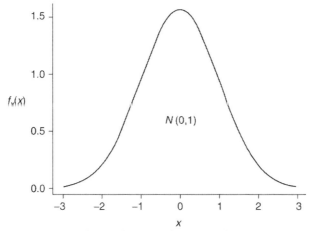

Figure 5.9a The shape of the standard normal curve

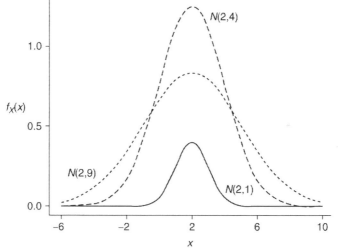

Figure 5.9b An illustration of the change in shape of a normal density function with changes in the variance.

Example 5.8 Let X and Y be two independent $N(\mu,\sigma^2)$ and $N(\gamma,\sigma^2)$ respectively. We are going to derive the joint density function of $X+Y$ and $X-Y$. The joint density function of X and Y is

$$f_{X,Y}(x,y) = \frac{1}{2\pi\sigma^2} e^{-\frac{1}{2\sigma^2}[(x-\mu)^2+(y-\gamma)^2]}, \quad -\infty < x, y < \infty$$

Define $U = X + Y$ and $V = X - Y$, then $X = \dfrac{U+V}{2}$ and $Y = \dfrac{U-V}{2}$

Using the change of variable technique and incorporating the Jacobian in the joint density function of U and V, we have

$$f_{U,V}(u,v) = \frac{1}{4\pi\sigma^2}\exp\left\{-\frac{1}{2\sigma^2}\left(\frac{u+v}{2}-\mu\right)^2\right\}\exp\left\{-\frac{1}{2\sigma^2}\left(\frac{u-v}{2}-\gamma\right)^2\right\}$$

$$= \frac{1}{4\pi\sigma^2}\exp\left\{-\frac{1}{2\sigma^2}\left(\frac{u^2}{2}-\frac{2u(\mu+\gamma)}{2}\right)\right\}\times$$

$$\exp\left\{-\frac{1}{2\sigma^2}\left(\frac{v^2}{2}-\frac{2v(\mu-\gamma)}{2}\right)\right\}\times\exp\left\{-\frac{1}{2\sigma^2}(\mu^2+\gamma^2)\right\}$$

After rearranging and completing the squares in the exponent of the right hand side of the above relationship we have

$$f_{U,V}(u,v) = \frac{1}{4\pi\sigma^2}\exp\left\{-\frac{1}{4\sigma^2}[u-(\mu+\gamma)]^2+\frac{1}{4\sigma^2}(\mu+\gamma)^2\right\}$$

$$\times\exp\left\{-\frac{1}{4\sigma^2}[v-(\mu-\gamma)]^2+\frac{1}{4\sigma^2}(\mu-\gamma)^2\right\}$$

$$\times\exp\left\{-\frac{1}{2\sigma^2}(\mu^2+\gamma^2)\right\} = \frac{1}{4\pi\sigma^2}\exp\left\{-\frac{1}{4\sigma^2}[u-(\mu+\gamma)]^2\right\}$$

$$\times\exp\left\{-\frac{1}{4\sigma^2}[v-(\mu-\gamma)]^2\right\}$$

Notice that the above expression is the product of two normal density functions one distributed as $N(\mu+\gamma,2\sigma^2)$ and the other as $N(\mu-\gamma,2\sigma^2)$. Therefore variables U and V are independent.

Example 5.9 Determine the distribution of the ratio of two independent normal variables X and Y which are $N(\mu_1,\sigma_1^2)$ and $N(\mu_2,\sigma_2^2)$ respectively. It is assumed that μ_2 is so large compared

to σ_2 that the range of Y is effectively positive. This example is from Kendall (1952).

Using the result from section 5.2.2 we know that if $V = \dfrac{X}{Y}$, the pdf of V is

$$f_V(v) = \int_{-\infty}^{\infty} y f_X(vy) f_Y(y) dy$$

Making the substitutions in the above relationship we have

$$f_V(v) = \frac{1}{2\pi\sigma_1\sigma_2} \int_{-\infty}^{\infty} y \exp\left[-\frac{1}{2\sigma_1^2}\left(y^2 v^2 + \mu_1^2 - 2\mu_1 vy\right)\right] \times$$

$$\exp\left[-\frac{1}{2\sigma_2^2}\left(y^2 + \mu_2^2 - 2\mu_2 y\right)\right] dy = \frac{1}{2\pi\sigma_1\sigma_2} \exp\left[-\frac{1}{2\sigma_1^2\sigma_2^2}\left(\mu_1^2\sigma_2^2 + \mu_2^2\sigma_1^2\right)\right]$$

$$\times \int_{-\infty}^{\infty} y \exp\left[-\frac{1}{2\sigma_1^2\sigma_2^2}\left\{y^2\left(v^2\sigma_2^2 + \sigma_1^2\right) - 2y\left(\mu_1 v\sigma_2^2 + \mu_2\sigma_1^2\right)\right\}\right] dy$$

$$= \frac{1}{2\pi\sigma_1\sigma_2} \exp\left[-\frac{1}{2}\frac{(\mu_1 - \mu_2 v)^2}{(v^2\sigma_2^2 + \sigma_1^2)}\right]$$

$$\times \int_{-\infty}^{\infty} y \exp\left[-\frac{(\sigma_2^2 v^2 + \sigma_1^2)}{2\sigma_1^2\sigma_2^2}\left\{y - \frac{(\mu_1 v\sigma_2^2 + \mu_2\sigma_1^2)}{v^2\sigma_2^2 + \sigma_1^2}\right\}^2\right] dy$$

The above integral is proportional to the mean of a

$$N\left[\frac{(\mu_1 v\sigma_2^2 + \mu_2\sigma_1^2)}{(v^2\sigma_2^2 + \sigma_1^2)}, \frac{\sigma_1^2\sigma_2^2}{(v^2\sigma_2^2 + \sigma_1^2)}\right]$$

variable and therefore

$$f_V(v) = \frac{1}{\sqrt{2\pi}} \exp\left[-\frac{(v\mu_2 - \mu_1)^2}{2(\sigma_1^2 + v^2\sigma_2^2)}\right] \frac{1}{\sqrt{(\sigma_2^2 v^2 + \sigma_1^2)}} \frac{(\mu_1 v\sigma_2^2 + \mu_2\sigma_1^2)}{v^2\sigma_2^2 + \sigma_1^2}$$

$$= \frac{1}{\sqrt{2\pi}} \exp\left[-\frac{1}{2}\frac{(v\mu_2 - \mu_1)^2}{(\sigma_1^2 + v^2\sigma_2^2)}\right] \frac{(\mu_1 v\sigma_2^2 + \mu_2\sigma_1^2)}{(v^2\sigma_2^2 + \sigma_1^2)^{\frac{3}{2}}}.$$

∎

5.4.1 Moment Generating Function of the Normal Distribution

The moment generating function (mgf) for a $N(\mu, \sigma^2)$ random variable is derived as follows

$$M_X(t) = E(e^{tX}) = \frac{1}{\sigma\sqrt{2\pi}} \int_{-\infty}^{\infty} e^{tx - \frac{1}{2\sigma^2}(x-\mu)^2} dx$$

$$= \frac{1}{\sigma\sqrt{2\pi}} \int_{-\infty}^{\infty} \exp\left[-\frac{1}{2\sigma^2}\left\{\left(x-(\mu+t\sigma^2)\right)^2 - \left(\mu+t\sigma^2\right)^2 + \mu^2\right\}\right]$$

$$= \exp\left(\frac{1}{2\sigma^2}(\mu+t\sigma^2)^2 - \frac{1}{2\sigma^2}\mu^2\right)\frac{1}{\sigma\sqrt{2\pi}} \int_{-\infty}^{\infty} \exp\left(-\frac{1}{2\sigma^2}\left(x-(\mu+t\sigma^2)\right)^2\right)$$

$$= e^{t\mu+\frac{1}{2}t^2\sigma^2}$$

Similarly the mgf of a $N(0,1)$ variable is $e^{\frac{1}{2}t^2}$. The results of Example 5.8 may be directly obtained using the mgf approach.

5.4.2 Distribution of a Linear Function of Normal Variables

The moment generating function of $\bar{X}\left(=\dfrac{X_1+X_2+\ldots+X_n}{n}\right)$ is

$$M_{\bar{X}}(t) = E\left(e^{t\bar{X}}\right) = E\left(e^{\frac{t}{n}\sum_{i=1}^{n}X_i}\right) \tag{5.6}$$

$$= \left(M_X\left(\frac{t}{n}\right)\right)^n$$

where $M_X(t)$ is the mgf of X_i. This relationship can be used to show

that as $n \to \infty$, the moment generating function of $\dfrac{(\bar{X}-\mu)}{\sigma}\sqrt{n}$ is the moment

generating function of a $N(0,1)$ variable. This is a consequence of the *Central Limit Theorem* discussed in greater detail in Chapter 7

Let $X_1,\ldots,X_n$ be a set of independently distributed $N(\mu_i,\sigma_i^2)$ $(i=1,2,\ldots,n)$ random variables
Define

$$Z = \sum_{i=1}^{n} a_i X_i$$

Then the mgf of Z is

$$M_Z(t) = E\left(e^{tZ}\right) = E\left(e^{ta_1X_1}\right)E\left(e^{ta_2X_2}\right)\ldots E\left(e^{ta_nX_n}\right)$$

$$= \prod_{i=1}^{n}\left(e^{ta_i\mu_i+\frac{1}{2}t^2a_i^2\sigma_i^2}\right) = e^{t\sum_{i=1}^{n}a_i\mu_i+\frac{1}{2}t^2\sum_{i=1}^{n}a_i^2\sigma_i^2}$$

The above relationship demonstrates a linear function of normal variables also follows a normal distribution. For the linear function Z under consideration, the distribution is that of a $N\left(\sum_{i=1}^{n} a_i \mu_i, \sum_{i=1}^{n} a_i^2 \sigma_i^2 \right)$ variable.

5.4.3 Binomial Approximation to the Normal

Consider a binomial variable X with parameters n and p_n then we know that X has a mean np_n and variance $np_n(1-p_n)$. Let $\mu = np_n$ and $\sigma = \sqrt{np_n(1-p_n)}$. Define

$$Z = \frac{X - np_n}{\sqrt{np_n(1-p_n)}} = \frac{X - \mu}{\sigma}$$

is a standardized binomial variable with mean zero and variance unity. It will be shown in Chapter 7 that for large n, Z has an approximate $N(0,1)$ distribution.

Example 5.10 Assume that 100 in 1000 individuals in the age group 35-55 develop hypertension. We have a group of 1000 individuals. What is the probability that at least 125 among them will develop hypertension.

In this case we want the probability that a binomial variable takes a value greater than 125. The required probability is

$$P(X > 125) = P\left(\frac{X - 1000(0.1)}{\sqrt{1000(0.1)(0.9)}} > \frac{125 - 1000(0.1)}{\sqrt{1000(0.1)(0.9)}} \right) = P(Z > 2.63)$$

$$= 0.0043$$

As we can see the likelihood of this occurring is very small

.

5.4.4 Poisson Approximation to Normal

Recall that the mgf of a Poisson variable X with parameter λ is $M_X(t) = e^{\lambda(e^t - 1)}$.

Consider a variable Z defined as $Z = \dfrac{X - \lambda}{\sqrt{\lambda}}$. Then Z is the standardized Poisson variable and has mean zero and variance unity. The mgf of Z is

$$\lim_{\lambda \to \infty} M_Z(t) = e^{\frac{1}{2}t^2}.$$

The above relationship is established in Chapter 7 and illustrates the fact that for large λ the Poisson distribution approximates the normal.

Example 5.11 Suppose we are interested in the hereditary susceptibility to colorectal cancer. We find from large population studies that 5 in 1000 individuals in the age group 40-65 will develop colorectal cancer. In a group of 500 individuals we find that 3 individuals have developed this disease. Is this an unusual event?

We are going to use the Poisson distribution argument first with $\lambda = 2.5$ (the population rate is 5 per 1000 therefore in a group of 500 individuals the rate should be 2.5).

$$P(X \geq 3) = 1 - [P(X = 0) + P(X = 1) + P(X = 2)] = 1 - (0.0821 + 0.2052$$
$$+ 0.2565) = 0.4562$$

Next we use the Poisson approximation to normal argument and obtain

$$P\left(\frac{X - 2.5}{\sqrt{2.5}} > \frac{3 - 2.5}{\sqrt{2.5}} \right) = P(Z > 0.3162) = 0.3745$$

In either case the probability of the observed value is not too low indicating that 3 individuals out of 500 developing the disease was purely by chance.

∎

5.4.5 Bivariate Normal Distribution

The joint distribution of variables X_1 and X_2 is said to follow a *bivariate normal distribution* if their joint pdf is of the form

$$f_{X_1, X_2}(x_1, x_2) = \frac{1}{2\pi\sigma_1\sigma_2\sqrt{1-\rho^2}} \exp\left[-\frac{1}{2(1-\rho^2)} \left(\frac{(x_1 - \mu_1)^2}{\sigma_1^2} + \frac{(x_2 - \mu_2)^2}{\sigma_2^2} \right) \right]$$

$$\times \exp\left[-\frac{1}{2(1-\rho^2)} \left(-2\rho\frac{(x_1 - \mu_1)(x_2 - \mu_2)}{\sigma_1\sigma_2} \right) \right] , -\infty < x_1, x_2 < \infty \qquad (5.7)$$

Equation (5.7) can be factored into two parts

$$f_{X_1, X_2}(x_1, x_2) = \frac{1}{2\pi\sigma_1\sigma_2\sqrt{1-\rho^2}} \exp\left[-\frac{1}{2(1-\rho^2)} \frac{(x_1 - \mu_1)^2}{\sigma_1^2} \right]$$

$$\exp\left[-\frac{1}{2(1-\rho^2)} \left[\frac{(x_2 - \mu_2)^2}{\sigma_2^2} - 2\rho\frac{(x_1 - \mu_1)(x_2 - \mu_2)}{\sigma_1\sigma_2} \right] \right] \qquad (5.8)$$

$$-\infty < x_1, x_2 < \infty$$

We are going to obtain the marginal density function of X_1 by integrating out x_2 in the above expression by making the substitution

$$y = \frac{x_1 - \mu_1}{\sigma_1} \quad \text{and} \quad z = \frac{x_2 - \mu_2}{\sigma_2} \qquad (5.9)$$

The Jacobian associated with the transformation in (5.9)) is $\sigma_1\sigma_2$. Hence (5.8) can be written as

$$g_{Y,Z}(y,z) = \frac{1}{2\pi\sqrt{1-\rho^2}} e\left(-\frac{1}{2(1-\rho^2)}y^2\right)$$

$$\exp\left[-\frac{1}{2(1-\rho^2)}\left[(z^2 - 2y\rho z + \rho^2 y^2) - \rho^2 y^2\right]\right] \quad -\infty < y, z < \infty \tag{5.10}$$

The marginal density function of Y may be obtained by integrating out for all values of z in (5.10). That is

$$f_Y(y) = \frac{1}{\sqrt{2\pi}} \frac{1}{\sqrt{2\pi(1-\rho^2)}} \exp\left(-\frac{1}{2(1-\rho^2)}y^2\right)$$

$$\times \int_{-\infty}^{\infty} \exp\left[-\frac{1}{2(1-\rho^2)}\left[(z^2 - 2y\rho z + \rho^2 y^2) - \rho^2 y^2\right]\right] dz \quad -\infty < y < \infty$$

$$f_Y(y) = \frac{1}{\sqrt{2\pi}} \exp\left(-\frac{1}{2}y^2\right)$$

$$\times \frac{1}{\sqrt{2\pi(1-\rho^2)}} \int_{-\infty}^{\infty} \exp\left[-\frac{1}{2(1-\rho^2)}\left[z^2 - 2y\rho z + \rho^2 y^2\right]\right] dz \quad -\infty < y < \infty$$

$$= \frac{1}{\sqrt{2\pi}} e^{-\frac{1}{2}y^2} \quad -\infty < y < \infty$$

Hence the marginal pdf of $Y = \dfrac{X_1 - \mu_1}{\sigma_1}$ is $N(0,1)$ implying that the marginal density function of X_1 is $N(\mu_1, \sigma_1^2)$. Similarly, it can be shown that the marginal probability density function of X_2 is $N(\mu_2, \sigma_2^2)$. If we were to integrate $f_Y(y)$ for all values of y we would obtain the following result

$$\int_{-\infty}^{\infty} f_Y(y)dy = 1$$

implying

$$\int_{-\infty}^{\infty}\int_{-\infty}^{\infty} f_{X_1,X_2}(x_1,x_2)dx_1 dx_2 = 1$$

This verifies that (5.7) is a valid pdf. We are now going to show that the correlation between X_1 and X_2 is ρ. Since

$$Corr(X_1, X_2) = E\left[\left(\frac{X_1 - \mu_1}{\sigma_1}\right)\left(\frac{X_2 - \mu_2}{\sigma_2}\right)\right]$$

We may obtain the correlation as

$$Corr(X_1, X_2) = \int_{-\infty}^{\infty} \int_{-\infty}^{\infty} \left[\left(\frac{x_1 - \mu_1}{\sigma_1} \right) \left(\frac{x_2 - \mu_2}{\sigma_2} \right) \right] f_{X_1 X_2}(x_1, x_2) dx_1 dx_2$$

Define

$$y = \frac{x_1 - \mu_1}{\sigma_1} \text{ and } z = \left(\frac{x_1 - \mu_1}{\sigma_1} \right) \left(\frac{x_2 - \mu_2}{\sigma_2} \right)$$

Then,

$$x_1 = \sigma_1 y + \mu_1 \text{ and } x_2 = \frac{\sigma_2 z}{y} + \mu_2$$

the Jacobian of this transformation is

$$J = \frac{\sigma_1 \sigma_2}{y}, \ y \neq 0$$

$$Corr(X_1, X_2) = \int_{-\infty}^{\infty} \int_{-\infty}^{\infty} \frac{z \sigma_1 \sigma_2}{2\pi \sigma_1 \sigma_2 y \sqrt{1 - \rho^2}} \exp \left[-\frac{1}{2(1-\rho^2)} \left(y^2 - 2\rho z + \frac{z^2}{y^2} \right) \right] dz dy$$

The term in the exponent above may be written as

$$y^2 - 2\rho z + \frac{z^2}{y^2} = \frac{z^2}{y^2} - 2\rho \frac{z}{y} y + \rho^2 y^2 - \rho^2 y^2 + y^2$$

$$= \left(\frac{z}{y} - \rho y \right)^2 + (1 - \rho^2) y^2$$

$$Corr(X_1, X_2) = \int_{-\infty}^{\infty} \int_{-\infty}^{\infty} \frac{1}{2\pi} e^{-\frac{1}{2} y^2} \frac{z}{y \sqrt{1 - \rho^2}} \exp \left\{ -\frac{\left(z - \rho y^2 \right)^2}{2(1-\rho^2) y^2} \right\} dz dy$$

Now

$$\int_{-\infty}^{\infty} \frac{z}{y \sqrt{(1 - \rho^2) 2\pi}} \exp \left\{ -\frac{\left(z - \rho y^2 \right)^2}{2(1 - \rho^2) y^2} \right\} dz = \rho y^2, \text{ since } Z \text{ is normal variate with}$$

mean ρy^2 an d variance $(1 - \rho^2) y^2$.

Hence

$$\int_{-\infty}^{\infty} \rho y^2 \frac{1}{\sqrt{2\pi}} e^{-\frac{1}{2} y^2} dy = \rho.$$

The above result is obtained using the fact that the left hand side in the above expression represents $\rho E(Y^2)$ which is simply ρ when Y has a $N(0,1)$ distribution. Thus we have shown that the coefficient of correlation between X_1 and X_2 is ρ

Example 5.12 Let Z_1 and Z_2 be two independent $N(0,1)$ variables and define X and Y as follows :

$$X = a_1 Z_1 + b_1 Z_2$$
$$Y = a_2 Z_1 + b_2 Z_2$$

Then, $E(X) = E(Y) = 0$ and $Var(X) = a_1^2 + b_1^2$ and $Var(Y) = a_2^2 + b_2^2$
$Cov(X,Y) = a_1 a_2 + b_1 b_2$

If we define ρ as the coefficient of correlation between X and Y, then

$$\rho = \frac{a_1 a_2 + b_1 b_2}{\sqrt{(a_1^2 + b_1^2)(a_2^2 + b_2^2)}}$$

Now,

$$f_{Z_1,Z_2}(z_1,z_2) = \frac{1}{2\pi} e^{-\frac{1}{2}(z_1^2 + z_2^2)} \quad -\infty < z_1, z_2 < \infty$$

and $z_1 = \dfrac{b_2 x - b_1 y}{a_1 b_2 - a_2 b_1}, z_2 = -\dfrac{a_2 x - a_1 y}{a_1 b_2 - a_2 b_1}$

Define

$$a_1 = \sigma_X \sqrt{\frac{1+\rho}{2}}, \, a_2 = \sigma_Y \sqrt{\frac{1+\rho}{2}}, \, b_1 = \sigma_X \sqrt{\frac{1-\rho}{2}}, \, b_2 = -\sigma_Y \sqrt{\frac{1-\rho}{2}}.$$

Then,

$Var(X) = \sigma_X^2, Var(Y) = \sigma_Y^2, Cov(X,Y) = \sigma_X \sigma_Y \rho$ for $-1 \le \rho \le 1$

The Jacobian J of the transformation is

$$\frac{1}{a_1 b_2 - a_2 b_1} = \frac{-1}{\sigma_X \sigma_Y \sqrt{1-\rho^2}}$$

$$f_{X,Y}(x,y) = \frac{1}{2\pi(a_1b_2 - a_2b_1)} \exp\left[-\frac{1}{2(a_1b_2 - a_2b_1)^2}\left[\{b_2x - b_1y\}^2 + \{a_2x - a_1y\}^2\right]\right]$$

$$= \frac{1}{2\pi(a_1b_2 - a_2b_1)} \exp\left[-\frac{1}{2\sigma_X^2\sigma_Y^2(1-\rho^2)}\left[\{b_2x - b_1y\}^2 + \{a_2x - a_1y\}^2\right]\right]$$

$$f_{X,Y}(x,y) = \frac{1}{2\pi}\left(\frac{1}{\sigma_X\sigma_Y\sqrt{1-\rho^2}}\right)$$

$$\times \exp\left[-\frac{1}{2(1-\rho^2)}\left\{\left(\frac{x}{\sigma_X}\right)^2 + \left(\frac{y}{\sigma_Y}\right)^2 - 2\rho\left(\frac{x}{\sigma_X}\right)\left(\frac{y}{\sigma_Y}\right)\right\}\right]$$

which is a bivariate normal density function with parameters and σ_X, σ_Y and ρ.
Note that in this case the means of both X and Y are zero.

5.4.6 The Moment Generating Function of the Bivariate Normal Distribution

$$f_{X_1,X_2}(x_1,x_2) = \frac{1}{2\pi\sigma_1\sigma_2\sqrt{1-\rho^2}} \exp\left[-\frac{1}{2(1-\rho^2)}\left\{\left(\frac{x_1-\mu_1}{\sigma_1}\right)^2 + \left(\frac{x_2-\mu_2}{\sigma_2}\right)^2\right\}\right]$$

$$\times\left[-\frac{1}{2(1-\rho^2)}\exp\left(-2\rho\left(\frac{x_1-\mu_1}{\sigma_1}\right)\left(\frac{x_2-\mu_2}{\sigma_2}\right)\right)\right]$$

$$M_{X_1,X_2}(t_1,t_2) = E\left(\exp(t_1X_1 + t_2X_2)\right) = \int_{-\infty}^{\infty}\int_{-\infty}^{\infty}\exp(t_1x_1 + t_2x_2)f_{X_1,X_2}(x_1,x_2)dx_1dx_2$$

Let $x = \left(\frac{x_1-\mu_1}{\sigma_1}\right)$ and $y = \left(\frac{x_2-\mu_2}{\sigma_2}\right)$

Jacobian of this transformation is $\sigma_1\sigma_2$

$$M_{X_1,X_2}(t_1,t_2) = \frac{e^{t_1\mu_1 + t_2\mu_2}}{2\pi\sqrt{1-\rho^2}}\int_{-\infty}^{\infty}\int_{-\infty}^{\infty}\exp\left[-\frac{1}{2(1-\rho^2)}\{x^2 - 2t_1\sigma_1x(1-\rho^2)\}\right]$$

$$\times\exp\left[-\frac{1}{2(1-\rho^2)}\{y^2 - 2y[\rho x + t_2\sigma_2(1-\rho^2)]\}\right]dydx$$

The second exponent in the above expression may be written as

$$= -\frac{1}{2(1-\rho^2)}\{y^2 - 2y[\rho x + t_2\sigma_2(1-\rho^2)] + [\rho x + t_2\sigma_2(1-\rho^2)]^2\}$$

$$+\frac{1}{2(1-\rho^2)}[\rho x + t_2\sigma_2(1-\rho^2)]^2$$

$$= \left[-\frac{1}{2(1-\rho^2)}\{y - [\rho x + t_2\sigma_2(1-\rho^2)]\}^2 + \frac{1}{2(1-\rho^2)}[\rho x + t_2\sigma_2(1-\rho^2)]^2 \right]$$

Replacing the second exponent in the expression of $M_{X_1,X_2}(t_1,t_2)$ by the above expression we have after integrating out for y

$$M_{X_1,X_2}(t_1,t_2) = \frac{e^{t_1\mu_1+t_2\mu_2}}{\sqrt{2\pi}} \int_{-\infty}^{\infty} \exp\left[-\frac{1}{2(1-\rho^2)}\{x^2 - 2t_1\sigma_1 x(1-\rho^2)\}\right]$$

$$\times \exp\left[\frac{1}{2(1-\rho^2)}[\rho x + t_2\sigma_2(1-\rho^2)]^2\right]dx$$

$$= \frac{e^{t_1\mu_1+t_2\mu_2}}{\sqrt{2\pi}} \int_{-\infty}^{\infty} \exp\left[-\frac{1}{2}\{x - (t_1\sigma_1 + \rho t_2\sigma_2)\}^2 + \frac{1}{2}(t_1^2\sigma_1^2 + t_2^2\sigma_2^2 + 2t_1 t_2\sigma_1\sigma_2\rho)\right]dx$$

$$= \exp\left[t_1\mu_1 + t_2\mu_2 + \frac{1}{2}(t_1^2\sigma_1^2 + t_2^2\sigma_2^2 + 2t_1 t_2\sigma_1\sigma_2\rho)\right]$$

Example 5.13 In the joint bivariate normal distribution of X_1 and X_2 of the form

$$f_{X_1,X_2}(x_1,x_2) = \frac{1}{2\pi\sqrt{1-\rho^2}}\exp\left(-\frac{1}{2(1-\rho^2)}\left[x_1^2 - 2x_1 x_2\rho + x_2^2\right]\right) \quad -\infty < x_1, x_2 < \infty$$

let $U = \dfrac{X_1 - \rho X_2}{\sqrt{1-\rho^2}}$ and $V = X_2$.

Then the Jacobian J of the transformation is

$$J = \sqrt{1-\rho^2} \text{ and }$$

$$f_{U,V}(u,v) = \frac{1}{2\pi}\exp\left[-\frac{1}{2}\{u^2 + v^2\}\right]$$

$$= g_U(u)h_V(v), \quad -\infty < u, v < \infty$$

Hence U and V are independent and $N(0,1)$ variables.

5.4.7 Conditional Distribution of $X_2 | X_1$

We can write (5.7) in a slightly different form (by completing the squares in the exponent)

$$f_{X_1,X_2}(x_1,x_2) = f_{X_2|X_1}(x_2 | x_1) f_{X_1}(x_1) = \frac{1}{\sigma_1 \sqrt{2\pi}} \exp\left[-\frac{1}{2}\left\{ \frac{x_1 - \mu_1}{\sigma_1} \right\}^2 \right]$$

$$\times \frac{1}{\sigma_2 \sqrt{2\pi(1-\rho^2)}} \exp\left[-\frac{1}{2\sigma_2^2(1-\rho^2)}\left\{ (x_2 - \mu_2) - \rho\frac{\sigma_2}{\sigma_1}(x_1 - \mu_1) \right\}^2 \right]$$

The second term on the right hand side of the above expression is the conditional distribution of X_2 given X_1 is $N(\mu_1,\sigma_1^2)$. Hence

$$E(X_2 | X_1) = \mu_2 + \rho\frac{\sigma_2}{\sigma_1}(x_1 - \mu_1)$$

$$Var(X_2 | X_1) = (1 - \rho^2)\sigma_2^2$$

(5.11)

The conditional distribution of X_1 given X_2 may be similarly obtained using considerations of symmetry as seen in (5.11). In *regression theory* terminology, the mean value of $X_2 | X_1$, that is the *regression function* of X_2 on X_1 is a linear function of X_1

5.5 Exponential Distribution

The *exponential distribution* (or the *negative exponential distribution*) is the continuous analog of the geometric distribution discussed in Chapter 4. A variable X is said to have an exponential distribution with parameter λ if its probability density function is of the form $f_X(x) = \lambda e^{-\lambda x}$, $0 < x < \infty$. This is commonly referred to as the *one parameter* exponential distribution. Furthermore, if $\lambda = 1$, the distribution is called the *standard* exponential distribution. In this book a variable following an exponential distribution with parameter λ will be referred to as *exponential(λ)*.

There are many situations in which exponential distribution may be applied. One of the most common applications of this distribution is in the *theory of recurrent events* or *renewal processes* where events occur at random points in time and the occurrence of an event has no bearing on the length of time it will take for the second event to occur. In *queuing theory* (or *waiting line theory*) where individuals arrive at a certain place (say a bank teller) independently of one another and each *customer's* arrival time has no influence on the time at which any other customer will arrive. The distribution of the time between arrival of customers called the *inter-arrival times* is assumed to have the exponential distribution. The moments of this distribution are straightforward and may be derived as follows.

$$E(X) = \int\limits_{x=0}^{\infty} \lambda x e^{-\lambda x}\, dx = \frac{1}{\lambda}$$

$$E(X^2) = \int\limits_{x=0}^{\infty} \lambda x^2 e^{-\lambda x}\, dx = \frac{2}{\lambda^2}$$

Hence

$$Var(X) = \frac{1}{\lambda^2}$$

The moment generating function is similarly obtained

$$M_X(t) = E(e^{tX}) = \lambda \int\limits_{x=0}^{\infty} e^{tx} e^{-\lambda x}\, dx = \lambda \int\limits_{x=0}^{\infty} e^{-x(\lambda-t)}\, dx = \frac{\lambda}{(\lambda-t)}, t < \lambda$$

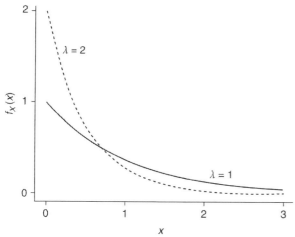

Figure 5.10a Exponential density functions for $\lambda = 1$ and $\lambda = 2$

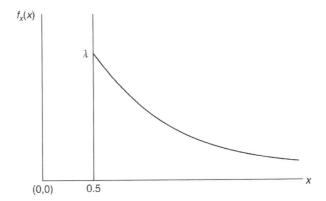

Figure 5.10b Exponential density with scale parameter λ and location parameter $\theta = 0.5$.

An exponential distribution with two parameters is of the form

$$f_X(x) = \lambda e^{-\lambda(x-\theta)}, \quad \theta < x < \infty \tag{5.12}$$

In (5.12) λ is the scale parameter and θ is the location parameter. The shape of such a curve would be as shown in Figure 5.10b above.

5.5.1 Sum of Exponential Variables

Consider two independent exponential variables X_1 and X_2 with parameters λ_1 and λ_2 ($\lambda_1 \neq \lambda_2$) respectively. Then the moment generating function of $Y = X_1 + X_2$ is

$$M_Y(t) = M_{X_1}(t)M_{X_2}(t) = \frac{\lambda_1\lambda_2}{(\lambda_1 - t)(\lambda_2 - t)}, t < \lambda_1, t < \lambda_2$$

Similarly for n independent variables $X_1, \ldots, X_n$ with parameters $\lambda_1, \ldots, \lambda_n$

$(\lambda_i \neq \lambda_j, i, j = 1, 2, \ldots, n)$ and $Z - \sum_{i=1}^{n} X_i$, the moment generating function of Z

obtained by induction is

$$M_Z(t) = \prod_{i=1}^{n} \frac{\lambda_i}{(\lambda_i - t)}, t < \lambda_i$$

The density function of Z as well as the mean and the variance of Z are derived in Chiang(1979) as follows

$$f_Z(z) = (-1)^{n-1} \lambda_1\lambda_2 \ldots \lambda_n \sum_{i=1}^{n} \frac{e^{-\lambda_i z}}{\prod_{\substack{j=1 \\ j \neq i}}^{n} (\lambda_i - \lambda_j)}$$

$$E(Z) = \sum_{i=1}^{n} \frac{1}{\lambda_i}$$

$$Var(Z) = \sum_{i=1}^{n} \frac{1}{\lambda_i^2}$$

For $\lambda_i = \lambda$ ($i = 1, 2, ..., n$)

$$M_Z(t) = \left(\frac{\lambda}{\lambda - t}\right)^n$$

which is the moment generating function of a *gamma* random variable with parameters n and λ. The *gamma distribution* is discussed in section 5.7.

■

Memoryless Property of Exponential Variable

Here we will show that like the geometric distribution, the exponential distribution is also memoryless.

Consider a variable X having an exponential distribution with parameter λ

Then

$$P(X > s | X > t) = \frac{P(X > s, X > t)}{P(X > t)}, \quad s > t$$

$$= \frac{P(X > s)}{P(X > t)} = \frac{\int_{x=s}^{\infty} e^{-\lambda x} dx}{\int_{x=t}^{\infty} e^{-\lambda x} dx} = e^{-(s-t)\lambda}$$

Hence as was the case for the geometric distribution, the above conditional probability shows that if the random variable X represents time to the occurrence of an event E, then the probability of an additional $(s-t)$ unit of time to the occurrence of E is the same as observing the occurrence of E during a time period of length $(s-t)$. What occurred during the first t time units has no bearing on occurrences between times t and s.

Example 5.14 Assume that time between arrival of cars at a car washing facility follows an exponential distribution at the rate of two cars per unit time. Suppose an arrival occurred at $t = 30$, what is the probability of an arrival by $t = 40$?
Since the exponential distribution is a memoryless distribution we need to calculate the probability of an arrival in 10 units of time which would be (for $\lambda = 2$)

$$P(t < 10) = \int_{t=0}^{10} 2e^{-2t} dt = 1 - e^{-20}$$

If we are asked to determine the probability that an arrival will occur before $t=40$ given it occured after $t=35$ then the required probability would be $(\lambda = 2)$

$$P(t < 40|t > 35) = \frac{P(35 < t < 40)}{P(t > 35)}$$

$$P(35 < t < 45) = \int_{35}^{40} 2e^{-2t}\,dt = e^{-70} - e^{-80}$$

and $P(t > 35) = \int_{35}^{\infty} 2e^{-2t}\,dt = e^{-70}$. Hence the required probability is

$$P(t < 40|t > 35) = \frac{P(35 < t < 40)}{P(t > 35)} = 1 - e^{-10}$$

■

Example 5.15 Assume that each of the three components in a ceiling fan must be in working condition in order for the fan to operate. Furthermore assume that if one of these components were to fail the fan would cease to operate (failure). Let X_1, X_2 and X_3 represent the lifetime of these three components and assume that these are independent random variables and that X_1, X_2, and X_3 follow an exponential distribution of the form $f_X(x) = e^{-x}$. In order to obtain the time to failure of the fan we need to derive the density function of $Y = Min(X_1, X_2, \ldots, X_n)$. In this case

$$F_Y(y) = P(Y \le y) = 1 - P(Y > y) \tag{5.13}$$

Now, if $Y > y$ each of the $X's$ must be greater than y and since the X's are independent of one another.

$$P(Y > y) = P(X_1 > y \text{ and } X_2 > y \text{ and } X_3 > y) = \left(\int_{y}^{\infty} e^{-x}\,dx \right)^3$$

$$F_Y(y) = 1 - \left(\int_{x=y}^{\infty} e^{-x}\,dx \right)^3 = 1 - e^{-3y}$$

The density function of Y may be obtained by differentiating $F_Y(y)$ as

$$f_Y(y) = 3e^{-3y}$$

which is clearly an exponential distribution with $\lambda = 3$. This concept will be discussed in detail in Chapter 6.

Example 5.16 Assume that the lifetime X of an electric bulb follows an exponential distribution with $\lambda = 10$. The manufacturer needs to determine t such that the probability that the bulb will burn out before time t is 0.8.

In this problem we need to determine t such that

$$\int_{x=0}^{t} 10 e^{-10x} dx = 1 - e^{-10t} = 0.8$$

$$t = 0.161$$

Example 5.17 Consider a random variable X having the following probability density function (The variable X is said to have a *doubly exponential distribution*).

$$f_X(x) = c e^{-|x|} \quad -\infty < x < \infty \tag{5.14}$$

We need to determine c such that $f_X(x)$ in (5.14) above is a proper density function. The integral of the right hand side of (5.14) may be written as

$$c \int_{x=-\infty}^{\infty} e^{-|x|} dx = c \int_{x=-\infty}^{0} e^{x} dx + c \int_{x=0}^{\infty} e^{-x} dx$$

$$= c e^{x} \Big|_{-\infty}^{0} - c e^{-x} \Big|_{0}^{\infty}$$

$$= 2c$$

Hence for (5.14) to be a proper probability density function $c = \dfrac{1}{2}$

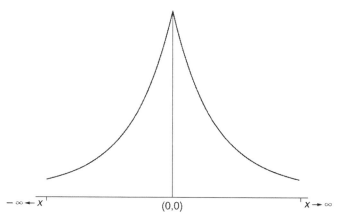

Figure 5.11 Doubly exponential distribution with $c = \dfrac{1}{2}$

The double exponential distribution with location parameter μ and scale parameter σ takes the form

$$f_X(x) = \frac{\sigma}{2} e^{-\sigma|x-\mu|} \quad -\infty < x, \mu < \infty, \sigma > 0$$

The moment generating function of the above distribution is obtained as follows

$$M_X(t) = \frac{\sigma}{2} \int_{-\infty}^{\infty} e^{tx} e^{-\sigma|x-\mu|} dx = \frac{\sigma}{2} \left(\int_{-\infty}^{\mu} e^{tx} e^{\sigma(x-\mu)} dx + \int_{\mu}^{\infty} e^{tx} e^{-\sigma(x-\mu)} dx \right)$$

$$= \frac{\sigma}{2} \left(\frac{e^{\mu t}}{\sigma + t} + \frac{e^{\mu t}}{\sigma - t} \right)$$

$$M_X(t) = \frac{e^{\mu t}}{1 - \left(\frac{t}{\sigma}\right)^2}, \quad |t| < \sigma$$

$$E(X) = \mu, Var(X) = \frac{2}{\sigma^2}$$

∎

5.6 Poisson Process

Moyé and Kapadia (2000) among others have discussed the *Poisson process* in the context of determining probability distribution of the number of individuals with a certain non contagious disease. They assume that the diseased individuals arrive into a population that at time zero has zero individuals with the disease. It is assumed that random effects influence the arrival of diseased individuals, that they arrive independently of one another, and that the forces and influences that determine their arrival into the population remain unchanged. Such a system is generally referred to as a *homogeneous Poisson process*. It is assumed that the probability of any particular arrival into the population is same for all time intervals of length t, and independent of where this time interval of length t is located. In stochastic processes in particular in *queuing theory,* what we are referring to as a population is referred to as "*the system*". We will use the two definitions interchangeably. Since we are assuming no departures from the system, this kind of model is often called the *immigration* model. A variation on this theme would be a model in which the disease is still noncontagious but the diseased individuals may leave the system due to death. Such models are referred to as the *immigration-emigration* models.

Define

$P_n(t) \equiv P$(at time t there are n diseased individuals in the population)

Also define

$$P_0(0) = 1 \qquad (5.15)$$

The boundary condition (5.15) implies that there were no diseased individuals in the population at time 0 (i.e., the time when we start observing the system)

Note that
$$P_0(t) \text{ and } 1 - P_0(t)$$
represent the probabilities of no new arrivals of diseased individuals into the system in time t and the probability of at least one new arrival in time t respectively.

      Consider Δt (a very small interval of time) in which at most one new non-contagious individual may enter the system. It is assumed that Δt is so small that the probability of more than one arrival in Δt is $o(\Delta t)$ (here the notation $o(\Delta t)$ implies a function of terms of power of Δt higher than the first. Since Δt is very small any function of the power of Δt other than the first power is set equal to zero). If on an average λ diseased individuals arrive in the system per unit of time, then in t units of time on an average λt individuals would have arrived. The probability of arrival in time Δt, would be $\lambda \Delta t$ and probability of no arrival in Δt would be $(1 - \lambda \Delta t)$.

      If we were to determine the probability of n arrivals in time $t + \Delta t$ then there are the following three ways in which such an event may occur (1) n arrivals by time t and no new arrivals in the interval $(t, t+\Delta t)$ (2) $(n-1)$ arrivals by time t and one arrival in the interval $(t, t+\Delta t)$ and (3) $(n-i)$ arrivals by time t and i $((i > 2)$ arrivals in $(t, t+\Delta t)$. Since the probability of the last event is $o(\Delta t)$ we can write
$$P_n(t + \Delta t) = P_n(t)(1 - \lambda \Delta t) + P_{n-1}(t)\lambda \Delta t + o(\Delta t)$$

or

$$\frac{P_n(t + \Delta t) - P(t)}{\Delta t} = -P_n(t)\lambda + P_{n-1}(t)\lambda + \frac{o(\Delta t)}{\Delta t} \qquad (5.16)$$

The left hand side of (5.16) approaches $P_n'(t)$ and the last term on the right hand side of (5.16) approaches zero as $\Delta t \to 0$. Hence we have

$$P_n'(t) = -\lambda P_n(t) + \lambda P_{n-1}(t), \quad 1 \leq n < \infty \qquad (5.17)$$

Note that

$$P_0(t + \Delta t) = P_0(t)(1 - \lambda \Delta t)$$
$$\frac{P_0(t + \Delta t) - P_0(t)}{\Delta t} = -\lambda P_0(t)$$

Hence,

$$P_0'(t) = -\lambda P_0(t) \qquad (5.18)$$

We may now solve for $P_n(t)$ by using probability generating functions. Define

$$G(s,t) = \sum_{n=0}^{\infty} P_n(t)s^n$$

Multiplying both sides of (5.17) by s^n and summing over all n from $n = 1$ to ∞, we have after incorporating (5.18)

$$\sum_{0}^{\infty} P_n^{'}(t)s^n - P_0^{'}(t) = -\lambda \sum_{0}^{\infty} P_n(t)s^n + \lambda P_0(t) + \lambda s \sum_{0}^{\infty} P_n(t)s^n$$

Rearranging terms in the above differential equation, and using (5.18) we have

$$\frac{\partial}{\partial t} G(s,t) = -\lambda(1-s)G(s,t)$$

$$\frac{\frac{d}{dt}G(s.t)}{G(s,t)} = -\lambda(1-s)$$

or

$$\log G(s,t) = -\lambda(1-s)t$$

$$G(s,t) = ke^{-\lambda(1-s)t}$$

Since

$$G(s,0) = 1, \text{ implies } k = 1 \text{ and therefore}, G(s,t) = e^{-\lambda(1-s)t}$$

Here, $P_n(t)$ is the coefficient of s^k in $G(s,t)$ and in this case is

$$P_n(t) = e^{-\lambda t} \frac{(\lambda t)^n}{n!}$$

which is a Poisson distribution with parameter λt. The model presented for $P_n(t)$ $(t > 0)$ above is an example of a stochastic process with a continuous time parameter t. This particular stochastic process is called the *Poisson process.*

Example 5.20 Suppose that the emergency room at a major medical center can handle no more than N patients per day. Once the emergency room has fulfilled its capacity for the day of seeing N patients, all arriving patients are triaged to other area hospitals. The incoming traffic is of the Poisson type with intensity λ but only as long as the capacity for the day is not exceeded. We need to determine the appropriate differential equations for $P_n(t)$ which represents the probability that exactly n patients have been through the system in time t
For $n = 0$ we have

$$P_0(t + \Delta t) = P_0(t)(1 - \lambda \Delta t)$$

or $\quad \lim_{\Delta t \to 0} \dfrac{P_0(t + \Delta t) - P_0(t)}{\Delta t} = -\lambda P_0(t)$ (5.19)

or $\qquad\qquad P_0'(t) = -\lambda P_0(t).$

For $n = N$ we obtain

$$P_N(t + \Delta t) = P_N(t) + P_{N-1}(t)\lambda \Delta t$$

or

$$P_N'(t) = \lambda P_{N-1}(t)$$ (5.20)

And for $1 \leq n < N$,

$$P_n'(t) = -\lambda P_n(t) + \lambda P_{n-1}(t)$$ (5.21)

The solution to (5.19) is easily obtained as

$$P_0'(t)/P_0(t) = -\lambda$$

$$P_0(t) = e^{-\lambda t}$$

Using the above result, equations (5.20) and (5.21) may be solved iteratively as non homogeneous differential equations.   .

Example 5.21 Customers arrive at a parking lot according to a Poisson process-distribution. Assume that the arrival rate is 20 per unit time and that at time $t = 0$ the parking lot was empty. (Assume the parking lot has infinite capacity and none of the cars leaves the parking lot). Here $\lambda = 20$

(a) To determine the average number of cars in the parking lot at $t{=}20$, we proceed as follows

$$E(\text{number of cars in 20 units of time}) = \sum_{n=0}^{\infty} nP_n(20) = \sum_{n=0}^{\infty} ne^{-20(20)} \frac{[20(20)]^n}{n!}$$

$$= e^{-20(20)}[20(20)]e^{20(20)} = 400$$

(b) To determine the probability that there are 20 cars in the parking lot at $t = 30$ given that there were 10 cars at $t = 15$. Since cars arrive at the rate of 20 per unit time it also means that on an average one car arrives every $\left(\dfrac{1}{20}\right)^{th}$ of a unit of time. This is the *average inter-arrival time*. Since the inter arrival distribution is exponential with its memoryless property, we need to determine the probability of 10 cars in $t{=}15$ units of time. Hence

$$P_{10}(15) = e^{-20(15)} \frac{[(20)(15)]^{10}}{10!}$$

∎

Gamma Distribution

A random variable X has a *gamma distribution* with parameters α, β and γ if its density function is of the form

$$f_X(x) = (x-\gamma)^{\alpha-1} e^{-\beta(x-\gamma)} \frac{\beta^{\alpha}}{\Gamma(\alpha)}, \quad \alpha, \beta > 0; x > \gamma$$

In the above density function, γ is the location parameter, β the scale parameter and α the shape parameter. The above distribution is similar to Type III of Pearson's system of distributions (Kendall 1952).

We will in our discussion of the gamma distribution assume that $\gamma = 0$ and that the density function is of the form

$$f_X(x) = \frac{\beta^{\alpha}}{\Gamma(\alpha)} x^{\alpha-1} e^{-\beta x}, \quad \alpha, \beta > 0; 0 < x < \infty \qquad (5.22)$$

For α an integer, (5.22) is the density function of the sum of α exponential variables each with parameter β (see section 5.5.1). The standard form of this distribution is obtained by setting $\beta = 1$. In queuing theory terminology the gamma distribution is referred to as the Erlang distribution because it was first used by Erlang in 1909 in connection with the congestion problem experienced by the Copenhagen Telephone Company. Note that for $\alpha = 1 = \beta$ the distribution in (5.22) is the exponential distribution discussed in section 5.5. In this book we will refer to a random variable that follows gamma distribution with parameters α and β as a *gamma*(α, β) variable.

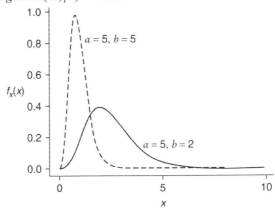

Figure 5.12 Graph of two gamma distributions

In section 5.5 we indicated that the exponential distribution is the continuous analog of the geometric distribution. Similarly, the gamma distribution (which is the sum of exponential random variables) is the continuous analog of the negative binomial distribution which is the sum of geometric random variables. The cumulative density function for a *gamma* (α, β) is

$$F_X(x) = \frac{\beta^\alpha}{\Gamma(\alpha)} \int_0^x t^{\alpha-1} e^{-\beta t} dt \quad \beta > 0, \alpha > 0 \tag{5.23}$$

If α is a positive integer, integration by parts can be used to reveal

$$F_X(x) = 1 - \left[1 + \beta x + \frac{1}{2!}(\beta x)^2 + \dots + \frac{1}{(\alpha-1)!}(\beta x)^{\alpha-1} \right] e^{-\beta x}, \quad x > 0$$

$$= 0 \text{ for } x \le 0$$

$F_X(x)$ can be easily evaluated using a computer program. For example for $\alpha = 50$ and $x = 10$, $F_X(10)$ for $\beta = 1, 2, 3,$ and 4 is 1, 0.96567, 0.20836 and 0.00256 respectively.

The cumulative density function of the gamma variable when α is not a positive integer is called the *incomplete gamma function* and tables for this function have been developed and are easily available. If α is not a positive integer the right hand side of (5.23) will have to be evaluated numerically.

5.7.1 *Moment Generating Function of the Gamma Distribution*
The moment generating function of the gamma distribution is obtained as follows

$$M_X(t) = \int_0^\infty \frac{\beta^\alpha}{\Gamma(\alpha)} x^{\alpha-1} e^{-x(\beta-t)} dx$$

$$= \beta^\alpha \int_0^\infty \frac{1}{\Gamma(\alpha)} x^{\alpha-1} e^{-x(\beta-t)} dx$$

$$= \left(\frac{\beta}{\beta-t} \right)^\alpha, t < \beta$$

which is the mgf of α exponential variables each with parameter β. The mean and variance of X is obtained as

$$E(X) = \frac{\alpha}{\beta}$$

$$Var(x) = \frac{\alpha}{\beta^2}$$

Example 5.22 Let $X_{\alpha,\beta}$ have the gamma distribution with parameters α and β as defined in (5.22). Let us denote the probability density function of $X_{\alpha,\beta}$ by $f_{X_{\alpha,\beta}}(x)$. It can be demonstrated that

$$P(l < X_{\alpha,\beta} < m) = \beta^{-1}[f_{X_{\alpha,\beta}}(l) - f_{X_{\alpha,\beta}}(m)] + P[l < X_{\alpha-1,\beta} < m]$$

$$P[l < X_{\alpha,\beta} < m] = \frac{\beta^\alpha}{\Gamma(\alpha)} \int_l^m x^{\alpha-1} e^{-x\beta} dx$$

$$= \frac{\beta^\alpha}{\Gamma(\alpha)} \left[-\frac{1}{\beta} m^{\alpha-1} e^{-m\beta} + \frac{1}{\beta} l^{\alpha-1} e^{-l\beta} + \frac{1}{\beta}(\alpha-1) \int_l^m x^{\alpha-2} e^{-x\beta} dx \right]$$

$$= \beta^{-1}[f_{X_{\alpha,\beta}}(l) - f_{X_{\alpha,\beta}}(m)] + P[l < X_{\alpha-1,\beta} < m]$$

Example 5.23 If X has a gamma probability density function as given in (5.22), then we are going to show that for a differentiable function $k(X)$ the following relationship holds

$$E\left[k(X)\left(X - \frac{\alpha}{\beta} \right) \right] = \beta^{-1} E[Xk'(X)] \qquad (5.24)$$

In (5.24), $k'(X)$ is the differential of $k(X)$ with respect to X and $E[Xk'(X)] < \infty$
The left hand side in (5.24) may be written as

$$E\left[k(X)\left(X-\frac{\alpha}{\beta}\right)\right]=\frac{\beta^{\alpha}}{\Gamma(\alpha)}\int_{0}^{\infty}k(x)(x-\alpha\beta^{-1})x^{\alpha-1}e^{-\beta x}dx$$

$$=\frac{\beta^{\alpha}}{\Gamma(\alpha)}\left[\int_{0}^{\infty}k(x)x^{\alpha}e^{-\beta x}dx-\alpha\beta^{-1}\int_{0}^{\infty}k(x))x^{\alpha-1}e^{-\beta x}dx\right]$$

$$=\frac{\beta^{\alpha}}{\Gamma(\alpha)}\left[\left\{-k(x)\beta^{-1}x^{\alpha}e^{-\beta x}\Big|_{0}^{\infty}\right\}\right]$$

$$+\frac{\beta^{\alpha}}{\Gamma(\alpha)}\beta^{-1}\int_{0}^{\infty}[k'(x)x^{\alpha}+\alpha k(x)x^{\alpha-1}]e^{-\beta x}dx$$

$$-\alpha\beta^{-1}\frac{\beta^{\alpha}}{\Gamma(\alpha)}\int_{0}^{\infty}k(x))x^{\alpha-1}e^{-\beta x}dx=\beta^{-1}E[Xk'(X)]$$

5.8 Beta Distribution

A random variable X is said to have a *beta distribution* if its density function is given by

$$f_X(x:\alpha,\beta)=\frac{\Gamma(\alpha+\beta)}{\Gamma(\alpha)\Gamma(\beta)}x^{\alpha-1}(1-x)^{\beta-1},\ 0<x<1 \tag{5.25}$$

$$=0\text{ elsewhere, }\alpha,\beta>0$$

.

The function (5.25)) represents a two parameter family of distributions and is often represented by *beta* (α,β). For $\alpha=\beta=1$ the distribution becomes the uniform distribution over the unit interval. The parameters α and β must both be positive real numbers. This distribution is also known as Pearson's Type I distribution (Kendall 1952).

We will first show that (5.25) is a proper density function by showing that

$$\int_{0}^{\infty}f_X(x:\alpha,\beta)dx=1$$

$$\int_{0}^{1}f_X(x:\alpha,\beta)dx=\frac{\Gamma(\alpha+\beta)}{\Gamma(\alpha)\Gamma(\beta)}\int_{0}^{1}x^{\alpha-1}(1-x)^{\beta-1}dx \tag{5.26}$$

Integrating (5.26) above we have

$$\int_0^1 x^{\alpha-1}(1-x)^{\beta-1}\,dx = \frac{(\beta-1)}{\alpha}\int_0^1 x^{\alpha}(1-x)^{\beta-2}\,dx$$

$$= \frac{(\beta-1)(\beta-2)}{\alpha(\alpha+1)}\int_0^1 x^{\alpha+1}(1-x)^{\beta-3}\,dx$$

$$= \frac{(\beta-1)(\beta-2)...3.2.1}{\alpha(\alpha+1)...(\alpha+\beta-1)} = \frac{\Gamma(\alpha)\Gamma(\beta)}{\Gamma(\alpha+\beta)}.$$

Hence, (5.25) is a proper probability distribution.

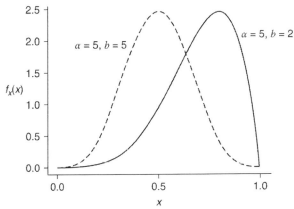

Figure 5.13 A graph of the beta distribution for two sets of parameter values.

The cumulative distribution for a *beta*(α, β) distribution is defined as

$$F_X(x) = \frac{\Gamma(\alpha+\beta)}{\Gamma(\alpha)\Gamma(\beta)}\int_0^x t^{\alpha-1}(1-t)^{\beta-1}\,dt \quad 0 < x < 1$$

$$= 1 \qquad\qquad\qquad\qquad x \geq 1$$

$$= 0 \qquad\qquad\qquad\qquad x \leq 0$$

The function defined above is called the *incomplete beta function*.
Suppose now we set $\alpha = \beta = \alpha$ in (5.26). Then

$$f_X(x:\alpha,\alpha) = \frac{\Gamma(\alpha+\alpha)}{\Gamma(\alpha)\Gamma(\alpha)} x^{\alpha-1}(1-x)^{\alpha-1}, 0 < x < 1$$

Notice that in this case $f_X(x)$ attains its maximum value at $x = 1/2$ and the curve is symmetric about that point and

$$\frac{[\Gamma(\alpha)]^2}{\Gamma(2\alpha)} = \int_0^1 x^{\alpha-1}(1-x)^{\alpha-1}\,dx = 2\int_0^{1/2}(x-x^2)^{\alpha-1}\,dx$$

Let $y = 4(x-x^2)$ for $0 < x < \dfrac{1}{2}$ or $x = \dfrac{1-\sqrt{1-y}}{2}$

We find using the technique of change of variables in the above integral

$$\frac{[\Gamma(\alpha)]^2}{\Gamma(2\alpha)} = 2^{1-2\alpha}\int_0^1 y^{\alpha-1}(1-y)^{-1/2}\,dy = 2^{1-2\alpha}\frac{\Gamma(\alpha)\Gamma\left(\dfrac{1}{2}\right)}{\Gamma\left(\alpha+\dfrac{1}{2}\right)}$$

Substituting $\sqrt{\pi}$ for $\Gamma\left(\dfrac{1}{2}\right)$ we have

$$\Gamma(2\alpha) = \frac{2^{2\alpha-1}}{\sqrt{\pi}}\Gamma(\alpha)\Gamma\left(\alpha+\frac{1}{2}\right) \tag{5.27}$$

The relationship (5.27) is called *Legendre's duplication formula for gamma functions* and the above result is due to Wald (1962). Stirling's formula may also be obtained by letting $\alpha \to \infty$.

5.8.1 Moments of the Beta Distribution

The moment generating function of *beta* (α, β) is (Casella and Berger 2002)

$$M_X(t) = 1 + \sum_{k=1}^{\infty}\left(\prod_{r=0}^{k-1}\frac{\alpha+r}{\alpha+\beta+r}\right)\frac{t^k}{k!}$$

Even though the moment generating function for this distribution is rather complex, determination of its moments is rather straightforward.

$$E(X^r) = \frac{\Gamma(\alpha+\beta)}{\Gamma(\alpha)\Gamma(\beta)}\int_0^1 x^{\alpha+r-1}(1-x)^{\beta-1},\ 0 < x < 1$$

$$= \frac{\Gamma(\alpha+\beta)\Gamma(\alpha+r)}{\Gamma(\alpha)\Gamma(\alpha+\beta+r)}$$

For $r = 1$ and $r = 2$ and using the properties of the gamma function presented in Chapter 1 section 1.2.3 we have

$$E(X) = \frac{\alpha}{\alpha+\beta} \text{ and } E(X^2) = \frac{\alpha(\alpha+1)}{(\alpha+\beta)(\alpha+\beta+1)}$$

$$Var(X) = \frac{\alpha\beta}{(\alpha+\beta)^2(\alpha+\beta+1)}$$

5.9 Chi-square Distribution

Consider the following *gamma* $\left(\dfrac{n}{2},\dfrac{1}{2}\right)$ distribution

$$f_X(x) = \frac{1}{2^{\frac{n}{2}}\Gamma\left(\dfrac{n}{2}\right)} x^{\frac{n}{2}-1} e^{-\frac{x}{2}}, \ 0 < x < \infty \tag{5.28}$$

If n is an even number greater than zero the above distribution is called the chi-square distribution with n degrees of freedom and is written as χ_n^2. Furthermore, it is easily seen that if X is a $N(0,1)$ variable, then X^2 is a chi-square variable with one degree of freedom it is also a *gamma*$(1/2,1/2)$ variable. Also the sum of squares of n independent $N(0,1)$ variables has a chi-square (see section 5.7).This distribution is particularly useful in studying goodness of fit between the observed data and the values obtained from the statistical model. It is also used in studying association between two categorical variables in the analysis of contingency tables. The chi-square distribution plays an important role in statistical inference especially since the assumption of normality is often made regarding the distribution of the variable of interest.

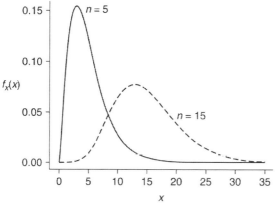

Figure 5.14 Chi-square distributions based on 5 and 15 degrees of freedom.

5.9.1 The Moment Generating Function of a χ_n^2 Variable

The mgf of (5.28) can be obtained from the mgf of a *gamma* $\left(\dfrac{n}{2},\dfrac{1}{2}\right)$ variable as

$$M_X(t) = \frac{1}{(1-2t)^{\frac{n}{2}}}, \quad t < \frac{1}{2}$$

Hence, $E(X) = n$ and $Var(X) = 2n$

Using the moment generating function it can be easily shown that the sum of n independent χ_1^2 variables is a χ_n^2 variable.

Consider now a set $X_1,\ldots,X_n$ of n independent variables, where X_i is a $N(\mu_i,\sigma_i^2)$ variable. Define $Z_i = \dfrac{X_i - \mu_i}{\sigma_i}$

Then, $Z_1,\ldots,Z_n$ is a set of $N(0,1)$ variables and each of the Z_i^2 ($i=1,2,..,n$) is a χ_1^2 variable. Hence the sum of squares of n $N(0,1)$ variable is a χ_n^2 variable.

Example 5.26 Let X and Y be each a χ_1^2 variable. We will obtain the distribution of $V = \sqrt{2(X+Y)}$

We know that $(X+Y)$ has a χ_2^2 distribution. Let $U = (X+Y)$. Then,

$V = \sqrt{2U}$

$$P(V \le v) = P\left(\sqrt{2U} \le v\right) = P\left(U \le \frac{v^2}{2}\right)$$

$$= F_{\chi_2^2}\left(\frac{v^2}{2}\right)$$

or

$$f_V(v) = v f_{\chi_2^2}\left(\frac{v^2}{2}\right) = \frac{1}{2} v e^{-\frac{v^2}{4}}, \quad 0 < v < \infty$$

Example 5.27 Let X be a random variable with pdf.

$$f_X(x) = \frac{1}{\Gamma\left(\dfrac{n-1}{2}\right) 2^{\frac{n-1}{2}}} x^{\frac{n-1}{2}-1} e^{-\frac{x}{2}} \quad 0 < x < \infty.$$

To determine the probability density function of $Y = \dfrac{K}{X}$ where K is a positive constant we proceed as follows

$$P(Y \le y) = F_Y(y) = P\left(\frac{K}{X} \le y\right)$$

$$= 1 - P\left(X \le \frac{K}{y}\right)$$

$$f_Y(y) = \frac{1}{\Gamma\left(\dfrac{n-1}{2}\right)2^{\frac{n-1}{2}}} e^{-\frac{K}{2y}} K^{\frac{n-1}{2}} \frac{1}{y^{\frac{n-1}{2}+1}}$$

$$= \left(\frac{K}{2}\right)^{\frac{n-1}{2}} \frac{1}{\Gamma\left(\dfrac{n-1}{2}\right)} e^{-\frac{K}{2y}} y^{-\left(\frac{n-1}{2}\right)-1} \qquad 0 < y < \infty.$$

The distribution of Y above is known as the *inverse gamma distribution* with parameters $\left(\dfrac{n-1}{2}, \dfrac{K}{2}\right)$

5.9.2 Distribution of $\dfrac{(n-1)S^2}{\sigma^2}$

If X_i's are iid are $N(\mu, \sigma^2)$ variables then the quantity $\sum_1^n \left(\dfrac{X_i - \mu}{\sigma}\right)^2$ is the sum
of squares of n independent $N(0,1)$ variables. Each of the terms in the sum is there-
fore a χ_1^2 variable. Hence $\sum_1^n \left(\dfrac{X_1 - \mu}{\sigma}\right)^2$ is a χ_n^2 variable.

We can write

$$\sum_1^n \left(\frac{X_i - \mu}{\sigma}\right)^2 = \frac{1}{\sigma^2}\sum_1^n (X_i - \mu + \bar{X} - \bar{X})^2 \tag{5.29}$$

$$= \frac{1}{\sigma^2}\sum_1^n (X_i - \bar{X})^2 + n\frac{(\bar{X} - \mu)^2}{\sigma^2}$$

The term on the left hand side of (5.29) is a χ_n^2 variable and the second term on
the extreme right hand side is a χ_1^2 variable being the square of a standard normal
variable. Let us now write

$W = \sum_1^n \left(\dfrac{X_i - \mu}{\sigma}\right)^2$, $Y = n\dfrac{(\bar{X} - \mu)^2}{\sigma^2}$ and $Z = \dfrac{1}{\sigma^2}\sum_1^n (X_i - \bar{X})^2$

Then, $W = Y + Z$

The mgf of W is $\dfrac{1}{(1-2t)^n}$, the mgf of Y is $\dfrac{1}{1-2t}$

Since mgf of W is the product of the mgf's of Y and Z, the mgf of Z is

$$M_W(t) = M_Y(t) \times M_Z(t)$$

$$M_Z(t) = \frac{M_W(t)}{M_Y(t)} = \left(\frac{1}{1-2t}\right)^{n-1}$$

The above result holds only when the X_i's are independent (as shown on Page 243) of one another.

We know that $\dfrac{1}{\sigma^2}\displaystyle\sum_1^n (X_i - \bar{X})^2 = \dfrac{(n-1)S^2}{\sigma^2}$

Hence we have shown that the variable $\dfrac{(n-1)S^2}{\sigma^2}$ has a χ^2_{n-1} distribution.

5.10 Student's *t* Distribution

Consider two independent variables Z and Y which are $N(0,1)$ and a χ^2_r variables respectively. Define

$$T = \frac{Z}{\sqrt{Y/r}}$$

The joint distribution of Z and Y is

$$f_{Z,Y}(z,y) = \frac{1}{\sqrt{2\pi}} e^{-\frac{1}{2}z^2} \frac{1}{2^{\frac{r}{2}}\Gamma\left(\frac{r}{2}\right)} y^{\frac{r}{2}-1} e^{-\frac{y}{2}}, \quad -\infty < z < \infty, 0 < y < \infty$$

We are now going to derive the pdf of T by using the following transformation

$t = \dfrac{z}{\sqrt{\dfrac{y}{r}}}$ and $v = y$ or $z = t\sqrt{\dfrac{y}{r}}$ and $y = v$. Then the Jacobian of the trans-

formation is $\sqrt{\dfrac{v}{r}}$

Then the joint pdf of T and V is

$$f_{T,V}(t,v) = \frac{1}{\sqrt{2\pi}\Gamma\left(\frac{r}{2}\right)2^{\frac{r}{2}}} \sqrt{\frac{v}{r}} v^{\frac{r}{2}-1} e^{\left[-\frac{v}{2}\left(1+\frac{t^2}{r}\right)\right]}, \quad -\infty < t < \infty, 0 < v < \infty$$

To get the marginal pdf of T we integrate the above expression for all values of

v. Hence, $\quad f_T(t) = \dfrac{1}{\sqrt{2\pi r}\,\Gamma\left(\dfrac{r}{2}\right)2^{\frac{r}{2}}} \displaystyle\int_0^\infty v^{\frac{r+1}{2}-1} e^{\left[-\frac{v}{2}\left(1+\frac{t^2}{r}\right)\right]} dv.$

Set $\qquad\qquad\qquad u = v\,\dfrac{1}{2}\left[1+\dfrac{t^2}{r}\right]$

After a little simplification we have

$$f_T(t) = \frac{\Gamma\left(\dfrac{r+1}{2}\right)}{\sqrt{\pi r}\,\Gamma\left(\dfrac{r}{2}\right)\left(1+\dfrac{t^2}{r}\right)^{\frac{r+1}{2}}}, \quad -\infty < t < \infty$$

This is the t distribution with r degrees of freedom and is donated by t_r.

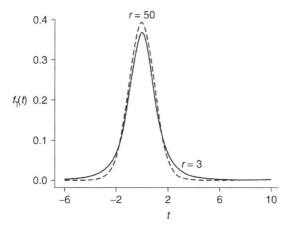

Figure 5.15 Graph of t distribution for 3 and 50 degrees of freedom

The shape of the t distribution is bell shaped, symmetric about zero, more spread out than the normal curve hence it is a flatter form of the normal curve. For $r=1$ it is the probability density function of the Cauchy distribution (see section 5.12). We will now show that for large r, the t distribution approximates the normal distribution. This distribution also known as the '*Student' t distribution* after the nom de plume of its discoverer Gossett in 1908. It was later modified by R.A. Fisher and is also sometimes called *Fisher's t distribution*.
Note that

$$\lim_{r \to \infty} f_T(t) = \lim_{x \to \infty} \frac{\Gamma\left(\dfrac{r+1}{2}\right)}{\sqrt{\pi r}\,\Gamma\left(\dfrac{r}{2}\right)\left(1+\dfrac{t^2}{r}\right)^{\frac{r+1}{2}}} = \lim_{r \to \infty} \frac{\Gamma\left(\dfrac{r+1}{2}\right)}{\sqrt{\pi r}\,\Gamma\left(\dfrac{r}{2}\right)} \lim_{r \to \infty}\left(1+\dfrac{t^2}{r}\right)^{-\frac{r+1}{2}}$$

$$= e^{-\frac{t^2}{2}} \lim_{r \to \infty} \frac{\Gamma\left(\dfrac{r+1}{2}\right)}{\sqrt{\pi r}\,\Gamma\left(\dfrac{r}{2}\right)}$$

Using Stirling's formula and taking the limits as $r \to \infty$

$$\lim_{r \to \infty} \frac{\Gamma\left(\dfrac{r+1}{2}\right)}{\sqrt{\pi r}\,\Gamma\left(\dfrac{r}{2}\right)} = \lim_{r \to \infty} \frac{\sqrt{2\pi}\,e^{-\frac{r+1}{2}}\left(\dfrac{r+1}{2}\right)^{\frac{r+1}{2}+\frac{1}{2}}}{\sqrt{2\pi}\,e^{-\frac{r}{2}}\left(\dfrac{r}{2}\right)^{\frac{r}{2}+\frac{1}{2}}\sqrt{\pi r}} = \lim_{r \to \infty} \frac{1}{\sqrt{\pi r}}\,e^{-\frac{1}{2}} \frac{\left(\dfrac{r+1}{2}\right)^{\frac{r}{2}+1}}{\left(\dfrac{r}{2}\right)^{\frac{r}{2}+\frac{1}{2}}}$$

The terms on the right hand side expression can be further simplified by writing them as

$$\frac{1}{\sqrt{\pi r}}\,e^{-\frac{1}{2}} \frac{\left(\dfrac{1}{2}\right)^{\frac{r}{2}+1}(r+1)^{\frac{r}{2}+1}}{\left(\dfrac{1}{2}\right)^{\frac{r}{2}+\frac{1}{2}}r^{\frac{r}{2}+\frac{1}{2}}} = \frac{1}{\sqrt{\pi}}\,e^{-\frac{1}{2}}\frac{1}{\sqrt{2}}\left(\dfrac{r+1}{r}\right)^{\frac{r}{2}+1}$$

Taking limits as r approaches infinity we have

$$= \lim_{r \to \infty} \frac{1}{\sqrt{\pi}}\,e^{-\frac{1}{2}}\frac{1}{\sqrt{2}}\left(1+\dfrac{1}{r}\right)^{\frac{r}{2}}\left(1+\dfrac{1}{r}\right)^{1} = \frac{1}{\sqrt{2\pi}}$$

Hence,

$$\lim_{r \to \infty} f_T(t) = \frac{1}{\sqrt{2\pi}}\,e^{-\frac{1}{2}t^2}$$

which is a $N(0,1)$ density function.

5.10.1 Moments of the t Distribution
We may identify the moments of the t distribution as follows

$$E(T) = \frac{\Gamma\left(\dfrac{r+1}{2}\right)}{\sqrt{\pi r}\,\Gamma\left(\dfrac{r}{2}\right)} \int_{-\infty}^{\infty} \frac{t}{\left(1+\dfrac{t^2}{r}\right)^{\frac{r+1}{2}}}\,dt = 0$$

The above integral is zero since the integrand is an odd function of t.

Obtaining higher moments of the t distribution is beyond the scope of the book. The following result is derived in Johnson and Katz (1970). All odd moments of the t distribution about zero are zero. If m is even, the mth central moment is

$$E(T^m) = r^{\frac{m}{2}} \frac{\Gamma\left(\frac{m+1}{2}\right)\Gamma\left(\frac{r-m}{2}\right)}{\Gamma\left(\frac{1}{2}\right)\Gamma\left(\frac{r}{2}\right)}$$

$$= r^{\frac{m}{2}} \frac{1.3...(m-1)}{(r-m)(r-m+2)...(r-2)}$$

For $m \geq r$, the mth moment is infinity

An alternative way of determining the mean and variance of the t distribution which will require the use of the moments of the normal and the chi square distributions is as follows

$$T = \frac{Z}{\sqrt{Y/r}}, \ E(T) = E(\sqrt{r}Z)E\left(\frac{1}{\sqrt{Y}}\right) = 0$$

Now T^2 is the ratio of the square of a $N(0,1)$ variable which is a χ_1^2 variable and a χ_r^2 variable divided by its degrees of freedom. Hence T^2 has a $F_{1,r}$ distribution (see section 5.11 below). The required variance is the expected value of a $F_{1,r}$ variable which is $\frac{r}{r-2}$, $r > 2$

5.10.2 Relationship with Sample Variance
We will prove that

$$T^* = \frac{(\bar{X} - \mu)\sqrt{n}}{S}$$

follows a t_{n-1} distribution, where

$$S^2 = \frac{1}{n-1}\sum_{i=1}^{n}(X_i - \bar{X})^2 .$$

We will first divide the numerator and the denominator by σ and obtain

$$T^* = \left(\frac{(\bar{X} - \mu)\sqrt{n}}{\sigma}\right)\frac{1}{\sqrt{\frac{1}{(n-1)\sigma^2}\sum_1^n (X_i - \bar{X})^2}} = \left(\frac{(\bar{X} - \mu)\sqrt{n}}{\sigma}\right)\frac{1}{\sqrt{\frac{(n-1)S^2}{\sigma^2}}/(n-1)}$$

The first term in the above expression is a $N(0,1)$ variable and the second (in the denominator) is a χ^2_{n-1} variable divided by its degrees of freedom. Hence the ratio is a t_{n-1} variable (Note that in section 5.16.5 it will be shown that $\bar{X}$ and S^2 are independent. Hence the numerator and the denominator in the expression for T^* are independent of one another).

5.11 *F* Distribution

If X and Y are independent χ^2_k and χ^2_m random variables then the F distribution or Snedecor's F is defined as the ratio

$$F = \frac{X/k}{Y/m}$$

The joint probability density function of X and Y is given by

$$f_{X,Y}(x,y) = \frac{1}{2^{\frac{k+m}{2}}\Gamma\left(\frac{k}{2}\right)\Gamma\left(\frac{m}{2}\right)}x^{\frac{k}{2}-1}y^{\frac{m}{2}-1}e^{-\frac{1}{2}(x+y)} \quad 0 < x, y < \infty$$

Now let

$$U = \frac{mX}{kY} \text{ and } V = Y \text{ then } X = \frac{kUV}{m}, \quad Y = V \quad \text{the Jacobian of the transfor-}$$

mation is $\frac{kv}{m}$.

Hence

$$f_{U,V}(u,v) = \frac{u^{\frac{k}{2}-1}k^{\frac{k}{2}-1}}{2^{\frac{k+m}{2}}\Gamma\left(\frac{k}{2}\right)\Gamma\left(\frac{m}{2}\right)m^{\frac{k}{2}-1}}v^{\frac{k+m}{2}-2}e^{-\frac{v}{2}\left(\frac{uk}{m}+1\right)}\frac{kv}{m} \quad 0 < u, v < \infty$$

Integrating over all values of v will give us the marginal distribution of U. This can be accomplished by recognizing the integrand as

$$gamma\left[\frac{k+m}{2}, \frac{1}{2}\left(\frac{uk}{m}+1\right)\right].$$

$$f_U(u) = \frac{\Gamma\left(\dfrac{k+m}{2}\right)}{\Gamma\left(\dfrac{k}{2}\right)\Gamma\left(\dfrac{m}{2}\right)}\left(\frac{k}{m}\right)^{\frac{k}{2}}\frac{u^{\frac{k}{2}-1}}{\left(1+\dfrac{uk}{m}\right)^{\frac{k+m}{2}}} \quad 0 < u < \infty$$

This is the F distribution with k and m degrees of freedom and is often written as $F_{k,m}$. Note that $\dfrac{1}{F}$ is the ratio of $\dfrac{Y/m}{X/k}$, and since Y and X are both chi-square variables with m and k degrees of freedom respectively $\dfrac{1}{F}$ will follow $F_{m,k}$ distribution.

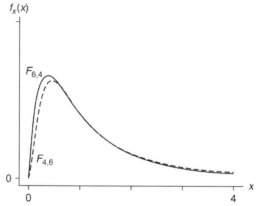

Figure 5.16 Graph of the F distribution for $F_{6,4}$ and $F_{4,6}$

5.11.1 *Moments of the* F *Distribution*

Let U be a $F_{k,l}$ variable. Then we need to obtain $E(U)$ and $Var(U)$. Since

$$U = \frac{X/k}{Y/m}$$

$$E(U) = E\left(\frac{X/k}{Y/m}\right) = E\left(\frac{X}{k}\right)E\left(\frac{m}{Y}\right) = \frac{m}{k}kE\left(\frac{1}{Y}\right)$$

Here we are using the fact that X and Y are independent and that $E(X)=k$

We are now going to obtain $E\left(\dfrac{1}{Y}\right)$ where Y is a χ^2_m variate

$$E\left(\frac{1}{Y}\right) = \int_0^\infty \frac{1}{y} \frac{y^{\frac{m}{2}-1} e^{-\frac{y}{2}}}{\Gamma\left(\frac{m}{2}\right) 2^{\frac{m}{2}}} dy = \frac{1}{\Gamma\left(\frac{m}{2}\right) 2^{\frac{m}{2}}} \int_0^\infty y^{\frac{m-2}{2}-1} e^{-\frac{y}{2}} dy = \frac{1}{m-2}$$

Hence $E(U) = \dfrac{m}{m-2}$

The variance of U may be derived as follows

$$E(U^2) = E\left(\frac{X/k}{Y/m}\right) = \frac{m^2}{k^2} E(X^2) E\left(\frac{1}{Y^2}\right) = \frac{m^2}{k^2}(2k + k^2) E\left(\frac{1}{Y^2}\right)$$

In the above derivation we used the mean and variance of a χ_k^2 variable as k and $2k$. Now $E\left(\dfrac{1}{Y^2}\right)$ may be derived directly from the χ_m^2 distribution and is

$$E\left(\frac{1}{Y^2}\right) = \int_0^\infty \frac{1}{y^2} \frac{y^{\frac{m}{2}-1} e^{-\frac{y}{2}}}{\Gamma\left(\frac{m}{2}\right) 2^{\frac{m}{2}}} dy = \frac{1}{\Gamma\left(\frac{m}{2}\right) 2^{\frac{m}{2}}} \int_0^\infty y^{\frac{m-2}{2}-2} e^{-\frac{y}{2}} dy = \frac{1}{(m-2)(m-4)}$$

Hence $E(U^2) = \dfrac{m^2(k+2)}{k(m-2)(m-4)}$

and $Var(U) = 2\left(\dfrac{m}{m-2}\right)^2 \left(\dfrac{m+k-2}{k(m-4)}\right)$, $m > 4$

5.11.2 Relationship with Sample Variances

Let $X_1, \ldots, X_n$ be a random sample from a $N(\mu_1, \sigma_1^2)$ distribution and $Y_1, \ldots, Y_m$ a random sample from a $N(\mu_2, \sigma_2^2)$ distribution. Then if S_1^2 and S_2^2 are the corresponding sample variances, then

$\dfrac{(n-1)S_1^2}{\sigma_1^2}$ and $\dfrac{(m-1)S_2^2}{\sigma_2^2}$ are distributed as χ_{n-1}^2 and χ_{m-1}^2 variables. And $\dfrac{S_1^2/\sigma_1^2}{S_2^2/\sigma_2^2}$ is the ratio of two chi square variables each divided by its degrees of freedom. Therefore $\dfrac{S_1^2/\sigma_1^2}{S_2^2/\sigma_2^2}$ follows an F distribution with $n-1$ and $m-1$ degrees of freedom.

Example 5.28 Let X be a random variable following the F distribution with k_1 and k_2 degrees of freedom. We will determine the pdf of

$$Y = \cfrac{1}{\left(1 + \cfrac{k_1}{k_2} X\right)}$$

The density function of X is

$$f_X(x) = \frac{\Gamma\left(\dfrac{k_1 + k_2}{2}\right)\left(\dfrac{k_1}{k_2}\right)^{\frac{k_1}{2}} x^{\frac{k_1}{2}-1}}{\Gamma\left(\dfrac{k_1}{2}\right)\Gamma\left(\dfrac{k_2}{2}\right)\left[1 + \dfrac{k_1}{k_2} x\right]^{\frac{k_1 + k_2}{2}}}$$

Let $\dfrac{k_1}{k_2} = r$. Then,

$$P(Y \le y) = F_Y(y) = P\left(\frac{1}{1+rX} \le y\right) = 1 - F_X\left(\frac{1-y}{ry}\right)$$

and after a little simplification

$$f_Y(y) = f_X\left(\frac{1-y}{ry}\right)\frac{1}{ry^2} = \frac{\Gamma\left(\dfrac{k_1 + k_2}{2}\right)}{\Gamma\left(\dfrac{k_1}{2}\right)\Gamma\left(\dfrac{k_2}{2}\right)} y^{\frac{k_2}{2}-1}(1-y)^{\frac{k_1}{2}-1}, \quad 0 < y < 1$$

The above is a $beta\left(\dfrac{k_2}{2}, \dfrac{k_1}{2}\right)$ distribution.

5.12 Cauchy Distribution

A random variable X is said to follow a Cauchy distribution with parameter θ if its probability density function is of the form

$$f_X(x) = \frac{1}{\pi[1+(x-\theta)^2]}, \quad -\infty < x < \infty$$

The most commonly used form of the Cauchy distribution is when $\theta = 0$. That is the form

$$f_X(x) = \frac{1}{\pi[1+x^2]}, \quad -\infty < x < \infty$$

Also

$$\int_{-\infty}^{\infty} \frac{dx}{\pi[1+(x-\theta)^2]} = \frac{1}{\pi}\tan^{-1}[(x-\theta)]\Big|_{-\infty}^{\infty} = 1$$

Hence, $f_X(x)$ is a proper probability density function. To obtain the expected value of X we proceed as follows

$$E(X) = \int_{-\infty}^{\infty} \frac{x\,dx}{\pi[1+(x-\theta)^2]} = \int_{-\infty}^{\infty} \frac{2(x-\theta+\theta)\,dx}{2\pi[1+(x-\theta)^2]}$$

$$= \int_{-\infty}^{\infty} \frac{2(x-\theta)\,dx}{2\pi[1+(x-\theta)^2]} + \theta \int_{-\infty}^{\infty} \frac{dx}{\pi[1+(x-\theta)^2]}$$

$$-\frac{1}{2\pi}\log[1+(x-\theta)^2]\Big|_{-\infty}^{\infty} + \theta$$

$$= \lim_{x\to\infty}\log[1+(x-\theta)^2] - \lim_{x\to-\infty}\log[1+(x-\theta)^2] + \theta.$$

Neither of the two limits in the above expression exist therefore, we can say that $E(X)$ does not exist.

5.12.1 *Sum of Cauchy Variables*
Define two independent random variables X and W, both of which are identically distributed as Cauchy variables such that

$$f_X(x) = \frac{1}{\pi(1+x^2)} \quad -\infty < x < \infty \text{ and } f_W(w) = \frac{1}{\pi(1+w^2)} \quad -\infty < w < \infty$$

Let $Z = X + W$.
We are going to obtain the probability density function of Z.
Using the convolution theorem discussed in Chapter 3, section 3.7 the probability density function of Z is

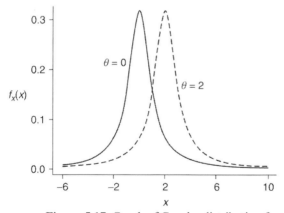

Figure 5.17 Graph of Cauchy distribution for $\theta = 0$ and $\theta = 2$

$$f_Z(z) = \int_{-\infty}^{\infty} f_X(x) f_W(z-x)\,dx = \frac{1}{\pi^2}\int_{-\infty}^{\infty} \frac{dx}{(1+x^2)[1+(z-x)^2]}.$$

Write

$$\frac{1}{(1+x^2)[1+(z-x)^2]} = \frac{1}{z(z-2x)(1+x^2)} - \frac{1}{z(z-2x)[1+(z-x)^2]}$$

Then,

$$f_Z(z) = \frac{1}{\pi^2 z} \int_{-\infty}^{\infty} \frac{dx}{(z-2x)(1+x^2)} - \frac{1}{\pi^2 z} \int_{-\infty}^{\infty} \frac{dx}{(z-2x)[1+(z-x)^2]}.$$

Let $y = z - 2x$ or, $x = \frac{1}{2}(z - y)$

Then,

$$\frac{1}{\pi^2 z} \int_{-\infty}^{\infty} \frac{dx}{(z-2x)(1+x^2)} = \frac{2}{\pi^2 z} \int_{-\infty}^{\infty} \frac{dy}{y(4+z^2-2zy+y^2)}.$$

Using the following formula obtained from the Standard Mathematical Tables (1971),

$$\int \frac{dx}{x(a+bx+cx^2)} = \frac{1}{2a} \log \frac{x^2}{a+bx+cx^2}$$

$$-\frac{b}{2a}\left[\frac{2}{\sqrt{4ac-b^2}} \tan^{-1} \frac{2cx+b}{\sqrt{4ac-b^2}}\right]$$

we have,

$$\frac{2}{\pi^2 z} \int_{-\infty}^{\infty} \frac{dy}{y(4+z^2-2zy+y^2)} = \frac{1}{\pi^2(4+z^2)} \tan^{-1} \frac{y-z}{2}\bigg|_{y=-\infty}^{\infty}$$

$$= \frac{1}{\pi(4+z^2)}$$

Similarly

$$-\frac{1}{\pi^2 z} \int_{-\infty}^{\infty} \frac{dx}{(z-2x)[1+(z-x)^2]} = -\frac{2}{\pi^2 z} \int_{-\infty}^{\infty} \frac{dy}{y\left[4+z^2+y^2+2zy\right]}$$

$$= \frac{1}{\pi(4+z^2)}$$

Hence the probability density function of Z is

$$f_Z(z) = \frac{2}{\pi(4+z^2)}$$

This suggests the general form for sum of n Cauchy variables as

$$f_Z(z) = \frac{n}{\pi(n^2+z^2)}$$

The pdf of W the mean of n Cauchy variable will then be

$$f_W(w) = \frac{1}{\pi(1+w^2)}$$

5.13 Exponential Family

A family of univariate probability distributions with k parameters is called an *exponential family* (or *Koopman Darmois family*) if

$$f_X(x|\theta) = \left\{ \exp\left(\sum_{i=1}^{k} c_i(\theta)T_i(x) + b(\theta) + S(x) \right) \right\}$$

for all possible range of values of x. Here $c_i(\theta)$ and $b(\theta)$ are functions of the vector θ only and $T(x)$ and $S(x)$ are functions of the x's only.

We will now show that some of the most commonly encountered distributions belong to this family.

5.13.1 Bernoulli Distribution

The probability mass function for this family with parameter p is

$P(X = x|p) = p^x(1-p)^{1-x}$, $x = 0,1$

which has the form of a member of an exponential family with one parameter

$S(x) = 0 = b(p), c_1(p) = \log(1-p), c_2(p) = \log(1-p), T_1(x) = x$

and $T_2(x) = (1-x)$

5.13.2 Gamma Distribution

$$f_X(x|\alpha, \beta) = \frac{\beta^\alpha}{\Gamma(\alpha)}(x)^{\alpha-1}e^{-\beta x}$$

has the required exponential form with

$b(\alpha, \beta) = \log\dfrac{\beta^\alpha}{\Gamma(\alpha)}, S(x) = 0, c_1 = \alpha - 1, c_2 = -\beta, T_1(x) = \log x$ and $T_2(x) = x$

5.13.3 Normal Distribution

For this case

$$f_X(x|\mu, \sigma^2) = \left(\frac{1}{\sigma\sqrt{2\pi}} \right) e^{-\frac{1}{2\sigma^2}(x-\mu)^2}$$

This is an example of a univariate member of the two parameter exponential family with

$$b(\mu,\sigma^2) = \log\frac{1}{\sigma\sqrt{2\pi}} - \frac{\mu^2}{2\sigma^2}, \; S(x) = 0, \; c_1(\mu,\sigma^2) = -\frac{1}{2\sigma^2}, T_1(x) = x^2$$

$$c_2(\mu,\sigma^2) = \frac{\mu}{\sigma^2} \text{ and } T_2(x) = x$$

The examples presented in this section, the beta, binomial, Poisson and negative binomial also belong to the family of exponential distributions as will be shown in Chapter 8.

5.14 Hierarchical Models-Mixture Distributions

As in Chapter 4 we will now discuss circumstances in which some parameters of a distribution may have their own probability density functions, which in Bayesian terminology are known as prior distributions. Using the prior distribution the adjusted distribution of the variable is obtained. This is then used to obtain its moments after the randomness in the parameters is adjusted for. Sometimes we may take this a step further and assume that the parameter in the prior distribution also has a prior. Hence we are then talking about three stage *hierarchical models*. We begin by considering a two stage hierarchical model. These distributions are also referred to as mixture distributions by some authors.

5.14.1 Beta- Binomial Hierarchy

Let X have a binomial distribution with parameters n and p where p in turn has a beta(α, β) distribution.. Then

$$P(X = r) = \int_0^1 P(X = r|p)f(p)dp = \int_0^1 \binom{n}{r}p^r(1-p)^{n-r}\frac{\Gamma(\alpha+\beta)}{\Gamma(\alpha)\Gamma(\beta)}p^{\alpha-1}(1-p)^{\beta-1}dp$$

$$= \binom{n}{r}\frac{\Gamma(\alpha+\beta)}{\Gamma(\alpha)\Gamma(\beta)}\int_0^1 p^{\alpha+r-1}(1-p)^{n+\beta-r-1}dp$$

$$P(X = r) = \binom{n}{r}\frac{\Gamma(\alpha+\beta)}{\Gamma(\alpha)\Gamma(\beta)}\frac{\Gamma(\alpha+r)\Gamma(n+\beta-r)}{\Gamma(\alpha+n+\beta)} \tag{5.30}$$

To obtain the moments of X we proceed as follows

$$E(X) = E_p[E_X(X|p)] = E_p[np] = nE(p) = \frac{n\alpha}{\alpha+\beta}$$

$$Var(X) = E_p[Var(X|p)] + Var_p[E(X|p]$$

$$= E_p[np(1-p)] + Var_p[np]$$

$$= nE[p(1-p)] + n^2 Var[p]$$

$$= nE(p) - nE(p^2) + n^2 Var(p)$$

$$= n\frac{\alpha}{\alpha + \beta} - n[Var(p) + \{E(p)\}^2] + n^2 Var(p)$$

$$= \frac{n\alpha\beta(n + \alpha + \beta)}{(\alpha + \beta)^2 (1 + \alpha + \beta)}$$

∎

We could have also obtained the mean and variance of X from (5.30) directly as follows

$$\sum_{r=0}^{n} rP(X = r) \text{ and } \sum_{r=0}^{n} [r - E(X)]^2 P(X = r).$$

5.14.2 Binomial-Poisson-Exponential Hierarchy (Three Stage Hierarchical Model)

Let Y represents the number of patients visiting the emergency room with SARS like symptoms in a County Hospital on a particular day and assume that X of these patients will eventually develop SARS. Furthermore, assume that X follows a binomial distribution with parameters Y and p and that Y follows a Poisson distribution with parameter λ where λ itself follows an exponential distribution with parameter β. Then,

$$P(X = x) = \sum_{y=x}^{\infty} \int_{0}^{\infty} \binom{y}{x} p^x (1-p)^{y-x} \frac{e^{-\lambda}}{y!} \lambda^y \beta e^{-\lambda\beta} d\lambda$$

$$= \sum_{y=x}^{\infty} \binom{y}{x} p^x (1-p)^{y-x} \frac{\beta}{y!} \int_{\lambda} e^{-\lambda} \lambda^y e^{-\lambda\beta} d\lambda$$

$$= \sum_{y=x}^{\infty} \binom{y}{x} p^x (1-p)^{y-x} \frac{\beta}{y!} \int_{0}^{\infty} \lambda^y e^{-\lambda(1+\beta)} d\lambda$$

$$= \sum_{y=x}^{\infty} \binom{y}{x} p^x (1-p)^{y-x} \frac{\beta}{(1+\beta)^{y+1} y!} \Gamma(y+1)$$

$$= \sum_{y=x}^{\infty} \binom{y}{x} p^x (1-p)^{y-x} \frac{\beta}{(1+\beta)^{y+1}}$$

$$= \sum_{y=x}^{\infty} \binom{y}{x} p^x (1-p)^{y-x} \frac{\beta}{(1+\beta)(1+\beta)^y}$$

The probability mass function of X can be written as

$$P(X = x) = \frac{\beta}{1+\beta}\left(\frac{p}{1-p}\right)^x \sum_{y=x}^{\infty} \binom{y}{x}\left[\frac{1-p}{1+\beta}\right]^y$$

$$P(X = x) = \frac{\beta}{1+\beta}\left(\frac{p}{1-p}\right)^x \left[\frac{1-p}{1+\beta}\right]^x \sum_{i=0}^{\infty} \binom{x+i}{i}\left[\frac{1-p}{1+\beta}\right]^i$$

Using the following formula due to Polya and Szegb (1925)

$$\sum_{k=0}^{\infty} \binom{a+k}{k} z^k = \frac{1}{(1-z)^{a+1}}, \quad |z| < 1$$

We have

$$P(X = x) = \frac{\beta}{1+\beta}\left(\frac{p}{1-p}\right)^x \left(\frac{1-p}{1+\beta}\right)^x \left(\frac{1}{1-\frac{1-p}{1+\beta}}\right)^{x+1} = \left(\frac{\beta}{\beta+p}\right)\left(\frac{p}{\beta+p}\right)^x$$

The above unconditional probability distribution of X has the form of the geometric distribution (discussed in Chapter 4, section 4.7) and represents the probability of x "failures" before the occurrence of the first "success" with probability of "success" and "failure" as:

$$\left(\frac{\beta}{\beta+p}\right) \text{ and } \left[\frac{p}{\beta+p}\right]$$

Hence the mean and variance of X are $E(X) = \dfrac{p}{\beta}$ and $Var(X) = \dfrac{p(\beta+p)}{\beta^2}$

∎

An alternative approach for determining $E(X)$ and $Var(X)$ is by using the formula for conditional mean and variance derived in Chapter 3 section 3.12 as follows

$$E(X) = E_Y[E_X(X|Y)] = E_Y[Yp] = pE(Y) = \frac{p}{\beta}$$

$$Var(X) = E[\text{var}(X|Y)] + Var[E(X|Y)]$$
$$= E[Yp(1-p)] + Var[Yp]$$

$$= p(1-p)\frac{1}{\beta} + p^2 Var(Y)$$

$$= p(1-p)\frac{1}{\beta} + p^2\frac{1+\beta}{\beta^2} = \frac{p(p+\beta)}{\beta^2}$$

5.15 Other Distributions

5.15.1 Lognormal Distribution

If X has a $N(\mu,\sigma^2)$ distribution and $X = \log Y$ then the distribution of Y is called the lognormal distribution. We will now derive the distribution of Y.

$$P(Y \le y) = F_Y(y) = P(e^X \le y) = P(X \le \log y)$$

$$f_Y(y) = f_X(\log y)\frac{1}{y} = \frac{1}{\sigma\sqrt{2\pi}\,y}e^{-\frac{1}{2\sigma^2}(\log y - \mu)^2} \qquad 0 < y < \infty$$

A more general form of this distribution is obtained if the variable Y is measured from λ

$$f_Y(y) = \frac{1}{\sigma\sqrt{2\pi}(y-\lambda)}e^{-\frac{1}{2\sigma^2}(\log y - \mu)^2} \qquad 0 < \lambda < y < \infty$$

This distribution in some cases provides a good fit to income distribution, weights of adults in a population and machine down times. The distribution is asymmetrical and positively skewed (or skewed to the right, that is, points above the median value tend to be further from the median in absolute value than the points below the median). Sometimes a logarithmic transformation of the data may allow us to treat the data as if it were drawn from a normal distribution.

The moments of this distribution are easy to obtain

$$E(Y) = E\left(e^X\right) = \frac{1}{\sigma\sqrt{2\pi}}\int_{-\infty}^{\infty} e^x e^{-\frac{1}{2\sigma^2}(x-\mu)^2}\,dx$$

$$= \frac{e^{\mu + \frac{\sigma^2}{2}}}{\sigma\sqrt{2\pi}}\int_{-\infty}^{\infty} e^{-\frac{(x-\mu-\sigma^2)^2}{2\sigma^2}}\,dx = e^{\mu + \frac{\sigma^2}{2}}$$

The above result could have also been obtained by noting that since $E(Y) = E(e^X)$, the value of $E(X)$ could be more easily obtained by setting $t = 1$ in the moment generating function of X. Similarly $E(Y^2)$ can be obtained by setting $t = 2$ in the moment generating function of X. Hence

$$E(Y^2) = E\left(e^{2X}\right) = e^{2\sigma^2 + 2\mu}$$

$$Var(Y) = E(Y^2) - [E(Y)]^2 = e^{\sigma^2 + 2\mu}\left(e^{\sigma^2} - 1\right).$$

5.15.2 Multivariate Normal Distribution

Consider variables X_i $(i = 1, 2, ..., n)$ with means μ_i and variances σ_i^2 and let $|\sigma_{ij}|$ represent the determinant of the variance co-variance matrix $\boldsymbol{\sigma}_{ij}$ and let $|\sigma^{ij}|$ represent the determinant of the inverse of σ_{ij}. Then the *multivariate normal distribution* is defined as

$$f_{X_1, ..., X_n}(x_1, ..., x_n) = \frac{\sqrt{|\sigma^{ij}|}}{(2\pi)^{\frac{n}{2}}} \exp\left[-\frac{1}{2}\sum_{i=1}^{n}\sum_{j=1}^{n}\left\{\sigma^{ij}(x_i - \mu_i)(x_j - \mu_j)\right\}\right],$$

$$-\infty < x_1, ..., x_n < \infty$$

In the above expression σ^{ij} is the $(i, j)^{th}$ element of the matrix of the inverse of the variance co-variance matrix (σ_{ij}). The above pdf may also be written in matrix notation as

$$f_X(x) = \frac{\sqrt{|\sigma^{ij}|}}{(2\pi)^{\frac{n}{2}}} \exp\left(-\frac{1}{2}(X - \mu)'(\sigma_{ij})^{-1}(X - \mu)\right)$$

Here the vector $(x-\mu)'$ contains elements $(x_1 - \mu_1, x_2 - \mu_2 ..., x_n - \mu_n)$ and is the transpose of the vector $(x - \mu)$ and $(\sigma_{ij})^{-1}$ is the inverse of the matrix σ_{ij}.

The mgf of the multivariate normal distribution is (Wilks 1962).

$$M_{X_1, ..., X_n}(t_1, t_2, ..., t_n) = E\left(e^{\sum_{1}^{n} t_i X_i}\right) = \exp\left[\sum_{1}^{n} t_i \mu_i + \frac{1}{2}\sum_{1}^{n}\sum_{1}^{n} t_i t_j \sigma_{ij}\right]$$

5.15.3 Weibull Distribution*

A variable X is said to have a *Weibull distribution* if its probability density is of the following form

$$f_X(x) = e^{-\left(\frac{x}{\alpha}\right)^{\beta}} \frac{\beta x^{\beta-1}}{\alpha^{\beta}}, \quad x > 0, \alpha, \beta > 0 \tag{5.31}$$

The Weibull distribution (5.31) is often used in reliability testing especially when modeling life time distributions.

Example 5.18 Consider an exponential variable X with parameter λ. We will derive the probability density function of $Y = X^{\frac{1}{\gamma}}$

$$P(Y < y) = F_Y(y) = P\left(X^{\frac{1}{\gamma}} < y \right) = P\left(X < y^\gamma \right) = F_X(y^\gamma)$$

$$f_Y(y) = f_X(y^\gamma)\left(\frac{d}{dy} y^\gamma \right) = \lambda\gamma e^{-\lambda y^\gamma} y^{\gamma-1} \quad 0 < y < \infty$$

(5.32)

The above distribution (5.32) is a variation on the distribution (5.22) and is known as the Weibull distribution with parameters γ and λ.

When the shape parameters γ and λ are set equal to unity, the above distribution is the same as the standard exponential distribution. The expected value and variance of Y in (5.32)

$$E(Y) = \frac{1}{\lambda^{\frac{1}{\gamma}}} \Gamma\left(\frac{1}{\gamma} + 1 \right)$$

$$Var(Y) = \frac{1}{\lambda^{\frac{2}{\gamma}}} \left[\Gamma\left(\frac{2}{\gamma} + 1 \right) - \left\{ \Gamma\left(\frac{1}{\gamma} + 1 \right) \right\}^2 \right].$$

5.15.4 Gumbell Distribution*

Let X have a standard exponential distribution with the probability distribution

$$f_X(x) = e^{-x} \quad x > 0$$

Define

$$Y = \alpha - \beta \log X$$

Then

$$P(Y < y) = F_Y(y) = P(\alpha - \beta \log X < y) = 1 - P\left(X < e^{\frac{\alpha-y}{\beta}} \right)$$

$$f_Y(y) = \frac{1}{\beta} e^{-e^{\frac{\alpha-y}{\beta}}} e^{\frac{\alpha-y}{\beta}} \quad -\infty < y < \infty$$

The above distribution is called the *Gumbell Distribution*.

5.15.5 Logistic Distribution

If a random variable X has the following probability density function

$$f_X(x) = \frac{e^{-(x-\alpha)\beta^{-1}}}{\beta[1+e^{-(x-\alpha)\beta^{-1}}]^2}, \quad -\infty < x < \infty \tag{5.33}$$

Then X is said to follow a *logistic distribution*. This distribution is commonly used in epidemiology for the assessment of risk.

The parameters α and $\beta(>0)$ are the location and scale parameters respectively.

Let $Y = \dfrac{X-\alpha}{\beta}$

Then,

$$f_Y(y) = \frac{e^{-y}}{(1+e^{-y})^2}, \quad -\infty < y < \infty$$

This distribution is called the *standard logistics distribution* and its moment generating function is

$$M_Y(t) = E(e^{ty}) = \int_{-\infty}^{\infty} e^{-(1-t)y}(1+e^{-y})^{-2}\, dy$$

Letting $z = (1+e^{-y})^{-1}$ in the above relationship and simplifying we have

$$M_Y(t) = \int_0^1 z^{-t}(1-z)^{-t}\, dz$$

$$= \pi t \cosec \pi t$$

The moment generating function for the distribution of X may be obtained as

$$M_X(t) = E\left(e^{t(\alpha+\beta y)}\right) = e^{t\alpha} E\left(e^{t\beta y}\right)$$

or

$$M_X(t) = \frac{\pi\beta t e^{\alpha t}}{\sin(\pi\beta t)}$$

The mean and variance for this distribution are

$$E(x) = \alpha$$

$$Var(X) = \frac{\pi^2 \beta^2}{3}$$

The density function is symmetric about the vertical axis. This distribution is used in modeling population growth where the growth is slow initially, reaches its peak, and then slows down.

Example 5.19 If two independent variables X_1 and X_2 follow standard exponential distributions, then we will derive the pdf of Z where

$$Z = -\log\frac{X_1}{X_2}.$$ (5.34)

The joint density function of X_1, X_2 is defined as

$$f_{X_1,X_2}(x_1,x_2) = e^{-(x_1+x_2)}, \quad 0 < x_1, x_2 < \infty \quad .$$

Let $U = \dfrac{X_1}{X_2}$ and $V = X_1 + X_2$,

Then, $X_1 = \dfrac{UV}{U+1}$ and $X_2 = \dfrac{V}{U+1}$

The joint density function of U and V is

$$f_{U,V}(u,v) = e^{-v}\frac{v}{(u+1)^2}, \quad 0 < u, v < \infty$$ (5.35)

The Jacobian of the transformation from (X_1, X_2) to (U,V) is incorporated in (5.35). From (5.35) we note that V and U are independent of one another and that V is the sum of two independent standard exponential variables and therefore has a *gamma*(2,1) distribution where as the probability density function of U is

$$f_U(u) = \frac{1}{(1+u)^2}, \quad 0 < u < \infty$$

If we define $Z = -\log U$
Then,

$$F_Z(z) = P(Z \le z) = P(-\log U \le z)$$
$$= P\left(e^{\log U} > e^{-z}\right)$$
$$= P\left(U > e^{-z}\right) = \int_{e^{-z}}^{\infty}\frac{du}{(1+u)^2} = -\frac{1}{1+u}\Big|_{e^{-z}}^{\infty} = \frac{1}{1+e^{-z}}$$

Hence the required density function is obtained by differentiating $F_Z(z)$ as

$$f_Z(z) = \frac{e^{-z}}{(1+e^{-z})^2}, \quad -\infty < z < \infty$$

Implying that Z has a standard logistic distribution.

5.15.6 *Inverted Gamma Distribution**

Let X be a *gamma* (α, β) random variable. Define

$$Y = \frac{1}{X}$$

We are going to derive the probability density function for Y

$$P(Y \le y) = F_Y(y) = P\left(\frac{1}{X} \le y\right)$$

$$= P\left(X > \frac{1}{y}\right) = 1 - F_X\left(X \le \frac{1}{y}\right)$$

$$f_Y(y) = f_X\left(\frac{1}{y}\right)\left|\frac{d}{dy}\left(\frac{1}{y}\right)\right|$$

$$= \frac{\beta^\alpha}{\Gamma(\alpha)} e^{-\frac{\beta}{y}}\left(\frac{1}{y}\right)^{\alpha+1} \quad y > 0 \tag{5.36}$$

The distribution (5.36) is known as the *inverted gamma distribution* and is often applied in the study of diffusion processes and the modeling of lifetime distributions. Its mean and variance are

$$E(X) = \frac{\beta}{\alpha - 1} \text{ for } \alpha > 1$$

$$Var(X) = \frac{\beta^2}{(\alpha - 1)^2(\alpha - 2)} \text{ for } \alpha > 2$$

∎

5.15.7 Inverted Beta Distribution*

We can prove that for two independent variable X_1 and X_2 where X_1 has the $gamma(\alpha, 1)$ distribution and X_2 has the $gamma(\beta, 1)$ distribution the distribution of the ration $W = \frac{X_1}{X_2}$ has what has come to be known as the *inverted beta distribution* with parameters α and β. The probability density function of the inverted beta distribution with parameters α and β is of the form

$$f_X(IB : x, \alpha, \beta) = \frac{\Gamma(\alpha)\Gamma(\beta)x^{\alpha-1}}{[\Gamma(\alpha + \beta)(1 + x)^{\alpha+\beta}]}, \quad x > 0 \tag{5.37}$$

If X has a *beta* (α, β) distribution then $Z = \frac{X}{1 - X}$ also follows the inverted beta distribution. The distribution in (5.37) is also known as the Pearson's Type VI distribution (Kendall 1952). Some statisticians (Keeping 1962) have called it the *beta-prime distribution*

*5.15.8 Dirichlet's Distribution**

Dirichlet's distribution is the multivariate analog of the univariate beta distribution.and is given by

$$f_{Y_1,\dots,Y_n}(y_1,\dots,y_n) = \frac{\Gamma(\alpha_1 + \alpha_2 + \dots \alpha_n + \alpha_{n+1})}{\Gamma(\alpha_1)\Gamma(\alpha_1)\dots\Gamma(\alpha_{n+1})} y_1^{\alpha_1-1} y_2^{\alpha_2-1}\dots y_n^{\alpha_n-1}\left(1 - \sum_1^n y_i\right)^{\alpha_{n+1}-1},$$

$$y_i > 0, \sum_1^n y_i) \le 1, l = 1,2,\dots,n$$

We will demonstrate that if a set of independent random variables $X_1,\dots,X_{n+1}$ has a *gamma*($\alpha_i,1$) ($i = 1,2,\dots,n+1$) distribution then the following transformation of variables

$$Y_i = \frac{X_i}{\sum_{i=1}^{n+1} X_i}, i = 1,2,\dots,n$$

$$Y_{n+1} = \sum_{i=1}^{n+1} X_i$$

will yields the Drichlet's distribution. The joint pdf of the $X's$ is

$$f_{X_1,\dots,X_{n+1}}(x_1,\dots,x_{n+1}) = \frac{1}{\prod_1^{n+1}\Gamma(\alpha_i)} \prod_1^{n+1} x_i^{\alpha_i-1} e^{-\sum_{i=1}^{n+1} X_i},$$

$$0 < x_i < \infty, i = 1,2,\dots n+1$$

The $X's$ may be expressed in terms of the $Y's$ as follows

$$X_i = Y_i Y_{n+1}, i = 1,2,\dots,n \text{ and } X_{n+1} = Y_{n+1}\left(1 - \sum_{i=1}^n Y_i\right)$$

The Jacobian of the transformation is a determinant several of whose terms are 0 and is

$$J = \begin{vmatrix} y_{n+1} & 0 & 0 & 0 \dots\dots y_1 \\ 0 & y_{n+1} & 0 & 0 \dots\dots y_2 \\ & & \dots\dots\dots\dots\dots & \\ -y_{n+1} & -y_{n+1} & \dots\dots 1 - \sum_1^n y_i \end{vmatrix} = y_{n+1}^n$$

Hence the joint distribution of the $Y's$ after incorporating the Jacobian is

.

$$f_{Y_1,\dots,Y_{n+1}}(y_1,\dots,y_{n+1}) = \frac{1}{\Gamma(\alpha_1)\Gamma(\alpha_2)\dots\Gamma(\alpha_{n+1})} y_1^{\alpha_1-1} y_2^{\alpha_2-1}\dots y_n^{\alpha_n-1}$$

$$\times \left(1-\sum_1^n y_i\right)^{\alpha_{n+1}-1} (y_{n+1})^{\sum_1^{n+1}\alpha_i-1} e^{-y_{n+1}} \qquad (5.38)$$

$$0 < y_{n+1} < \infty,\ y_i > 0, \sum_{i=1}^n y_i \le 1$$

The joint density function (5.38) is a product of two density functions of the form

$$f_{Y_1,\dots,Y_{n+1}}(y_1,\dots,y_{n+1}) = f_{Y_1,\dots,Y_n}(y_1,\dots,y_n) f_{Y_{n+1}}(y_{n+1})$$

Hence Y_{n+1} is independent of $Y_1,\dots,Y_n$ and has a *gamma*$(\sum_1^{n+1}\alpha_i,1)$ distribution.

Integrating over for all values of y_{n+1} in (5.38) we obtain

$$f_{Y_1,\dots,Y_n}(y_1,\dots,y_n) = \frac{1}{\Gamma(\alpha_1)\Gamma(\alpha_2)\dots\Gamma(\alpha_{n+1})} y_1^{\alpha_1-1} y_2^{\alpha_2-1}\dots y_n^{\alpha_n-1}\left(1-\sum_1^n y_i\right)^{\alpha_{n+1}-1}$$

$$\times \int_0^\infty (y_{n+1})^{\sum_1^{n+1}\alpha_i-1} e^{-y_{n+1}} dy_{n+1}$$

$$= \frac{\Gamma(\sum_1^{n+1}\alpha_i)}{\Gamma(\alpha_1)\Gamma(\alpha_2)\dots\Gamma(\alpha_{n+1})} y_1^{\alpha_1-1} y_2^{\alpha_2-1}\dots y_n^{\alpha_n-1}\left(1-\sum_1^n y_i\right)^{\alpha_{n+1}-1}$$

Hence the joint distribution of $Y_1,\dots,Y_n$ is the multivariate Dirichlet distribution.

5.15.9 Hazard Function*

The *hazard function* (*or force of mortality or failure rate*) associated with any random variable X is defined as

$$h(x) = \frac{f_X(x)}{1-F_X(x)}$$

The numerator is the probability density function of X and the denominator is the probability that X takes a value greater than x i.e., $P(X > x)$. The latter is also called the *survival function*. The hazard function associated with the logistic distribution in (5.33) is given by

$$\frac{1}{\beta[1+e^{-(x-\alpha)\beta^{-1}}]}$$

Similarly, the hazard function for the standard exponential is

$$\frac{e^{-x}}{e^{-x}} = 1$$

and the hazard function for the Weibull distribution in (5.31) is

$$\frac{\beta x^{\beta-1}}{\alpha^\beta}$$

5.15.10 Pareto Distribution*

A random variable X which has the following density

$$f_X(x) = \frac{\alpha}{x_0} \left(\frac{x_0}{x} \right)^{\alpha+1}, \quad x > x_0, \alpha > 0$$

is known as having a *Pareto distribution* with scale parameter α and location parameter x_0. This distribution is often used to model income distributions of individuals whose incomes exceeds x_0 as well as modeling the dynamics of stock movements in warehouses.

The mean and variance of this distribution are

$$E(X) = \frac{\alpha x_0}{\alpha - 1}, \, \alpha > 1$$

$$Var(X) = \frac{\alpha x_0^2}{(\alpha-1)^2 (\alpha-2)}, \, \alpha > 2$$

5.16 Distributional Relationships

5.16.1 Gamma-Poisson Relationship

As was mentioned in section 5.6 that the number of customers that arrive at a service facility in time t follows the Poisson distribution, and that the time until the nth arrival follows the gamma distribution (section 5.5.1). Consider now a random-variable X which has a *gamma*$(\eta + 1, 1)$, (where η is a positive integer) distribution, and a random variable Y which follows a Poisson distribution with parameter λ. We will show that

$$P(X > \lambda) = P(Y \le \eta) \quad\quad\quad (5.39)$$

The left hand side of (5.39) may be written as

$$P(X > \lambda) = \frac{1}{\eta!} \int_{x=\lambda}^{\infty} x^{\eta} e^{-x} dx = \frac{1}{\eta!} \left(\lambda^{\eta} e^{-\lambda} + \eta \int_{x=\lambda}^{\infty} x^{\eta-1} e^{-x} dx \right)$$

$$= \frac{1}{\eta!} \left[\lambda^{\eta} e^{-\lambda} + \eta(\lambda^{\eta-1} e^{-\lambda}) + \eta(\eta-1)(\lambda^{\eta-2} e^{-\lambda}) + \ldots + \eta! e^{-\lambda} \right]$$

$$= e^{-\lambda} \sum_{i=0}^{\eta} \frac{\lambda^{i}}{i!} = P(Y \leq \eta)$$

5.16.2 Gamma-Normal Relationship

Let X be a $N(0,1)$ variable. Define $Y = X^2$. We are going to obtain the density function for Y

$$P(Y \leq y) = P(X^2 \leq y) = P\left(-\sqrt{y} \leq X \leq \sqrt{y}\right)$$

$$F_Y(y) = F_X\left(\sqrt{y}\right) - F_X\left(-\sqrt{y}\right)$$

Differentiating both sides of the above relationship and using the results from Chapter 3 section 3.8 we obtain

$$f_Y(y) = f_X\left(\sqrt{y}\right) \frac{1}{2} y^{-\frac{1}{2}} + f_X\left(-\sqrt{y}\right) \frac{1}{2} y^{-\frac{1}{2}}$$

$$= \frac{e^{-\frac{1}{2}y} y^{-\frac{1}{2}}}{2^{\frac{1}{2}} \Gamma\left(\frac{1}{2}\right)}$$

which is a *gamma*(1/2,1/2) variable. This is also a chi-square distribution with one degree of freedom. In the above derivation we use the fact that $\sqrt{\pi} = \Gamma\left(\frac{1}{2}\right)$

5.16.3 Beta-Binomial Relationship

If X has the binomial distribution with parameters n and p, and Y has a beta distribution with parameters a and $n - a + 1$ (a is a positive integer $\leq n$), then it can be shown that

$$P(X \geq a) = P(Y \leq p) \tag{5.40}$$

We will first derive the expression for the right hand side of (5.40)

$$P(Y \le p) = \frac{\Gamma(n+1)}{\Gamma(a)\Gamma(n-a+1)} \int_{y=0}^{p} y^{a-1}(1-y)^{n-a}\, dy$$

$$= \frac{\Gamma(n+1)}{\Gamma(a)\Gamma(n-a+1)} \left[-\frac{(1-y)^{n-a+1}}{n-a+1} y^{a-1} \Big|_{y=0}^{p} \right]$$

$$+ \frac{a}{(n-a+1)} \frac{1}{} \frac{\Gamma(n+1)}{\Gamma(a)\Gamma(n-a+1)} \int_{y=0}^{p} y^{a-2}(1-y)^{n-a+1}\, dy$$

On repeated integration in the above expression we obtain

$$P(Y \le p) = \frac{n!}{(a-1)!(n-a)!} \left[-\sum_{i=1}^{a-1} \frac{p^{a-i}(1-p)^{n-(a-i)}}{\prod_{j=1}^{i} \frac{(n-a+j)}{[a-(j-1)]}} \right]$$

$$+ \frac{1}{\prod_{j=1}^{a-1} \frac{(n-a+j)}{[a-(j-1)]}} \frac{n!}{(a-1)!(n-a)!} \int_{y=0}^{p} (1-y)^{n-1}\, dy$$

After a little simplification we obtain

$$P(Y \le p) = -\sum_{i=1}^{a} \binom{n}{a-i} p^{a-i}(1-p)^{n-a+i} + 1$$

$$= -\sum_{i=1}^{a} \binom{n}{a-i} p^{a-i}(1-p)^{n-a+i} + [p+(1-p)]^{n}$$

$$= \sum_{i=a}^{n} \binom{n}{i} p^{i}(1-p)^{n-i} = P(X \ge a)$$

∎

Example 5.24 let X have a *beta*(α, β) distribution. Define $Y = \dfrac{X}{(1-X)}$

Then

$$P(Y \le y) = F_Y(y) = P\left(\frac{X}{1-X} < y \right) = P\left(X < \frac{y}{y+1} \right)$$

$$f_Y(y) = \frac{d}{dy} F_Y(y) = \frac{\Gamma(\alpha+\beta)}{\Gamma(\alpha)\Gamma(\beta)} \left(\frac{y}{y+1} \right)^{\alpha-1} \left(1 - \frac{y}{1+y} \right)^{\beta-1} \left| \frac{d}{dy}\left(\frac{y}{y+1} \right) \right|$$

$$f_Y(y) = \frac{\Gamma(\alpha+\beta)}{\Gamma(\alpha)\Gamma(\beta)} \left(\frac{y}{y+1} \right)^{\alpha-1} \left(1 - \frac{y}{1+y} \right)^{\beta-1} \frac{1}{(y+1)^2}, \; y > 0$$

5.16.4 Beta-Gamma Relationship

In this section we will show that the beta distribution is related to the gamma distribution.

Consider two independent gamma variables X_1 and X_2 distributed as $gamma(\alpha, 1)$ and $gamma(\beta, 1)$ respectively. We are going to show that

$$U = \frac{X_1}{X_1 + X_2} \tag{5.41}$$

has a $beta(\alpha, \beta)$ distribution.

Define $V = X_2$

Then, X_1 and X_2 in terms of U and V are

$$X_1 = \frac{UV}{1-U} \text{ and } X_2 = V$$

Since X_1 and X_2 are independently distributed, their joint probability density function is the product of the two density functions

$$f_{X_1, X_2}(x_1, x_2) = \frac{1}{\Gamma(\alpha)\Gamma(\beta)} x_1^{\alpha-1} x_2^{\beta-1} e^{-(x_1+x_2)}$$

.and the Jacobian of the transformation is

$$J = \begin{vmatrix} \dfrac{v}{(1-u)^2} & \dfrac{u}{1-u} \\ 0 & 1 \end{vmatrix} = \frac{v}{(1-u)^2}$$

Hence the joint probability density function of U and V is given by

$$g_{U,V}(u,v) = \frac{1}{\Gamma(\alpha)\Gamma(\beta)} u^{\alpha-1} v^{\alpha+\beta-1} e^{-\frac{v}{1-u}} \frac{1}{(1-u)^{\alpha+1}}, \ 0 < u < 1, \ 0 < v < \infty$$

Integrating $g_{U,V}(u,v)$ over all values of v in the interval $(0, \infty)$, we obtain

$$g_U(u) = \frac{1}{\Gamma(\alpha)\Gamma(\beta)} \frac{1}{(1-u)^{\alpha+1}} u^{\alpha-1} \int_0^\infty v^{\alpha+\beta-1} e^{-\frac{v}{1-u}} dv$$

Set $w = \dfrac{v}{1-u}$ in the above integral.

$$g_U(u) = \frac{1}{\Gamma(\alpha)\Gamma(\beta)} \frac{1}{(1-u)^{\alpha+1}} u^{\alpha-1} \int_0^\infty w^{\alpha+\beta-1} (1-u)^{\alpha+\beta-1} (1-u) e^{-w} dw$$

$$= \frac{1}{\Gamma(\alpha)\Gamma(\beta)} u^{\alpha-1} (1-u)^{\beta-1} \int_0^\infty w^{\alpha+\beta-1} e^{-w} dw = \frac{\Gamma(\alpha+\beta)}{\Gamma(\alpha)\Gamma(\beta)} u^{\alpha-1} (1-u)^{\beta-1}$$

The above result shows that U has a *beta*(α, β) distribution. It can be similarly shown that if X_1 has the *gamma* (α, r) distribution and X_2 has the *gamma* (β, r) distribution, then U defined in (5.41) will have a *beta*(α, β) distribution.

We may now obtain the density function of U for the case when all the Y's have a *gamma*(1,1) distribution and U is defined as

$$U = \frac{Y_1}{Y_1 + Y_2 + Y_3}$$

Since sum of two gamma (1,1) variables is a *gamma*(2,1) variable, the denominator of U may be written as the sum of Y_1 and $(Y_2 + Y_3)$ where Y_1 is *gamma*(1,1) variable and $(Y_2 + Y_3)$ is a *gamma*(2,1) variable. Hence, using the result derived earlier in this section U is a *beta*(1,2) variable. This relationship is important and is often used in deriving the F (for Fisher) distribution.

5.16.5 *Sampling Distribution of* $\bar{X}$ *and* S^2 *for the Normal Case*

The mgf of $\bar{X}$ when each X_i has a $N(\mu, \sigma^2)$ distribution is easily obtained from section 5.4.1 as

$$M_{\bar{X}}(t) = \left[\exp\left\{ \mu\left(\frac{t}{n}\right) + \left(\frac{t}{n}\right)^2 \sigma^2 \right\} \right]^n = \exp\left(\mu t + t^2 \left(\frac{\sigma^2}{n}\right) \right)$$

Hence when X_i is distributed as $N(\mu, \sigma^2)$

the mgf of $\bar{X}$ is that of a $N\left(\mu, \dfrac{\sigma^2}{n}\right)$ variable. We will now derive the distribu-

tion of $S^2 = \dfrac{\sum\limits_{1}^{n}(x_i - \bar{x})^2}{n-1}$ when the X's follow a $N(\mu, \sigma^2)$ distribution

Define the following transformation

$$Y_1 = \bar{X}$$
$$Y_i = X_i \ (i = 2, 3, ..., n)$$

With this transformation, the X's may be written as

$$X_1 = nY_1 - Y_2 - ... - Y_n, \text{ and } X_i = Y_i \ (i = 2, 3, ..., n).$$

The Jacobian of this transformation is n.

The Joint density function of $X_1, X_2, ..., X_n$ may be written as

$$f_{X_1, ..., X_n}(x_1, ..., x_n) = \left(\frac{1}{2\pi\sigma^2}\right)^{\frac{n}{2}} e^{-\frac{1}{2\sigma^2}\sum_1^n (x_i - \mu)^2} \qquad -\infty < x_1, ..., x_n < \infty \qquad (5.42)$$

Now, $\displaystyle\sum_1^n (x_i - \mu)^2 = \sum_1^n (x_i - \bar{x} + \bar{x} - \mu)^2 = \sum_1^n (x_i - \bar{x})^2 + n(\bar{x} - \mu)^2$

Hence the joint pdf is then

$$f_{X_1, ..., X_n}(x_1, ..., x_n) = \left(\frac{1}{2\pi\sigma^2}\right)^{\frac{n}{2}} \exp\left[-\left\{\frac{1}{2\sigma^2}\left(\sum_1^n (x_i - \bar{x})^2 + n(\bar{x} - \mu)^2\right)\right\}\right],$$

$$-\infty < x_1, ..., x_n < \infty$$

Converting the x's into y's and incorporating the Jacobian we have

$$f_{Y_1, ..., Y_n}(y_1, ..., y_n) = n\left(\frac{1}{2\pi\sigma^2}\right)^{\frac{n}{2}} \exp\left[-\frac{1}{2\sigma^2}\left((n-1)y_1 - \sum_2^n y_i\right)^2\right]$$

$$\times \exp\left[-\frac{1}{2\sigma^2}\left(\sum_2^n (y_i - y_1)^2 + n(y_1 - \mu)^2\right)\right] \qquad (5.43)$$

$$-\infty < y_1, ..., y_n < \infty$$

The joint density in (5.43) can be written as the product of three exponentials terms as follows

$$f_{Y_1, ..., Y_n}(y_1, ..., y_n) = \sqrt{n}\frac{1}{\sqrt{2\pi}\sigma} \exp\left(-\frac{n(y_1 - \mu)^2}{2\sigma^2}\right)$$

$$\times \sqrt{n}\left(\frac{1}{\sqrt{2\pi}\sigma}\right)^{n-1} \exp\left[-\frac{1}{2\sigma^2}\left((n-1)y_1 - \sum_2^n y_i\right)^2\right] \qquad (5.44)$$

$$\times \exp\left[-\frac{1}{2\sigma^2}\sum_2^n (y_i - y_1)^2\right]$$

Now

$$\left((n-1)y_1 - \sum_2^n y_i\right)^2 = \left[\sum_2^n (y_1 - y_i)\right]^2 = \left[\sum_2^n (\bar{x} - x_i)\right]^2 \qquad (5.45)$$

Also

$$\sum_2^n (y_i - y_1)^2 = \sum_2^n (x_i - \bar{x})^2 \qquad (5.46)$$

Adding (5.45) and (5.46) we obtain

$$\left[(n-1)y_1 - \sum_{2}^{n} y_i\right]^2 + \sum_{2}^{n}(y_i - y_1)^2 = \left(\sum_{2}^{n}(y_i - y_1)\right)^2 + \sum_{2}^{n}(y_i - y_1)^2 = (n-1)S^2$$

Hence (5.44) may now be written as

$$f_{Y_1,\ldots,Y_n}(y_1,\ldots,y_n) = \sqrt{n}\,\frac{1}{\sigma\sqrt{2\pi}}\exp\left(-\frac{n(y_1 - \mu)^2}{2\sigma^2}\right)$$

$$\times \frac{\sqrt{n}}{2^{\frac{n-1}{2}}\sigma^{n-1}\left[\Gamma\left(\frac{1}{2}\right)\right]^{n-1}}\exp Q$$

Where (5.47)

$$Q = \left(-\frac{1}{2\sigma^2}\left\{\sum_{2}^{n}(y_1 - y_i)\right\}^2 + \sum_{2}^{n}(y_i - y_1)^2\right)$$

The right hand side of (5.47) is the product of two functions. The first is the pdf of y_1 and the second is the conditional pdf of $y_2,\ldots,y_n$ given y_1. But y_1 represents $\bar{x}$ and the exponent in the second expression is $\frac{(n-1)S^2}{\sigma^2}$. Hence $\bar{x}$ is independent of S^2. We also know from section 5.9.2 that $\frac{(n-1)S^2}{\sigma^2}$ follows a χ^2_{n-1} distribution independent of the value of $\bar{x}$. Hence S^2 is independent of $\bar{x}$. We have therefore shown that $\bar{x}$ and S^2 are independent of each other.

5.16.6 Beta-t Relationship*
If T has the t-distribution with k degrees of freedom, we show that

$$X = \frac{1}{1 + \dfrac{T^2}{k}}$$

has a $beta\left(\dfrac{k}{2},\dfrac{1}{2}\right)$ distribution

$$P(X \le x) = P\left(\frac{k}{k+T^2} \le x\right) = P\left(T^2 > \frac{k(1-x)}{x}\right) = 1 - P\left(T^2 \le \frac{k(1-x)}{x}\right)$$

$$= 1 - P\left(-\sqrt{\frac{k(1-x)}{x}} \le T \le \sqrt{\frac{k(1-x)}{x}}\right) = 1 - 2F_T\left[\sqrt{\frac{k(1-x)}{x}}\right] - 1$$

The above result uses the fact that $F_T(-y) = 1 - F_T(y)$. Differentiating both sides of the above relationship and using the fact that the t-distribution is symmetric, we have

$$f_X(x) = 2f_T\left(\sqrt{\frac{k(1-x)}{x}}\right)\frac{\sqrt{k}}{2x\sqrt{x(1-x)}}$$

After a little simplification

$$f_X(x) = \frac{\Gamma\left(\dfrac{k+1}{2}\right)}{\Gamma\left(\dfrac{k}{2}\right)\Gamma\left(\dfrac{1}{2}\right)} x^{\frac{k}{2}-1}(1-x)^{\frac{1}{2}-1}, \ \ 0 < x < 1$$

which is a $beta\left(\dfrac{k}{2},\dfrac{1}{2}\right)$ distribution.

5.16.7 F-t Relationship*

If T has a t-distribution with r degrees of freedom then T^2 follows the $F_{1,r}$ distribution.

$$T = \frac{Z}{\sqrt{\dfrac{Y}{r}}}$$

where Z is $N(0,1)$ and Y is χ_r^2, and Z and Y are independent. Therefore

$$T^2 = \frac{Z^2}{Y/r} = \frac{(X/1)}{Y/r}$$

where X is Z^2 which is a χ_1^2 distribution. Hence T^2 is a $F_{1,r}$ variable.

5.16.8 Cauchy-Normal Relationship*

Let X and Y have $N(0,1)$ distributions. Define $Z = \dfrac{X}{Y}$. Then,

$$P(Z \leq z) = P\left(\frac{X}{Y} \leq z \right)$$

We will use the formula developed in section 5.3 for the density function of the ratio of two variables. The feasible region for the variables X and Y will be the same as defined in Figure 5.1. Hence

$$f_Z(z) = \int_0^{\infty} y f_X(zy) f_Y(y) dy - \int_{-\infty}^0 y f_X(zy) f_Y(y) dy, \; -\infty < y < \infty$$

Substituting for $f_X(zy)$ and $f_Y(y)$ in the above density function for Z we have

$$f_Z(z) = \frac{1}{2\pi} \int_0^{\infty} y e^{-\frac{1}{2}z^2 y^2} e^{-\frac{1}{2}y^2} dy - \frac{1}{2\pi} \int_{-\infty}^0 y e^{-\frac{1}{2}z^2 y^2} e^{-\frac{1}{2}y^2} dy$$

$$= \frac{1}{2\pi} \int_0^{\infty} y e^{-\frac{1}{2}(z^2+1)y^2} dy - \frac{1}{2\pi} \int_{-\infty}^0 y e^{-\frac{1}{2}(z^2+1)y^2} dy$$

$$= \frac{1}{2\pi(1+z^2)} + \frac{1}{2\pi(1+z^2)} = \frac{1}{\pi(1+z^2)}, \; -\infty < z < \infty$$

Hence the ratio of two $N(0,1)$ variables is a Cauchy Variable.

5.16.9 Cauchy-Uniform Relationship*

Let X have a $U\left(-\frac{\pi}{2}, \frac{\pi}{2} \right)$ distribution. We will demonstrate that $Y = \tan X$ has a Cauchy distribution. The pdf of X is

$$f_X(x) = \frac{1}{\pi}, \; -\frac{\pi}{2} < x < \frac{\pi}{2}$$

$$P(Y \leq y) = F_Y(y) = P(\tan X \leq y) = P(X \leq \tan^{-1} y)$$

$$f_Y(y) = f_X(\tan^{-1} y)\frac{1}{1+y^2} = \frac{1}{\pi(1+y^2)}, \; -\infty < y < \infty$$

5.17 Additional Distributional Findings*

5.17.1 Truncated Normal Distributions

We will first prove that if $f_X(x)$ is a pdf of a random variable X, then for $F_Y(a) < 1$

$$\int_a^{\infty} \frac{f_X(x)}{1 - F_X(a)} dx = 1$$

$$\int_a^\infty \frac{f_X(x)}{1-F_X(a)}dx = \frac{1}{1-F_X(a)}\int_a^\infty f_X(x)dx$$

$$= \frac{1}{1-F_X(a)}\left(1-F_X(a)\right) = 1$$

The random variable with the pdf of the following form (for $x \geq a$)

$$\frac{f_X(x)}{1-F_X(a)}$$

is said to be *truncated*. The above is an example of a distribution that is truncated from below. A distribution may also be truncated from above or be *doubly truncated*. Recall that in Chapter 4 Example 4.11 we had introduced the truncated Poisson distribution.

A random variable X has a truncated normal distribution if its density function is of the form

$$f_X(x) = \frac{1}{2\pi\sigma}e^{-\frac{1}{2\sigma^2}(x-\mu)^2}\left[\frac{1}{2\pi\sigma}\int_A^\infty e^{-\frac{1}{2\sigma^2}(y-\mu)^2}dy\right]^{-1}, A \leq x \leq \infty \qquad (5.48)$$

The distribution in (5.48) is said to be truncated from below.

If the distribution of X is of the form

$$f_X(x) = \frac{1}{2\pi\sigma}e^{-\frac{1}{2\sigma^2}(x-\mu)^2}\left[\frac{1}{2\pi\sigma}\int_{-\infty}^B e^{-\frac{1}{2\sigma^2}(y-\mu)^2}dy\right]^{-1}, -\infty < x < B$$

then, the distribution of X is said to be truncated from above.

For X to have a doubly truncated normal distribution, the probability density function will be of the form

$$f_X(x) = \frac{1}{2\pi\sigma}e^{-\frac{1}{2\sigma^2}(x-\mu)^2}\left[\frac{1}{2\pi\sigma}\int_A^B e^{-\frac{1}{2\sigma^2}(y-\mu)^2}dy\right]^{-1} A \leq x \leq B$$

It needs to be pointed out that if the degree of truncation is large the distribution of X may not resemble a normal distribution.

5.17.2 *Truncated Gamma Distribution**

Gamma distribution is often used in life testing and often truncation from above is recommended. In order that the random variable representing duration of life may not take values exceeding a fixed number, say B. The gamma truncated distribution used in such cases has the form

$$f_X(x) = \frac{x^{\alpha-1}e^{-\beta x}}{\int_0^B x^{\alpha-1}e^{-\beta x}dx}, 0 < x < B.$$

The moments of a truncated gamma distribution are expressed in terms of incomplete gamma functions.

■

5.17.3 Noncentral χ^2

If $X_1,\ldots,X_n$ are independent $N(0,1)$ variables and $b_1,\ldots,b_n$ are some constants

then the distribution of W defined as $W = \sum_{1}^{n}(X_i + b_i)^2$

is called the *noncentral χ^2 distribution* with n degrees of freedom and with

a non-centrality parameter defined as $\delta = \sum_{1}^{n} b_i^2$. We will represent a noncentral

chi square variable with n degrees of freedom and noncentrality parameter δ by
$\chi^2(n,\delta)$. We know that sum of squares of n $N(0,1)$ variables is a χ_n^2 variable.
Hence for $b_i = 0$ $(i = 1,2,\ldots n)$ a non central chi square variable is the same as the
chi square variable. Hence $\chi_n^2 = \chi^2(n,0)$

      The distribution of W was originally obtained by Fisher (1928b) and it
has applications in mathematical physics and communications theory. It is also
used for describing the sizes of groups of species in a predetermined area when the
species are subject to movement to other areas. The distribution is also known as
the *generalized Raleigh distribution.*
The pdf of W is given by

$$f_W(w) = \frac{e^{-\frac{1}{2}(w+\delta)}}{2^{\frac{n}{2}}} \sum_{j=0}^{\infty} \frac{w^{\frac{n}{2}+j-1} \delta^j}{\Gamma\left(\frac{n}{2}+j\right) 2^{2j} j!}, \quad 0 < w < \infty$$

The mgf of this distribution is

$$M_W(t) = E(e^{tW}) = \frac{e^{-\frac{1}{2}\delta}}{2\left(\frac{1}{2}\right)^n} \sum_{0}^{\infty} \frac{\delta^j}{2^{2j} j!\Gamma\left(\frac{1}{2}n+j\right)} \int_0^{\infty} e^{-\frac{1}{2}(1-2t)w} (w)^{\frac{1}{2}n+j-1} dw$$

$$= e^{-\frac{1}{2}\delta} e^{\frac{\delta/2}{(1-2t)}} \frac{1}{(1-2t)^{\frac{n}{2}}} = \frac{1}{(1-2t)^{\frac{n}{2}}} e^{\delta t(1-2t)^{-1}} \quad t < \frac{1}{2}$$

From the above equation we observe that the noncentral χ^2 has the reproductive
property with respect to n. Thus if X_i is a $\chi^2(n_i,\delta_i)$ variable and that $X_1,\ldots,X_m$
are all independent , then

$$Z = \sum_{i=1}^{n} X_i \text{ is a } \chi^2\left(\sum_{1}^{m} n_i, \sum_{1}^{m} \delta_i\right) \text{ variable}$$

5.17.4 Noncentral t Distribution*

If $n, \bar{X}, S^2$ are the size, mean and variance of a sample from $N(\mu, \sigma^2)$ then the distribution of

$$T = \frac{(\bar{X} - \mu + \delta)\sqrt{n}}{S}$$

is known as the *noncentral t-distribution* with $(n-1)$ degrees of freedom and non-centrality parameter δ.

5.17.5 Noncentral F Distribution*

Define independent random variables X and Y such that X has a noncentral $\chi^2(n, \delta)$ distribution and Y has a χ_m^2 distribution. Then F defined as $F = \dfrac{X/n}{Y/m}$ has a noncentral F distribution with n and m degrees of freedom and *noncentrality parameter* δ. If the numerator and denominator variables in the above ratio are both noncentral chi squares, then the corresponding distribution of F is the *doubly noncentral F distribution* with n and m degrees of freedom.

Problems

1. Let $X_1, ..., X_n$ denote a random sample from a population with density function $f_x(x) = 3x^2, \ 0 < x < 1$
    a) Write down the joint pdf of $X_1, ..., X_n$
    b) Find the probability that the first observation is less than 0.5 i.e., $P(X_1 < 0.5)$.
    c) Find the probability that all the observations are less than 0.5.

2. A random variable X has the probability density function defined by $f_X(x) = 2x, \ 0 < x < 1$. Obtain the density function of $Y = 8X^2$

3   A random variable X has density function $f_X(x) = 2xe^{-x^2}, \ x > 0$. find the probability density function of $Y = X^2$

4. For a variable X with moment generating function,
    $(pe^t)^k (1 - pe^t)(1 - e^t + p^k qe^{(k+1)t})^{-1}$
    find $E(X)$ and $Var(X)$.

5. Let X be a random variable whose cumulative density function is given by

$$F_X(x) = \begin{cases} 0 & \text{for } x < 0 \\ X/3 & \text{for } 0 \le x < 1 \\ X/2 & \text{for } 1 \le x < 2 \\ 1 & \text{for } x \ge 2 \end{cases}$$

Find $P(1/2 \le x \le 3/2)$.

6. Let the variable Y have a Poisson distribution with parameter λ and λ has a *gamma*(α, β) distribution. Find the marginal distribution of Y.

7. Let X and Y be distributed as $beta(\alpha, \beta)$ and $beta(\alpha + \beta, \gamma)$ respectively. De rive the distribution of XY.

8. For any three random variables X, Y, and Z with finite variances, prove that
   a) X and $Y - E(Y|X)$ are uncorrelated
   b) $Var[Y - E(Y|X)] = E[Var(Y|X)]$
   c) $Cov[Z, E(Y|Z)] = Cov(Z, Y)$

9. Show that if X and Y are independent and $N(0,1)$ then
   $$Z = \frac{X}{X + Y} \text{ has a Cauchy distribution}$$

10. A random variable X has the following Pareto density
    $$f_X(x) = \frac{(1+\alpha)}{x^{2\alpha+1}} \text{ if } 1 \le x < \infty, \ \alpha > 0$$
    $$= 0 \text{ otherwise}$$
    Find the variance of X

11. If X and Y have a joint distribution of the form
    $$f_{X,Y}(x, y) = 2x, \ 0 \le x \le 1, 0 \le y \le 1$$
    Find $P(X^2 < Y < X)$

12. Suppose X and Y have joint density function
    $$f_{X,Y}(x, y) = x + y, \ 0 < x, y < 1$$
    Determine
    a) $E[X|Y = y]$.
    b) $E\left(Xe^{\left[Y+\frac{1}{Y}\right]} \Big| Y = y \right).$

13. Suppose that X and Y are distributed as $N(\alpha, \sigma^2)$ and $N(\beta, \tau^2)$ respectively. Define $Z = X+Y$. Find
   a) The conditional distribution of X given $Z = z$
   b) The conditional distribution of Z given $X = x$

14. Suppose that the distribution of Y conditional on $X = x$ is $N(x, x^2)$ and that the marginal distribution of X is U(0,1).
   Find $E(Y)$, $Var(Y)$, $Cov(X,Y)$.

15. For the binomial variable X with parameters n and p, let the success probability p have a *beta* (α, β) distribution. Show that the following relationship holds

$$Var(X) = nE(p)[1 - E(p)] + n(n-1)Var(p)$$

16. For the negative binomial random variable X with parameters r and p, let the success probability p have a *beta* (α, β) distribution. Derive the marginal distribution of X and its variance.

17. Consider a set of binomial variables with parameters n_i and p_i. Let p_i follow a *beta* (α, β) distribution. Show that

$$Var(X_i) = n_i \frac{\alpha\beta(\alpha + \beta + n_i)}{(\alpha + \beta)^2(\alpha + \beta + 1)}$$

18. In Problem 11 define $Y = \sum_{1}^{n} X_i$. Obtain the mean and variance of Y.

19. If the joint distribution of X_1 and X_2 is the bivariate normal with the fol-lowing probability density function

$$f_{X_1, X_2}(x_1, x_2) = \frac{1}{2\pi\sqrt{1-\rho^2}} \exp\left(-\frac{1}{2(1-\rho^2)}\left[x_1^2 - 2x_1x_2\rho + x_2^2\right]\right)$$

$$-\infty < x_1, x_2 < \infty$$

show that the $Corr(X_1^2, X_2^2) = \rho^2$.

20. Let X and Y be distributed as *beta* (α, β) and *beta* $(\alpha + \beta, \gamma)$ respectively. Derive the distribution of XY.

21. Let $X_1, \ldots, X_{n+1}$ be a sample from an exponential distribution with pdf $f_X(x) = e^{-x}$. Let $S_m = \sum_1^m X_i, m \le n+1$. Show that joint probability density function of

$$Y = \left(\frac{S_1}{S_{n+1}}, \ldots, \frac{S_n}{S_{n+1}} \right) \text{ is}$$

$$f_Y(y_1, y_2, \ldots y_n) = n! \text{ if } 0 < y_1 < y_2 < \ldots < y_n < 1$$

22. Suppose that X_1 and X_2 are independent exponential random variables with parameter λ. Let $Y_1 = X_1 - X_2$ and $Y_2 = X_2$.
    a) Find the joint probability density of Y_1 and Y_2.
    b) Show that the pdf of Y_1 is $f_{Y_1}(y_1) = \frac{1}{2}\lambda e^{-\lambda|y_1|}, \quad -\infty < y_1 < \infty$

23. Suppose that X and Y are independent random variables with probability density functions $gamma(p,1)$ and $gamma(p+1/2,1)$ respectively. Show that $z = \sqrt{XY}$ has a $gamma(2p,1)$ distribution.

24. Let $(X_1, Y_1), \ldots (X_n, Y_n)$ be a sample from the following bivariate normal distribution

$$f_{X,Y}(x,y) = \frac{1}{2\pi\sigma_1\sigma_2\sqrt{1-\rho^2}} \exp\left[-\frac{1}{2(1-\rho^2)} Q \right]$$

where, $Q = \left\{ \frac{(x-\mu_1)^2}{\sigma_1^2} + \frac{(y-\mu_2)^2}{\sigma_2^2} - 2\rho\frac{(x-\mu_1)(y-\mu_2)}{\sigma_1\sigma_2} \right\}$

$$-\infty < x_1, x_2 < \infty$$

Define

$$\bar{X} = \frac{1}{n}\sum_{i=1}^n X_i, \ \bar{Y} = \frac{1}{n}\sum_{i=1}^n Y_i, \ S_1^2 = \sum_{i=1}^n (X_i - \bar{X})^2,$$

$$S_2^2 = \sum_{i=1}^n (Y_i - \bar{Y})^2, S_{12} = \sum_{i=1}^n (X_i - \bar{X})(Y_i - \bar{Y}), \ R = \frac{S_{12}}{S_1 S_2}$$

a) Show that S_1^2, S_2^2 and S_{12} are independent.

b) Show that when $\rho = 0$, $T = \frac{R\sqrt{n-2}}{\sqrt{1-R^2}}$ has a t-distribution with n-2

degrees of freedom.

25. Let (X,Y) follow a bivariate normal distribution

$$f_{X,Y}(x,y) = \frac{1}{2\pi\sigma_1\sigma_2\sqrt{(1-\rho^2)}}\exp\left[\frac{-1}{(1-\rho^2)}\left\{\frac{(x-\mu_1)^2}{\sigma_1^2}+\frac{(y-\mu_2)^2}{\sigma_2^2}\right\}\right]$$

$$\times\exp\left[\frac{2}{(1-\rho^2)}\left(\frac{x-\mu_1}{\sigma_1}\right)\left(\frac{x-\mu_2}{\sigma_2}\right)\right]$$

$$-\infty < x,y < \infty$$

with parameters $\mu_1,\mu_2,\sigma_1^2,\sigma_2^2,\rho$, show that if $\sigma_1,\sigma_2 > 0, |\rho| < 1$

Show that if $\sigma_1^2 = \sigma_2^2$, then $X+Y$ and $X-Y$ are independent.

26. Let $X_1,\ldots,X_n$ be independent random variables having the pdf

$$f_{X_i}(x_i) = \frac{\beta^{\alpha_i}x^{\alpha_i-1}e^{-\beta x}}{\Gamma(\alpha_i)}$$

Find the pdf of the sum $Y = \sum_{i=1}^{n}X_i$.

27. If $X_1,\ldots,X_n$ are independent and distributed as $gamma\,(1,\lambda)$ show that $\sum_{i=1}^{n}X_i$ has a $gamma\,(n,\lambda)$ distribution and that $2\lambda\sum_{i=1}^{n}X_i$ has the χ_{2n}^2 distribution.

28. Prove that for the probability density function

$$f_Y(y) = \sqrt{\frac{2}{\pi}}\exp\left(-\frac{y^2}{2}\right), 0 < y < \infty$$

$$E[(Y)] = \sqrt{\frac{2}{\pi}} \text{ and the } Var[(Y)] = 1-\frac{2}{\pi}$$

29. Let X have a $N(\theta,\sigma^2)$ distribution and let $h(x)$ be a differentiable function such that $E[h'(X)] < \infty$. Show that (Stein's formula)
$$E[h(X)(X-\theta)] = \sigma^2 E[h'(X)]$$

30. Show that if
a) X is a Poisson variable with parameter λ, then
$$E[\lambda h(X)] = E[Xh(X-1)]$$
b) X is a negative binomial variable with parameters r and p, then
$$E[(1-p)h(X)] = \left[\frac{X}{r+X-1}h(X-1)\right].$$

31. Obtain the mean and variance of a $U(a,b)$ distribution using the moment generating function.

32. For
    X_1 and X_2 iid $U(\theta, \theta + 1)$ variables obtain the distribution of their sum.

33. Let X be a random variable pdf belonging to the one-parameter exponential family in natural form and ξ is an interior point of Ψ.

    a) Show that the moment generating function of $T(X)$ exists and is given by

    $$m(s) = \exp\left(d^*(\xi) - d^*(s + \xi)\right) \text{ for s in some neighborhood of } 0.$$

    b) $E\left(T(X)\right) = -d^{*\prime}(\xi)$

    c) $Var\left(T(X)\right) = -d^{*\prime\prime}(\xi)$.

34. Let $X_1, \ldots, X_n$ be iid $N(\mu, \sigma^2)$ random variables. Find the conditional distribution of X_1, given $\sum_{i=1}^{n} X_i$.

References

Bartoszynski, R. and Niewiadomska-Bugaj, M. (1996), *Probability and Statistical Inference*, John Wiley and Sons, New York.

Casella, G. and Berger, R.L. (2000), *Statistical Inference*. 2nd ed. Duxbury Press. Belmont

Chung, K.L. (2000), *A Course in Probability Theory*, Academic Press, New York..

Colton, T. (1974), *Statistics in Medicine*, Little, Brown and Company, Boston.

Cox, D.R. and Hinkley, D.V. (1974), *Theoretical Statistics*, Chapman and Hall, London.

Dudewicz, E.J. and Mishra, S.N. (1988), *Modern Mathematical Statistics*, John Wiley and Sons, New York.

Erlang, A.K. (1909), *Probability and Telephone Calls*, Nyt Tidsskr, Mat Series b, Vol 20.

Evans, M., Nicholas, H. and Peacock, B. (2000), *Statistical Distribution*, 3rd ed., Wiley Series in Probability and Statistics, New York.

Feller, W. (1965), *An Introduction to Probability Theory and Its Applications*, Vol I. 2nd ed., John Wiley and Sons, New York.

Feller, W. (1971), *An Introduction to Probability Theory and Its Applications*, Vol II, John Wiley and Sons, New York.

Freund, J.F. (1971), *Mathematical Statistics*, 2nd ed. Prentice Hall, Englewood Cliffs, New Jersey.

Hogg, R.V. and Craig, A.T. (1978), *Introduction to Mathematical Statistics*, 4th ed., Macmillan Publishing Company, New York.

Johnson, N.L. and Kotz, S, (1970), *Continuous Univariate Distributions-1*, The Houghton Mifflin Series in Statistics, New York.

Johnson, N.L. and Kotz, S. (1970), *Continuous Univariate Distributions-2*, The Houghton and Mifflin Series in Statistics, New York.

Keeping, E.S. (1962), *Introduction to Statistical Inference*, D.Van Nostrand.

Kendall, M.G. (1952) *The Advanced Theory of Statistics*, Vol I, 5th ed., Charles Griffin and Company, London.

Mood, A.M. Graybill, F.A. and Boes D.C. (1974), *Introduction to the Theory of Statistics*, 3rd ed., McGraw-Hill, New York.

Polya, G. and Szegb, G. (1925), *Aufgaben und Lehrsatze*, Vol I, part 2. Berlin.

Rao, C.R. (1952), *Advanced Statistical Methods in Biometric Research*, John Wiley and Sons, New York.

Rao, C.R. (1977), *Linear Statistical Inference and Its Applications*, 2nd ed., John Wiley and Sons, New York.

Rice, J.A. (1995), *Mathematical Statistics and Data Analysis*, 2nd ed., Duxbury Press, Belmont, CA.

Rohatgi, V.K. (1975), *An Introduction to Probability Theory and Mathematical Statistics*, John Wiley and Sons, New York.

Wilks, S.S. (1962), *Mathematical Statistics*, John Wiley and Sons, New York.

6

Distributions of Order Statistics

6.1 Introduction

The study of order statistics focuses on functions of random variables that have been rank ordered (i.e., ordered from the smallest to the largest). For example, the minimum of a collection of n observations is the first order statistic, the next largest is the second order statistic, the third largest is the third order statistic, etc. The examination of the characteristics of order statistics have important implications in many areas in statistics. Of particular interest are the distribution of the minimum and maximum values in a collection of random variables. These two values are the objects of close inspection because the minimum and maximum represent two extremes. In addition, the median (or the observation that represents the 50th percentile value from a collection of observations) often receives careful consideration because it represents an alternative measure to the mean as a measure of central tendency.

The abilities of 1) careful observation and 2) the successful application of random variable transformation techniques are the main skills needed to depict the relationships between a collection of random variables and their order statistics. We will begin with some simple examples to help develop the reader's intuition in this new area, and then identify some very useful results.

6.2 Rank Ordering

The hallmark of order statistics is the use of relative ranking. Thus far in this text, we have considered a finite collection of random variables $X_1, X_2, X_3, \ldots, X_n$. Each member of this collection has typically followed the same probability distribution, and has been independent of the other members in the collection. This has actually provided all the structure that we have needed in order to compute some elementary, but nevertheless important probabilities.

Example 6.1 For example, consider two saplings planted in the same ground. The height of each sapling after three years of growth is normally distributed with mean 20 cm. and variance 5 cm. What is the probability that the height of the first sapling is greater than the height of the second at three years post planting?

We begin by simply letting X_1 be the height of the first tree, and X_2 the height of the second tree. Then, the $P[X_1 > X_2]$ reduces to a short collection of helpful probability statements using the underlying normal distribution, as follows

$$P[X_1 > X_2] = P[X_1 - X_2 > 0] = P[Y = X_1 - X_2 > 0] = 0.50, .$$

where the random variable Y follows a normal distribution with $\mu = 0$ and $\sigma^2 = 10$. This result satisfies our intuition that the answer should be ½. Another, more intuitive approach that we might have taken would have been to consider the event as being one of three events; $X_1 > X_2$, $X_1 < X_2$, or $X_1 = X_2$. Since the probability that a random variable that follows a continuous distribution is equal to a single value from that distribution is zero, we can say that $P[X_1 = X_2] = 0$ when X_1 and X_2 are normally distributed. This realization allows us to focus on the two inequalities. Thus when X_1 and X_2 are independent and identically distributed, then these two events should be equally likely. Since there are no other possibilities, and the probability of equal heights is zero, the probability that one sapling is larger than the other is $P[X_1 > X_2] = P[X_2 > X_1] = \dfrac{1}{2}$.

■

6.2.1 Complications of Distribution Heterogeneity

Some computations about the relative magnitude of two random variables are more complicated.

Example 6.2 Consider the case of a patient who has chest pain related to an impending myocardial infarction (heart attack). When the patient has chest pain, they call the ambulance for immediate transportation to a hospital. It is important that they get to the hospital before the heart attack occurs, because early use of an intervention may reduce the size of (or actually prevent) the heart attack. Assume that the length of time from the occurrence of the chest pain to the patient's arrival in the hospital emergency room and the immediate initiation of therapy follows an exponential distribution with parameter λ. Also, assume that (based on research data) the probability distribution of the time from the chest pain to the actual heart attack follows a gamma distribution with parameter α and r. We are interested in determining the probability that the patient arrives in the emergency room and begins therapy before the occurrence of the heart attack. We begin to solve this problem by letting S be the duration of time from the occurrence of the chest pain to the patient's arrival at the emergency room, and T be the time from chest pain to the occurrence of the heart attack. We will assume that the hospital arrival time is independent of the timing of the heart attack. We may write the probability density functions of S and T respectively as

$$f_S(s) = \lambda e^{-\lambda s} 1_{[0,\infty)}(s) : f_T(t) = \frac{\alpha^r}{\Gamma(r)} t^{r-1} e^{-\alpha t} 1_{[0,\infty)}(t),$$

where $1_A(x)$ is the indicator function, taking on the value one when X is contained in the set A. We must find P[$S < T$]. We may write this as

$$P[S < T] = \iint\limits_{s<t} f_S(s) f_T(t) = \iint\limits_{0 \le s < t < \infty} \lambda e^{-\lambda s} ds \frac{\alpha^r}{\Gamma(r)} t^{r-1} e^{-\alpha t} dt.$$

Proceeding,

$$P[S < T] = \left[\int_0^\infty \frac{\alpha^r}{\Gamma(r)} t^{r-1} e^{-\alpha t} \left[\int_0^t \lambda e^{-\lambda s} ds \right] dt \right] = \int_0^\infty \frac{\alpha^r}{\Gamma(r)} t^{r-1} e^{-\alpha t} \left(1 - e^{-\lambda t}\right) dt$$

$$= 1 - \int_0^\infty \frac{\alpha^r}{\Gamma(r)} t^{r-1} e^{-(\alpha+\lambda)t} dt.$$

(6.1)

We can simplify further by writing

$$\int_0^\infty \frac{\alpha^r}{\Gamma(r)} t^{r-1} e^{-(\alpha+\lambda)t} dt = \frac{\alpha^r}{(\alpha+\lambda)^r} \int_0^\infty \frac{(\alpha+\lambda)^r}{\Gamma(r)} t^{r-1} e^{-(\alpha+\lambda)t} dt = \left[\frac{\alpha}{(\alpha+\lambda)} \right]^r.$$

This final simplification is a consequence of the fact that

$$\int_0^\infty \frac{(\alpha+\lambda)^r}{\Gamma(r)} t^{r-1} e^{-(\alpha+\lambda)t} dt = 1.$$

Thus $P[S < T] = 1 - \left[\frac{\alpha}{\alpha+\lambda} \right]^r$. The fact that the random variables S and T were not identically distributed, required us to carry out this more complex calculation. ∎

Example 6.3 As another example of our current ability to explore the probability of events that involve order statistics, consider the case of three independent random variables where X_1 follows an $\exp(\lambda_1)$, X_2 follows an $\exp(\lambda_2)$, and X_3 follows an $\exp(\lambda_3)$ probability distribution. We seek P [$X_1 \le X_2 \le X_3$]. We may proceed directly as follows.

$$P\left[X_1 \leq X_2 \leq X_3\right] = \iiint\limits_{x_1 \leq x_2 \leq x_3} \lambda_1 e^{-\lambda_1 x_1} \lambda_2 e^{-\lambda_2 x_2} \lambda_3 e^{-\lambda_3 x_3} \, dx_1 dx_2 dx_3$$

$$= \int_0^\infty \lambda_1 e^{-\lambda_1 x_1} \int_{x_1}^\infty \lambda_2 e^{-\lambda_2 x_2} \left[\int_{x_2}^\infty \lambda_3 e^{-\lambda_3 x_3} \, dx_3\right] dx_2 dx_1$$

$$= \int_0^\infty \lambda_1 e^{-\lambda_1 x_1} \int_{x_1}^\infty \lambda_2 e^{-\lambda_2 x_2} e^{-\lambda_3 x_2} \, dx_2 dx_1$$

$$= \int_0^\infty \lambda_1 e^{-\lambda_1 x_1} \int_{x_1}^\infty \lambda_2 e^{-(\lambda_2 + \lambda_3) x_2} \, dx_2 . dx_1$$

Writing

$$\int_{x_1}^\infty \lambda_2 e^{-(\lambda_2 + \lambda_3) x_2} \, dx_2 = \lambda_2 \int_{x_1}^\infty e^{-(\lambda_2 + \lambda_3) x_2} \, dx_2 = \frac{\lambda_2}{(\lambda_2 + \lambda_3)} \int_{x_1}^\infty (\lambda_2 + \lambda_3) e^{-(\lambda_2 + \lambda_3) x_2} \, dx_2$$

$$= \frac{\lambda_2}{(\lambda_2 + \lambda_3)} e^{-(\lambda_2 + \lambda_3) x_1}.$$

and substituting this result into the last line of the previously expression, we may write

$$P\left[X_1 \leq X_2 \leq X_3\right] = \int_0^\infty \lambda_1 e^{-\lambda_1 x_1} \frac{\lambda_2}{(\lambda_2 + \lambda_3)} e^{-(\lambda_2 + \lambda_3) x_1} \, dx_1$$

$$= \frac{\lambda_1 \lambda_2}{(\lambda_2 + \lambda_3)} \int_0^\infty e^{-(\lambda_1 + \lambda_2 + \lambda_3) x_1} \, dx_1 \tag{6.2}$$

$$= \frac{\lambda_1 \lambda_2}{(\lambda_2 + \lambda_3)(\lambda_1 + \lambda_2 + \lambda_3)} \int_0^\infty (\lambda_1 + \lambda_2 + \lambda_3) e^{-(\lambda_1 + \lambda_2 + \lambda_3) x_1} \, dx_1$$

$$= \frac{\lambda_1 \lambda_2}{(\lambda_2 + \lambda_3)(\lambda_1 + \lambda_2 + \lambda_3)}.$$

Note that when $\lambda_1 = \lambda_2 = \lambda_3$, then, the solution reduces to $P\left[X_1 \leq X_2 \leq X_3\right]$

$= \frac{1}{6}$. In this case, the random variables X_1, X_2, and X_3 are identically distributed.

However, a different, and simpler approach would take full advantage of the observation that the three random variables were independent and identically distributed, we might have followed an alternative process. Recognizing that there are only three random variable, we can enumerate six and only six ways to order them

$$X_1 \leq X_2 \leq X_3$$
$$X_1 \leq X_3 \leq X_2$$
$$X_2 \leq X_1 \leq X_3$$
$$X_2 \leq X_3 \leq X_1 \qquad (6.3)$$
$$X_3 \leq X_1 \leq X_2$$
$$X_3 \leq X_2 \leq X_1$$

If the random variables are i.i.d., then the six orderings of these random variables are equally likely. Since the occurrence of the union of these six events exhausts the event space, we know that the probability of any one of them must be 1/6. ∎

6.2.2 New Notation

Up until this point, we paid scant attention to the relative magnitudes of the observations, and our notational devices were not focused on the observations' relative positions in the collection. Thus, we can currently assess the $P[X_3 > 0]$, but we did not embed the notational structure that would allow us to ask about the likelihood that X_3 was greater than the minimum of X_1 and X_2. In the problems that we have solved in section 6.2.1, the event whose probability we sought produced the ordering, and we embedded this relationship into the regions of integration as we solved the problem.

However, a very simple maneuver will induce the structure that we need to evaluate the probabilities of more complicated events involving the ordered observations. Given a collection of observations X_1, X_2, X_3,..., X_n, we will now *rank order* them as $X_{(1)}$, $X_{(2)}$, $X_{(3)}$,...,$X_{(n)}$. In this new structure, $x_{(1)}$ is the minimum of the collection X_1, X_2, X_3,...,X_n, $X_{(2)}$ is the second largest observation in this collection, and, continuing until we exhaust the sample, we let $X_{(n)}$ which is the largest. Thus, the transformation of the original sequence X_1, X_2, X_3,..., X_n to the rank ordered sequence $X_{(1)}$, $X_{(2)}$, $X_{(3)}$,...,$X_{(n)}$ requires the completion of a transformation of n random variables to a new collection of n random variables. The result of this transformation is the reordering of the sequence.

We can examine how this transformation works for a very simple example. Let X_1, and X_2 be independent and identically distributed random variables that follow an exponential distribution with parameter λ. We wish to find the joint distribution of $X_{(1)}$, $X_{(2)}$, or the joint distribution of min (X_1, X_2), and max (X_1, X_2). Let $W = \max (X_1, X_2)$, and $V = \min (X_1, X_2)$. Then by distributional theory of the transformation of variables we know that

$$f_{V,W}(v,w) = f_{X_1,X_2}(v,w) J\left[(x_1,x_2) \to (v,w)\right]. \tag{6.4}$$

where $J\left[(x_1,x_2) \to (v,w)\right]$ is the notation that represents the Jacobian that governs the transformation of $(X_1, X_2) \to (V,W)$ mapping. Since the random variables X_1 and X_2 are exponentially distributed random variables, we may write their joint density function as $f_{X_1,X_2}(x_1,x_2) = \lambda^2 e^{-\lambda(x_1+x_2)} \mathbf{1}_{[0,\infty)}(x_1) \mathbf{1}_{[0,\infty)}(x_2)$.

The transformation may be written as two functions:

$$
\begin{aligned}
X_1 &= V\mathbf{1}_{X_1 \le X_2}(X_1,X_2) + W\mathbf{1}_{X_1 > X_2}(X_1,X_2) \\
X_2 &= V\mathbf{1}_{X_2 < X_1}(X_1,X_2) + W\mathbf{1}_{X_2 \ge X_1}(X_1,X_2).
\end{aligned}
\tag{6.5}
$$

These functions link X_1 and X_2 to V and W. If $X_1 < X_2$, then X_1 is equal to V and X_2 is W. Conversely, if $X_1 > X_2$, then X_1 is W and X_2 is equal to V. The Jacobian of this transformation is 1. We observe that the region in the (X_1, X_2) plane denoted by $0 \le X_1 < \infty$, $0 \le X_2 < \infty$ is equivalent to the set of (V, W) values for which $0 \le V \le W < \infty$. However, there are two equivalent mappings from (X_1, X_2) to (V, W). Thus,

$$f_{V,W}(v,w) = f_{X_1,X_2}(v,w) J\left[(x_1,x_2) \to (v,w)\right] = 2!\lambda^2 e^{-\lambda(v+w)} \mathbf{1}_{0 \le v \le w < \infty}(v,w). \tag{6.6}$$

To demonstrate that the right hand side of the previous expression is a probability density function, we simply integrate over the region $\Omega_{V,W} = \{(v,w): 0 \le v \le w < \infty\}$ as follows

$$
\iint\limits_{\Omega_{V,W}} f_{V,W}(v,w)\,dvdw = \iint\limits_{\Omega_{V,W}} 2\lambda^2 e^{-\lambda(v+w)} \mathbf{1}_{0 \le v \le w < \infty}\,dvdw
$$

$$
= 2\int_0^\infty \lambda e^{-\lambda v} \int_v^\infty \lambda e^{-\lambda w}\,dwdv = 2\int_0^\infty \lambda e^{-\lambda v}\,dv = \int_0^\infty 2\lambda e^{-2\lambda v}\,dv = 1.
$$

Using this procedure, we can find the joint distribution of (V, M, W) where $V = \min(X_1, X_2, X_3)$, $W = \max(X_1, X_2, X_3)$, and $M = \text{median}(X_1, X_2, X_3)$. Adapting equation (6.6), we find

$$f_{V,M,W}(v,m,w) = 3!\lambda^3 e^{-\lambda(v+m+w)} \mathbf{1}_{0 \le v \le m \le w < \infty}(v,m,w) \tag{6.7}$$

where the six in the previous expression represents the fact that there are $3! = 6$ ways to order the (X_1, X_2, X_3) sequence.[1] Now generalizing, for n random variables, there are $n!$ ways to order the sequence. We may therefore write the joint distribution of the order statistics of a collection of independent and identically distributed random variables $(x_1, x_2, x_3, \ldots, x_n)$ as

$$f_{X_{(1)}, X_{(2)}, X_{(3)}, \ldots, X_{(n)}} \left(x_{(1)}, x_{(2)}, x_{(3)} \ldots x_{(n)} \right)$$
$$= n! \lambda^n e^{-\lambda \left(x_{(1)} + x_{(2)} + x_{(3)} + \ldots + x_{(n)} \right)} \mathbf{1}_{0 \leq x_{(1)} \leq x_{(2)} \leq x_{(3)} \leq \ldots \leq x_{(n)} \leq \infty} \left(x_{(1)}, x_{(2)}, x_{(3)} \ldots x_{(n)} \right). \tag{6.8}$$

6.3 The Probability Integral Transformation

One very useful tool that is commonly used in probability and statistical developments is the random variable that follows the uniform distribution on the [0, 1] interval. We will now see that this probability distribution has an important application to the generation of other random variables and ultimately, to order statistics.

Recall from Chapter 3 that, if X is a random variable with cumulative distribution function $F_X(x)$ and we define $Y = F_X(x)$, then $f_Y(y) = \mathbf{1}_{[0,1]}(y)$ or y is uniformly distributed on the [0,1] interval. This transformation is termed the *probability integral transformation*.

One of the most useful applications of this tool is the generation of random variables that follow a pre-specified probability distribution. For example, consider the researcher who wishes to estimate the failure time to a circuit component in the space shuttle fleet. She anticipates that the component's failure time follows an exponential distribution with parameter λ. Although $U(0, 1)$ random variables are easy to generate, she wishes to generate random variables from the desired exponential distribution. She begins with writing the cumulative distribution function for this random variable

$$F_T(t) = 1 - e^{-\lambda t}$$

Now, she knows from the probability integral transformation that $Y = F_T(t)$ follows a $U(0,1)$, and also that uniformly distributed random variables are easy to generate. She only needs to apply the inverse function F_T^{-1} to Y in order to

---

[1] There are three possible values for the minimum, then, given that the minimum has been chosen, there are two possibilities for the median, and finally, given that both the maximum and the minimum have been chosen, there is only one possibility for the minimum of the three observations.

generate a random variable that follows an exponential distribution. She proceeds as follows.

$$Y = 1 - e^{-\lambda T}$$

$$1 - Y = e^{-\lambda T} \quad\quad (6.9)$$

$$-\frac{1}{\lambda} \log(1 - Y) = T$$

Thus, in order to generate a random value from an exponential(λ) distribution, she needs to only generate a uniformly distributed random variable, and then to apply the relationship between T and Y in (6.9) to generate a value of T.

Another example of interest that exemplifies the concept of the cumulative distribution function of a random variable as a random variable itself derives from the concept of expectation. For example, we are interested in computing

$$E[F_Z(Z)] \quad\quad (6.10)$$

where Z is a standard normal random variable. Recall that we have already demonstrated that the random variable $Y = F_Z(Z)$ follows a uniform distribution on the [0,1] interval, so we know at once that $E[F_Z(Z)] = E[Y] = \frac{1}{2}$. An alternate approach to this problem would be to let X and Y be independent and identically distributed $N(0,1)$ random variables. How would we compute $P[X < Y]$? One way to express this event is

$$P[X < Y] = \iint\limits_{x < y} f_{X,Y}(x,y)dxdx = \int_{-\infty}^{\infty} \int_{-\infty}^{y} f_X(x)f_Y(y)dxdy. \quad\quad (6.11)$$

However $\int_{-\infty}^{y} f_x(x)dx - F_X(y)$ so we may proceed with

$$P[X < Y] = \int_{-\infty}^{\infty} F_X(y)f_Y(y)dy = E[F_X(Y)] = E[F_Y(Y)] = E[F_Z(Z)] \quad (6.12)$$

where Z is a standard normal random variable. However, we also know that

$$P[X < Y] = P[X - Y < 0] = P[W < 0] = \frac{1}{2}, \quad \text{where } W \text{ follows a } N(0,2)$$

distribution. We have therefore shown that $\frac{1}{2} = P[X < Y] = E[F_Z(Z)]$.

We can use this device to evaluate many expectations that involve the cumulative distribution function of the random variable of interest. For example, in order to find the $E[1 - F_Z(Z + A)]$ where Z is a standard normal random variable, consider $P[X - Y > A]$ where X and Y are independent standard normal random variables. Then we can write

$$P[X - Y > A] = \iint\limits_{x-y>A} f_{X,Y}(x,y)dxdy = \int\limits_{-\infty}^{\infty} \int\limits_{y+A}^{\infty} f_X(x)f_Y(y)dxdy$$

$$= \int\limits_{-\infty}^{\infty} (1 - F_X(Y+A))f_Y(y)dy = \int\limits_{-\infty}^{\infty}(1 - F_Y(Y+A))f_Y(y)dy \quad (6.13)$$

$$= E[1 - F_Y(Y+A)] = E[1 - F_Z(Z+A)].$$

Alternative, we also know that

$$P[X - Y > A] = P[W > A] = 1 - F_Z\left(\frac{A}{\sqrt{2}}\right). \qquad (6.14)$$

where W follows a $N(0,2)$ distribution. Combining (6.13) with (6.14) we find

$$E[1 - F_Z(Z+A)] = 1 - F_Z\left(\frac{A}{\sqrt{2}}\right), \text{ or } E[F_Z(Z+A)] = F_Z\left(\frac{A}{\sqrt{2}}\right).$$

6.4 Distributions of Order Statistics in i.i.d. Samples

When the collection of observations X_1, X_2, X_3, ..., X_n are independent and identically distributed, identifying the specific distributions of order statistics proceeds in a straightforward manner. We will first consider the probability distribution of single, individual order statistics (such as the minimum and the maximum), and then move to the more complex circumstance of the joint distributions of order statistics.

6.4.1 The First Order Statistic (The Minimum)

The probability distribution of V, the minimum value of the collection of X_1, X_2, X_3, ..., X_n i.i.d. observations can be identified easily, once we take advantage of the unique property of the minimum, i.e., it is the smallest value in the sample. Our plan is to first identify $F_V(v)$, and then, if possible, find its derivative to obtain the probability density function $f_V(v)$.

In order to work directly with $F_V(v)$ we need to relate the event $V \leq v$ with an event involving X_1, X_2, X_3, ..., X_n. However, the event that the *Min* $(X_1, X_2, X_3, ..., X_n) \leq v$ implies a very complicated event for the collection of the n individual observations. The fact that $V \leq a$ could imply, that for some samples, all of the observations are less than a. On the other hand, for some values of a, V

$\leq a$ implies that only some of the individual X_i's are less than a. Since the event that Min $(X_1, X_2, X_3, \ldots, X_n) \leq v$ is quite complicated and raises a complicated enumeration problem, we will consider the alternative event denoted by $V > v$. If the minimum is greater than v, then, since each of the individual observations must be greater than or equal to the minimum, then all of the observations must be greater than v as well. We can now write for $-\infty \leq v \leq \infty$, .

$$1 - F_V(v) = P[V > v]$$
$$= P\left[\{X_1 \geq v\} \cap \{X_2 \geq v\} \cap \{X_3 \geq v\} \cap \ldots \cap \{X_n \geq v\}\right]. \tag{6.15}$$

We may now use the i.i.d. property of the sample to write

$$P\left[\{X_1 \geq v\} \cap \{X_2 \geq v\} \cap \{X_3 \geq v\} \cap \ldots \cap \{X_n \geq v\}\right]$$
$$= \prod_{i=1}^{n} P[X_i \geq v] = [1 - F_X(v)]^n. \tag{6.16}$$

where $F_X(x)$ is the cumulative probability distribution function of the random variable X. Combining expression (6.15) with (6.16) we now find

$$1 - F_V(v) = [1 - F_X(v)]^n$$
$$F_V(v) = 1 - [1 - F_X(v)]^n. \tag{6.17}$$

Thus, if $F_X(x)$ is differentiable, we can compute

$$f_V(v) = \frac{dF_V(v)}{dv} = \frac{d\left(1 - [1 - F_X(v)]^n\right)}{dv} = -n[1 - F_X(v)]^{n-1} \frac{d(-F_X(v))}{dv} \tag{6.18}$$
$$f_V(v) = n[1 - F_X(v)]^{n-1} f_X(v).$$

Thus the probability density function of the minimum v of n i.i.d. observations with cumulative distribution function $F_X(x)$ and probability density function $f_X(x)$ is $f_V(v) = n[1 - F_X(v)]^{n-1} f_X(v)$.

6.4.2 Components in Sequence

Consider an electronic component that is constructed from a sequence of electric elements. Each of the elements must function for the assembled component to function properly. If any single one of the elements fails, a fault indicator lights, and the assembled unit must be replaced (Figure 6.1 shows a component with three elements).

Assume that the ith element of the n elements in the sequence has a lifetime t_i that is a random variable that follows an exponential distribution with parameter λ. We will also assume that the lifetimes of each of the n elements are independent random variables that follow the same probability distribution. Our goal is to identify the expected lifetime of the electronic assembly.

Consideration of this problem's construction reveals that the assembly fails when any one of its elements fails. Another way to say this is that the assembly fails with the very first component failure. Thus, the expected lifetime of the unit is the minimum lifetime of the n elements. If V is the minimum lifetime of the component and X is the lifetime of an element in the assembly, then from expression (6.17),

$$F_V(v) = 1 - \left[1 - F_X(v)\right]^n. \tag{6.19}$$

Since $f_X(x) = \lambda e^{-\lambda x} \mathbf{1}_{[0,\infty)}(x)$, we may write the cumulative probability distribution function of the component is $F_V(v) = 1 - e^{-n\lambda v}$, and write the density function as $f_V(v) = n\lambda e^{-n\lambda v} \mathbf{1}_{[0,\infty)}(v) dv$. Thus, the lifetime of the component is a random variable that follows an exponential distribution with parameter $n\lambda$. The expected lifetime of this component is $E[v] = \dfrac{1}{n\lambda}$.

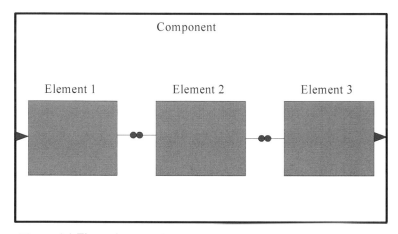

Figure 6.1 Three elements that must work in series in order for the component to work

Thus the average lifetime of the component is inversely proportional to the number of elements from which it is constructed. If complication is defined by the number of elements that are required to function in sequence (or in series) for the system to function, then the more complicated the system is, the shorter will be its expected lifetime.

6.4.3 The Last Order Statistic (The Maximum)

The distribution of the maximum of a collection of i.i.d. random variables X, X_2, X_3, ..., X_n may also be easily computed. In fact, since the observation that the maximum W of a collection of random variables is less than a specific value w implies that each random variable in the collection must also be less than that value w allows us to write

$$
\begin{aligned}
F_W(w) &= P\left[W \leq w\right] \\
&= P\left[\{X_1 \leq w\} \cap \{X_2 \leq w\} \cap \{X_3 \leq w\} \cap ... \cap \{X_n \leq w\}\right] \quad (6.20) \\
&= \prod_{i=1}^{n} F_{X_i}(w) = \left(F_X(w)\right)^n.
\end{aligned}
$$

If $F_X(w)$ has a derivative, we may write $f_W(w)$ as

$$
f_W(w) = \frac{dF_W(w)}{dw} = \frac{d\left[\left(F_X(w)\right)^n\right]}{dw} = n\left(F_X(w)\right)^{n-1} f_X(w). \quad (6.21)
$$

If for example, X_1, X_2, X_3, ..., X_n are i.i.d. random variables that follow an exponential distribution with parameter λ, then the distribution of the maximum W is

$$
f_W(w) = n\left(F_X(w)\right)^{n-1} f_X(w) = n\left(1 - e^{-\lambda w}\right)^{n-1} \lambda e^{-\lambda w} 1_{[0,\infty)}(w).
$$

We can demonstrate that this is a proper probability density function, i.e. that $\int_{\Omega_w} f_W(w) = 1$ (where Ω_W is the support of the random variable W) by writing

$$
\int_{\Omega_W} f_W(w) = \int_0^\infty n\left(1 - e^{-\lambda w}\right)^{n-1} \lambda e^{-\lambda w} dw. \quad (6.22)
$$

The right hand side of equation (6.22) can be evaluated through a one-to-one transformation of variables. Let $y = 1 - e^{-\lambda w}$. Then, for this transformation,

$dy = \lambda e^{-\lambda w} dw$.[2] An examination of the regions of integration reveals that, when $0 \le w \le \infty$, then $0 \le y \le 1$. Thus

$$\int_0^\infty n\left(1-e^{-\lambda w}\right)^{n-1} \lambda e^{-\lambda w} \mathbf{1}_{[0,\infty)} w dw = \int_0^1 ny^{n-1} dy = y^n \big]_0^1 = 1$$

One other way to demonstrate that $f_W(w)$ is a proper density function introduces a simple tool that will be very useful for the discussions of order statistics to follow. We may write the probability density function

$f_W(w) = n\left(1-e^{-\lambda w}\right)^{n-1} \lambda e^{-\lambda w} \mathbf{1}_{[0,\infty)}(w)$. Rather than attempt to integrate $f_W(w)$ directly, we can first recall from the binomial theorem (introduced in chapter 1)

that $(a+b)^n = \sum_{k=0}^{n} \binom{n}{k} a^k b^{n-k} = \sum_{k=0}^{n} \binom{n}{k} a^{n-k} b^k$. We may therefore write

$$\left(1-e^{-\lambda w}\right)^{n-1} = \sum_{k=0}^{n-1} \binom{n-1}{k} (-1)^k e^{-k\lambda w}. \tag{6.23}$$

Integrating $f_W(w)$, we see that

$$\int_{\Omega_W} f_W(w) = \int_0^\infty n\left(1-e^{-\lambda w}\right)^{n-1} \lambda e^{-\lambda w} dw$$

$$= \int_0^\infty n \sum_{k=0}^{n-1} \binom{n-1}{k} (-1)^k e^{-k\lambda w} \lambda e^{-\lambda w} dw. \tag{6.24}$$

Reversing the order of the sum and integral sign, write

---

[2]   This is identical to the probability integral transformation, in which $y = 1 - e^{-\lambda w} = F_W(w)$. However, in this circumstance, we may proceed by writing

$$\int_0^\infty n\left(1-e^{-\lambda w}\right)^{n-1} \lambda e^{-\lambda w} \mathbf{1}_{[0,\infty)}(w) dw = \int_0^1 n\left(F_W(w)\right)^{n-1} f_W(w) = F_W(w) \big]_0^1 = 1.$$

$$\int_{\Omega_W} f_W(w)dw = \int_0^{\infty} n \sum_{k=0}^{n-1} \binom{n-1}{k}(-1)^k e^{-k\lambda w} \lambda e^{-\lambda w} dw = n \sum_{k=0}^{n-1} \binom{n-1}{k}(-1)^k \int_0^{\infty} \lambda e^{-(k+1)\lambda w} dw$$

$$= n \sum_{k=0}^{n-1} \binom{n-1}{k}(-1)^k \frac{1}{k+1} \int_0^{\infty} \lambda(k+1) e^{-(k+1)\lambda w} dw$$

$$= n \sum_{k=0}^{n-1} \binom{n-1}{k}(-1)^k \frac{1}{k+1}$$

$$(6.25)$$

by recognizing that $\int_0^{\infty} \lambda(k+1) e^{-(k+1)\lambda w} dw = 1$. We now note that

$$n \sum_{k=0}^{n-1} \binom{n-1}{k}(-1)^k \frac{1}{k+1} = \sum_{k=0}^{n-1} \frac{n(n-1)!}{k!(n-1-k)!(k+1)}(-1)^k$$

$$= \sum_{k=0}^{n-1} \frac{n!}{(k+1)!(n-(k+1))!}(-1)^k = \sum_{k=1}^{n} \binom{n}{k}(-1)^{k-1} = (-1)\sum_{k=1}^{n} \binom{n}{k}(-1)^k$$

$$= (-1)\left[(1-1)^n - 1\right]$$

(See Chapter 1, section 1.4). Therefore

$$\int_{\Omega_W} f_W(w) = \int_0^{\infty} n\left(1 - e^{-\lambda w}\right)^{n-1} \lambda e^{-\lambda w} 1_{w \geq 0} dw = n \sum_{k=0}^{n-1} \binom{n}{k}(-1)^k \frac{1}{k+1}$$

$$= (-1)\sum_{k=1}^{n} \binom{n}{k}(-1)^k = (-1)(-1) = 1.$$

Example 6.4 U.S. Air Force air traffic controllers must commonly plan for air attack scenarios against targets that are protected by ground based and air based anti-aircraft weapons. In this scenario, two squads of aircraft are sent to attack a suspected chemical weapons depot. The target depot is surrounded by anti-aircraft weapons designed to destroy the swiftly approaching bombers. In order to suppress this antiaircraft weapons fire, the attack planners send a group of radar suppression and jamming aircraft. The idea is for the group of radar suppression aircraft to arrive at the chemical weapons factory before the bombers and destroy the antiaircraft weapons that would be used against the bombers, so that the bombing fleet can complete its mission without itself coming under attack. However, in order for this plan to succeed, the radar suppression aircraft must strike before the bombers arrive. The timing must be precise. If the bombers arrive too soon, before the radar suppressors finish, the bombers will be vulnerable to fire from the unsuppressed anti-aircraft missile batteries. If the bombers lag too far behind, the enemy will have ample time to detect the exposed bombers. What is the probability that the first of the bombers arrives after the last of the radar suppressors?

To solve this problem we will let X_i be the arrival time of the ith bomber and Y_j be the arrival plus attack time of the jth radar suppressor aircraft. Assume X_1, X_2, X_3,...X_n are i.i.d. observations from an exponential distribution with parameter λ_B, and Y_1, Y_2, Y_3, ... Y_m are i.i.d. observations from an exponential distribution with parameter λ_R. Then the probability that the first bomber arrives after the last radar suppressor is the probability that the $V = Min(X_1, X_2, X_3,...X_n)$, is greater than the $W = Max (Y_1, Y_2, Y_3, ... Y_m)$. Since each X_i is independent of Y_j for all $i = 1$ to n and all $j = 1$ to m. then V is independent of W. In order to solve this problem in airspace management, we only have to find the $P[V \geq W]$, where the probability density function for V is defined as

$$f_{V_n}(v) = n\lambda_B e^{-n\lambda_B v} \mathbf{1}_{[0,\infty)}(v) \text{ and } f_{W_m}(w) = m\left(1-e^{-\lambda_R w}\right)^{m-1} \lambda_R e^{-\lambda_R w} \mathbf{1}_{[0,\infty)}(w). \text{ Thus}$$

$$P[V \geq W] = \iint_{0<w<v<\infty} f_{V_n W_m}(v, w) dv dw$$

$$= \int_0^\infty \left[\int_w^\infty n\lambda_B e^{-n\lambda_B v} dv \right] m\left(1-e^{-\lambda_R w}\right)^{m-1} \lambda_R e^{-\lambda_R w} dw \qquad (6.26)$$

$$= \int_0^\infty m\left(1-e^{-\lambda_R w}\right)^{m-1} e^{-n\lambda_B w} \lambda_R e^{-\lambda_R w} dw.$$

Writing $\left(1-e^{-\lambda_R w}\right)^{m-1} = \sum_{k=0}^{m-1} \binom{m-1}{k}(-1)^k e^{-\lambda_R k w}$ as

$$P[V \geq W] = \int_0^\infty m\left(1 - e^{-\lambda_R w}\right)^{m-1} e^{-n\lambda_B w} \lambda_R e^{-\lambda_R w} dw$$

$$= \int_0^\infty m \sum_{k=0}^{m-1} \binom{m-1}{k} (-1)^k e^{-\lambda_R k w} e^{-n\lambda_B w} \lambda_R e^{-\lambda_R w} dw$$

$$= m \sum_{k=0}^{m-1} \binom{m-1}{k} (-1)^k \int_0^\infty e^{-\lambda_R k w} e^{-n\lambda_B w} \lambda_R e^{-\lambda_R w} dw$$

$$= m \sum_{k=0}^{m-1} \binom{m-1}{k} (-1)^k \left[\int_0^\infty \lambda_R e^{-\left(\lambda_R(k+1)+n\lambda_B\right)w} dw \right]$$

$$= m \sum_{k=0}^{m-1} \binom{m-1}{k} (-1)^k \left[\frac{\lambda_R}{\lambda_R(k+1) + n\lambda_B} \right].$$

For example, if the air controllers are working to vector in two radar suppression planes to be followed by three bombers, the probability that the bombers arrive after the radar suppression aircraft is ($n = 3$, $m = 2$)

$$P[V \geq W] = 2 \sum_{k=0}^{1} \binom{1}{k} (-1)^k \left[\frac{\lambda_R}{\lambda_R(k+1) + 3\lambda_B} \right]$$

$$= 2\lambda_R \left[\frac{1}{\lambda_R + 3\lambda_B} - \frac{1}{2\lambda_R + 3\lambda_B} \right].$$

(6.27)

∎

6.5 Expectations of Minimum and Maximum Order Statistics

Since order statistics, being functions of random variables, have probability distributions, we can compute their moments. For example, the work that we have completed thus far permits us to compute the expected value of the minimum and maximum order statistics. Let $V = Min\ (X_1, X_2, X_3,\ldots,X_n)$ and $W = Max\ (X_1, X_2, X_3,\ldots,X_n)$ where X_i's are i.i.d. with cumulative distribution function $F_X(x)$ and probability density function $f_X(x)$. Then,

$$E[V] = \int_{\Omega_V} v f_V(v) dv = \int_{\Omega_V} v n \left(1 - F_X(v)\right)^{n-1} f_X(v) dv$$

$$E[W] = \int_{\Omega_W} w f_W(w) dw = \int_{\Omega_W} w n \left(F_X(v)\right)^{n-1} f_X(w) dw.$$

(6.28)

For the special case in which $f_X(x) = \lambda e^{-\lambda x} \mathbf{1}_{[0,\infty)}(x)$, we may write

$$E[V] = \int_{\Omega_V} v f_V(v) dv = \int_{\Omega_V} vn \left(1 - F_X(v)\right)^{n-1} f_X(v) dv = \int_0^\infty vn e^{-\lambda(n-1)v} \lambda e^{-\lambda v} dv$$

$$= \int_0^\infty v \lambda n e^{-\lambda n v} dv = \frac{1}{n\lambda}.$$

Recall the problem of the electronics elements in sequence from section 6.4.2. In that problem, the minimum lifetime of the five elements was the lifetime of the assembled component. Suppose that the technicians need to be sure that a component with five elements in series is to have an average lifespan of 5 months. In this case, $\dfrac{1}{5\lambda} = 5$, or $\dfrac{1}{\lambda} = 25$. Therefore, for the assembled component to have an average life time of five months, a computation that is based on the minimum lifetime of each of the five components, then the average lifetime of each of the components must be 25 months, or much longer. By insisting that the minimum lifetime be sufficiently long, we are requiring that the average lifetime be very long.

To find the expected value of the maximum W, we write

$$E[W] = \int_{\Omega_W} w f_W(w) dw = \int_{\Omega_W} wn \left(F_X(v)\right)^{n-1} f_X(w) dw$$

$$= \int_0^\infty wn \left(1 - e^{-\lambda w}\right)^{n-1} \lambda e^{-\lambda w} dw. \tag{6.29}$$

To help evaluate this last integral, we invoke the binomial formula to write $\left(1 - e^{-\lambda w}\right)^{n-1} = \sum_{k=0}^{n-1} \binom{n-1}{k} (-1)^k e^{-\lambda k w}$ and

$$E[w] = \int_0^\infty wn \sum_{k=0}^{n-1} \binom{n-1}{k} (-1)^k e^{-\lambda k w} \lambda e^{-\lambda w} dw. \tag{6.30}$$

Interchanging the order of the integration and the summation signs we find

$$E[w] = n \sum_{k=0}^{n-1} \binom{n-1}{k} (-1)^k \int_0^\infty w e^{-\lambda k w} \lambda e^{-\lambda w} dw$$

$$= n \sum_{k=0}^{n-1} \binom{n-1}{k} (-1)^k \int_0^\infty \lambda w e^{-\lambda(k+1)w} dw$$

$$= n \sum_{k=0}^{n-1} \binom{n-1}{k} (-1)^k \frac{1}{k+1} \int_0^\infty \lambda(k+1) w e^{-\lambda(k+1)w} dw.$$

Now, recognize that $\int_0^\infty \lambda(k+1) w e^{-\lambda(k+1)w} dw = \dfrac{1}{\lambda(k+1)}$. Therefore

$$E[w] = n \sum_{k=0}^{n-1} \binom{n-1}{k} (-1)^k \frac{1}{\lambda(k+1)^2}.$$

We may now complete the computation

$$E[w] = n \sum_{k=0}^{n-1} \binom{n-1}{k} (-1)^k \frac{1}{\lambda(k+1)^2}$$

$$= \frac{n}{\lambda} \left[\frac{2^n - 1}{n} - 1 \right].$$

(6.31)

Example 6.5 Traffic Flow One of the major issues in urban life is automobile congestion, and traffic engineers work to design traffic patterns to improve the flow of moving traffic. In this particular circumstance, engineers are considering increasing the number of lanes on a road connecting two commonly used locales. The goal of the engineers is to decrease the automobile transit time for this stretch. While an obvious approach would be to increase the number of lanes, these experienced engineers recognize that increasing the number of lanes also increases the use of the highway, and the increased number of cars may offset the decrease in transit time produced by the additional lanes.

One parameterization of this is as follows. Define the transit time (i.e. the time it takes a car to travel between two points A and B) for a highway with k lanes of traffic be $X(k)$. Then assume that $X(k)$ follows a Gamma distribution with parameters αk and k. The probability density function of $X(k)$ is

$$f_{X(k)}(x) = \frac{(\alpha k)^k}{\Gamma(k)} x^{k-1} e^{-\alpha k x} 1_{[0,\infty)}(x).$$

Note the mean transit time $E[X(k)] = \dfrac{k}{\alpha k} = \dfrac{1}{\alpha}$. Thus the mean is independent of the number of lanes, a reflection of the observation that the increase in the number of traffic lanes may not decrease mean transit time in the long run. However, the shape of the distribution changes as a function of k (Figure 6.2) An additional question is how likely is it that the transit time on a k lane highway is less than the transit time of a highway with $k-1$ lanes of traffic. Let $X = X(k)$ and $Y = X(k-1)$. We need to compute $P[X < Y]$, the probability that the transit time on a k lane highway is less than the transit time on a $k-1$ lane highway. In order to prove this assertion, we will use.

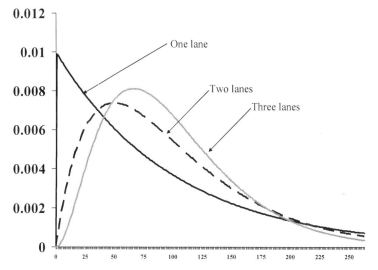

Figure 6.2 Probability density function of highway transit times as function of the number of lanes and population density.

Lemma 6.1 Let X be a random variable that follows a gamma distribution with parameters α and n where n is an integer. Denote its cumulative probability density function by $F_X(x)$. Then

$$F_X(c) = 1 - e^{-\alpha c} \sum_{j=0}^{n-1} \frac{\alpha^{n-j} c^{n-(j+1)}}{j!\,\Gamma(n-j)}.$$

The proof of the lemma is straightforward. We begin by writing that

$F_X(c) = 1 - \int_c^\infty \dfrac{\alpha^n}{\Gamma(n)} x^{n-1} e^{-\alpha x} dx$. We now evaluate this integral by creating a new

intermediate variable $Y = X - c$. For $c \le x \le \infty$, the range of the value of Y becomes $0 \le Y \le \infty$. We also note that $dx = dy$. Thus, we may write

$$\int_c^\infty \dfrac{\alpha^n}{\Gamma(n)} x^{n-1} e^{-\alpha x} dx = \int_0^\infty \dfrac{\alpha^n}{\Gamma(n)} (y+c)^{n-1} e^{-\alpha(y+c)} dy$$

$$= \dfrac{\alpha^n}{\Gamma(n)} e^{-\alpha c} \int_0^\infty (y+c)^{n-1} e^{-\alpha y} dy.$$

We may evaluate this integral by using the binomial formula by writing

$$(y+c)^{n-1} = \sum_{j=0}^{n-1} \binom{n-1}{j} y^j c^{n-(j+1)} .$$

Substituting   this   expression   into   the   integrand   of

$\dfrac{\alpha^n}{\Gamma(n)} e^{-\alpha c} \int_0^\infty (y+c)^{n-1} e^{-\alpha y} dy$ and simplifying, we obtain

$$\dfrac{\alpha^n}{\Gamma(n)} e^{-\alpha c} \int_0^\infty (y+c)^{n-1} e^{-\alpha y} dy = \dfrac{\alpha^n}{\Gamma(n)} e^{-\alpha c} \int_0^\infty \sum_{j=0}^{n-1} \binom{n-1}{j} y^j c^{n-(j+1)} e^{-\alpha y} dy =$$

$$\dfrac{\alpha^n}{\Gamma(n)} e^{-\alpha c} \sum_{j=0}^{n-1} \dfrac{\Gamma(n)}{\Gamma(j+1)\Gamma(n-j)} c^{n-(j+1)} \int_0^\infty y^j e^{-\alpha y} dy.$$

Since $\int_0^\infty y^j e^{-\alpha y} dy = \dfrac{\Gamma(j+1)}{\alpha^j}$. Therefore

$$\dfrac{\alpha^n}{\Gamma(n)} e^{-\alpha c} \int_0^\infty (y+c)^{n-1} e^{-\alpha y} dy = \dfrac{\alpha^n}{\Gamma(n)} e^{-\alpha c} \sum_{j=0}^{n-1} \dfrac{\Gamma(n)}{\Gamma(j+1)\Gamma(n-j)} c^{n-(j+1)} \dfrac{\Gamma(j+1)}{\alpha^j}$$

$$= e^{-\alpha c} \sum_{j=0}^{n-1} \dfrac{\alpha^{n-j} c^{n-(j+1)}}{\Gamma(n-j)}.$$

and $F_X(c) = 1 - \int_c^\infty \dfrac{\alpha^n}{\Gamma(n)} x^{n-1} e^{-\alpha x} dx = 1 - e^{-\alpha c} \sum_{j=0}^{n-1} \dfrac{\alpha^{n-j} c^{n-(j+1)}}{\Gamma(n-j)}$ ∎

Note that the integral in the above expression is a form of the incomplete gamma. We can now compute the probability that the transit time on a k lane highway is less than that on a $k-1$ lane highway. Begin with

$$P[X<Y]= \int_{X<Y} f_X(x)f_Y(y)dxdy$$

$$= \int_0^{\infty}\left[\int_0^{y}\frac{\left(\alpha(k-1)\right)^{k+1}}{\Gamma(k-1)}x^k e^{-\alpha(k+1)x}dx\right]\frac{(\alpha k)^{k}}{\Gamma(k)}y^{k-1}e^{-\alpha ky}dy$$

$$= \int_0^{\infty}\left[1-e^{-\alpha(k+1)y}\sum_{j=0}^{k}\frac{\left[\alpha(k-1)\right]^{k+1-j}y^{k-j}}{\Gamma(k-1-j)}\right]\frac{(\alpha k)^{k}}{\Gamma(k)}y^{k-1}e^{-\alpha ky}dy$$

$$= 1-\sum_{j=0}^{k}\frac{\left[\alpha(k-1)\right]^{k+1-j}}{\Gamma(k-1-j)}\frac{(\alpha k)^{k}}{\Gamma(k)}\int_0^{\infty}y^{2k-j-1}e^{-\alpha(2k+1)y}dy.$$

(6.32)

Since

$$\int_0^{\infty}y^{2k-j-1}e^{-\alpha(2k+1)y}dy=\frac{\Gamma(2k-j)}{\left[\alpha(2k+1)\right]^{2k-j}},$$

by inserting the value of this integral into the last line of expression (6.32) and obtain

$$P[X<Y]=1-\sum_{j=0}^{k}\frac{\left[\alpha(k+1)\right]^{k+1-j}}{j!\Gamma(k+1-j)}\frac{(\alpha k)^{k}}{\Gamma(k)}\frac{\Gamma(2k-j)}{\left[\alpha(2k+1)\right]^{2k-j}},$$

which, after a little simplification becomes

$$P[X<Y]=1-\frac{\alpha}{(k-1)!}\sum_{j=0}^{k}\frac{(2k-j-1)!(k+1)^{k+1-j}k^k}{j!(k-j)!}\frac{k^k}{(2k+1)^{2k-j}}$$

6.5 Distributions of Single Order Statistics

We have successfully identified the distribution of the minimum and maximum order statistics for a collection of independent and identically distributed random variables. Ultimately, we will identify the joint distribution of the minimum and maximum order statistics. In order to do this, we will need to devote some time to developing a heuristic tool. This device will permit us to write the joint probability density function for a collection of order statistics very easily under certain conditions. We will begin this development with an examination of the distribution of a single order statistic, and from there proceed to an examination of the joint distribution of order statistics from a sample of n independent and identically distributed observations.

6.5.1 The Heuristic Approach

As before, we start with a selection of n i.i.d. observations X_1, X_2, X_3, ..., X_n which are the realizations of a random variable with a cumulative distribution function $F_X(x)$, and the corresponding probability density function $f_X(x)$. Recall that we derived the probability density function of W = maximum $(X_1, X_2, X_3,$ ..., X_n) as $f_W(w) = n\left(F_X(w)\right)^{n-1} f_X(w)\mathbf{1}_{\Omega_W}(w)$. We were able to identify this density by taking advantage of the fact that if the maximum was less than a particular value w, then all of the observations in the sample must be less than w. However, it is possible to visualize this process from another perspective. We choose the maximum from n observations by selecting one observation from the n observations. We know that if this is the maximum W, then the remaining $n-1$ observations must be less than or equal to W. Thus, there are two classes of observation 1) the maximum and 2) all other values. If the maximum falls at a particular location in Ω_W, then all of the other observations can only take on values less than this (Figure 6.3)

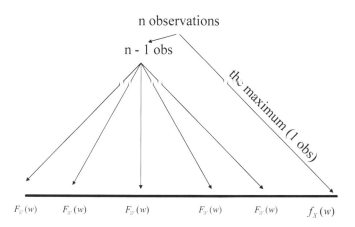

Figure 6.3 Dispersion of sample data points on the real line to determine maximum (w).

In this process, we are not particularly concerned about the relative order of these remaining $n - 1$ observations. Our only concern is that these $n - 1$ observations less than the maximum value in the sample. This observation is the key to the simplification of the process by which the probability density function of the nth order statistic is identified. We also know that there are $\binom{n}{n-1}$ ways to choose these two classes of observations. Once they are chosen, then the probability that the maximum is in the range $(w, w+dw)$ is $f_X(w)dw$. When we know W, the probability that the other $n-1$ observations are less than w is simply $\left[F_W(w)\right]^{n-1}$ (Figure 6.3).

We may now assemble the probability density function of W, the maximum of n i.i.d. observations as

$$f_W(w) = \binom{n}{1 \quad n-1}\left[F_X(w)\right]^{n-1} f_X(w)$$

$$= n\left[F_X(w)\right]^{n-1} f_\Lambda(w),$$

which is the result we derived earlier.

What aided us in this assembly of the probability density function of the maximum was the observation that we need not be concerned with the exact location of the $n - 1$ observations that were not the maximum. We only need focus on the fact that each of these $n - 1$ observations was less than or equal to W.

6.5.2 Distribution of the Sample Median

In this section, we will now derive the probability distribution of the sample median, M from a sample of an odd number of observations. Again, we assume that these observations are i.i.d. with a cumulative distribution function $F_X(x)$, and probability density function $f_X(x)$. Using the work from the previous examples, we demonstrate how we might compute the probability density function of the median (Figure 6.4). From Figure 6.4, we demonstrate that the identification of the median M has several possibilities. One is represented by a combinatorial component, and the second might be best considered as a probability component. The combinatorial $\binom{n}{\frac{n-1}{2} \quad 1 \quad \frac{n-1}{2}}$ counts the number

of ways one can choose n observations (assume n is odd) that

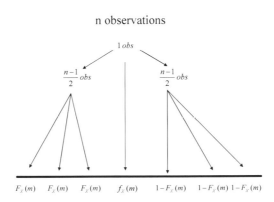

n observations

Figure 6.4 Dispersion of sample data points on the real line to determine the median (M).

fall into three classes 1) $\dfrac{n-1}{2}$ observations that are less than the median 2) one

observation that is the median and 3) $\dfrac{n-1}{2}$ observations whose values are

greater than the median.

The probability component computes the likelihood that wherever the median has fallen (an event that we consider to have occurred with "probability", $f_X(m)dm$, $\dfrac{n-1}{2}$ observations occur with values less than m, with

probability $F_X(m)$, and $\dfrac{n-1}{2}$ observations occur at values greater than m, each

with probability $1 - F_X(m)$. Thus, we may write the probability distribution of the median M as

$$f_M(m) = \begin{pmatrix} n \\ \dfrac{n-1}{2} \quad 1 \quad \dfrac{n-1}{2} \end{pmatrix} F_X(m)^{\frac{n-1}{2}} \left(1 - F_X(m)\right)^{\frac{n-1}{2}} f_X(m).$$

If the probability density function of x is exponential with parameter λ, then we can write

$$f_M(m) = \begin{pmatrix} n \\ \dfrac{n-1}{2} \quad 1 \quad \dfrac{n-1}{2} \end{pmatrix} \left(1 - e^{-\lambda m}\right)^{\frac{n-1}{2}} \left(e^{-\lambda m}\right)^{\frac{n-1}{2}} \lambda e^{-\lambda m} \mathbf{1}_{[0,\infty)}(m).$$

This is a probability distribution that we can work with fairly easily. For example to identify the expected value of the median of n i.i.d. random variables that follow an exponential distribution, with parameter λ, we find

$$E[M] = \int_{\Omega_S} m f_M(m) = \int_0^\infty m \begin{pmatrix} n \\ \dfrac{n-1}{2} \quad 1 \quad \dfrac{n-1}{2} \end{pmatrix} \left(1 - e^{-\lambda m}\right)^{\frac{n-1}{2}} \left(e^{-\lambda m}\right)^{\frac{n-1}{2}} \lambda e^{-\lambda m} \, dm$$

$$= \begin{pmatrix} n \\ \dfrac{n-1}{2} \quad 1 \quad \dfrac{n-1}{2} \end{pmatrix} \int_0^\infty m \left(1 - e^{-\lambda m}\right)^{\frac{n-1}{2}} \left(e^{-\lambda m}\right)^{\frac{n-1}{2}} \lambda e^{-\lambda m} \, dm.$$

Before proceeding, we can write

$$\left(1-e^{-\lambda m}\right)^{\frac{n-1}{2}} = \sum_{k=0}^{\frac{n-1}{2}} \binom{\frac{n-1}{2}}{k} e^{-k\lambda m} (-1)^k,$$

and

$$E[M] =$$

$$= \binom{n}{\frac{n-1}{2} \quad 1 \quad \frac{n-1}{2}} \int_0^\infty \left[m \sum_{k=0}^{\frac{n-1}{2}} \binom{\frac{n-1}{2}}{k} e^{-k\lambda m} (-1)^k \left(e^{-\lambda \frac{n-1}{2} m} \right) \lambda e^{-\lambda m} dm \right.$$

$$= \binom{n}{\frac{n-1}{2} \quad 1 \quad \frac{n-1}{2}} \sum_{k=0}^{\frac{n-1}{2}} \binom{\frac{n-1}{2}}{k} (-1)^k \int_0^\infty \lambda m e^{-\left((k+1)+\frac{n-1}{2}\right)\lambda m} dm$$

This integral can be easily simplified as follows

$$\int_0^\infty \lambda s e^{-\left((k+1)+\frac{n-1}{2}\right)\lambda s} ds = \frac{\lambda}{(k+1)+\frac{n-1}{2}} \int_0^\infty \left((k+1)+\frac{n-1}{2}\right) s e^{-\left((k+1)+\frac{n-1}{2}\right)\lambda s} ds =$$

$$= \frac{\lambda}{\left((k+1)+\frac{n-1}{2}\right)}.$$

Continuing, we obtain

$$E[M] = \binom{n}{\frac{n-1}{2} \quad 1 \quad \frac{n-1}{2}} \sum_{k=0}^{\frac{n-1}{2}} \binom{\frac{n-1}{2}}{k} \frac{(-1)^k \lambda}{\left((k+1)+\frac{n-1}{2}\right)}. \qquad (6.33)$$

For the solution when n is even, assume $\frac{(n+2)}{2}$ observations are less than the median and follow the same development as above.

6.5.3 Distribution of Any Percentile Value

The techniques discussed thus far in this section can be easily adapted to the instances of finding the distribution of any of the percentile values of a sample of observations that are i.i.d. with cumulative distribution function $F_X(x)$ and probability density function $f_X(x)$. We want to find the probability density function of the pth percentile value from the sample of n observations. Begin by defining the random variable $S = p^{th}$ percentile value and use the heuristic approach to identifying the distribution of $f_S(s)$. To simply this consideration, then, if there are a total of 100 observations (Figure 6.5).

n observations

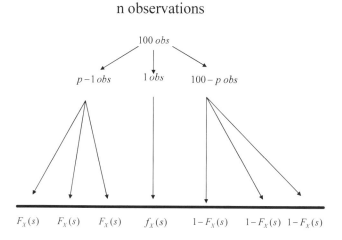

Figure 6.5 Dispersion of sample data points on the real line to determine the distribution of the pth percentile value (s)

Then, there are $p-1$ observations that take on a value less than s, and $100-p$ observations that have values that are greater than s. We can now write the density function of the pth order statistic as

$$f_S(s) = \binom{100}{p-1 \quad 1 \quad 100-p}\left(F_X(s)\right)^{p-1} f_X(s)\left(1-F_X(s)\right)^{100-p}. \qquad (6.34)$$

If, for example, X_i follows a $U(0,1)$ distribution for X_i, $i = 1$ to 100, then

$$f_S(s) = \binom{100}{p-1 \ \ 1 \ \ 100-p} s^{p-1} (1-s)^{100-p} \mathbf{1}_{[0,1]}(s).$$

We can rewrite the density of S as

$$f_S(s) = \frac{\Gamma(101)}{\Gamma(p)\Gamma(100-p+1)} s^{p-1} (1-s)^{100-p} \mathbf{1}_{[0,1]}(s).$$

This is a beta distribution. A graph of the beta distribution for different percentile values is provided (Figure 6.6).

6.6 Joint Distributions of Order Statistics

The tool that we have found so useful in developing the probability distribution function for a collection of independent and identically distributed random

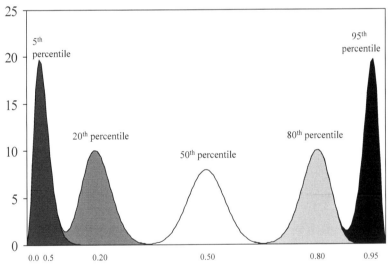

Figure 6.6 Probability density functions for each of the 5th, 20th, 50th, 80th, and 95th percentile values for observations from a U[0,1] distribution

variables can be generalized to identify the joint distributions of order statistics. These joint distributions can be particularly useful if the statistician wishes to evaluate the relationship between two order statistics, or to identify the distribution of a function of the order statistics. An example of this last concept is the identification of the distribution of the range, R, of collection of observations, where $R = X_{[n]} - X_{[1]}$.

6.6.1 Joint Distribution of the Minimum and Maximum Order Statistics

Assume that we have a collection of observations X_1, X_2, X_3, ..., X_n which are independent and identically distributed. Let their common cumulative distribution function be $F_X(x)$, and their common probability density function be $f_X(x)$. If V is the minimum $(X_1, X_2, X_3, ..., X_n)$, and W is the maximum $(X_1, X_2, X_3, ..., X_n)$, we seek the joint probability density function of V and W, $f_{V,W}(v,w)$. For a sample of n observations, we know that v falls in some interval $(v, v + dv)$ with "probability" $f_X(v)dv$, W falls in an interval $(w, w + dw)$ with "probability" $f_X(w)dw$ and the remaining $n - 2$ observations fall in between the values V and W (Figure 6.8).

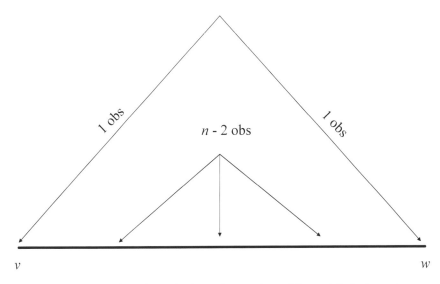

Figure 6.7 Dispersion of sample data points on the real line to jointly determine the minimum (v) and maximum (w).

Out of n observations we can count the ways that we select three classes of observations 1) the minimum (1 observation) 2) the maximum (one observation and 3) the $n - 2$ remaining observations whose only criterion for selection is that they must fall between v and w. There are $\begin{pmatrix} n \\ 1 \quad n-2 \quad 1 \end{pmatrix}$ ways to select these three classes of observations. We also know that each of these selections occurs with probability $f_X(v)[F_X(w)-F_X(v)]^{n-2} f_X(w)$. Of course we must add the additional constraint that $V \le W$. Thus the joint probability density function of the minimum and maximum of a collection of i.i.d. observations is

$$f_{V,W}(v,w) = \begin{pmatrix} n \\ 1 \quad n-2 \quad 1 \end{pmatrix} f_X(v)[F_X(w)-F_X(v)]^{n-2} f_X(w)\mathbf{1}_{v \le w}$$
$$= n(n-1)f_X(v)[F_X(w)-F_X(v)]^{n-2} f_X(w)\mathbf{1}_{v \le w}. \tag{6.35}$$

The use of the joint distribution of the minimum and maximum of a collection of observations is illustrated by the following example:

Example 6.6 Sunday mornings are among the least busy times in a hospital's emergency room. Assume that the patient arrivals after midnight Sunday morning (12AM) follow an exponential distribution with parameter λ. The issue that concerns the emergency room administrator is the possibility of clustering. Clustering is when many patients appear in the emergency room in a very short period of time. It is possible that there will be absolutely no arrivals to the emergency room for several hours, followed by a cluster of emergency room visits. This event can overwhelm a staff that has not been overworked during a shift with an anticipated low patient visit rate. One way to characterize this event is in terms of order statistics.

We focus on the eight hours between 12AM midnight and 8AM. During this time period, n patients arrive at the emergency room with arrival times $X_1, X_2, X_3, \ldots, X_n$. Let V be the minimum $(X_1, X_2, X_3, \ldots, X_n)$, and W the maximum $(X_1, X_2, X_3, \ldots, X_n)$. What is the probability that all patients arrive between 3 AM and 7AM? We can compute this probability directly from expression (6.35). We simply seek $P[3 \le V \le W \le 7]$. This can be written as

$$P[3 \le V \le W \le 7] = \iint_{3 \le v \le w \le 7} f_{V,W}(v,w)dvdw$$

Since the X_i's are i.i.d. exponential with parameter λ, we can utilize expression (6.35) to write

$$f_{V,W}(v,w) = n(n-1)f_X(v)\left[F_X(w)-F_X(v)\right]^{n-2} f_X(w)\mathbf{1}_{[0\leq v\leq w<\infty]}(v,w)$$

$$= n(n-1)\lambda e^{-\lambda v}\mathbf{1}_{0\leq v\leq\infty}\left[\left(1-e^{-\lambda w}\right)-\left(1-e^{-\lambda v}\right)\right]^{n-2}\lambda e^{-\lambda w}\mathbf{1}_{[0\leq v\leq w<\infty]}(v,w)$$

$$= n(n-1)\lambda e^{-\lambda v}\left[e^{-\lambda v}-e^{-\lambda w}\right]^{n-2}\lambda e^{-\lambda w}\mathbf{1}_{0\leq w\leq v<\infty}(v,w).$$

Applying the binomial theorem to $\left[e^{-\lambda v}-e^{-\lambda w}\right]^{n-2}$, we obtain

$$\left[e^{-\lambda v}-e^{-\lambda w}\right]^{n-2} = \sum_{k=0}^{n-2}\binom{n-2}{k}(-1)^{n-2-k}e^{-\lambda kv}e^{-\lambda(n-2-k)w}. \qquad (6.36)$$

Substituting this expression reveals

$$f_{V,W}(v,w) = n(n-1)\lambda e^{-\lambda v}\sum_{k=0}^{n-2}\binom{n-2}{k}(-1)^{n-2-k}e^{-\lambda kv}e^{-\lambda(n-2-k)w}\lambda e^{-\lambda w}\mathbf{1}_{0\leq w\leq v<\infty}(v,w)$$

$$= n(n-1)\sum_{k=0}^{n-2}\binom{n-2}{k}(-1)^{n-2-k}\lambda e^{-\lambda(k+1)v}\lambda e^{-\lambda(n-k-1)w}\mathbf{1}_{0\leq w\leq v<\infty}(v,w). \qquad (6.37)$$

Thus,

$$P[3\leq V\leq W\leq 7] = \iint_{3\leq v\leq w\leq 7} f_{V,W}(v,w)dvdw$$

$$= \iint_{3\leq v\leq w\leq 7} n(n-1)\sum_{k=0}^{n-2}\binom{n-2}{k}(-1)^{n-2-k}\lambda e^{-\lambda(k+1)v}\lambda e^{-\lambda(n-k-1)w}dvdw$$

$$= n(n-1)\sum_{k=0}^{n-2}\binom{n-2}{k}(-1)^{n-2-k}\iint_{3\leq v\leq w\leq 7}\lambda e^{-\lambda(k+1)v}\lambda e^{-\lambda(n-k-1)w}dvdw$$

$$= n(n-1)\sum_{k=0}^{n-2}\binom{n-2}{k}(-1)^{n-2-k}\int_3^7\left[\int_v^7\lambda e^{-\lambda(n-k-1)w}dw\right]\lambda e^{-\lambda(k+1)v}dv.$$

Writing

$$\int_3^7 \left[\int_v^7 \lambda e^{-\lambda(n-k-1)w} dw \right] \lambda e^{-\lambda(k+1)v} dv$$

$$= \int_3^7 \left[\frac{e^{-\lambda(n-k-1)v} - e^{-7\lambda(n-k-1)}}{n-k-1} \right] \lambda e^{-\lambda(k+1)v} dv$$

$$= \frac{1}{n-k-1} \left[\int_3^7 \lambda e^{-n\lambda v} dv - e^{-7\lambda(n-k-1)} \int_3^7 \lambda e^{-\lambda(k+1)v} dv \right]$$

$$= \frac{1}{n-k-1} \left[\frac{1}{n} \int_3^7 \lambda n e^{-n\lambda v} dv - \frac{e^{-7\lambda(n-k-1)}}{k+1} \int_3^7 \lambda(k+1) e^{-\lambda(k+1)v} dv \right]$$

$$= \frac{1}{n-k-1} \left[\frac{e^{-3n\lambda} - e^{-7n\lambda}}{n} - \frac{e^{-7\lambda(n-k-1)} \left(e^{-3\lambda(k+1)} - e^{-7\lambda(k+1)} \right)}{k+1} \right].$$

We may now compute

$$P[3 \le V \le W \le 7] = n(n-1) \sum_{k=0}^{n-2} \binom{n-2}{k} (-1)^{n-2-k} \int_3^7 \left[\int_v^7 \lambda e^{-\lambda(n-k-1)w} dw \right] \lambda e^{-\lambda(k+1)v} dv$$

$$- n(n-1) \sum_{k=0}^{n-2} \binom{n-2}{k} (-1)^{n-2-k} \left[\frac{1}{n-k-1} \left[\frac{e^{-3n\lambda} - e^{-7n\lambda}}{n} - \frac{e^{-7\lambda(n-k-1)} \left(e^{-3\lambda(k+1)} - e^{-7\lambda(k+1)} \right)}{k+1} \right] \right]. \qquad (6.38)$$

To identify the effect of clustering of emergency room visits, we may find the probability of clustering during any time period. Consider the event that $V \ge \dfrac{L}{L+1} W$, where L is an arbitrary positive integer. Here, $L = 1$ implies that $V \ge \dfrac{1}{2} W$; For example, if W = 4:00AM (or 4 hours after midnight), then $2 \le V \le W = 4$, or all of the patients arrival times cluster between 2:00AM and 4:00AM. Similarly, $W = 6$ and $L = 5$ is the event that all of the patient arrival times cluster between 5AM and 6AM. In order to measure the probability of clustering regardless of the time during the first eight hours of Sunday morning, we wish to find $P\left[V \ge \dfrac{L}{L+1} W \right]$ for $0 \le W \le 8$ and fixed value L. We begin as before

$$P\left[V \geq \frac{L}{L+1}W\right]$$

$$= n(n-1)\sum_{k=0}^{n-2}\binom{n-2}{k}(-1)^{n-2-k}\iint\limits_{\frac{L}{L+1}w \leq v \leq w \leq 8} \lambda e^{-\lambda(k+1)v}\lambda e^{-\lambda(n-k-1)w}\,dvdw. \tag{6.39}$$

which may be written as

$$P\left[V \geq \frac{L}{L+1}W\right]$$

$$= n(n-1)\sum_{k=0}^{n-2}\binom{n-2}{k}(-1)^{n-2-k}\int_0^8\left[\int_{\frac{L}{L+1}w}^{w}\lambda e^{-\lambda(k+1)v}\,dv\right]\lambda e^{-\lambda(n-k-1)w}\,dw. \tag{6.40}$$

We proceed as before, this time evaluating $\displaystyle\int_0^8\left[\int_{\frac{L}{L+1}w}^{w}\lambda e^{-\lambda(k+1)v}\,dv\right]\lambda e^{-\lambda(n-k-1)w}\,dw$

$$\int_0^8\left[\int_{\frac{L}{L+1}w}^{w}\lambda e^{-\lambda(k+1)v}\,dv\right]\lambda e^{-\lambda(n-k-1)w}\,dw$$

$$= \int_0^8\left[\frac{1}{k+1}\left(e^{\frac{-\lambda(k+1)L}{L+1}w}-e^{-\lambda(k+1)w}\right)\right]\lambda e^{-\lambda(n-k-1)w}\,dw \tag{6.41}$$

$$= \frac{1}{k+1}\left[\int_0^8\lambda e^{-\lambda\left[\frac{L(k+1)}{L+1}+n-(k+1)\right]w}\,dw - \int_0^8\lambda e^{-n\lambda w}\,dw\right]$$

$$= \frac{1}{k+1}\left[\frac{L+1}{L(k+1)+(n-k-1)(L+1)}\left(1-e^{-8\lambda\left[\frac{L(k+1)}{L+1}+n-(k+1)\right]}\right) - \frac{1}{n}\left(1-e^{-8\lambda n}\right)\right].$$

Substituting the last line of expression (6.41) into the last line of expression (6.40), we have

$$P\left[v \geq \frac{L}{L+1} w\right]$$

$$= n(n-1) \sum_{k=0}^{n-2} \binom{n-2}{k} (-1)^{n-2-k} \frac{1}{k+1} \left[\frac{L+1}{L(k+1)+(n-k-1)(L+1)} \left(1 - e^{-8\lambda \left[\frac{L(k+1)}{L+1} + n-(k+1) \right]} \right) - \frac{1}{n} \left(1 - e^{-8\lambda n} \right) \right].$$

(6.42)

6.6.2 Distribution of the Range

The range of a collection of random variables is simply the difference between the maximum value and the minimum value of a set of observations. It is an important tools for practicing statisticians, providing a useful measure of dispersion of the sample of observations. Using the results of the previous section, we can find the probability density function of this important quantity for certain commonly used probability distributions.

As a first example, let X_i, $i = 1,\ldots, n$ be a collection of i.i.d. observations from the $U[0,1]$ distribution. We seek the probability density function of $R = W - V$, where v is the minimum observation of the sample and w is the maximum observation of the sample. Our plan will be as follows. We will first identify the joint distribution of V, W, and then transform the two variables V and W to the (R, W) space. then carry out a two variable-to-two variable transformation in order to identify the joint distribution function of R and W. We will then integrate over the range of W, producing the marginal probability density function of R.

We know from the previous section that the joint probability density function of V and W, $f_{V,W}(v,w)$ is

$$f_{V,W}(v, w) = n(n-1) f_X(v) \left(F_X(w) - F_X(v) \right)^{n-2} f_X(w) \mathbf{1}_{v \leq w}.$$ (6.43)

In this example the random variable X follows a uniform $(0,1)$ distribution. Therefore

$$f_{V,W}(v, w) = n(n-1)(w-v)^{n-2} \mathbf{1}_{0 \leq v \leq w \leq 1}(v, w).$$

We now carry out the (v, w) to (r, s) transformation where $r = w - v$ and $s = w$. The Jacobian of this transformation is easily seen to be 1. The relevant region (r, s) is, as we anticipated $0 \leq r \leq s \leq 1$. Thus we have

$$f_{R,S}(r, s) = f_{V,W}(r, s) J[(V, W) \rightarrow (R, S)]$$
$$= n(n-1) r^{n-2} \mathbf{1}_{0 \leq r \leq s \leq 1}(r, s),$$

and if Ω_s is the set of all s such that $0 \le r \le s \le 1$, then we find the distribution of R as

$$f_R(r) = \int_{\Omega_S} f_{R,S}(r,s)ds$$

$$= n(n-1)r^{n-2} dr \int_r^1 ds \qquad (6.44)$$

$$= n(n-1)r^{n-2}(1-r)\mathbf{1}_{[0,1]}(r),$$

which we recognize as a Beta $(n-1, 2)$ distribution.

     We can also derive the distribution of R, when the observation are i.i.d. exponential (λ). As we have seen in the previous section

$$f_{V,W}(v,w) = n(n-1)\lambda e^{-\lambda v}\left(e^{-\lambda v} - e^{-\lambda w}\right)^{n-2}\lambda e^{-\lambda w}\mathbf{1}_{0 \le v \le w < \infty}, \qquad (6.45)$$

which we have shown can be written as

$$f_{V,W}(v,w) = n(n-1)\sum_{k=0}^{n-2}\binom{n-2}{k}(-1)^{n-2-k}\lambda e^{-\lambda(k+1)v}\lambda e^{-\lambda(n-k-1)w}\mathbf{1}_{0 \le v \le w < \infty}. (6.46)$$

by applying the binomial theorem to $\left(e^{-\lambda v} - e^{-\lambda w}\right)^{n-2}$. We now carry out the same transformation as in the previous example, allowing $R = W - V$ and $S = W$. Again, the Jacobian of this transformation is one. We now write

$$f_{R,S}(r,s) = f_{V,W}(r,s)J[(v,w) \to (r,s)]$$

$$= n(n-1)\sum_{k=0}^{n-2}\binom{n-2}{k}(-1)^{n-2-k}\lambda e^{-\lambda(k+1)(s-r)}\lambda e^{-\lambda(n-k-1)s}\mathbf{1}_{0 \le r \le s < \infty}(r,s)$$

$$= n(n-1)\sum_{k=0}^{n-2}\binom{n-2}{k}(-1)^{n-2-k}\lambda e^{-\lambda(-k-1)r}\lambda e^{-\lambda ns}\mathbf{1}_{0 \le r \le s < \infty}(r,s).$$

We now integrate over Ω_s to find the marginal distribution of R.

$$f_R(r) = \int_{\Omega_S} f_{R,S}(r,s)dr$$

$$= (n-1)\sum_{k=0}^{n-2}\binom{n-2}{k}(-1)^{n-2-k}\lambda e^{-\lambda(-k-1)r}\mathbf{1}_{0\leq r\leq\infty}\int_r^{\infty}\lambda n e^{-\lambda ns}\,ds$$

$$= (n-1)\sum_{k=0}^{n-2}\binom{n-2}{k}(-1)^{n-2-k}\lambda e^{-\lambda(-k-1)r}\left(1-e^{-\lambda nr}\right)\mathbf{1}_{[0,\infty)}(r) \qquad (6.47)$$

$$= (-1)^{n-2}\lambda(n-1)\left(1-e^{-\lambda nr}\right)\sum_{k=0}^{n-2}\binom{n-2}{k}e^{-\lambda(-k-1)r}\mathbf{1}_{[0,\infty)}(r)$$

$$= (-1)^{n-2}\lambda(n-1)e^{\lambda r}\left(1-e^{-\lambda r}\right)^{n-2}\mathbf{1}_{[0,\infty)}(r)$$

6.6.3 Interquartile Ranges

We have identified a measure of dispersion from a collection of observations that are independent and follow the same distribution from two order statistics. We can continue this process by examining, not the minimum and maximum, but the $V = 25^{\text{th}}$ and $W = 75^{\text{th}}$ percentile values. The interquartile range $T = W - V$, like the range, is a measure of dispersion. However, it is not quite so variable. If either the minimum or the maximum is unusually extreme, then the range can be uncharacteristically large. However, the interquartile range is protected from the influence of these extreme observations. Thus, the interquartile range is likely to be somewhat more stable than the range.

Our goal is to identify the distribution of T, the interquartile range from a collection of 100 independent observations sampled from U(0,1) distribution. We will proceed as before, first identifying the joint distribution of V and W, then carrying out a two-variable-to-two-variable transformation. Finally, we will integrate out the auxiliary variable in order to identify $f_T(t)$, the probability distribution function for the interquartile range.

Using our heuristic rule, we can write at once the joint probability density function for V and W, the 25th and 75th interquartile range from a collection of 100 i.i.d. observations from the $U(0,1)$ distribution.

$$f_{V,W}(v,w) =$$

$$\binom{100}{24\ \ 1\ \ 50\ \ 1\ \ 24}\left(F_X(v)\right)^{24}f_X(v)\left(F_X(w)-F_X(v)\right)^{50}f_X(w)\left(1-F_X(w)\right)^{24}\mathbf{1}_{0\leq v\leq w\leq 1}(v,w) \qquad (6.48)$$

$$= \binom{100}{24\ \ 1\ \ 50\ \ 1\ \ 24}v^{24}\left(w-v\right)^{50}\left(1-w\right)^{24}\mathbf{1}_{0\leq v\leq w\leq 1}(v,w).$$

We can now use a simple transformation of variable procedure. As before, we define $T = S - V$. and $W = S$. The region in the (s, t) plane that corresponds to $0 \leq V \leq W \leq 1$ is $0 \leq T \leq S \leq 1$. We can therefore write $f_{S,T}(s,t)$ as

$$f_{S,T}(s,t) = \begin{pmatrix} 100 \\ 24 \ 1 \ 50 \ 1 \ 24 \end{pmatrix}(s-t)^{24} t^{50}(1-s)^{24} \mathbf{1}_{0 \leq s \leq t \leq 1}(s,t). \quad (6.49)$$

Applying the binomial formula to $(s-t)^{24}$, we can write

$$f_{S,T}(s,t) = \begin{pmatrix} 100 \\ 24 \ 1 \ 50 \ 1 \ 24 \end{pmatrix}\sum_{k=0}^{24}\binom{24}{k}s^k t^{74-k}(-1)^{74-k} t^{50}(1-s)^{24}\mathbf{1}_{0 \leq t \leq s \leq 1}(s,t).$$

It now remains for us to identify the marginal distribution $f_T(t)$. If Ω_s is the set of all s such that $0 \leq r \leq s \leq 1$, we may write

$$f_T(t) = \int_{\Omega_s} f_{S,T}(s,t)$$

$$= \begin{pmatrix} 100 \\ 24 \ 1 \ 50 \ 1 \ 24 \end{pmatrix}\sum_{k=0}^{24}\binom{24}{k}t^{74-k}(-1)^{74-k}dt\mathbf{1}_{0 \leq t \leq 1}(t)\int_t^1 s^k(1-s)^{24}ds. \quad (6.50)$$

and it remains for us to integrate the incomplete beta function $\int_t^1 s^k(1-s)^{24}ds.$ We proceed by carrying out a second random variable transformation. The difficulty posed by this integral is the range of integration. If we could transform from the s space to the u space where $T \leq S \leq 1$ implied that $0 \leq U \leq 1$, we would be able to complete the integration. We start with the simplest type of linear transformation. The fact that $S = T$ implies $u = 0$ suggests that we begin with the transform $U = S - T$. Continuing to develop and explore this transformation reveals that $U = \dfrac{S-T}{1-T}$, or $S = T + U(1-T)$ is precisely the transformation we seek. The Jacobian of this transformation is 1 and the integral becomes

$$\int_t^1 s^k(1-s)^{24}ds = \int_0^1 (t+u(1-t))^k((1-t)(1-u))^{24}du$$

$$= (1-t)^{24}\int_0^1 (t+u(1-t))^k(1-u)^{24}du.$$

This development will continue with another application of the binomial formula, this time $(t + u(1 - t))^k = \sum_{j=0}^{k} \binom{k}{j} u^j t^{k-j} (1-t)^j$. We can now rewrite the incomplete beta integral as

$$\int_t^1 s^k (1-s)^{24} \, ds = (1-t)^{24} \int_0^1 \sum_{j=0}^{k} \binom{k}{j} u^j t^{k-j} (1-t)^j (1-u)^{24} \, du$$

$$= \sum_{j=0}^{k} \binom{k}{j} t^{k-j} (1-t)^{24+j} \int_0^1 u^j (1-u)^{24} \, du$$

$$= \sum_{j=0}^{k} \binom{k}{j} \frac{\Gamma(j+1)\Gamma(25)}{\Gamma(26+j)} t^{k-j} (1-t)^{24+j} \int_0^1 \frac{\Gamma(26+j)}{\Gamma(j+1)\Gamma(25)} u^j (1-u)^{24} \, du$$

$$= \sum_{j=0}^{k} \binom{k}{j} \frac{\Gamma(j+1)\Gamma(25)}{\Gamma(26+j)} t^{k-j} (1-t)^{24+j},$$

and substituting this result into expression (6.50) we have

$$f_T(t)$$

$$= \binom{100}{24 \ \ 1 \ \ 50 \ \ 1 \ \ 24} \sum_{k=0}^{24} \binom{24}{k} t^{74-k} (-1)^{74-k} \mathbf{1}_{0 \le t \le 1}(t) \sum_{j=0}^{k} \binom{k}{j} \frac{\Gamma(j+1)\Gamma(25)}{\Gamma(26+j)} t^{k-j} (1-t)^{24+j} \quad (6.51)$$

$$= \binom{100}{24 \ \ 1 \ \ 50 \ \ 1 \ \ 24} \sum_{k=0}^{24} \sum_{j=0}^{k} \binom{24}{k} \binom{k}{j} \frac{\Gamma(j+1)\Gamma(25)}{\Gamma(26+j)} t^{74-j} (1-t)^{24+j} (-1)^{74-k} \mathbf{1}_{0 \le t \le 1}(t),$$

and simplification of the factorial expressions reveals that the probability density function of the interquartile range is

$$f_T(t) = \frac{100!}{50!} \sum_{k=0}^{24} \sum_{j=0}^{k} \frac{(-1)^{74-k} t^{74-j} (1-t)^{24+j}}{(24-k)!(k-j)!(25+j)!} \mathbf{1}_{0 \le t \le 1}(t). \qquad (6.52)$$

We can use this same approach to identify the interquartile range for a collection of 100 random variables that are independent and follow the exponential distribution with parameter λ. Let P be the 25th percentile and Q be the 75th percentile. Then we have

$$f_{P,Q}(p,q) =$$

$$\begin{pmatrix} & 100 & \\ 24 & 1 & 50 & 1 & 24 \end{pmatrix} \left(F_X(p)\right)^{24} f_X(p)\left(F_X(q)-F_X(p)\right)^{50} f_X(q)\left(1-F_X(q)\right)^{24} \mathbf{1}_{p \leq q} \quad (6.53)$$

$$= \begin{pmatrix} & 100 & \\ 24 & 1 & 50 & 1 & 24 \end{pmatrix} \left(1-e^{-\lambda p}\right)^{24} \lambda e^{-\lambda p}\left(e^{-\lambda p}-e^{-\lambda q}\right)^{50} \lambda e^{-\lambda q} e^{-24\lambda q} \mathbf{1}_{0 \leq p \leq q \leq \infty}$$

In order to proceed, we will invoke the binomial formula twice and note

$$\left(1-e^{-\lambda p}\right)^{24} = \sum_{j=0}^{24} \binom{24}{j}(-1)^j e^{-\lambda jp}$$

$$\left(e^{-\lambda p}-e^{-\lambda q}\right)^{50} = \sum_{k=0}^{50} \binom{50}{k}(-1)^{50-k} e^{-\lambda kp} e^{-\lambda(50-k)q}.$$

Inserting these summations into the last line of expression (6.53) we find

$$f_{P,Q}(p,q) =$$

$$= \begin{pmatrix} & 100 & \\ 24 & 50 & 24 \end{pmatrix} \left(\sum_{j=0}^{24} \binom{24}{j}(-1)^j e^{-\lambda jp} \right) \lambda e^{-\lambda p} \left(\sum_{k=0}^{50} \binom{n}{k}(-1)^{50-k} e^{-\lambda kp} e^{-\lambda(50-k)q} \right) \lambda e^{-\lambda q} e^{-24\lambda q} \mathbf{1}_{0 \leq p \leq q \leq \infty}$$

$$= \begin{pmatrix} & 100 & \\ 24 & 50 & 24 \end{pmatrix} \sum_{k=0}^{50}\sum_{j=0}^{24} \binom{n}{k}\binom{24}{j}(-1)^j(-1)^{50-k} \lambda e^{-\lambda(k+j+1)p} \lambda e^{-\lambda(74-k+1)q} \mathbf{1}_{0 \leq p \leq q \leq \infty}$$

As before, we define the interquartile range as $T = Q - P$ and $S = Q$. The Jacobian of this transformation is 1, and the range function $\mathbf{1}_{0 \leq p \leq q < \infty}$ is equivalent to $\mathbf{1}_{0 \leq t \leq s < \infty}$ in the (s, t) space. We may therefore write

$$f_{S,T}(s,t) = f_{P,Q}(s,t)J(p,q) \to (s,t))$$

$$= \begin{pmatrix} & 100 & \\ 24 & 50 & 24 \end{pmatrix} \sum_{k=0}^{50}\sum_{j=0}^{24} \binom{n}{k}\binom{24}{j}(-1)^j(-1)^{50-k} \lambda e^{-\lambda(k+j+1)(s-t)} \lambda e^{-\lambda(74-k+1)s} \mathbf{1}_{0 \leq t \leq s < \infty} \quad (6.54)$$

$$= \begin{pmatrix} & 100 & \\ 24 & 50 & 24 \end{pmatrix} \sum_{k=0}^{50}\sum_{j=0}^{24} \binom{n}{k}\binom{24}{j}(-1)^j(-1)^{50-k} \lambda e^{-\lambda((k+j+1)(t))} \lambda e^{-\lambda(74+j+2)s} \mathbf{1}_{0 \leq t \leq s < \infty}.$$

We find $f_T(t)$ by integrating the joint density over $\Omega_s = 0 \leq t \leq s < \infty$.

$f_T(t)$

$$= \left(\frac{100}{24}\frac{1}{50}\frac{1}{24}\right)\sum_{k=0}^{50}\sum_{j=0}^{24}\binom{24}{j}\binom{n}{k}(-1)^j(-1)^{50-k}\,\lambda e^{-\lambda(-(k+j+1))t}\,\mathbf{1}_{0\leq t\leq\infty}(t)\int_t^{\infty}\lambda e^{-\lambda(76+j)s}\,ds \qquad (6.55)$$

$$= \left(\frac{100}{24}\frac{1}{50}\frac{1}{24}\right)\sum_{k=0}^{50}\sum_{j=0}^{24}\binom{24}{j}\binom{n}{k}(-1)^j(-1)^{50-k}\,\frac{\lambda}{76+j}e^{-\lambda(-(k+j+1))t}\left(1-e^{-\lambda(76+j)t}\right)\mathbf{1}_{0\leq t\leq\infty}.$$

Problems

1. If S is a random variable that follows a Gamma (α_S, n_S) distribution, and T is a random variable that follows a Gamma (α_T, n_T) independent of S, where n_s and n_r are each integers greater than one, than compute the probability that S is less than or equal to T.

2. Let x_i i= 1 to 4 be a collection of independent random variables following the exponential distribution with parameter λ_i, i = i...4. Compute the probability that $x_1 \leq x_2 \leq x_3 \leq x_4$. Show that this probability is $\frac{1}{4!}=\frac{1}{24}$ when $\lambda_i = \lambda$, i = i...4.

3. Let x_1, and x_2 be two random variables, and $v = \min(x_1, x_2)$, and $w = \max(x_1, x_2)$. Show that the Jacobian of the transformation (x_1, x_2) to (v,w) is one.

4. Show that $f_{V,M,W}(v, m, w) = 6\lambda^6 e^{-\lambda(v+m+w)}\mathbf{1}_{0\leq v\leq m\leq w\leq\infty}\,dvdmdw$ is a probability density function.

5. Show that there are $n!$ ways to order a collection of n i.i.d. random variables.

6. Using the result from problem 6.5, find the joint probability density function for all n order statistics of n i.i.d. random variables.

7. How could one generate a sample of n random variables that i.i.d. follow a Gamma distribution with shape parameter k (an integer) and scale parameter λ?

8. If z is a standard normal random variable and a, b and c are arbitrary constants such that $a \neq 0$ and $b \neq 0$, then

$$E\left[1 - F_Z\left(\frac{bY + c}{a}\right)\right] = 1 - F_Z\left[\frac{c}{\sqrt{a^2 + b^2}}\right].$$

9. Consider a circuit which has four identical components in parallel, each constructed from four identical elements in series. The time to failure for each of the elements is i.i.d. exponential. What is the mean survival time for the circuit if the mean time to failure for each element in the circuit is 1 month?

10. Assume each of the elements in the circuit depicted in Figure 6.8 has a lifetime that follows an exponential distribution with parameter λ. Compute the expected lifetime of the circuit.

11. Assume each of the elements in the circuit depicted in Figure 6.9 has a lifetime that follows an exponential distribution with parameter λ. Compute the expected lifetime of the circuit.

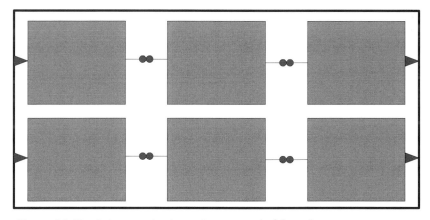

Figure 6.8 Circuit A: two redundant units composed of three elements

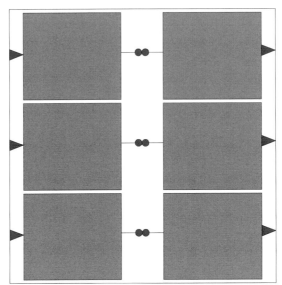

Figure 6.9 Circuit B: three redundant components each composed
of two elements

12.   Assume that the circuit in Figure 6.8 and the circuit in Figure 6.9 carry
      out the same function. How much longer must the survival time for
      elements in circuit A be in order for Circuit A to have the same lifetime
      as Circuit B?

13.   Let x_i be a collection of i.i.d. random variables following a uniform
      distribution on the interval $[0, a]$. Find the probability density function of
      the range.

14.   Let x_i be a collection of i.i.d. random variables following a uniform
      distribution on the interval $[0, 1]$. Find the probability density function of
      the truncated range $x_{[n-1]} - x_{[2]}$.

7

Asymptotic Distribution Theory

7.1 Introduction

Asymptotic distribution theory is the study of the properties of infinite sequences of random variables. While this aspect of statistical theory is a rigorous and intellectually stimulating body of knowledge in and of itself, its utility in the realm of statistical applications is a question worthy of a detailed answer. There is no doubt that statistical distribution theory with its concepts of limits and weak vs. strong convergence is challenging. The question is whether rising to this challenge produces any real dividends for the applied statistician.

Since the use of probability and statistics is a daily occurrence, we fortunately do not have to look very far to identify the benefits derived from the use of limiting theory. We only need to casually peruse a local newspaper on any given day to appreciate the role of statistics in conveying information about the world in which we live. Projections commonly appear concerning the impact of global warming on our ecosystem. Pollsters attempt to learn the moods of a populace by taking a survey. Estimates of the financial and human costs of a future war are calculated and reported. Clinical investigators work to learn the effect of a new therapy that has been developed to prevent the occurrence of strokes in patients with diabetes mellitus.

For each of these circumstances, the forecasters, the pollsters, the predictors, or the experimenters attempt to discern facts based upon a relatively small database whose primary utility is its availability. For example, the environmental scientist works to learn about future temperatures on earth by estimating past temperatures and measuring future ones. The pollster cannot obtain the responses of the entire national population, but he can gauge the mood of a few hundred respondents. The forecaster cannot know the impact of a future conflict with all of its uncertainties. Instead, she attempts to project findings about current economic systems and the destructive power of modern weapons. Finally the clinical experimenter is interested in generalizing a research finding that is based on 600 patients to a population of twenty million patients with diabetes mellitus. In each of these circumstances, the worker is attempting to generalize results from samples they can measure to populations that they cannot measure.

The obvious question one must ask is how reliable are the estimates of the effects that these scientists hope to calculate from these relatively small samples? The answers to these questions calibrate us, telling us how to judge the utility of the ongoing work. If the estimates are not accurate, then the consumer of the statistical information must be appropriately wary of the conclusions, and any actions taken based on these estimates must be disciplined and tightly restrained.

While there are several reasons why an estimate obtained from a sample can be inaccurate, an ultimate explanation lies in the fact that the estimate is based not on the population, but instead on only a sample of the population. One useful style of intuition informs us that, if the estimate is a good one, then the larger the sample becomes, the closer the estimate based on the sample will be to the true population value. Even the gambler, who stubbornly argues that his string of bad luck must turn around, is (perhaps unwittingly), using a property of an estimate from a large sample in order to help guide his persistent actions. Thus, statisticians, when confronted with several different and competing estimators, commonly focus on its *large sample, or asymptotic properties*.

These asymptotic properties are used in very comprehensible contexts. For example, a researcher may ask whether the estimator under consideration consistently overestimates or underestimates the population parameter that it attempts to measure. If it does, then that estimator is asymptotically biased, suggesting that perhaps another estimator should be sought. In another case, if there are several estimators that purport to measure the same population parameter, then a comparison of the large sample variances of these estimators might be a useful way to select from among the competitors. These are each important questions that asymptotic theory addresses. In order to examine these issues, we must understand the use of limits in probability and statistics, and learn some of the classic results of the utilization of this theory. These will be the topic of this chapter.

7.2 Introducing Probability to the Limit Process

In this chapter, we must jointly consider two complicated concepts in mathematics — the idea of a limit, and the notion of probability. It is best to start this process by taking them one at a time. A review of the limit process in mathematics has been provided in Chapter 1. Having become sufficiently reacquainted with the idea of limits, we can now consider how this concept has been adapted for use in probability. Certainly, the incorporation of the concept of probability into our ongoing conversations about limits and converging sequences is a complication. However, it is exactly this complication that we seek if these tools are to be helpful to us when we gauge the reliability of sample based parameter estimates.

Ultimately, based on a sample of size n, we will have an estimator, $T_n(X_1, X_2, X_3,...,X_n)$ of a population parameter $q(\theta)$. In that complicated situation, we will wish to know the properties of $T_n(X_1, X_2, X_3,...,X_n)$ as n increases. However, during this elementary stage of our discussions, we will begin with the simple consideration of a sequence of random variables. We are interested in the ultimate behavior of these random variables as n increases, seeking an answer to the question, "if n is large enough, what will be the behavior of X_n?" Since the possible values of X_n are determined by its probability distribution function, then the answer to this question depends on the probability distribution we assume for X_n. Thus, learning about the behavior of a random variable is, at its root, developing an understanding of the characteristics of the probability distribution function that governs it values. If, for example we assume that each of the X_n's are independent and has the same probability distribution, all that is available to learn could be learned from an examination of the behavior of X_1.

The learning process becomes much more interesting if the probability distribution functions for the random variables that comprise the sequence $\{X_n\}$ are permitted to change as a function of n. In fact, this is the technique that is commonly used. We will see that in some circumstances, a sequence of random variables that have a binomial distribution can be approximated by a Poisson distribution. Similarly, sequences of binomial random variables, and sequences of Poisson random variables can each have their probability distributions approximated by the normal distribution. Other examples of convergence will lead us to Chebyshev's inequality, the weak and strong law of large numbers, and the central limit theorem.

In order to introduce the properties of convergence that are associated with random variables, we will start with a trivial example. Our plan will be to begin with some simple, transparent examples, and then add successive layers of complexity to the theoretical development. This will permit us to differentiate features of random variables that exhibit different types of convergence. The four types of convergence that we will cover are 1) convergence in distribution (convergence in law), 2) convergence in probability, 3) convergence almost surely (convergence with probability one), and 4) convergence in r^{th} mean.

7.3 Introduction to Convergence in Distribution

How then might random variables converge? One useful combination of the concept of probability and the use of the limit process is as follows. Let $\{X_n\}$ be a sequence of random variables, for which each random variable has a cumulative distribution function $F_{X_n}(x)$. This provides a sequence of distribution functions $\left\{ F_{X_n}(x) \right\}$. Then, the simple convergence of these distribution functions to a single cumulative distribution $F_X(x)$ is a simple, intuitive, and useful first approach to embedding the notion of probability into the concept of limiting processes. This type of convergence is called *convergence in distribution*, or *convergence in law*.

Definition 7.1. Convergence in Distribution (convergence in law)
A sequence of random variables $\{X_n\}$ with associated cumulative distribution functions $F_{X_n}(x)$ is said to converge in distribution (or converge in law) to the random variable X that has a cumulative distribution function identified as $F_X(x)$ if $\lim_{n\to\infty} F_{X_n}(x) = F_X(x)$ for every x at which the cumulative distribution function $F_X(x)$ is continuous.

Thus, if the cumulative distribution functions of a sequence of random variables converges to a cumulative distribution function, then the random variables themselves converge in distribution.

It is important to see how the introduction of probability has altered our definition of convergence. In Chapter 1's discussion of the limiting process, the elements of the sequence $\{X_n\}$ became closer to one another as n increases. In the current setting, the elements of the sequence $\{X_n\}$ are random — they are unknown. What converges is not the random variable values themselves, but the distributions which govern the values (and their associated probabilities) that these random variables can assume. The importance of $\lim_{n\to\infty} F_{X_n}(x) = F_X(x)$ for only those points x for which the cumulative distribution function is continuous is discussed in detail elsewhere(Rao, 1984).

Let's now consider a simple example to illustrate convergence in distribution. Consider the sequence of random variables $\{X_n\}$ for which X_n follows a uniform distribution on the interval $\left[0, \frac{n}{n+1}\right]$. Thus X_1 is uniformly distributed between 0 and $\frac{1}{2}$, $X_2 \sim U\left[0, \frac{2}{3}\right]$, $X_3 \sim U\left[0, \frac{3}{4}\right]$, and so on. In this circumstance, it is easy to write the probability density function of the random variable X_n as $f_{X_n}(x) = \frac{n+1}{n} 1_{\left[0, \frac{n}{n+1}\right]}(x)dx$. The question that we wish to address is whether this sequence of random variables converges in distribution.

Our intuition suggests that convergence should be to the uniform distribution on the [0, 1] interval since the range of the uniformly distributed random variable X_n approaches [0, 1] as n increases. We can verify this suspicion by computing the cumulative distribution function for X_n as

$$F_{X_n}(x) = \int_0^x f_{X_n}(w) = \int_0^x \frac{n+1}{n} \mathbf{1}_{\left[0, \frac{n}{n+1}\right]}(w)dw$$

There are two cases here. The first is when $0 \le x \le \dfrac{n}{n+1}$, and the second is for

the circumstance in which we find $x > \dfrac{n}{n+1}$. For the case where

$0 \le x \le \dfrac{n}{n+1}$, then $F_{X_n}(x) = \displaystyle\int_0^x \frac{n+1}{n} dw = x\frac{n+1}{n}$. In the special situation where

$x > \dfrac{n}{n+1}$, we have $F_{X_n}(x) = \displaystyle\int_0^{\frac{n}{n+1}} \frac{n+1}{n} dw = 1$. The evaluation of these two cases

reveals that the formula for $F_{X_n}(x)$ depends on the value of x. We may write the solution as

$$F_{X_n}(x) = \frac{n+1}{n} x \mathbf{1}_{\left[0, \frac{n}{n+1}\right]}(x) + \mathbf{1}_{\left[\frac{n}{n+1}, 1\right]}(x)$$

We now need to observe what happens to this cumulative distribution function

as n increases to infinity. For large n, $\displaystyle\lim_{n\to\infty} \frac{n+1}{n} = \lim_{n\to\infty} 1 + \frac{1}{n} = 1$. Thus, for values

of X that lie in the interval $0 \le x < 1$, the limit of the cumulative distribution

function $\displaystyle\lim_{n\to\infty} F_{X_n}(x) = \lim_{n\to\infty} \frac{n+1}{n} x = x$. For $x \ge 1$, $F_X(x) = 1$. Thus, the limit of the

sequence of distribution functions $\left\{F_{X_n}(x)\right\}$ is, as we suspected, simply the
cumulative density function of the uniform distribution on the $[0, 1]$ interval.
We can therefore conclude that the sequence of random variables $\{X_n\}$
converges to the U(0,1) random variable in distribution.

It is important to note what is actually converging in this example. The
random variables are not converging to any one point, or to any one particular
value. The value of X_n remains unpredictable. What has converged is the
probability formula that governs the range of X_n and their probabilities. We will
see later that this concept of convergence in distribution produces some of the
most important and productive results in probability theory for the applied
statistician.

7.4 Non-convergence

The concept of convergence in distribution is a new one for us. Up until section 7.3, convergent sequences were sequences in which the individual values of the sequence members approach a limiting value. Convergence in distribution focused not on the limiting value of a random variable, but a limiting cumulative distribution function for that random variable. To distinguish these two concepts, we will now consider an infinite sequence of random variables X_1, X_2, X_3, ..., X_n,... which are independent, and identically distributed, each of which follows the same uniform (0, 1) distribution. Is there any limiting behavior that we can observe or predict in this sequence of random variables?

Certainly there is a (trivial) convergence in distribution for this infinite sequence of random variables to a $U(0,1)$ distribution. However, is there any one limiting value that this sequence of random variables is increasingly likely to take? The concept of limit that we reviewed in Chapter 1 required that if a sequence of variables $\{X_n\}$ has a limiting value X, then the farther out in that sequence we go, the closer the individual X_n must be to that value X. For example, is $\lim_{n \to \infty} x_n = 0.50$? Is there any single value that the X_n approach as n increases to infinity?

The answer to this question is certainly no. Our sequence of independent and identically distributed random variables all have the same, even disbursement of probability on the [0, 1] interval. The idea that any one value of X_n becomes more likely as n increases is defeated by the realization that each of these random variables can widely vary across the [0,1] interval in an unpredictable fashion. Thus convergence in distribution, while it will be very useful for us, can be relatively uninformative about the values the random variables in the sequence can actually take.

7.5 Introduction to Convergence in Probability

Recall that the core concept of the limiting process that we considered in Chapter 1 was the link between the elements of the convergent sequence and their limiting value. The farther out we went in the sequence $\{X_n\}$, the closer the sequence members came to their limiting value X. Another way of stating this is that as n was allowed to increase, the actual size of the deviations $|X_n - X|$ got progressively smaller. If we were to apply the concept of probability to this limiting process, we could focus, not on $|X_n - X|$, but instead on the probability of events involving $|X_n - X|$. A very useful concept for applying probability to the limiting process of a sequence of random variables $\{X_n\}$ is to require that, if the random variable X_n is considered to have a limit X, then the probability of large values of $|X_n - X|$ should become smaller as n increases. Another way of saying this is that when challenged with an $\varepsilon > 0$, then $P\big[|X_n - X| \geq \varepsilon\big]$ should go to zero as n increases to infinity if the random variables X_n approaches some limiting value X.

How does this concept work for our sequence of $U(0,1)$ random variables? In this case, when challenged with an $\varepsilon > 0$, and a candidate limit random variable X then $P\big[|X_n - X| > \varepsilon\big]$ is always a constant that is never a function of n for this probability distribution. For example, for any value $X = c$ contained on the $[0, 1]$, we find

$$P\big[|X_n - X| \geq \varepsilon\big] = P\big[|X_n - c| \geq \varepsilon\big] = 1 - P\big[|X_n - c| \leq \varepsilon\big],$$

using the cumulative distribution function for the U(0,1) random variable we find that

$$P\big[|X_n - c| \leq \varepsilon\big] = P\big[c - \varepsilon \leq X_n \leq \varepsilon + c\big] = \min(\varepsilon + c, 1) - \max(c - \varepsilon, 0).$$

Since $\varepsilon + c$ can be greater than 1 (for example $\varepsilon = 0.001$, $c = 0.9999$) and the probability density function is defined only defined on $[0, 1]$, we see that the upper bound of the cumulative distribution function is $\min(\varepsilon + c, 1)$. Analogously, it is also possible that $\varepsilon - c$ is less than 0. Since the probability density function is defined only on $[0, 1]$ and the lower bound of the cumulative density function is 0, we write $\max(c - \varepsilon, 0)$.

Therefore

$$P\big[|X_n - X| \geq \varepsilon\big] = 1 - \big[\min(\varepsilon + c, 1) - \max(c - \varepsilon, 0)\big]$$

This quantity is not a function of n for the probability distribution that we have selected and we cannot find an X for which we can say that $\lim_{n \to \infty} P\big[|X_n - X| \geq \varepsilon\big] = 0$. Thus, the sequence of independent and identically distributed U(0,1) random variables does not meet out new criteria for random variable convergence, even though the sequence converges in distribution.

We have to be clear about the preceding example. What guided the demonstration in the previous paragraph was the fact that the random variables take values on $[0, 1]$. Random variables can take other values but have the probability of those values mapped to U(0,1). In this circumstance, the convergence result would be different.

However, we can further illuminate properties of convergence by altering the sequence of random variables. Let the sequence $\{X_n\}$ now follow a beta $(n, 1)$ distribution. Thus $f_{X_n}(x) = nx^{n-1}\mathbf{1}_{[0,1]}(x)$. In this circumstance, the probability density functions of the random variables in the sequence does vary as a function of n, and varies in a predictable fashion (Figure 7.1). We see that as n increases the probability that the random variable X_n will fall close to the value one also increases. Since the distribution is a continuous one, we know that we can not expect to ever observe $X_n = 1$, i.e., $P[X_n = 1] = 0$; however, the probability that X_n takes on a value far from one decreases as n increases.

In fact, in this example, we can easily compute the probability that a large deviation occurs. Hence, the P $[|X_n - 1| \geq \varepsilon]$ may be obtained as follows

$$P\left[|X_n - 1| \geq \varepsilon\right] = P\left[1 - X_n \geq \varepsilon\right] = P\left[-X_n \geq \varepsilon - 1\right] = P\left[X_n \leq 1 - \varepsilon\right]. \quad (7.1)$$

For the beta $(n, 1)$ distribution, this becomes

$$P\left[|X_n - 1| \geq \varepsilon\right] = \int_0^{1-\varepsilon} nx^{n-1}dx = x^n \Big]_0^{1-\varepsilon} = (1 - \varepsilon)^n \quad (7.2)$$

As n increases the quantity $(1 - \varepsilon)^n$ approaches zero. Thus we have demonstrated that the probability of large deviations of X_n from the value one become less likely as n goes to infinity. Now, observe that the only restriction on ε, the size of the deviation, was that it was greater than zero. Thus, the above argument holds for every value of ε with which we are challenged. Therefore, we have demonstrated that we can reduce the probability of a deviation of any magnitude to as close as we desire to zero merely by going far enough out in this sequence of random variables $X_1, X_2, X_3, \ldots, X_n$. This type of convergence is defined as *convergence in probability*, which we now define.

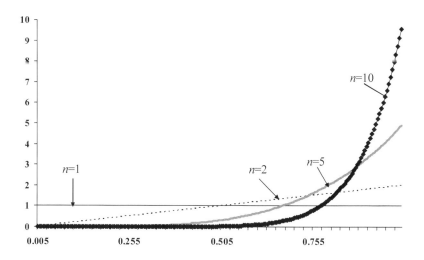

Figure 7.1 Probability density function of the beta distribution for $\alpha=n$; $\beta = 1$. As n increases, the distribution is shifted to the right.

Definition 7.2. Convergence in Probability.
Let $\{X_n\}$ be a sequence of random variables. Then the sequence is said to converge in probability to a random variable X, if for each $\varepsilon >$ $0 \lim_{n\to\infty} P\left[\left|X_n - X\right| \geq \varepsilon\right] = 0$. This is equivalent to demonstrating that for each $\varepsilon >$ 0, $\lim_{n\to\infty} P\left[\left|X_n - X\right| \leq \varepsilon\right] = 1$

Note that convergence in probability does not imply that all values of X_n in the sequence must be close to the limit X. We can only say the probability that discrepant values of the random variables occur approaches zero. These discrepant values are unlikely, but we cannot preclude their occurrence. For this reason, convergence in probability is commonly referred to as *weak* convergence. However, we will see later in the chapter that weak convergence can produce very powerful and useful results.

7.6 Convergence Almost Surely (with Probability One)

We will have much more to say about convergence in probability in the next few sections of this chapter. However, using the term "weak" to describe this manner of convergence begs the question "is there a stronger convergence modality involving random variables?" The answer to this question is yes.

One natural way we might consider tightening the definition of convergence would be to completely remove the possibility of large deviations. When a sequence of random variables $\{X_n\}$ converges in probability to X, we know that the probability that $X_n - X$ is large goes to zero, but we also must concede that large deviations can occur. Are there any circumstance in which we could be assured that large deviations are impossible?

Consider the sequence of random variables $\{X_n\}$ for which each random variable has the probability density function

$$f_{X_n}(x) = (n+1)\mathbf{1}_{\left[\frac{n}{n+1},1\right]}(x). \tag{7.3}$$

This collection of probability density functions actually represents a family of uniform distributions whose range of values decreases as n increases. For example, we observe that for $n = 1$, X_1 follows a $U\left(\frac{1}{2},1\right)$ distribution. The distribution of X_2 is $U\left(\frac{2}{3},1\right)$, the distribution of X_3 is $U\left(\frac{3}{4},1\right)$, and so on. Note that the probability that X_n is close to one increases as was the case for the previous sequence of random variables that followed a beta distribution. However, for that sequence, the probability that X_n was close to zero, although steadily decreasing, was always positive. That is not the case for this sequence of random variables that follow uniform distributions. In this latter circumstance, the range of values of X_n for which X_n has positive probability is steadily and consistently decreasing (Figure 7.2). In this case, the values of the random variable fall in $[0, 1]$.

From the discussion of convergence in probability, our intuition tells us that this sequence of random variables should converges in probability to one. We can easily demonstrate this. We must show that for $\varepsilon > 0 \lim_{n \to \infty} P\left[|X_n - 1| \geq \varepsilon\right] = 0$. We start, as we did in the example that was provided in the previous section, with the observation that for any value of X_n, $P\left[|X_n - 1| \geq \varepsilon\right] = P\left[1 - X_n \geq \varepsilon\right] = P\left[X_n \leq 1 - \varepsilon\right]$. This final probability can be evaluated easily. The minimum possible value of X_n is $\frac{n}{n+1}$. If $1 - \varepsilon \leq \frac{n}{n+1}$, then P[$X_n \leq 1 - \varepsilon$] = 0. If $1 - \varepsilon \geq \frac{n}{n+1}$, then

$$P\left[x_n \le 1-\varepsilon\right] = \int_{\frac{n}{n+1}}^{1-\varepsilon} f_{X_n}(x) = (n+1)\left[1-\varepsilon - \frac{n}{n+1}\right] = 1-(n+1)\varepsilon \qquad (7.4)$$

Thus $\lim_{n\to\infty} P\left[\left|x_n - 1\right| \ge \varepsilon\right] = \lim_{n\to\infty}\left[\max\left(0, 1-(n+1)\varepsilon\right)\right] = 0$, and we have shown that X_n converges in probability to one.

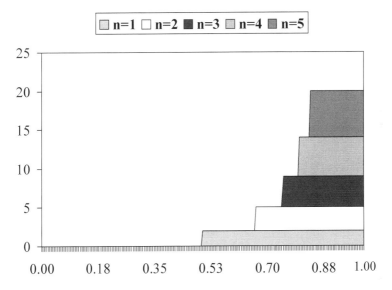

Figure 7.2 Probability concentration near the value of one in a sequence of uniformly distributed random variables.

However there is a stronger form of convergence available here. Since the greatest lower bound of X_n is $\dfrac{n}{n+1}$ and this greatest lower bound increases as n increases, small values of X_n $\left(i.e.\ X_n < \dfrac{n}{n+1}\right)$ cannot occur. We therefore observe that as n increases to infinity, the greatest lower bound of X_n increases to one as well. The random variable X_n is firmly trapped between the value one on the right, and a greatest lower bound on the left that is itself consistently increasing to the value one. Thus, as n increases, we are assured that X_n must both become close and stay close to the value one. Another way to say this is that, for any $\varepsilon > 0$, X_n is within ε of 1 all but finitely many times (sometimes abbreviated as a.b.f.m.t.). A condition occurs all but finitely many times when there exist an N such that for all $n > N$, the condition holds true.

We can demonstrate this formally. Choose $\varepsilon > 0$. We must demonstrate that $|X_n - 1| \leq \varepsilon$ all but finitely many times. Thus, our task is to find an N such that for all $n > N$, $|X_n - 1| \leq \varepsilon$. We first note that the condition $|X_n - 1| \leq \varepsilon$ is equivalent to $X_n \geq 1 - \varepsilon$. When must $X_n \geq 1 - \varepsilon$? Recall that the greatest lower bound for $X_n = \dfrac{n}{n+1}$ so X_n must be greater than or equal to $1 - \varepsilon$ when the inequality $\dfrac{n}{n+1} \geq 1 - \varepsilon$ is true. This is the case when $n \geq \dfrac{1-\varepsilon}{\varepsilon}$. Thus by choosing $N = \dfrac{1-\varepsilon}{\varepsilon}$, we are assured that for all $n > N$, X_n will be within ε of one.

This type of convergence is known as convergence *almost surely*, or *convergence with probability one*.

Definition 7.3. Convergence almost surely (also known as convergence with probability one).

A sequence of random variables $\{X_n\}$ converges almost surely to X, when for every $\varepsilon > 0$, X_n is within ε of X all but finitely many times.

An important distinction between weak convergence and strong convergence is the likelihood of events that do not satisfy the convergence condition of X_n being within ε of **X.** In the sequence $\{X_n\}$ converges weakly to X, then $|X_n - X| \geq \varepsilon$ can occur for any value of n. The probability of this event decreases with increasing n; however, although this probability reaches a point of negligibility, it is always positive. Thus even though X_n converges to X, for every n it is possible to observe an X_n such that $|X_n - X| \geq \varepsilon$. On the other hand, if the convergence in which we are interested is almost surely convergence, then there exists an n such that for all $n > N$, the probability of the occurrence of an X_n such that $|X_n - X| \geq \varepsilon$ zero.

Example 7.1 To further differentiate between the roles of convergence in probability and almost surely convergence, consider the following failure model. Let $\{X_n\}$ be a sequence of outcomes of an experiment testing the successful operation of an increasingly complicated electronic component. The value of X_n is based on the successful operation of n smaller circuits that themselves may be considered a sequence of n independent and identically distributed Bernoulli trials $(y_{n1}, y_{n2}, y_{n3}, \ldots, y_{nn})$ with probability of success p. The first component's operation is described by the random variable X_1. If the component works successfully, then $X_1 = 1$, an event that occurs only if $y_{11} = 1$. If $y_{11} = 0$, then $X_1 = 0$. The result of the second component's test is denoted by X_2. The event that the second component is successful is described by $X_2 = 1$; this occurs when $y_{21} = y_{22} = 1$. If either $y_{21} = 0$ or $y_{22} = 0$, then the second component fails. The third component's successful operation is described by $X_3 = 1$, which occurs iff $y_{31} = y_{32} = y_{33} = 1$; $X_3 = 0$ otherwise, etc. Does the sequence of increasingly complex components ever get to the point where they are guaranteed to either always fail or always work?

Another way to state this question is "In what sense and to what value does the sequence of random variables $\{X_n\}$ converge?" We observe that for all n, X_n that we have chosen is such that it can only take on the value 0 or 1. The probability that $X_n = 1$ decreases over time since it is based on a sequence of consecutive successes. Does this selection of random variables $\{X_n\}$ converge in probability to zero? If so, we must show that the limit as n goes to infinity of the probability that X_n is equal to zero is one. This is a straightforward task, since

$$\lim_{n \to \infty} P[X_n = 0] = \lim_{n \to \infty} 1 - p^n = 1$$

Thus, the sequence of X_n's converges to 0 in probability.

However, is this failure certain? Does the sequence $\{X_n\}$ converge almost surely? If so, then there must be some N such that for all $n > N$, X_n is guaranteed to be zero. However, no matter how large n is, there is a positive probability that the value of X_n will be based on a string of n successes. Even though this probability almost vanishes, it is never zero. This extremely remote probability, or "outside chance" of a string of n successes for any value of n blocks the occurrence of almost sure convergence. No matter how complex the component is, it is never guaranteed to fail.

With this distinction, it is no surprise that almost sure convergence implies convergence in probability. This observation leads to the descriptor of almost sure convergence as *strong convergence*, and convergence in probability (as well as convergence in distribution) is described as *weak convergence*.

7.7 Convergence in r^th Mean

Finally, we will define convergence in mean square

Definition 7.4. Convergence in mean square

Let $\{X_n\}$ be a sequence of random variables. Then if $\lim_{n \to \infty} E\left[\left|X_n - X\right|^r\right] = 0$, then X_n converges to X in rth mean. In particular, if $r = 2$, we say that $\{X_n\}$ converges to X in mean square.

Convergence in rth mean may seem to be a somewhat unintuitive definition, but in fact it is quite useful, very powerful, and oftentimes is easy to prove. For example, to show that the sample mean $\overline{X}_n$ of a collection of n independent observations from a normal distribution with known mean μ and finite variance σ^2 converges to μ in mean square , we need merely write

$$\lim_{n \to \infty} E\left[\overline{X}_n - \mu\right]^2 = \lim_{n \to \infty} Var\left[\overline{X}_n\right] = \lim_{n \to \infty} \frac{\sigma^2}{n} = 0$$

Example 7.1 Home Inspections
A newly recognized problem in many dwellings in the southeastern United States is fungal or mold infestation. Homes that have been subjected to flooding rains, followed by exposure to warm and humid air have been shown to be a breeding ground for mold. A minority of these molds are toxic, producing byproducts that poison the interior air of a house. These poison gases can make, the home uninhabitable. The cost of both relocating and treating the affected homes' residents, as well as the cost of drying, treating, and sometimes rebuilding the home are substantial and have important implications for the home insurance industry.

To gain an assessment of the magnitude of this problem in a subdivision of homes, a survey team selects n homes at random from this subdivision for a mold evaluation. This evaluation is thorough and exhaustive, producing a mold score which is a measure of mold infestation and toxicity. The mold score is a continuous measure that lies between 0 and 100. A score of 0 means there is no evidence of mold. A score of 100 means the dwelling is toxic and represents a health hazard. The surveyors wish to estimate the probability that a home has a mold score less than or equal to 25. Stated another way, if X is a random variable reflecting the result of a mold score evaluation, then the researchers wish to estimate $F_X(25)$, the probability that a mold score evaluation is less than or equal to 25.

These scientists would like to use the proportion of homes in their survey with mold scores ≤ 25 as an estimate of $F_X(25)$. However, is this estimate a reliable one? If we can demonstrate that this estimate converges in mean square to the cumulative distribution function evaluated for a mold score of 25, then we also know that this estimate converges in probability and in distribution.

We begin this demonstration by introducing some terminology. The survey scientists begin with a collection of random variables $X_1, X_2, X_3, \ldots X_n$, representing the mold scores for each of the n homes in their sample. They intend to define a new variable W_i, where $W_i = \mathbf{1}_{x_i \leq 25}$ (i.e., $W_i = 1$ if $x_i \leq 25$, and

$W_i = 0$ otherwise. Then we see that, in this case, $\overline{W}_n$ is the same as the proportion of x_i's with values less than or equal to 25, and it is reasonable to estimate $F_X(25)$ by $F_n(25) = \overline{W}_n$. We are going to apply the concept of convergence in mean square here. We must show that $\lim_{n\to\infty} E\left[F_n(25) - F_X(25)\right]^2 = 0$. We first identify the expected value of $F_n(25)$.

$$E[W_i] = E\left[\mathbf{1}_{[0<X<25]}\right] = 1P[X_i \le 25] + 0P[X_i > 25] = P[X_i \le 25]$$

We now write

$$E[F_n(25)] = E\left[\frac{\sum_{i=1}^{n} W_i}{n}\right] = \frac{1}{n}nP[X_i \le 25] = P[X_i \le 25] = F_X(25)$$

So, examining $\lim_{n\to\infty} E\left[F_n(25) - F_X(25)\right]^2 = \lim_{n\to\infty} E\left[F_n(25) - E[F_n(25)]\right]^2$, which is simply the $Var[F_n(25)]$. We identify this quantify by first writing

$$E[W_i^2] = E\left[\mathbf{1}_{x_i \le 25}\right] = 1P[X_i \le 25] + 0P[X_i > 25] = P[X_i \le 25] = F_X(25)$$

and observe that

$$Var[W_i] = E[W_i^2] - E^2[W_i] = F_X(25) - F_X(25)^2 = F_X(25)[1 - F_X(25)].$$

Since $F_n(25) = \overline{W}_n$, we may then write

$$Var[F_n(25)] = Var\left[\frac{\sum_{i=1}^{n} W_i}{n}\right] = \frac{nVar[W_i]}{n^2} = \frac{F_X(25)[1 - F_X(25)]}{n}$$

We may now conclude that

$$\lim_{n\to\infty} E\left[F_n(25) - F_X(25)\right]^2 = \lim_{n\to\infty} Var[F_n(25)] = \lim_{n\to\infty}\left[\frac{F_X(25)[1 - F_X(25)]}{n}\right] = 0$$

and researchers use of the proportion of dwellings with a mold score less than or equal to 25 converges in mean square to the desired cumulative distribution. The fact that the empirical measure of the distribution function converges to the

theoretical cumulative distribution function is sometimes described as the
Fundamental Theorem of Statistics. ∎

7.8 Relationships between Convergence Modalities

As we suspected, the introduction of probability into our understanding of
limiting process has been complicated. We have discussed four different modes
of convergence that involve the concept of probability. These are convergence in
distribution, convergence in probability, convergence almost surely (or
convergence with probability 1), and convergence in mean square. Each of these
modes has been the foundation of important theoretical work in probability, but
they have different implications.

A sequence of random variables that converges in probability also must
converge in distribution. A sequence of random variables that converges almost
surely, also converges in probability, and therefore converges in distribution.
Similarly, a sequence of random variables that converges in mean square also
converges in probability, and must converge in distribution as well.

The relationship between convergence almost surely and convergence
in mean square is complex. Each of them implies convergence in probability and
convergence in distribution but if a sequence of random variables $\{x_n\}$
converges almost surely to X, it need not converge in mean square to X. Also,
convergence in mean square does not imply convergence with probability one.

Finally, as pointed out above, a sequence of random variables that
converges almost surely must also converge in probability. However, if $\{X_n\}$ is a
sequence of random variables that converges in probability to a random variable
X, although $\{X_n\}$ does not converge to X with probability one, it contains a sub-
sequence $\left\{X_{n_k}\right\}$ that does converges to the value X with probability one.

7.9 Application of Convergence in Distribution

Convergence in distribution is among the "weakest" of the convergence types
that we will consider in this chapter. However, it has produced several useful
and illuminating expressions. We will review two of them in this section.

7.9.1 The Poisson Approximation to Binomial Probabilities

One of the most important features of convergence in distribution is the ability
to identify useful rules and an occasional shortcut for the computation of some
probabilities. This process begins at once for us when we consider two of the
most ubiquitous discrete probability distributions, the binomial distribution and
the Poisson distribution.

Recall that the binomial distribution is based on the sums of
independent Bernoulli trials, each occurring with a constant probability of

success, p. In this case, a failure is represented by zero and a success is represented by a one. Let X be the number of successes in n such Bernoulli trials, then

$$P[X = k] = \binom{n}{k} p^k (1-p)^{n-k} \mathbf{1}_{I[0,n]}(k) \tag{7.5}$$

where $I_{[0,n]}$ represents the set of integers on the $[0, n]$ interval.

The Poisson distribution assigns probability for all non-negative integers. It is commonly used to compute the probability of events that occur relatively infrequently and independently of one another. In this circumstance if X follows a Poisson distribution, then

$$P[X = k] = \frac{\lambda^k}{k!} e^{-\lambda} \mathbf{1}_{I[0,n]}(k)$$

Each of these probability mass functions describe probabilities of discrete events. Bernoulli trials are described as successes or failures, and Poisson events can be linked to the set of integers. In addition, we described the Poisson random variable as an outcome that depicts the occurrence of a relatively rare event. This circumstance arises in the binomial distribution when n is large and p is small. It is therefore natural to suspect that there might be some circumstances in which these probabilities approximate each other. Our goal then is to closely examine the binomial mass function under these circumstances. A useful lemma that we will need for this analysis is

Lemma 7.1. If $\{x_n\}$ is a sequence of real numbers that converges to x, then

$$\lim_{n \to \infty} \left(1 + \frac{x}{n} \right)^n = e^x \tag{7.6}$$

This useful result was proved in Chapter 1. There are other forms of this equality, namely

$$\lim_{n \to \infty} \left(1 - \frac{x}{n} \right)^n = e^{-x} \tag{7.7}$$

and, more generally,

$$\lim_{n \to \infty} \left(1 + \frac{x_n}{n} \right)^n = e^{\lim_{n \to \infty} x_n} \tag{7.8}$$

It is useful to be very familiar with these concepts. We are now ready to begin our demonstration. We will let $\lambda_n = np_n$, and we write equation (7.5) as

$$P[X=k] = \binom{n}{k} p_n^{\ k} \left(1-p_n\right)^{n-k} \mathbf{1}_{k \in I_{[0,n]}} = \binom{n}{k} \left(\frac{\lambda_n}{n}\right)^k \left(1-\left(\frac{\lambda_n}{n}\right)\right)^{n-k} \mathbf{1}_{I_{[0,n]}}(k)$$

$$= \binom{n}{k} \frac{1}{n^k} \lambda_n^{\ k} \left(1-\left(\frac{\lambda_n}{n}\right)\right)^{n-k} \mathbf{1}_{I_{[0,n]}}(k)$$

Now rewrite

$$\binom{n}{k} \frac{1}{n^k} = \frac{n!}{k!(n-k)!n^k} = \left(\frac{1}{k!}\right) \frac{n(n-1)(n-2)(n-3)\ldots 1}{(n-k)(n-k-1)(n-k-2)\ldots 1} \left(\frac{1}{n^k}\right)$$

$$= \left(\frac{1}{k!}\right) \frac{n(n-1)(n-2)(n-3)\ldots(n-k+1)}{n^k} \qquad (7.9)$$

$$= \left(\frac{1}{k!}\right)\left(\frac{n}{n}\right)\left(\frac{n-1}{n}\right)\left(\frac{n-2}{n}\right)\ldots\left(\frac{n-k+1}{n}\right)$$

So

$$P[X=k] = \left(\frac{1}{k!}\right)\left(\frac{n}{n}\right)\left(\frac{n-1}{n}\right)\left(\frac{n-2}{n}\right)\ldots\left(\frac{n-k}{n}\right) \lambda_n^{\ k} \left(1-\left(\frac{\lambda_n}{n}\right)\right)^{n-k} \mathbf{1}_{I_{[0,n]}}(k) \quad (7.10)$$

We can now take the limit of expression (7.10) as n approaches infinity. We see that

$$\lim_{n \to \infty} \left(\frac{n}{n}\right)\left(\frac{n-1}{n}\right)\left(\frac{n-2}{n}\right)\ldots\left(\frac{n-k+1}{n}\right) = 1 \qquad (7.11)$$

and $\lim\limits_{n \to \infty} \left(1-\left(\frac{\lambda_n}{n}\right)\right)^{n-k} = e^{-\lambda}$. We now need to examine the limiting condition.
Clearly n will increase to infinity, but we also require p_n to decrease in such a fashion such that np is a constant. We will denote this latter condition as $p \to \varepsilon$

$$\lim_{\substack{n \to \infty \\ p \to \varepsilon}} \binom{n}{k} p_n^{\ k} \left(1-p_n\right)^{n-k} \mathbf{1}_{I_{[0,n]}}(k) = \frac{\lambda^k}{k!} e^{-\lambda} \mathbf{1}_{I_{[0,n]}}(k)$$

where the limiting conditions are that n goes to infinity, and p decreases in such a fashion that np is a constant. Under these conditions, the binomial distribution converges in distribution to the Poisson distribution.

7.9.2 Example of Convergence in Probability

As we saw in the previous section, an infinite sequence of random variables converges in probability to a single value when the random variables are very likely to be close to the limiting value. However convergence in probability does not assure us that the random variable must be close to the limiting value; only that the probability of being close to the limiting value gets increasingly large, and approaches one.

Consider the following example. An inexperienced but talented young football player works patiently to improve her ability to kick a football a distance of 30 yards between two poles that are 10 yards apart. She can always kick the ball at least 40 yards, however, she must work to improve her ability to kick the ball between the two goal posts. If the center of the goal post is marked as $X = 0$, then the ball must fall in the interval denoted by $-5 \le X \le 5$ or $(-5, 5)$ before the goal can be counted. Early in her career, the distribution of the ball's position as the result of her kick is uniformly distributed on the interval $(-20, 20)$. However, as she refines her ability, she is eventually able to narrow the target area of her kick to $(-8, 8)$. In what sense does the accuracy of her kicks converge?

In this example, we may consider the position of the ball after the nth kick as X_n, and the sequence of these kicks over her career can be easily represented by a (seemingly) infinite sequence of random variables. Early in her career (i.e., for n small) the X_n's follow a $U(-20, 20)$ distribution. As she improves approaching the peak of her ability the distribution of the X_n's follow a $U(-6, 6)$ distribution. Is there any convergence that takes place in this setting? Although, as her experience increases, her kicks undeniably become more accurate, it must also be acknowledged that no one kick, regardless of the point in her career at which it was made, is guaranteed to go through the goal post. In fact, the $n+1^{st}$ attempt is not guaranteed to be any better than the nth kick. This suggest that the kicking accuracy does not converge in an almost sure sense. However, the notion of convergence in probability is particularly attractive here. As n increases, and her kicks become increasingly accurate, the probability of a kick resulting in an X_n that is not within $(-6, 6)$ becomes very small over time. It is not too farfetched to say that the accuracy of her kicks converge in probability to a $U(-6, 6)$ distribution.

Even in the absence of a guarantee that X_n will be close to its limiting value X, convergence in probability has proven to be an extremely useful concept in probability and in statistics. We will first explore some properties of this mode of convergence, and then directly apply it in several useful settings.

7.10.1 Properties of Convergence in Probability

We demonstrated some of the properties of convergent sequences in Chapter 1. There we showed that certain useful combinations (e.g., sums and products) of convergent sequences themselves converge. The same is true for sequences that converge in probability, demonstrated below.

Lemma 7.2 If $\{X_n\}$ is a sequence of random variables that converges to X in probability and a is a scalar constant, then the sequence $\{W_n\}$ defined by $W_n = aX_n$, then w_n converges in probability to $W = aX$.

The proof is straightforward and follows the proof provided for lemma 7.1. We must show that $\lim_{n\to\infty} P\left[\left|W_n - W\right| \le \varepsilon\right] = 1$. We write, for $a > 0$

$$P\left[\left|W_n - W\right| \le \varepsilon\right] = P\left[\left|aX_n - aX\right| \le \varepsilon\right] = P\left[a\left|X_n - X\right| \le \varepsilon\right] = P\left[\left|X_n - X\right| \le \frac{\varepsilon}{a}\right]$$

and we can now write that $\lim_{n\to\infty} P\left[\left|W_n - W\right| \le \varepsilon\right] = \lim_{n\to\infty} P\left[\left|X_n - X\right| \le \frac{\varepsilon}{a}\right] = 1$ since

$\{X_n\}$ converges in probability to X. We will relegate the proof of the lemma that the convergence of probability of $\{X_n\}$ to X implies that $\{-X_n\}$ converges to $-X$ to the problems and therefore the sequence $\{aX_n\}$ converges to aX when a is negative. ∎

Lemma 7.3. If $\{X_n\}$ is a sequence of random variables that converges to X in probability and $\{Y_n\}$ is a sequence of random variables that converge to Y in probability then the random variable $\{W_n\}$ where $W_n = X_n + Y_n$ converges in probability to $W = X + Y$.

We must show that, when we are challenged with an $\varepsilon \ge 0$, then we can show that the $\lim_{n\to\infty} P\left[\left|W_n - W\right| \ge \varepsilon\right] = 0$. Begin by writing

$$P\left[\left|W_n - W\right| \ge \varepsilon\right] = P\left[\left|X_n + Y_n - \left(X + Y\right)\right| \ge \varepsilon\right] = P\left[\left|X_n - X + Y_n - Y\right| \ge \varepsilon\right]$$

Now identify ε_x and ε_y such that $\varepsilon_x > 0$, $\varepsilon_y > 0$ and $\varepsilon = \varepsilon_x + \varepsilon_y$. Then

$$P\left[\left|X_n - X + Y_n - Y\right| \ge \varepsilon\right] \le P\left[\left|X_n - X\right| \ge \varepsilon_x \ \cup \ \left|Y_n - Y\right| \ge \varepsilon_y\right]$$

$$\le P\left[\left|X_n - X\right| \ge \varepsilon_x\right] + P\left[\left|Y_n - Y\right| \ge \varepsilon_y\right]$$

Thus

$$\lim_{n\to\infty} P\left[\left|W_n - W\right| \ge \varepsilon\right] \ \le \ \lim_{n\to\infty} P\left[\left|X_n - X\right| \ge \varepsilon_x\right] + \lim_{n\to\infty} P\left[\left|Y_n - Y\right| \ge \varepsilon_y\right] \ = \ 0$$

∎

7.10.2 Slutsky's Theorem

There are some very useful properties of random variables that converge in probability that are readily available to us (from Dudedwicz and Mishra).

Theorem 7.1. Slutsky's Theorem (general form)
Consider a sequence of random variables $\{X_n\}$, and a collection of functions h_1, h_2, h_k, defined on the sequence $\{X_n\}$ so that $h_i(\{X_n\})$ converges to a constant a_i, $i = 1,2,3,...,k$. Define $g(a_1, a_2, a_3, ..., a_k) < \infty$. Then $g(h_1(\{X_n\}), h_2(\{X_n\}), h_3(\{X_n\}) ... h_k(\{X_n\}))$ converges to $g(a_1, a_2, a_3, ..., a_k)$ in probability. In particular, if g is a rational function (i.e., the ratio of two polynomials), then this result is known as Slutsky's Theorem.

This seems quite complicated on first appearance, so we will first consider a simple example. Begin with a sequence of random variables such that $\{X_n\}$ converges to a constant μ in probability. Define the family of functions h_k as

$$h_1\left(\{x_n\}\right) = \{x_n\}$$
$$h_2\left(\{x_n\}\right) = \{2x_n\}$$
$$h_3\left(\{x_n\}\right) = \{3x_n\} \tag{7.12}$$
$$\vdots$$
$$h_k\left(\{x_n\}\right) = \{kx_n\}$$

We observe that $h_i(\{X_n\})$ converges in probability to the constant $i\mu$. As an example of a function g, we next let the function $g\left(h_1,h_2,h_3,...,h_k\right) = \sum_{i=1}^{k} h_i$. We are now in a position to invoke Slutsky's theorem and obtain

$$\lim_{n\to\infty} g\left(h_1,h_2,h_3,...,h_k\right) = \sum_{i=1}^{k} h_i = \frac{k(k+1)}{2}\mu \tag{7.13}$$

Slutsky's theorem does not have as many applied applications as it does theoretical ones. However, in this latter circumstance, Slutsky's theorem is very powerful theorem for two reasons. First, it allows functions of random variables to have their convergent properties explored directly. However, perhaps more importantly, the application of Slutsky's theorem in a very simple form illuminates critical properties of random variables that converge in probability. For example, consider the case where we have a sequence of random variables $\{X_n\}$ that converges in probability to X, and a separate sequence of random variables $\{Y_n\}$ that converges in probability to Y. Define two functions h_1, and h_2 as

$$h_1\left(\{x_n\}\right) = \{x_n\}; h_2\left(\{y_n\}\right) = \{y_n\}$$

Thus the functions h_1 and h_2 are simply identity functions. Finally, define the function $g(h_1, h_2)$ as $h_1 - h_2$ (i.e., the difference of the two identity functions). Then, from Slutsky's theorem, we find that the function $g(h_1, h_2)$ converges to the value $X - Y$ in probability. Another way to write this result is

Lemma 7.4 If $\{X_n\}$ is a sequence of random variables that converges in probability to X and $\{Y_n\}$ is a sequence of random variables that converges in probability to **y** then the random variable $\{W_n\}$ where $W_n = X_n - Y_n$ converges in probability to the random variable $W = X-Y$.

Slutsky's theorem can also be used to prove lemmas 7.5, and 7.6.

Lemma 7.5. If $\{X_n\}$ is a sequence of random variables that converges in probability to X and $\{Y_n\}$ is a sequence of random variables that converge in probability to Y then the random variable $\{W_n\}$ where $W_n = X_n Y_n$ converges in probability to $W = XY$.

Lemma 7.6 If $\{X_n\}$ is a sequence of random variables that converges to X in probability and $\{Y_n\}$ is a sequence of random variables that converge to Y in probability $(Y \neq 0)$ then the random variable $\{W_n\}$ where $W_n = \dfrac{X_n}{Y_n}$ converges in

probability to $W = \dfrac{X}{Y}$.

The statement of Slutsky's theorem above may appear to be complicated. A much more common and useful form of this theorem follows.

Theorem 7.1a. Slutsky's Theorem (common form):
If a random variable W_n converges in distribution to W, and another random variable U_n converges in probability to the constant U, then
1) the sequence $\{W_n + U_n\}$ converges in distribution to W + U.
2) the sequence $\{W_n U_n\}$ converges in distribution to WU.

While the devices suggested by the preceding lemmas are helpful in demonstrating convergence in probability, it is sometimes useful to prove that a sequence of random variables converge in probability from first principles. Consider the following example.

A sequence of random variables $\{X_n\}$ follows a $U(0,1)$ distribution. We have already demonstrated that this sequence does not converge. However, what happens to the max $\{X_n\}$ as n goes to infinity? In this case, we create a new sequence of random variables $W_1 = \max(X_1)$, $W_2 = \max(X_1, X_2)$, $W_3 = \max(X_1, X_2, X_3)$. Does the sequence $\{W_n\}$ converge in probability to 1?

We must show that $\lim_{n \to \infty} P\left[|W_n - 1| > \varepsilon\right] = 0$. We note as before that

$\lim_{n \to \infty} P\left[|W_n - 1| > \varepsilon\right] = P\left[1 - W_n > \varepsilon\right] = P\left[W_n < 1 - \varepsilon\right]$. The probability distribution of the w_n is also easily identified. We may write

$$P\left[W_n < 1 - \varepsilon\right] = P\left[Max\left(X_1, X_2, X_3, \ldots X_n\right) < 1 - \varepsilon\right]$$
$$= P\left[X_1 < 1 - \varepsilon,\ X_2 < 1 - \varepsilon,\ X_3 < 1 - \varepsilon, \ldots\ X_n < 1 - \varepsilon\right]$$
$$= \prod_{i=1}^{n} P\left[X_i < 1 - \varepsilon\right] = \left(1 - \varepsilon\right)^n$$

We can now write

$$\lim_{n \to \infty} P\left[|W_n - 1| > \varepsilon\right] = \lim_{n \to \infty} \left(1 - \varepsilon\right)^n = 0.$$

and we have shown that the maximum of a sequence of independent $U(0,1)$ random variables converges in probability to one.

Finally, we can demonstrate that continuous functions of a sequences of random variables that converge in probability themselves converge in probability, as demonstrated in the following lemma.

Lemma 7.5 Let $\{X_n\}$ be a sequence of random variables that converge in probability to X. If $f(X_n)$ is a continuous function at all points X_n in the sequence, then the sequence $\{f(X_n)\}$ converges to $f(X)$.

Lemma 7.5 is yet another implication of Slutsky's theorem. However, a direct demonstration of this lemma is also available. We know that the concept of a continuous function might be expressed in the observation that as two points on the real line get closer to each other, then the value of the continuous function of those two points must also get closer together. The convergence in probability assumption allows us to say that, if it is very likely that those two points X and a are close to each other, then it is very likely that $f(X)$ will be close to $f(a)$.

More formally, we must show that $\lim_{n\to\infty} P\Big[\big|f(X_n)-f(X)\big|\le\varepsilon\Big]=1$ We know that, since X_n converges to X in probability, we can go far enough out in the sequence $\{X_n\}$ so that $P(|X_n-X|\le\delta]$ is large. However, the continuity of the function f assures us that if $|X_n-X|\le\delta^*$, then it must be true that $|f(X_n)-f(X)|\le\varepsilon$. Thus, $|X_n-X|\le\delta$ implies that $\big|f(X_n)-f(X)\big|\le\varepsilon$. But if this implication is true that the inequality

$$P\Big[\big|f(X_n)-f(X)\big|\le\varepsilon\Big] \;\ge\; P\Big[|X_n-X|\le\delta\Big]$$

must   also   be   true.   Therefore   we   know   that   the   quantity $P\Big[\big|f(X_n)-f(X)\big|\le\varepsilon\Big]$ can be brought arbitrarily close to one by choosing n large enough, proving $\lim_{n\to\infty} P\Big[\big|f(X_n)-f(X)\big|\le\delta\Big]=1$

We may succinctly write this as

$$\lim_{n\to\infty} P\Big[\big|f(X_n)-f(X)\big|\le\varepsilon\Big] \;\ge\; \lim_{n\to\infty} P\Big[|X_n-X|\le\delta\Big]=1$$

Thus, just as limit function passed though the continuous function argument for real numbers, we find that the limit in probability function passes through continuous functions of random variables.

7.11 The Law of Large Numbers and Chebyshev's Inequality

The law of large numbers is a statement about the limiting behavior of the sample mean of a collection of observations. Sometimes described in the vernacular as the "law of averages", this tenet states that, under suitable conditions, as n increases, the sample mean $\overline{X}_n$ approaches the population mean μ.

7.11.1 The Universality of the "Law of Averages"

Unlike many laws and theorems in probability and statistics, non-probabilists in the population at large have a common and intuitive understanding of the underlying meaning of this law. Consider the ubiquitous example of flipping a fair coin. After ten flips, it is not so uncommon to have seven heads and three tails. Yet, we all expect that, after a very large number of flips, the number of heads and tails will "even out", i.e., we can expect approximately the same number of heads as tails. This expectation can be placed in the setting of the law

---

* This argument is actually based on the property of uniform continuity of the function f, in which the value of ε does not depend on the value of X.

of large numbers. Let the experiment be the result of a single coin toss, and let X_i be the result of that toss. Define $X_i = 1$ if that coin toss results in a head, and $X_i = 0$ if the result is a tail. Then in this case $\sum_{i=1}^{n} x_i$ is just the total number of heads in n tosses, and $\overline{X}_n$ is the relative frequency of the occurrence of heads

   The law of large numbers has both a weak version and a strong version We will focus in this section on the weak version.

Theorem 7.3. The Weak Law of Large Numbers
Let $X_1,..,X_n$ be a collection of independent random variables with common mean μ and a common finite variance σ^2. Then the sample mean, $\overline{X}_n$ converges to μ in probability.

We will find that this most useful result will be very easy to prove. To facilitate this process, we will begin with a brief discussion of "tail probabilities"

7.11.2 Inequalities and the Probability of Tail Events
There are important and useful observations concerning the probability of special events, termed *tail events.* A tail event for a random variable X is the event that $X \geq c$ for a positive, real number c. The event that $X \leq -c$ is of course, also a tail event. It is the value of c that makes tail events so useful, since the likelihood of a tail event can be an important indicator for gauging the reliability of an estimator.

   Consider the circumstance in which we have a collection of observations X_1, X_2, X_3, ... X_n (which we will summarize as $\{X_n\}$) from the same probability distribution which has a parameter θ. In this example, it has been suggested that a function of the sample values $T(\{X_n\})$ is a useful estimator for a function of the probability distribution's parameter $q(\theta)$. As we study the properties of the estimator $T(\{X_n\})$ a core component of this evaluation will of course be whether $T([X_n])$ is actually close to $q(\theta)$. If this is the case, then we would expect that the difference $T(\{X_n\}) - q(\theta)$ or, more appropriately, its absolute value, would be small. Thus $P\left[\left|T(\{X_n\}) - q(\theta)\right| \geq c\right]$, a probability based on the tail event that $\left|T(\{X_n\}) - q(\theta)\right| \geq c$, can provide useful information about the likelihood that the proposed estimator $T(\{X_n\})$ will deviate substantially from the parameter $q(\theta)$. For example, the occurrence of large values of $P\left[\left|T(\{X_n\}) - q(\theta)\right| \geq c\right]$ for relatively small values of c would suggest that the estimator may not perform as well as we hoped. A detailed discussion of this subject is presented in Chapter 8.

There are some remarkably informative inequalities that can make it easy to identify the probability of these tail events. Let X be a random variable with probability density function $f_X(x)$. Then we may write

$$E\left[|X|\right] = \int_{-\infty}^{\infty} |x| f_x(x) dx$$

Now, choose any value $c > 0$. Then

$$E\left[|X|\right] = \int_{-\infty}^{\infty} |x| f_x(x) dx = \int_{-\infty}^{c} |x| f_x(x) dx + \int_{c}^{\infty} |x| f_x(x) dx$$

Now we note that since $E\left[|X|\right]$, $\int_{-\infty}^{c} |x| f_x(x) dx$, and $\int_{c}^{\infty} |x| f_x(x) dx$ are all positive

terms, and $E\left[X\right] \geq \int_{c}^{\infty} |x| f_x(x) dx$.

Next, examine the integrand and the region of integration contained in this last expression. On the region of integration where $c \leq X \leq \infty$, then we may

write $\int_{c}^{\infty} |x| f_x(x) dx \geq \int_{c}^{\infty} c\, f_x(x) dx$. But $\int_{c}^{\infty} c f_X(x) = c \int_{c}^{\infty} f_X(x) = cP\left[x \geq c\right]$. Thus we

have

$$P\left[X \geq c\right] \leq \frac{E\left[|X|\right]}{c}, \tag{7.14}$$

an expression that relates a tail probability to a function of the expectation of a random variable. It is easy to show that if X is a non negative random variable,

then $P\left[X \geq c\right] \leq \frac{E\left[X\right]}{c}$.

7.11.3 Markov's Inequality

There are some remarkably informative inequalities that can make it easy to identify the probability of these tail events. Let X be a random variable with probability density function $f_x(x)$. Then we may write

$$E\left[|X|\right] = \int_{-\infty}^{\infty} |x| f_x(x) dx \tag{7.15}$$

Choose any value $c > 0$. Then we have shown that $E\left[|X|\right] \geq \int_{c}^{\infty} |x| f_x(x) dx$.

or $P[X \geq c] \leq \dfrac{E\big[|X|\big]}{c}$.

A generalization of Markov's inequality is presented below. This effort expands the evaluation in which we are currently involved to positive powers of |x|. In this case

$$E\Big[|X|^r\Big] = \int_{-\infty}^{\infty}|x|^r f_x(x)dx = \int_{-\infty}^{c}|x|^r f_x(x)dx + \int_{c}^{\infty}|x|^r f_x(x)dx$$

$$\geq \int_{c}^{\infty}|x|^r f_x(x)dx \geq \int_{c}^{\infty}c^r f_x(x)dx = c^r \int_{c}^{\infty}f_x(x)dx = c^r P[x \geq c]$$

or $P\Big[X^r \geq c\Big] \leq \dfrac{E\Big[|X|^r\Big]}{c^r}$

7.11.4 Chebyshev's Inequality

The most famous of this series of inequalities is that attributed to P.L. Chebyshev, a Russian probabilist who lived in the 19th century. Chebyshev's inequality, alluded to in Chapter 3, is a probability statement about the deviation of a random variable from its mean. While the original proof of this inequality was quite complex, we can develop it from a simple application of Markov's inequality. Let X be a random variable with mean μ and finite variance σ^2. We wish to find an upper bound for the probability $P\big[|X - \mu| \geq c\big]$.

The direct application of Markov's equality with r = 2 produces the immediate result

$$P\big[|X - \mu| \geq c\big] \leq \frac{E\Big[|X - \mu|^2\Big]}{c^2} = \frac{E\Big[(X - \mu)^2\Big]}{c^2} = \frac{\sigma^2}{c^2} \tag{7.16}$$

The degree to which we can expect X to deviate from its mean is related to the variance σ^2 and the degree of departure $c > 0$.

Chebyshev's inequality does not always produce an accurate bound for the probability of a tail event. If, for example, we attempted to use Chebyshev's inequality to find the probability that a random variable that followed a Poisson probability with parameter $\lambda = 12$ was within one unit of its mean, we would compute from equation (7.16)

$$P\left[\left|X-\mu\right|\geq1\right]\leq\frac{\lambda}{1}=12?$$

While true, this is certainly not very helpful. A more useful example follows.

Example 7.2. A national bank has agreed to set up an ATM machine in a location that is relatively safe and is likely to experience a certain level of use by the bank's customers. The number of customers who visit the ATM machine is anticipated to follow a Poisson distribution with parameter $\lambda = 12$ customers per day. If 22 or more customers use the machine in a day, the ATM's store of money will be rapidly depleted and the machine will need to be re-supplied. If use of the machine drops to two or less customers per day, the delicate internal mechanism that self calibrates with use will need to be serviced more frequently. What is the probability that the ATM machine will require attention?

The machine requires attention when usage drops to either two or less customers, or, alternatively 22 or more customers a day use the ATM. This may easily be recast as the probability $P\left[\left|X-12\right|\geq10\right]$. Applying Chebyshev's inequality in this circumstance reveals

$$P\left[\left|X-12\right|\geq10\right]\leq\frac{12}{100}=0.12$$

This suggests that it is unlikely that this machine will require servicing on any particular day. What contributed to the utility of the inequality in this circumstance is that we were interested in a tail event that was extreme.

Example 7.3

As another example, consider the task of a computer chip manufacturer who produces electronic chips in lots of 1000. If the variance of the computing parameter for each chip is too great, the percentage of chips that are considered defective chips or "floor sweepings" (meaning that the chip is discarded, figuratively thrown on the floor to be swept up and placed in the trash) is too great. If $\sigma^2 = 1$, what is the probability that the process variance will deviate from σ^2 by more than one unit.

In this circumstance, we need to compute $P\left[\left|S^2-\sigma^2\right|\geq1\right]$. From Chebyshev's equality, we know that $P\left[\left|S^2-\sigma^2\right|\geq1\right]\leq\frac{Var\left(S^2\right)}{1}=Var\left(S^2\right)$. If we can assume that the quantity $\frac{(n-1)S^2}{\sigma^2}$ follows a χ^2_{n-1}, then we know the

$Var\left[\dfrac{(n-1)S^2}{\sigma^2}\right] = 2(n-1).$ We may therefore find the variance of S^2

as $2(n-1) = Var\left[S^2\dfrac{(n-1)}{\sigma^2}\right] = \dfrac{(n-1)^2}{\sigma^4}Var\left(S^2\right).$

Or

$$\dfrac{(n-1)^2}{\sigma^4}Var\left(S^2\right) = 2(n-1)$$

$$Var\left[S^2\right] = \dfrac{2(n-1)\sigma^4}{(n-1)^2} = \dfrac{2\sigma^4}{(n-1)}$$

Thus we may write

$$P\left[\left|S^2 - \sigma^2\right| \geq 1\right] \leq Var\left(S^2\right) = \dfrac{2\sigma^4}{(n-1)} = \dfrac{2}{(n-1)} \text{ for } \sigma^2 = 1.$$

For example, in a sample of 1000 chips the $P\left[\left|S^2 - \sigma^2\right| \geq 1\right] \leq 0.002.$

■

Attempts have been made to generalize Chebyshev's inequality; several of them are provided (Parzen, page 228). However, the inequality is not as well known for its computational utility. Its most important use has been in the demonstration of the law of large numbers.

7.11.5 The Weak Law of Large Numbers
The law of large numbers has an important history, but can be simply stated. This law asserts that, as the sample size increases, the sample means of an independent sample gets closer to the population mean. We will formalize this statement, defining the law of large numbers first in terms of convergence in probability.

Our proof for this will be short and straightforward. In order to demonstrate that the quantity $\overline{X}_n$ converges in probability to μ, we must show that $\lim\limits_{n\to\infty} P\left[\left|\overline{X}_n - \mu\right| \geq \varepsilon\right] = 0.$ However, we know from Chebyshev's inequality that $P\left[\left|\overline{X}_n - \mu\right| \geq \varepsilon\right] \leq \dfrac{Var\left[\overline{X}_n\right]}{\varepsilon^2}$ We may therefore write

$$\lim\limits_{n\to\infty} P\left[\left|X_n - \mu\right| \geq \varepsilon\right] \leq \lim\limits_{n\to\infty}\dfrac{Var\left[\overline{X}_n\right]}{\varepsilon^2} = \lim\limits_{n\to\infty}\dfrac{\sigma^2}{n\varepsilon} = 0. \qquad (7.17)$$

We note the role of the finite variance. If the variance was infinite we have no guarantee that the final limit expression in (7.17) would converge to zero.

The result that the sample mean converges to the population mean may appear as no surprise to the modern student in statistics. However this result was by no means clear to 16th century probabilists who were confronted with several competing estimators for the probability of a successful Bernoulli trial. The weak law of large numbers was first demonstrated by Bernoulli in 1713 for independent Bernoulli trials (Feller). However, this original proof did not incorporate Chebyshev's inequality, but instead involved a laborious evaluation of $P\left[\left|\overline{X}_n - \mu\right| \geq \varepsilon\right]$. It was the demonstration of Bernoulli's version of the law of large numbers that supported the natural belief that the proportion of successful Bernoulli trials is an accurate assessment of the probability of a Bernoulli trial. However, while this intuitive sense that relative frequency computations can be a good measure of the probability of a successful Bernoulli trial is a correct interpretation of this version of the law of large numbers, there remain today some common interpretations of this law that are also intuitive but incorrect.

A false interpretation of the law of large numbers is all too easily translated into the world of gambling. A gambler who is experiencing a run of bad luck commonly believes that continuing to play will assure victory at the gambling table. If we parameterize his situation by letting $X_n = 1$ if his n^{th} gamble earns him money, and $X_n = 0$ if he loses money, then his conviction might be more mathematically stated as a belief that, in the long run, $\overline{X}_n$ must be close to p, the probability that the gambler will win his bet. Therefore, he reasons that the string of X_n's that he has observed for which $X_n = 0$ cannot continue to go on, and will soon be reversed by the occurrence of a compensatory collection of X_n's for which X_n will be one. Thus, he believes that by continuing to play, his "luck will turn around".

However, there are two difficulties with this approach. The first, more obvious one is that, all too commonly, the value of p is lower than he anticipates. This low value would require a far longer appearance at the gambling table than his diminishing fortune will allow. Secondly, and perhaps less obviously, a change in the gambler's luck occurs far less commonly than our gambler might anticipate. The gambler assumes that the number of times the gambler is winning should increase in proportion to the number of times he gambles. This is not the case. Feller (1) has demonstrated that the number of times the lead changes in a sequence of Bernoulli trials increases not as a function of n, but as a function of $\sqrt{n}$. Thus the number of gambles the gambler has to make before he regains his losses is likely to be far longer than he can tolerate.

7.12 The Central Limit Theorem

7.12.1 Basic Concepts

One of the most fascinating observations in probability is the finding that so many random variables appear to follow a normal distribution. Actually, however, only a relatively small number of these random variables are truly normally distributed. Some random variables can be transformed (e.g. by taking their cube roots) into new random variables that follow a normal distribution. However, the overwhelming majority of random variables are not normally distributed at all. Nevertheless, probabilities of events based on these random variables are very nicely approximated by the normal distribution. This may seem counterintuitive because of the complexity of the density function of the normal distribution. It is this latter class of random variables that will be the focus of our attention.

The use of normal probabilities in these cases is governed by the central limit theorem. We will begin with a simple statement of this powerful observation.

Theorem 7.4. The Central Limit Theorem

Let $\{X_n\}$ be a collection of random variables which are independent and identically distributed, following a distribution with a known finite mean μ and finite variance σ^2. Then for any fixed c

$$P\left[\frac{S_n - n\mu}{\sqrt{n\sigma^2}} \leq c \right] converges\ to\ \Phi_Z(c) \tag{7.18}$$

where $\Phi_Z(c)$ is the cumulative distribution function of the standard normal distribution (with mean zero and variance one).

7.12.2 Examination of Uniformly Distributed Random Variable

Before we attempt to prove this theorem, we should examine this remarkable statement in detail. According to this expression, the original random variables need not themselves be normally distributed. As long as they are independent of one another, have a mean, and have finite variances, the probabilities of events that involve the sums of these random variables can be approximated by a normal distribution. The wide applicability of this theorem is one of the reasons why the cumulative distribution function $F_X(x) = \int\limits_{-\infty}^{x} \frac{1}{\sqrt{2\pi}} e^{-\frac{x^2}{2}} dx$ is called the normal distribution; it is the distribution that is most usually or "normally" used.

In order to gain a sense of how quickly the movement toward a normal distribution begins, consider the following exercise. Let the random variable X_1 follow a uniform distribution on [0, 1]. Let $y_1 = X_1$ We know that the density of

y_1 is $f_{Y_1}(y) = \mathbf{1}_{y \in [0,1]} dy$ and we can draw this density readily (Figure 7.3). Now, define the random variable X_2 as a U(0,1) distribution that is independent of X_1 and create the random variable $y_2 = X_1 + X_2$. What does the distribution of the sum of two of these random variables look like? It has been demonstrated in Chapter 5 that $f_{Y_2}(y) = y\mathbf{1}_{y \in [0,1]} + (2 - y)\mathbf{1}_{y \in (1,2]}$. The graph of this density function demonstrates a striking trend to concentration of the probability in the center to the distribution (Figure 7.3).

While we cannot say that the density function of $y = X_1 + X_2$ is that of a normal random variable, we cannot deny that the location of the random variable that is the sum of two $U(0,1)$ random variables is far less likely to be close to 0 or close to the 2 than it is to be in the center of the distribution. This is a striking finding and might be considered to be counterintuitive. Since a uniformly distributed random variable is equally likely to fall in an interval of equal widths, it is hard to imagine why sums of these random variables are more likely to cluster nearer to the value one. We will explore this property of sums shortly.

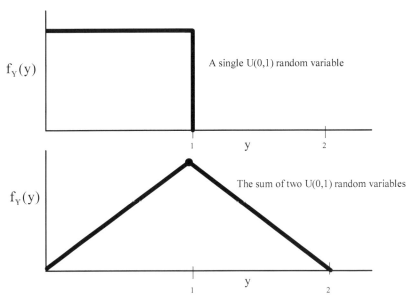

Figure 7.3 Comparison of the probability density functions for the sum of one and the sum of two U(0.1) random variables.

The applicability of the central limit theorem is amplified if we add one more independent U(0,1) random variable to our sum. If we now let y be the sum of three independent and identically distributed random variables, then we can identify the density of y, $f_Y(y)$ can be readily identified (Chapter 5, section 5.3.3)

$$f_Y(y) = \frac{y^2}{2}\mathbf{1}_{0 \le y \le 1} + \left(3y - y^2 - \frac{3}{2}\right)\mathbf{1}_{1 < y \le 2} + \left(\frac{y^2}{2} - 3y + \frac{9}{2}\right)\mathbf{1}_{2 < y \le 3} \qquad (7.19)$$

The derivation of the density function requires us to divide the region of y into several different sub regions. In each of these sub regions, we can find the cumulative density function $F_Y(y)$. After completing this series of integrations, we next check to insure that $F_Y(y)$ is continuous at the juncture of these regions. Finally, we take derivatives of the cumulative density function in each of these regions, arriving at the probability density function of y. In this circumstance, the density has three different formulas, which depend on the interval in which y lies. The shape of this complicated density function is remarkably "normal" (Figure 7.4).

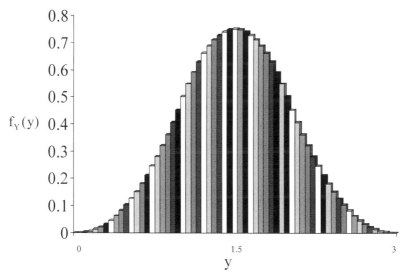

Figure 7.4 Exact function of the sum of three U(0,1) random variables

The sum of these three independent, uniformly distributed random variables is tightly constrained to be within the interval [0,3]. However, within this region, the probability distribution already takes on a remarkably familiar shape. The straight lines which delimited the density function have been replaced by curves which smoothly reflect the change in slope of the probability density function as the value of y increases from $y = 0$ to $y = 3$. In addition, the values of this random variable center around 1.5, which is the midpoint of the [0, 3] range. The shape of the density function has been transformed. No longer made up of lines parallel to the X and y axes (as was the case for y_1) or by straight lines (as was the case for y_2), the combination of parabolic shapes from the density (7.19) is what will be approximated by a normal distribution. However a comparison of this exact density to that of the normal density function that would be used to approximate the density function of y_3 reveals that computations based on the approximation would be quite inaccurate for the sum of three i.i.d. U(0,1) random variables (Figure 7.5). Nevertheless, we can see how the approximation of a normal distribution to the density function of the sum of U(0,1) random variables is produced.

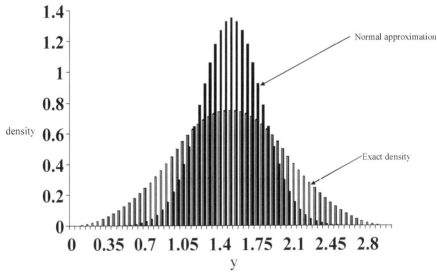

Figure 7.5 Comparison of the exact density of the sum of three U(0,1) random variables and that of the normal distribution.

7.12.3 Dichotomous Random Variables

The preceding demonstration of the convergence of the sum of independent and identically distributed random variables to a "normal like" distribution demonstrated the applicability of the central limit theorem. However, it provided no insight into why there should be movement to a central value in the sum of random variables at all. A simple example will provide this important insight.

A dichotomous random variable is simply a random variable that can only assume one of two values. For example, a simple Bernoulli trial is an example of a dichotomous random variable. For our purposes, we define the random variable X as dichotomous in the following manner: it can assume the value $X = 0.01$ with probability 0.50, or the value $X = 0.99$ with probability 0.50. We may write the density of X, $f_X(x)$ as

$$f_X(x) = 0.50\mathbf{1}_{x=0.01} + 0.50\mathbf{1}_{x=0.99}. \tag{7.20}$$

There is certainly no indication of central tendency for this random variable that can only be one of two values that are relatively far apart on the [0,1] interval. A second such random variable from the same distribution will also not demonstrate any movement to central tendency. However, the evaluation of the sum of two such random variables demonstrates very different behavior. We may quite easily write the possible values of the sum of two random variables, i.e., $V = X + Y$, and compute the probabilities of these values. (Table 7.1).

Table 7.1 Distribution of Sum of Two Dichotomous Random Variables

		P(X)	0.50	0.50
		X	0.01	0.99
P(Y)	Y			
0.50	0.01		0.020	1.000
0.50	0.99		1.000	1.980

There are four possible values for v. Of these four values, two are extreme (0.020 and 1.980), and the other two cluster in the center. Fifty percent of the possible values this new random variable can take are centered in the middle of the probability distribution. If we proceed one step further and ask why do half of the possible values of V demonstrate central tendency, we see that it is because there are more ways to combine small values of one of the summands with large values of the other (0.01, 0.99), (0.99, 0.01), than there are to create either an extremely small pair of summands (0.01, 0.01), or an extremely large pair (0.99, 0.99).

Proceeding further, we can create a new random variable, W, that is the sum of three independent and identically distributed random variables that has the same distribution as (7.20)

Table 7.2   Distribution of Sum of Three Dichotomous Random Variables

P(V)	P(X) X	0.50 0.01	0.50 0.99
	V		
0.25	0.02	0.030	1.010
0.50	0.01	1.010	1.990
0.25	1.98	1.990	2.970

In this case, there are only four possible values for the random variable W; two of these demonstrate central tendency, and two do not. However, the probability of the two central values of W, 1.010 and 1.990 (P[W = 1.010] = P[W = 1.990]) = 0.375) are much greater than the probabilities of the extreme values (P[W = 0.030] = P[W = 2.970] = 0.125). This probability is greater because there are more combinations of the three random variables that produce these central values than there are combinations that produce extreme values. This same point is made in the construction of the sum of four random variables (Table 7.3).

Table 7.3   Distribution of Sum of Four Dichotomous Random Variables

P(W)	P(X) X	0.50 0.01	0.50 0.99
	W		
0.125	0.03	0.040	1.120
0.375	1.01	1.020	2.000
0.375	1.99	2.000	2.980
0.125	2.97	2.980	3.960

Thus, the move to central tendency increases as the number of random variables included in the summand increases. This is because the summation process produces more moderate than extreme values from the original random variables (Figure 7.6).

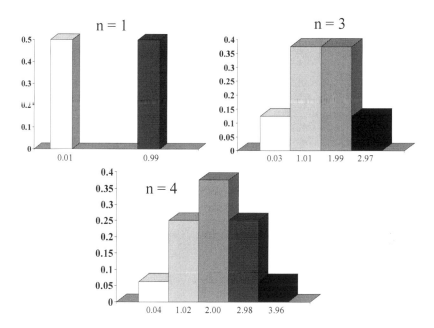

Figure 7.6 Distribution of sums of dichotomous variables

7.13 Proof of the Central Limit Theorem

The central limit theorem is sometimes called the Lindburg-Levy Theorem. Its proof is based on a property of convergence in distribution (convergence in law) that we have not yet discussed, namely the use of moment generating functions to demonstrate that random variables converge.

7.13.1 Convergence in Distribution and $M_n(t)$

In section 7.3 we defined the concept of convergence in distribution as one that involves a sequence of random variables $\{X_n\}$ and a corresponding sequence of cumulative distribution functions, $\{F_{X_n}(x)\}$ that, for the sake of notational brevity, we will describe as $\{F_n(x)\}$. We said that the sequence of random variables $\{X_n\}$ converged in distribution to the random variable X if $\lim_{n \to \infty} F_n(x) = F(x)$. However, we might also use the link between moment generating functions and cumulative distribution functions to discuss convergence in distribution in terms of these helpful devices. Recall that a cumulative distribution function has one and only one moment generating function. Therefore, since the moment generating functions are unique, the convergence of $F_n(X)$ to $F(X)$ suggests that if $M_n(t)$ is the moment generating function of the random variable X_n, then $M_n(t)$ must converge to $M(t)$ where $M_X(t) = E\left[e^{tx}\right]$. This is called Levy's Continuity Theorem for Moment Generating Functions, which we present without proof.

Theorem 7.5.   Levy's Continuity Theorem for Moment Generating Functions. Let $\{X_n\}$ be a sequence of random variables. Denote the cumulative distribution function of X_n as $F_n(X)$, and the moment generating function of X_n by $M_n(t)$. Then

$\lim_{n \to \infty} F_n(x) = F(\mathbf{x})$ *iff* $\lim_{n \to \infty} M_n(t) = M(t)$ for every point t such that $|t| < 1$ and

$\lim_{t \to 0} M(t) = 1$.

7.13.2 Comment about Levy's Theorem

Our goal is to prove the central limit theorem by relying on Levy's criteria for the convergence of moment generating functions. We will find that, in general the moment generating function is a useful tool in demonstrating the convergence of random variables. However useful moment generating functions may be, it must be recognized, that working with moment generating functions can at first be frustrating. The explorations of the manner in which the sum of uniformly distributed random variables on the unit interval produced probability mass in the center of the distribution in section 7.12 provided intuition into why central tendency emerged from sums of variables which by themselves presented no such tendency. However, frequently there is no such insight when one works with moment generating functions. This is not to suggest that arguments based solely on moment generating functions are impenetrable — only that the intuition for the underlying argument lies in a deeper branch of mathematics, that of limits and continuity.

We will find in our approach that a useful tool on which moment generating functions commonly rely is the following statement.

$$\lim_{n \to \infty} \left(1 + \frac{x}{n}\right)^n = e^x \qquad\qquad (7.21)$$

We proved this finding earlier in Chapter 1 by applying two familiar findings 1) $\sum_{n=0}^{\infty} \frac{x^n}{n!} = e^x$, and 2) $(a+b)^n = \sum_{k=0}^{n} \binom{n}{k} a^k b^{n-k} = \sum_{k=0}^{n} \binom{n}{k} b^k a^{n-k}$.

7.13.3 Approximation of Binomial Probabilities by Poisson Probabilities — Moment Generating Function Approach

We have demonstrated in an earlier section of this chapter that under certain conditions, we can approximate the probability of an event involving a binomial variable by using the Poisson probability distribution. This was accomplished by demonstrating that a binomial (n, p) mass function approaches that of the Poisson (λ) mass function when p, the probability of a success on a Bernoulli trial is small and n, the number of trials goes to infinity in such a manner that $\lambda = np$ is constant. We can now demonstrate this result using a moment generating function argument. Recall that the moment generating function for a Bernoulli trial is with probability of success p is $q + pe^t$, and $M_n(t)$ the moment generating function for the sum of n of these random variables is

$$E\left[e^{\sum_{k=1}^{n} x_k t}\right] = \prod_{k=1}^{n}\left(Ee^{(x_k t)}\right) = \prod_{k=1}^{n}\left(M_{x_k}(t)\right) = \left(M_x(t)\right)^n = \left(q + pe^t\right)^n \quad (7.22)$$

For the Poisson distribution with parameter n, the moment generating function is

$$\sum_{k=0}^{\infty} e^{kt}\frac{\lambda^k}{k!}e^{-\lambda} = e^{-\lambda}\sum_{k=0}^{\infty}\frac{\left(\lambda e^t\right)^k}{k!} = e^{-\lambda}e^{\lambda e^t} = e^{\lambda\left(e^t - 1\right)} \quad (7.23)$$

We must therefore show that $\lim_{\substack{n\to\infty \\ np\to\lambda}}\left(q + pe^t\right)^n = e^{\lambda\left(e^t - 1\right)}$. Our development thus far allows us to complete this demonstration easily. We can first write that

$\left(q + pe^t\right)^n = \left(1 - p + pe^t\right)^n = \left(1 + p(e^t - 1)\right)^n$. If we now let $p = \dfrac{\lambda}{n}$ we have

$$\left(q + pe^t\right)^n = \left(1 + \frac{\lambda(e^t - 1)}{n}\right)^n \quad (7.24)$$

From here we simply observe that

$$\lim_{n\to\infty}\left(q + pe^t\right)^n = \lim_{n\to\infty}\left(1 + \frac{\lambda(e^t - 1)}{n}\right)^n = e^{\lambda(e^t - 1)} \quad (7.25)$$

the desired result. We have demonstrated the weak convergence of the binomial distribution to the Poisson distribution by demonstrating that both the distribution function and the moment generating function of the binomial distribution converge to these respective functions of the Poisson distribution.

7.13.4 Weak Convergence of Poisson Random Variables to the Normal Distribution

The use of the normal distribution as an approximation to the distribution of other distributions will begin with an examination of the manner in which the Poisson distribution produces probabilities that are approximately Gaussian. We will again use moment generating functions in this demonstration, but this time, we will rely on a different, established property of these functions. Recall that, by taking successive derivatives of the moment generating function of a random variable X, we can write

$$M_X(t) = \sum_{k=0}^{\infty} \frac{E[X^k]}{k!} t^k$$

$$M_X(t) = 1 + \frac{E[X]}{1} t + \frac{E[X^2]}{2} t^2 + \frac{E[X^3]}{3!} t^3 + \dots + \frac{E[X^k]}{k!} t^k + o(t^k)$$

(7.26)

where $o(t^k)$ reflects additional terms that contain powers of t that are greater than k (i.e., terms that are of order t^{k+1}). A careful analysis of the expansion of the moment generating function is $\lim_{t \to 0} \dfrac{o(t^k)}{t^k} = 0$.

Our goal is to demonstrate the utility of this expansion in proving that the probability distribution of a random variable converges to the probability distribution of another. Specifically, we will show that the expansion of the moment generating function of the Poisson distribution, when expanded as in expression (7.26) will converge to that of the normal distribution. Recall that if X is a Poisson random variable with mean λ, then

$$M_X(t) = e^{\lambda(e^t - 1)}$$

(7.27)

Now consider the random variable Z defined as given by $Z = \dfrac{X - \lambda}{\sqrt{\lambda}}$, which is the Poisson random variable with its mean subtracted and divided by its standard deviation (i.e., a standardized Poisson random variable). Then the mgf of Z is

$$M_Z(t) = E\left[e^{\frac{(X-\lambda)t}{\sqrt{\lambda}}} \right] = e^{-t\sqrt{\lambda}} E\left[e^{\frac{t}{\sqrt{\lambda}} X} \right].$$

Since $M_X(t) = e^{\lambda(e^t - 1)}$, we may write $E\left[e^{\frac{t}{\sqrt{\lambda}}X}\right] = e^{\lambda\left(e^{\frac{t}{\sqrt{\lambda}}} - 1\right)}$, and continue as

$M_Z(t) = e^{-t\sqrt{\lambda}}\left[e^{\lambda\left(e^{t/\sqrt{\lambda}} - 1\right)}\right]$. It now remains for us to write

$$\lambda\left(e^{\frac{t}{\sqrt{\lambda}}} - 1\right) = \lambda\left[1 + \frac{t}{\sqrt{\lambda}} + \frac{t^2}{2!\lambda} + \frac{t^3}{3!\lambda^{3/2}} + \ldots + \frac{t^n}{n!\lambda^{n/2}} + \ldots - 1\right]$$

$$= \lambda\left[1 + \frac{t}{\sqrt{\lambda}} + \frac{t^2}{2!\lambda} + o(t) - 1\right] = \lambda\left[\frac{t}{\sqrt{\lambda}} + \frac{t^2}{2!\lambda} + o(t)\right]$$

Thus

$$M_Z(t) = E\left(e^{\frac{(X-\lambda)t}{\sqrt{\lambda}}}\right) = e^{-t\sqrt{\lambda}}\left[e^{\lambda\left(e^{t/\sqrt{\lambda}} - 1\right)}\right] = e^{-t\sqrt{\lambda}}\left[e^{t\sqrt{\lambda} + \frac{1}{2}t^2 + o(t)}\right] = e^{\frac{1}{2}t^2 + o(t)}$$

And

$$\lim_{\lambda \to \infty} M_Z(t) = e^{\frac{1}{2}t^2}.$$

However, another demonstration of this result illustrates the importance of a finite variance in the limiting process. In this circumstance, we are interested in identifying the moment generating function for the standardized sum of n Poisson random variables that we will write as

$$T_n = \frac{\sum_{i=1}^{n} x_i - n\lambda}{\sqrt{n\lambda}} \tag{7.28}$$

We will first use standard moment generating function techniques to find the moment generating function of $M_{T_n}(t)$. Begin by writing expression (7.28)

$$T_n = \frac{\sum_{i=1}^{n} X_i - n\lambda}{\sqrt{n\lambda}} = \frac{1}{\sqrt{n\lambda}}\sum_{i=1}^{n}(X_i - \lambda) \tag{7.29}$$

We may now write

$$M_{T_n}(t) = M_{\sum_{i=1}^{n}(x_i - \lambda)}\left(\frac{t}{\sqrt{n\lambda}}\right) = \left[M_{(x_i - \lambda)}\left(\frac{t}{\sqrt{n\lambda}}\right)\right]^n \tag{7.30}$$

We can now use the result of (7.26) to evaluate $M_{(x_i-\lambda)}\left(\dfrac{t}{\sqrt{n\lambda}}\right)$.

$$M_{(x_i-\lambda)}\left(\frac{t}{\sqrt{n\lambda}}\right) = 1 + \frac{E[X_i - \lambda]}{1}\left(\frac{t}{\sqrt{n\lambda}}\right) + \frac{E[X_i - \lambda]^2}{2}\left(\frac{t}{\sqrt{n\lambda}}\right)^2 + o\left(\frac{t^2}{n\lambda^2}\right)$$

$$= 1 + \frac{\lambda}{2}\left(\frac{t^2}{n\lambda}\right) + o\left(\frac{t^2}{n\lambda^2}\right) = 1 + \left(\frac{t^2}{2n}\right) + o\left(\frac{t^2}{n\lambda^2}\right) \qquad (7.31)$$

Relationship (7.30) shows that $M_{T_n}(t) = \left[M_{(x_i-\lambda)}\left(\dfrac{t}{\sqrt{n\lambda}}\right)\right]^n$, we may write

$$M_{T_n}(t) = \left[1 + \left(\frac{t^2}{2n}\right) + o\left(\frac{t^2}{n\lambda^2}\right)\right]^n = \left[1 + \frac{\left(\dfrac{t^2}{2}\right) + no\left(\dfrac{t^2}{n\lambda^2}\right)}{n}\right]^n \qquad (7.32)$$

and now it only remains for us to take a limit

$$\lim_{n\to\infty} M_{T_n}(t) = \lim_{n\to\infty}\left[1 + \frac{\left(\dfrac{t^2}{2}\right) + no\left(\dfrac{t^2}{n\lambda^2}\right)}{n}\right]^n = e^{\lim\limits_{n\to\infty}\left[\left(\frac{t^2}{2}\right)+no\left(\frac{t^2}{n\lambda^2}\right)\right]} = e^{\frac{t^2}{2}} \qquad (7.33)$$

Recognizing that $e^{\frac{t^2}{2}}$ is the moment generating function for the standard normal distribution, we have our desired result.

It would be worthwhile to take a moment to review the details of this proof because the techniques we used to demonstrate convergence of the Poisson distribution (in law) to the normal distribution will be used repeatedly. We had to first be able to identify the moment generating function for each of the two distributions. Secondly, writing the moment generating function in terms of the moments of the distribution allows us to include the variance of the random variable. If the variance of the random variable does not exist (i.e., is infinite) then our proof would break down. Additionally, by working with the sum of independent random variables, we were able to take the product of the generating functions. Finally, we invoked the result $\lim\limits_{n\to\infty}\left(1 + \dfrac{x_n}{n}\right)^n = e^{\lim\limits_{n\to\infty} x_n}$. These steps we will use again to demonstrate the weak convergence of a binomial random variable.

7.13.5 Weak Convergence of Binomial Distribution to the Normal Distribution

In a previous section, we demonstrated the weak convergence of the binomial distribution to the Poisson distribution, first by working directly with the distribution functions of the two random variables, and secondly, through an examination of their moment generating functions. In this section, we will demonstrate the weak convergence of the binomial distribution to the normal distribution.

You may have already anticipated this result. Since we already know that binomial random variables converge in distribution to Poisson random variables, and Poisson random variables converge (again, in distribution) to normal random variables, we might have suspected that binomial random variables would converge in distribution "through" the Poisson distribution to normal random variables. Here is the demonstration of that fact.

Let X_n be a sequence of independent Bernoulli random variable with parameter p. Then $S_n = \sum_{i=1}^{n} X_i$ is a binomial random variable. We need to demonstrate that $T_n = \dfrac{S_n - np}{\sqrt{np(1-p)}}$ converges in distribution to a random variable that is normally distributed with mean zero and variance one. We will use the moment generating function approach. Begin by recognizing that

$$T_n = \frac{S_n - np}{\sqrt{np(1-p)}} = \sum_{i=1}^{n} \frac{X_i - p}{\sqrt{np(1-p)}} \tag{7.34}$$

This observation permits us to write $M_{T_n}(t) = \left[M_S(t)\right]^n$ where $S = \dfrac{x_i - p}{\sqrt{np(1-p)}}$.

We now proceed to evaluate $M_{\frac{x_i - p}{\sqrt{np(1-p)}}}(t)$. Begin by writing

$$M_{\frac{x_i - p}{\sqrt{np(1-p)}}}(t) = \sum_{k=0}^{\infty} \frac{E[X_1 - p]^k}{k!} \left(\frac{t}{\sqrt{np(1-p)}}\right)^k$$

$$= 1 + \frac{E[X_1 - p]}{1}\left(\frac{t}{\sqrt{np(1-p)}}\right) + \frac{E[X_1 - p]^2}{2}\left(\frac{t}{\sqrt{np(1-p)}}\right)^2 + o\left(\left(\frac{t}{\sqrt{np(1-p)}}\right)^2\right)$$

$$\tag{7.35}$$

which may be simplified to

$$M_{\frac{X_i-p}{\sqrt{np(1-p)}}}(t) = 1 + \frac{p(1-p)}{2}\frac{t^2}{np(1-p)} + \left(o\left(\frac{t}{\sqrt{np(1-p)}}\right)^2\right)$$

$$= 1 + \frac{t^2}{2n} + o\left(\frac{t^2}{np(1-p)}\right) = 1 + \frac{\frac{t^2}{2} + no\left(\frac{t^2}{np(1-p)}\right)}{n}$$

(7.36)

Now it remains for us to take advantage of the fact that the moment generating function of the sum of independent random variables is the product of the moment generating function of the summands. Thus

$$M_{T_n}(t) = \left[M_{\frac{X_i-p}{\sqrt{np(1-p)}}}(t)\right]^n = \left(1 + \frac{\frac{t^2}{2} + no\left(\frac{t^2}{np(1-p)}\right)}{n}\right)^n$$

(7.37)

and, taking limits, we conclude

$$\lim_{n\to\infty} M_{T_n}(t) = \lim_{n\to\infty}\left(1 + \frac{\frac{t^2}{2} + no\left(\frac{t^2}{np(1-p)}\right)}{n}\right)^n = \lim_{n\to\infty} e^{\frac{t^2}{2}+no\left(\frac{t^2}{np(1-p)}\right)}$$

(7.38)

$$= e^{\frac{t^2}{2}+\lim_{n\to\infty} no\left(\frac{t^2}{np(1-p)}\right)} = e^{\frac{t^2}{2}}$$

The demonstration that a binomial distribution could be usefully approximated by the normal distribution was first proved by De'Moivre and Laplace. The De'Moivre-Laplace theorem is one of the most important and useful theorems in probability, as in the following example.

Example 7.3. An important threat in modern warfare is the targeting of civilian population by artillery and surface to surface missiles. An adversary is suspected of possessing 1000 such mobile sites from which attacks against civilian populations can be unleashed. If the probability of identifying any particular one of these launch sites is 0.85, what is the probability that between 86% and 86.5% of these 1000 sites could be identified for destruction.

If we treat the attempt to identify each of the launching sites as an independent Bernoulli trial with parameter $p = 0.85$, the problem reduces to the probability that in 1000 Bernoulli trials, there are between 860 and 865 successes. This exact probability is

$$\sum_{k=860}^{865} \binom{1000}{k} (0.85)^k (0.15)^{n-k}$$

However, even with modern computing tools, this will take time to compute. Using the DeMoivre-Laplace theorem, a quicker solution is available.

$$P\left[860 \le S_n \le 865\right]$$
$$= P\left[\frac{860 - np}{\sqrt{np(1-p)}} \le \frac{S_n - np}{\sqrt{np(1-p)}} \le \frac{865 - np}{\sqrt{np(1-p)}}\right]$$
$$= P\left[\frac{860 - 850}{11.29} \le \frac{S_n - 500}{11.29} \le \frac{865 - 850}{11.29}\right]$$
$$= P\left[0.89 \le N(0,1) \le 1.33\right] = 9.5\%$$

Other probabilities are also readily obtained. The probability that between 700 and 900 of the sites are destroyed is approximately 100%. However, the probability that all sites are identified $= 0.85^{1000}$ or, essentially, zero. Thus, although there is high probability of getting most of the sites, it is also quite likely that at least one of the missile sites will survive and remain a continuing threat to the civilian population. ∎

7.13.6 Proof of the Central Limit Theorem

With our new experience in demonstrating, understanding, and using the central limit theorem, its proof will be surprisingly easy. We begin with its restatement.

If X_i, i = 1,2,3,..., n, if a family of independent random variable each with a moment generating function $M_X(t)$, mean μ, and variance σ^2, then the random variable

$$\frac{\sum\limits_{i=1}^{n} X_i - n\mu}{\sqrt{n\sigma^2}}$$ converges in distribution to a $N(0,1)$ random variable as n goes to infinity.

We can proceed directly to the proof. Let us start as we did in the previous two sections when we worked with Poisson, and binomial random variables, by beginning with the observation that

$$T_n = \frac{S_n - n\mu}{\sqrt{n\sigma^2}} = \sum_{i=1}^{n} \frac{X_i - \mu}{\sqrt{n\sigma^2}} \tag{7.39}$$

Thus, the random variable of interest is the sum of n independent random variables. We may use this fact to write

$$M_{T_n}(t) = \left[M_{\frac{X_i - \mu}{\sqrt{n\sigma^2}}}(t) \right]^n \tag{7.40}$$

We now examine the moment generating function $M_{\frac{X_i - u}{\sqrt{n\sigma^2}}}(t)$. We can start by writing

$$M_{\frac{X_i - \mu}{\sqrt{n\sigma^2}}}(t) = \sum_{k=0}^{\infty} \frac{E[X_1 - \mu]^k}{k!} \left(\frac{t}{\sqrt{n\sigma^2}} \right)^k$$

$$= 1 + \frac{E[X_1 - \mu]}{1} \left(\frac{t}{\sqrt{n\sigma^2}} \right) + \frac{E[X_1 - \mu]^2}{2} \left(\frac{t}{\sqrt{n\sigma^2}} \right)^2 + \left(o \left(\frac{t}{\sqrt{n\sigma^2}} \right)^2 \right) \tag{7.41}$$

Since $E[x_1] = \mu$, and $E[x_1 - \mu]^2 = \sigma^2$, we may substitute these quantities into expressions (7.41) to find that

$$M_{\frac{X_i - \mu}{\sqrt{n\sigma^2}}}(t) = 1 + \frac{\sigma^2}{2} \frac{t^2}{n\sigma^2} + \left(o \left(\frac{t}{\sqrt{n\sigma^2}} \right)^2 \right)$$

$$= 1 + \frac{t^2}{2n} + o \left(\frac{t^2}{n\sigma^2} \right) = 1 + \frac{\frac{t^2}{2} + no \left(\frac{t^2}{n\sigma^2} \right)}{n} \tag{7.42}$$

Now it remains for us to take advantage of the fact that the moment generating function of the sum of independent random variables is the product of the moment generating function of the summands. Thus

$$M_{T_n}(t) = \left[M_{\frac{X_i - \mu}{\sqrt{n\sigma^2}}}(t) \right]^n = \left(1 + \frac{\frac{t^2}{2} + no\left(\frac{t^2}{n\sigma^2}\right)}{n} \right)^n \qquad (7.43)$$

and, taking limits, we conclude

$$\lim_{n \to \infty} M_{T_n}(t) = \lim_{n \to \infty} \left(1 + \frac{\frac{t^2}{2} + no\left(\frac{t^2}{n\sigma^2}\right)}{n} \right)^n = \lim_{n \to \infty} e^{\frac{t^2}{2} + no\left(\frac{t^2}{n\sigma^2}\right)} \qquad (7.44)$$

$$= e^{\frac{t^2}{2} + \lim_{n \to \infty} no\left(\frac{t^2}{n\sigma^2}\right)} = e^{\frac{t^2}{2}}$$

This is our desired result ∎

7.13.7 "Within Normal Limits"

In health care, one of the most commonly used expressions among physicians, nurses, and other health care providers in the process of studying the characteristics and findings of their patients is "within normal limits". This term is applied to heart diameter measurements, blood sugar levels, plasma hormone assessments, serum cholesterol evaluations, and blood pressure measurements to name just a few. Essentially, a measurement is considered to be within normal limits if it falls within a range of values commonly seen in healthy people. Generally, the normal distribution is used to determine the values of these normal limits.

However, just how applicable is the normal distribution in this setting? As an example, let's consider one of the parameters that is assumed to follow a normal distribution, an individual's white blood cell count. White cells are particular cells that inhabit the blood and the lymphatic system. These cells are part of the body's line of defense against invasion. Unlike red blood cells (erythrocytes) which are carried along passively by the currents and eddies of the blood stream, white blood cells are capable of independent motion, and can freely choose their own direction of movement. However, they are especially attracted to toxins released by invading organisms and to compounds that are released by damaged cells. When sensitized by these substances, these white blood cells react aggressively, attaching and moving through blood vessel walls, as they leave the blood stream and negotiate their way to the region of injury. Once they arrive at the site of cellular disruption, they produce substances that kill the invading organism (e.g. bacteria), or destroy the foreign body (e.g. a wooden splinter). White blood cells are short-lived, typically not surviving for longer than 48 hours. However, their counts can dramatically increase with important systemic infections occur (e.g. pneumonia). White blood cells can also be produced in astonishing huge and damaging numbers when they are the product of cancer.

Clearly, there are many factors that affect the white blood cell count, and it would be difficult to see why the precise probability distribution of this count would be normal. Nevertheless, this is the probability distribution that is used to describe the white blood cell count. While the use of the normal distribution is a natural consequence of the central limit theorem, the applicability of this theorem can be examined from another perspective in this example. There are many factors that influence the white blood cell count. While we can think of a few (e.g. presence of infection, foreign bodies, cancer producing substances, hormone levels, compounds that are elaborated by other white cells) there are undoubtedly many, many more of these influences. By and large, the impact of any single one of these influence is to either increase or decrease the white blood cell count by a small amount. Thus the white blood cell count is the result of the combined effect of all of these factors, each of which exerts only a small effect. This is essentially what the central limit theorem states. The impact of the sum of many independent influences individually have a small effect, when suitably normalized, follows a standard normal distribution.

We can go one step further. Although the assumption of a normal distribution for the white blood cell count admits a wide range of possible values, 95% of the population whose white blood cell counts are healthy will have their count fall within 1.96σ of the population mean. Therefore, one could compute the mean μ and standard deviation σ in a population of subjects who have healthy white blood cell counts. From this computation, the lower 2.5 percentile value (μ – 1.96σ) and the upper 97.5 percentile value (μ + 1.96σ) could be calculated. This region (μ– 1.96σ , μ + 1.96σ), commonly described as the 95% confidence interval, is the range of white blood cell counts that are "within normal limits". The construction and use of this region represents an attempt to incorporate the observation that, while variability is a routine occurrence in nature, too much variability, while possibly normal, is the hallmark of an abnormality.

7.14 The Delta Method

From the preceding section, we have seen that, under commonly occurring conditions, the probability distribution of the sample mean of a random variable (suitable normalized) can be very reasonably approximated by a normal distribution. In the examples that we have developed thus far, the mean and variance of the random variable have been directly available. However, frequently, we are interested in the distribution of functions of these random variable that do not have an easily calculated mean and variance. A procedure, commonly known as the delta method provides a very helpful result.

Let $\{X_i\}$ i = 1 to n be a collection of independent and identically distributed random variables with mean μ and variance σ^2. If the random variable Y is defined as $Y = g(X)$ where g is a differentiable function of X then, the delta method provides the following finding:

$$\sqrt{n}\left(g\left(\overline{X}_n\right) - g\left(\mu\right)\right) \text{ is approximately } N\left(0, \left[g'\left(\mu\right)\right]^2 \sigma^2\right) \qquad (7.45)$$

The normal distribution that appears on the right hand side of expression (7.45) follows from an application of the central limit theorem. It is the computation of the mean and variance of $g\left(\overline{X}_n\right)$ that are the special product of the delta method. Let X be a random variable with expectation μ and variance σ^2. Our goal is to identify the mean and variance of g(Y). The first idea to examine the variance might be to use a direct approach i.e.

$$Var[Y] = \int_{\Omega_x} [g(x)]^2 f_X(x) - \left[\int_{\Omega_x} [g(x)] f_X(x) dx \right]^2 \tag{7.46}$$

However, on many occasions, the integrals in equation (7.46) can not be directly evaluated. In such circumstances, a more indirect approach is available to us, an approach that, at its heart, involves a very simple application of an elementary Taylor series expansion. We can begin by writing $Y = g(X)$ as a simple Taylor series approximation around the point $X = \mu$. This provides

$$g(x) = g(\mu) + \left[\frac{dg(\mu)}{dx} \right](x-\mu) + \left[\frac{d^2g(\mu)}{dx^2} \right] \frac{(x-\mu)^2}{2} + \left[\frac{d^3g(\mu)}{dx^3} \right] \frac{(x-\mu)^3}{3!} + \dots \tag{7.47}$$

Since the higher power terms will be negligible, we will truncate expression (7.47) so that $g(X)$ is a linear function of X, or

$$g(x) \approx g(\mu) + \left[\frac{dg(x)}{dx} \right]_{x=u} (x-\mu). \tag{7.48}$$

Substituting $\overline{X}$ for X in equation (7.48) reveals

$$g(\overline{X}_n) = g(\mu) + \left[\frac{dg(x)}{dx} \right]_{x=u} (\overline{X}_n - \mu). \tag{7.49}$$

Taking expectations of both sides of equation (7.49) reveals

$$E\left[g(\overline{X}_n) \right] = E[g(\mu)] + \left[\frac{dg(x)}{dx} \right]_{x=u} E\left[(\overline{X}_n - \mu) \right] = g(\mu) \tag{7.50}$$

Thus, we have identified an approximation to $E\left[g(\overline{X}_n) \right]$. The variance of $g\left(\overline{X} \right)$ also follows from a similar argument

$$Var\left[g(\bar{x}_n)\right] = Var\left[g(\mu) + \left[\frac{dg(\mu)}{dx}\right](\bar{x}_n - \mu)\right].$$

$$= Var\left[\left[\frac{dg(x)}{dx}\right]_{x=u}(\bar{X}_n - \mu)\right]$$

$$= \left[\frac{dg(x)}{dx}\right]_{x=u}^2 Var\left[(\bar{X}_n - \mu)\right] \qquad (7.51)$$

$$= \left[\frac{dg(x)}{dx}\right]_{x=u}^2 \frac{\sigma^2}{n}$$

7.14.1 Assessing the Relative Risk of Poisons

Epidemiology is the study of the causes and distribution of diseases in the population. Carefully planned clinical studies are commonly used to quantify the relationship between a potentially toxic agent to which the human subjects were exposed and the development of subsequent disease in those subjects. These types of studies are designed to detect the effect of a potential environmental toxin in the environment that is believed to be linked to a particular disease. In such an evaluation, the experience of subjects who have been exposed to the possible toxin is compared to the experience of patients who have not been exposed to the toxin as follows. Disease-free patients who have been subjected to the potentially poisonous agent are observed for a pre-specified period of time in order to determine the number of these patients who contract the disease after their exposure. Subjects who have not been exposed to the toxic agent are followed for the same period of time; new occurrences of the disease in these patients are also noted. The epidemiologist computes the proportion of patients who contract the disease in the exposed group, p_e, and the proportion of patients who contract the disease in the unexposed group p_u. From these quantities the relative risk of exposure R$_e$ can be computed as $R_e = \frac{p_e}{p_u}$.

For example, in one study, 100 patients without the disease are exposed to the potential toxic agent at time t = 0 and followed for one year, at which time, 30 patients are identified to have developed the disease. In addition, 100 unexposed patients without the disease at time t = 0 are also followed simultaneously for one year. At the end of this time period, 10 patients who were not exposed are found to have developed the disease. In this circumstance, the estimate of the relative risk is $R_e = \frac{p_e}{p_u} = \frac{0.30}{0.10} = 3.0$. Patients in the exposed group are three times as likely as the unexposed group to contract the disease in one year.

Epidemiologists have long recognized that the relative risk has a skewed probability distribution. In order to counteract this innate asymmetry, these scientists characteristically compute the natural logarithm of the relative risk and focus on the mean and variance of this distribution. To compute the mean and variance of this transformed relative risk, they commonly turn to the Taylor series expansion. In this specific circumstance, the epidemiologist chooses $y = \log(R_e) = \log\left(\dfrac{p_e}{p_u}\right) = \log(p_e) - \log(p_u)$. From here we can see directly that $E[\log R_e] = E[\log p_e] - E[\log p_u]$. In addition, since one patient's experience is independent of the experience of any other patient, we may also write that $Var\left[\log(R_e)\right] = Var\left[\log(p_e)\right] + Var\left[\log(p_u)\right]$. An examination of p_e reveals that it is the proportion of failures in a sequence of patient Bernoulli trials. We may therefore represent $E[p_e] = \theta_e$. We now use a Taylor series expansion to write

$$\log p_e \approx \log(\theta_e) + \left[\frac{d \log p_e}{d p_e}\right](p_e - \theta_e) \qquad (7.52)$$

From equation (7.52) we may compute

$$E[\log p_e] \approx \log(\theta_e) + \left[\frac{d \log p_e}{d p_e}\right] E\left[(p_e - \theta_e)\right] = \log(\theta_e) \qquad (7.53)$$

Proceeding to the variance computation, we find that

$$Var[\log p_e] \approx Var\left[\log(\theta_e) + \left[\frac{d \log p_e}{d p_e}\right](p_e - \theta_e)\right]$$

$$= \left[\frac{d \log p_e}{d p_e}\right]^2 E\left[(p_e - \theta_e)^2\right] \qquad (7.54)$$

$$= \left[\frac{d \log p_e}{d p_e}\right]^2 Var[p_e]$$

We observe that $\left[\dfrac{d \log p_e}{d p_e}\right]^2 = \dfrac{1}{p_e^2}$, and the estimate of the variance of p_e,

$\widehat{Var[p_e]} = \dfrac{p_e(1 - p_e)}{n_e}$. Substituting these quantities into expression (7.54) reveals

$$Var\left[\log p_e\right] \approx \left[\frac{d\log p_e}{dp_e}\right]^2 Var\left[p_e\right]$$

$$= \frac{1}{p_e^2}\frac{p_e\left(1-p_e\right)}{n_e} = \frac{\left(1-p_e\right)}{p_e n_e} \tag{7.55}$$

An analogous computation produces $Var\left[\log p_u\right] \approx \dfrac{\left(1-p_u\right)}{p_u n_u}$. Combining these

results we have

$$Var\left[\log\left(R_e\right)\right] = Var\left[\log\left(p_e\right)\right] + Var\left[\log\left(p_u\right)\right]$$

$$\approx \frac{1-p_e}{n_e p_e} + \frac{1-p_u}{n_u p_u} \tag{7.56}$$

■

7.15 Convergence Almost Surely (With Probability One)

The demonstration of convergence in distribution of the binomial distribution to the Poisson distribution, and convergence of each of these distributions in law to the normal distribution each represent examples of the utility of convergence in distribution. The continued, vibrant applicability of these sturdy results no doubt cause confusion when these results are described as "weak" in some tracts. However, we must keep in mind that, within mathematics, weak is not synonymous with the terms "fragile", "pathetic" or puny". To the mathematician, describing results as weak merely means that the finding implies less than other types of findings. It does not imply that the implications of weak results are useless. As we have seen, the consequences of weak convergence are quite useful and powerful.

Another type of convergence that we saw earlier in this chapter is that of convergence almost surely. Recall that convergence almost surely implied that it was impossible for any possible value of the random variable to be far from the limit point when n was large enough (as opposed to "weaker" forms of convergence where it is possible, but unlikely for the random variable to be far from its limit point for large n).

In some cases, almost sure convergence produces its own probability law. For example, the strong law of large numbers tells us that not only does the sample mean converge to the population mean in probability (i.e., weakly), but also with probability one (i.e., strongly). These can be very useful results in probability theory.

In addition, probabilists have worked to relax some of the assumptions that underlie the law of large numbers. The most common form of this law is when the sequence of observations is independent and identically distributed. Relaxations of these criteria are available, but they require new assumptions about the existence of variances and assurances that the variances do not grow too fast. Some of these are due to Chebyshev and Kolmogorov (Parzen). Similar development work has taken place with the central limit theorem. The version of this law that we have discussed in detail in this chapter is due to Lindeberg and Lévy. Feller(1) contains an examination of the degree to which the assumptions that support the central limit theorem can be relaxed. There are also multivariable extensions of this very useful theorem (Sterling).

7.16 Conclusions

Asymptotic distribution is central to the study of the properties of infinite sequences of random variables. The use of probability and statistics is a daily occurrence, and we must have reliable estimates of statistical estimators on which important scientific and public health decisions reside.

There are several reasons why an estimate obtained from a sample can be inaccurate, an ultimate explanation lies in the fact that the estimate is based not on the population, but instead on only a sample of the population. Thus, statisticians, when confronted with several different and competing estimators, commonly focus on its asymptotic properties. Our intuition serves us well here. One useful style of intuition informs us that, if the estimate is a good one, then the larger the sample becomes, the closer the estimate based on the sample will be to the true population value.

These asymptotic properties are used in very comprehensible contexts. For example, a researcher may ask whether the estimator under consideration consistently overestimates or underestimates the population parameter that it attempts to measure. If it does, then that estimator is asymptotically biased, suggesting that perhaps another estimator should be sought. The properties of convergence in distribution, convergence in probability, convergence in law and convergence almost surely help us to establish different useful metrics to assess the utility of these estimators. When there are several estimators that purport to measure the same population parameter, then a comparison of the large sample variances of these estimators might be a useful way to select from among the competitors. These are each important questions that asymptotic theory addresses. In order to examine these issues, we must understand the use of limits in probability and statistics, and learn some of the classic results of the utilization of this theory.

Problems

1.   What utility does asymptotic theory have for the practicing statistician?

2.   Prove the assertion made in section 7.6 that if $\{X_n\}$ is a sequence of independent and identically distributed uniform $[0, 1]$ random variables, then the members of the sequence do not get arbitrarily close to one another as n increases.

3.   Show that if $\{X_n\}$ is a sequence of independent and identically distributed random variables, then, for any $\varepsilon > 0$, $P\left[\left|X_n - X\right| \geq \varepsilon\right]$ is either always zero, or is a constant that is not a function of n.

4   It $\{X_n\}$ converges in probability to X, then show that $\{-X_n\}$ converges in probability to $-X$.

5   Show that the sequence of random variables $\{X_n\}$ whose probability density function is $f_{X_n}(x) = (n+1)\mathbf{1}_{\left[0, \frac{1}{n+1}\right]}$ actually converges almost surely to zero.

6.   Let $\{X_n\}$ be a sequence of independent and identically distributed random variables which has a probability density function that is indexed by n.
$$f_{x_n}(x) = \frac{1}{2}\mathbf{1}_{x=0} + \frac{n+1}{2}\mathbf{1}_{x\in\left(\frac{n}{n+1}, 1\right)}.$$
In what sense does the sequence $\{X_n\}$ converge?

7.   Let $\{X_n\}$ be a sequence of independent and identically distributed random variables which has a probability density function that is indexed by n.
$$f_{x_n}(x) = \frac{1}{2}\mathbf{1}_{(0, 1)}(x) + \frac{n+1}{2}\mathbf{1}_{\left(\frac{n}{n+1}, 1\right)}(x).$$
In what sense does the sequence $\{X_n^*\}$ converge to the value one?

8.   Let $\{X_n\}$ be a sequence of independent and identically distributed random variables   with   the   probability   density

function $f_{x_n}(x) = \dfrac{1}{n+1}\mathbf{1}_{(0,1)}(x) + n\mathbf{1}_{\left(\frac{n}{n+1},1\right)}(x)$. In what sense does the sequence $\{X_n\}$ converge?

9. Let $\{X_n\}$ be a sequence of uniformly distributed random variables. The probability distribution function of X_1 distributed as $U(0,1)$. The probability distribution function of X_n is $U\left(\dfrac{n-1}{n},1\right)$. In what sense does the sequence of random variables converge?

10. Show that if X is a non negative random variable, then
$$P[x \geq c] \leq \frac{E[x]}{c}$$

11. Prove the following version of the weak law of large numbers. Let X_i, i $=1 \dots n$ be a collection of independent random variables where $E(X_i) = \mu_i$ and $Var(x_i) = \sigma_i^2$ where $\sigma_i^2 \leq \infty$. Then the sample mean $\overline{x}_n$ converges in probability to μ if it is true that $\lim\limits_{n \to \infty} \dfrac{\sum\limits_{i=1}^{n} \sigma_i^2}{n^2} = 0$.

12. Prove the Bernoulli law of large numbers. Let X_i, i $=1 \dots n$ be the results of a collection of independent and identically distributed Bernoulli trials. Then the sample mean $\overline{x}_n$ converges in probability to p, the probability of a successful Bernoulli trial.

13. Let X_1, X_2, X_3 be three independently distributed U(0, 1) random variables. Verify the probability density for the sum of three U(0,1) random variables is as expressed in (7.19).

14. Use Slutsky's theorem to prove that if $\{X_n\}$ and $\{Y_n\}$ are a sequence of random variables that converge to X and $\mathbf{y}$ respectively, then the sequence of random variables $\{w_n\}$ where $w_n = X_n - Y_n$ converges to $W = X - Y$.

15. Use Slutsky's theorem to prove that if $\{X_n\}$ and $\{Y_n\}$ are a sequence of random variables that converge to X and Y respectively where $Y \neq 0$ then the sequence of random variables $\{W_n\}$ where $W_n = X_n / Y_n$ converges to $W = X / Y$.

16. Let X_n be a chi square distribution with n degrees of freedom. Show using the central limit theorem that the probability distribution function of the

quantity $\dfrac{X_n - n}{\sqrt{n}}$ may be approximated by the standard normal distribution.

17. Let $\{X_n\}$ be a sequence of independent and identically distributed exponential distributions such that $E[X_n] = \dfrac{1}{\lambda}$. Show using the central limit theorem that the probability distribution function of the quantity

$$\dfrac{\sum_{i=1}^{n} X_i - \dfrac{n}{\lambda}}{\sqrt{\dfrac{n}{\lambda}}}$$ may be approximated by the standard normal distribution.

18. A business's gross receipts on a given day can either grow by $2000 shrink by $5000, or remain unchanged (i.e., the gross receipts neither grow by at least $2000 nor shrink by at least $5000). Let the probability that gross receipts are greater than $2000 be p, the probability that gross receipts shrink by at least $5000 is q and the probability that changes in gross receipts remain between these two regions (i.e., are unchanged) is r, where $p + q + r = 1$. Let the gross daily receipts be independent from day to day. A month consists of 28 business days and a year is 336 business days. Compute the following probabilities.
    a) The gross receipts for the month have grown more that $50,000.
    b) The gross receipts for the month have grown more than $100,000.
    c) The gross receipts for the year remain unchanged.
    d) The gross receipts for the year have shrunk by more than $168,000.
    Your answers should be in terms of p, q, and r.

19. Hourly arrivals to a clinic during a 10 hour night shift follow a Poisson distribution with parameter $\lambda = 3$ visits per hour. Compute the following probabilities.
    a) The total number of patients seen in 7 nights is between 200 and 220 patients.
    b) The total number of patients seen in 10 nights is less than 325 patients
    c) In a month, more than 950 patients are seen in clinic during the night shift.

References

Dudewizc, E.J., Mishra, S.N. (1988) *Modern Mathematical Statistics*. New York. John Wiley and Sons.

Feller, W. (1968) *An Introduction to Probability Theory and Its Applications –* vol 1. Third Edition   New York. John Wiley & Sons. p 84.

Parzen, E. (1960) *Modern Probability Theory and Its Applications..* New York. John Wiley & Sons. p 372.

Rao, M.M. (1984) *Probability Theory with Applications.* Orlando. Academic Press. p 45.

Stirling, R.J. (1980) *Approximation Theorems of Mathematical Statistics.* New York. John Wiley and Sons.

8

Point Estimation

8.1 Introduction

In the nano-technology era, new probability distributions and new estimators are being identified. The major problem of the estimation is that commonly more than one estimator is offered to estimate a mathematical parameter. In other cases, there is a new estimator which is offered to supplant an established estimator. In these circumstances, controversies can rage around the choice of the best estimator. There is no doubt that the estimators have different properties. The question often asked is which property works best for the problem at hand.

There have been several early attempts to estimate parameters using data. One of the earliest attempts to identify the value of mathematical parameters from observations is the work carried out by the Babylonian astronomers 500-300 B.C.[1] These workers developed mathematical formulas that they believed governed the motion of the planets. These formulas contained parameters, and the astronomers used some simple arithmetic schemes to carry out these computations to produce predictions for the appearance of the sun and other heavenly bodies. The estimations that they attempted were a combination of parameterization and the use of computation to fill in the missing pieces of the formula. No written record that tracks whether or how arithmetic measures were used to actually estimate parameters survived them.

Several hundred years later, the Greek Hipparchus became involved in an attempt to estimate variability of the duration of day.[2] A portion of his astronomical endeavors included estimating the time intervals of successive passages of the sun through the solstice. In this work he found that there was year-to-year variability. Hipparchus deduced that this annual variability meant that the length of the

---

[1]These historical accounts are taken from Plackett, R.L. (1958). The principle of the arithmetic mean. *Biometrika* 45 130-135.

[2] This is in contradistinction to the measurement of the length of daylight which was well known to exhibit seasonal variability.

year was itself not constant, suggesting that day length itself was not a constant 24 hours but exhibited variability. He estimated the maximum variance of the length of a year as 0.75 days, an estimate that he constructed by taking one half of the range of his observations. This represents an early attempt to estimate sample variance.

One of the most striking examples of the activity and controversy that can surround an estimator is the development of the sample mean. For example, as pointed out by Lehman [3], we accept that the best estimate of the population mean μ of the normal distribution based on n observations $x_1, x_2, x_3, \ldots, x_n$ is $\bar{X} = \dfrac{\sum\limits_{i=1}^{n} x_i}{n}$.

However, this was not always the case.

The idea of repeating and combining observations made on the same quantity appears to have been introduced as a scientific method by Tycho Brae towards the end of the 16th century. He used the arithmetic mean to replace individual observations as a summary measurement. The demonstration that the sample mean was a more precise value than a single measurement did not appear until the end of the seventeenth century, and was based on the work by the astronomer Flamsten.

In the meantime, additional work was being carried out by a French expedition under Maupertuis that was sent to Lapland to measure the length of a degree of longitude. The goal of this effort was to determine if the Earth was flattened at either the poles or the equator. Each observer in the expedition was required to make their own estimate, but that estimate was required to be based on several observations. Specifically, each observer was asked to make his own observation of the angles, write them separately, and then report their arithmetic mean to the group.

At this point, the simultaneous organization of the formulation of discrete probability laws and the development of the differential calculus permitted an analytical examination of the probability distribution of the mean. In 1710, Simpson's work proved that the arithmetic mean has a smaller variance than a single observation. In a letter written from Lagrange to Laplace in December 1776, Lagrange provided a derivation of an estimator from the multinomial distribution that had some useful mathematical properties.[4] In the same letter, Lagrange extended this work to continuous distributions. Laplace incorporated this work into his own arguments as he worked on the problem of combining observations.

While the problem of the sample mean has been solved, the field of estimation continues to be a dynamic one to this day. Occasionally, commonly accepted estimators such as least square regression parameter estimators are chal-

---

[3] Lehmann E.L. 1983. *Theory of Point Estimation.* New York. John Wiley. p 3.

[4] We will be introduced to these *maximum likelihood estimators* later in this chapter.

lenged with other estimators, such as ridge estimators. Bayesian estimators compete with traditional estimators in many fields including clinical research. Estimation theory continues to be a dynamic field and requires the attention of the serious statistician.

In each of Chapter 4 and Chapter 5, we discussed probability distributions of random variables. Each distribution is associated with one or more unknown parameter. These parameters play important roles in understanding the behavior of the random variable and in making decision related to the distribution. For example, a parent-teacher organization of an elementary school is to give a designed T-shirt to each of its several-hundred enrolled students. Rather than measure the height of each child, they assume the height of the children is normally distributed with a mean and a variance. However, if they do not know these two parameters, it would be difficult to decide on the number of T-shirts of each size that are required. To know these parameters, they can certainly measure the height of every student and calculate the mean and the variance. However, it is cumbersome or sometimes impossible to complete the measurement of the height for all students, especially when the student body is large. If the organization could randomly choose a sample of reasonable size, they may be able to logically "approximate" the mean and the variance from this sample. There are several ways to "guesstimate" parameters based on observations from a random sample. These "guesstimates" are called estimators. Traditionally, any function of a random sample, for example $\bar{X} = \frac{1}{n} \sum_{i=1}^{n} X_i$, is called an *estimator*. Any realization of an estimator, for example $\bar{x} = 39$, is called an *estimate*. In this chapter we shall discuss several methods of estimation.

8.2 Method of Moments Estimators

Let $X_1, X_2, \cdots, X_n$ be a random sample from a population with a continuous or discrete distribution with k parameters $\theta_1, \theta_2, \cdots, \theta_k$. Assume that the k population moments are finite. The method of moments estimator can be found by equating the first k sample moments to the corresponding k population moments, and solving the resulting system of simultaneous equations. More specifically, consider the system of equations

$$\mu'_j(\theta_1, \theta_2, \cdots, \theta_k) = \frac{1}{n} \sum_{i=1}^{n} X_i^j, \text{ for } j = 1, 2, \cdots, k, \tag{8.1}$$

where $\mu'_j(\theta_1, \theta_2, \cdots, \theta_k) = E(X^j)$ and the jth population moment of all $X_i, i = 1, \cdots, n$, is typically a function of parameters associated with the distribution. The method of moments estimators $\tilde{\theta}_1, \tilde{\theta}_2, \cdots, \tilde{\theta}_k$ for $\theta_1, \theta_2, \cdots, \theta_k$ is the solution of the system of equations in (8.1).

Example 8.1 Let $X_1, X_2, \cdots, X_n$ be a iid random sample from a population distributed as $N(\theta_1, \theta_2)$, where θ_1 and θ_2 are both unknown. The equations for solving method of moments estimators are

$$E(X_1) = \theta_1 = \frac{1}{n}\sum_{i=1}^{n} X_i \text{ and } E(X_1^2) = \theta_1^2 + \theta_2 = \frac{1}{n}\sum_{i=1}^{n} X_i^2.$$

The solution for these equations gives the method of moments estimators

$$\tilde{\theta}_1 = \frac{1}{n}\sum_{i=1}^{n} X_i = \bar{X} \text{ and } \tilde{\theta}_2 = \frac{1}{n}\sum_{i=1}^{n} X_i^2 - \bar{X}^2 = \frac{1}{n}\sum_{i=1}^{n} (X_i - \bar{X})^2.$$

Example 8.2 Systolic blood pressure is a measure of the force of blood flow. In order to understand the distribution of the systolic blood pressure (SBP) of a cohort of adolescents in a study, a random sample of 10 subjects was chosen and their SBP measured. The measurements were 102.5, 117.0, 94.0, 103.0, 97.0, 96.0, 98.0, 104.0, 102.0 and 95.0 (mmHg). If we assume that the distribution of SBP in this cohort is normal with unknown parameters μ and σ^2, we can estimate these parameters as $\tilde{\mu} = 100.85$ and $\tilde{\sigma}^2 = 40.7$.

Note that the method of moment estimate for σ^2 is not the same as the sample variance discussed in Chapter 5.

Example 8.3 It has been observed that the number of phone calls received in a fixed time period follows a Poisson distribution. The following is the number of phone calls arriving at a professor's office during lunch time (12 to 1 pm) on 14 consecutive Wednesdays: 4, 2, 1, 3, 1, 1, 1, 3, 3, 4, 1, 4, 0 and 3. If we are to estimate the parameter λ (representing the rate of calls per hour) of this Poisson distribution using these data, we should first find the sample mean $\frac{1}{14}\sum_{i=1}^{14} x_i = \frac{31}{14} \approx 2.21$. Since the population mean of a Poisson distribution is equal to its parameter λ, the method of moment estimate can be calculated as $\tilde{\lambda} = 2.21$.

Example 8.4 The delay time for patients with potential coronary heart disease is referred as the duration between the onset of an acute cardiovascular symptom and the arrival at a hospital emergency room. It is assumed that the delay time follows an exponential distribution with parameter θ. The delay times of the first 20 patients recorded in REACT study published by Luepker and et al. [2000] were recorded (in minutes) as 525, 719, 2880, 150, 30, 251, 45, 858, 15 , 47, 90, 56, 68, 6, 139, 180, 60, 60, 294, 747.

If we are to find the method of moment estimate for θ using these observations, we will derive the mean of the exponential distribution $E(X) = \frac{1}{\theta}$ and set equal to the sample mean. It is then obvious that

$$\tilde{\theta} = \frac{1}{\bar{x}} = \frac{1}{361} = 0.00277.$$

8.3 Maximum Likelihood Estimators

The method of maximum likelihood is the most commonly used procedure for estimating parameters. It selects as the estimate the parameter value that maximizes the probability (likelihood) that the given data will be observed. This principle can be easily understood by a simple example. Consider a box containing four balls. We are told that it either has three green balls and one blue ball or three blue balls and one green ball. One ball is randomly chosen from the box and found to be green. We would be more likely to believe that the box has three green balls. Of course, this method may not always give us a correct answer. It will certainly improve our chances of making a correct guess.

We can further understand the maximum likelihood principle by examining observations from a discrete random variable. Consider a random sample $X_1, \ldots, X_n$ obtained from a population with probability mass function $p(x|\theta)$. Let $x_1, x_2, \cdots, x_n$ be a set of n such observations. For a given parameter value, we can calculate the probability that a particular combination will be observed as $P(X_1 = x_1, X_2 = x_2, \cdots, X_n = x_n | \theta)$. The maximum likelihood principle will lead us towards finding the parameter value that maximizes this probability among all possible values of the parameters.

Definition 8.1 Let $X_1, X_2, \cdots, X_n$ be a random sample (i.i.d random variables) obtained from a population with either pdf $f(x|\theta_1, \theta_2, \cdots, \theta_k)$ or pmf $p(x|\theta_1, \theta_2, \cdots, \theta_k)$. The *likelihood function* $L(\theta|x_1, x_2, \cdots, x_n)$ is defined as

$$L(\theta|x_1, x_2, \cdots, x_n) = f(x_1|\theta)f(x_2|\theta)\cdots f(x_n|\theta) = \prod_{i=1}^{n} f(x_i|\theta)$$

for the continuous case, or

$$L(\theta|x_1, x_2, \cdots, x_n) = p(x_1|\theta)p(x_2|\theta)\cdots p(x_n|\theta) = \prod_{i=1}^{n} p(x_i|\theta),$$

for the discrete case, where $\theta = (\theta_1, \theta_2, \cdots, \theta_k)$.

Definition 8.2 Let $X_1, X_2, \cdots, X_n$ be a random sample obtained from a population with either pdf $f(x|\theta_1, \theta_2, \cdots, \theta_k)$ or pmf $p(x|\theta_1, \theta_2, \cdots, \theta_k)$, where $\theta = (\theta_1, \theta_2, \cdots, \theta_k) \in \Theta$, and Θ is the set of all possible parameter values and is called the parameter space. The *maximum likelihood estimator* (MLE) is the parameter value that maximizes the likelihood function on Θ.

8.3.1 MLE with Complete Observations

In this section, we will present some examples that assume the data are all observable and require either an analytical method or a numerical method to find the MLE.

Example 8.5 Let $x_1, x_2, \cdots, x_n$ be a sequence of random observations from a binomial distribution $B(m, \theta)$ with known m and unknown parameter θ, where $0 \le \theta \le 1$. The likelihood function associated with this sample is given by

$$L(\theta|x_1, x_2, \cdots, x_n) = \prod_{i=1}^{n}\left\{\binom{m}{x_i}\theta^{x_i}(1-\theta)^{m-x_i}\right\} = \left\{\prod_{i=1}^{n}\binom{m}{x_i}\right\}\theta^{\sum_{i=1}^{n}x_i}(1-\theta)^{mn-\sum_{i=1}^{n}x_i}.$$

To maximize the likelihood function $L(\theta|x_1, x_2, \cdots, x_n)$, we should first differentiate the above function with respect to θ. We would then set the first derivative equal to zero and solve the resulting equation. However, we can also perform these mathematical operations on the logarithmic transformation of $L(\theta|x_1, x_2, \cdots, x_n)$, since the logarithmic function is a monotone increasing one. The first derivative of this function is

$$\left(\sum_{i=1}^{n}x_i\right)\frac{d\log(\theta)}{d\theta} + \left(mn - \sum_{i=1}^{n}x_i\right)\frac{d\log(1-\theta)}{d\theta}$$

$$= \left(\sum_{i=1}^{n}x_i\right)\frac{1}{\theta} + \left(mn - \sum_{i=1}^{n}x_i\right)\frac{-1}{1-\theta}.$$

The only solution to the above equation is $\hat{\theta} = \dfrac{\sum_{i=1}^{n}x_i}{mn}$, if $\sum_{i=1}^{n}x_i \ne mn$ and

$\sum_{i=1}^{n}x_i \ne 0$. Since the second derivative is negative for all possible values of x,

$\hat{\theta}$ maximizes the log-transformation of the likelihood function. For the two ex-treme cases when $\sum_{i=1}^{n} x_i = mn$ or $\sum_{i=1}^{n} x_i = 0$, the likelihood function becomes ei-ther θ^{mn} or $(1-\theta)^{mn}$ respectively. These two monotone functions of θ have maximum at 1 and 0 respectively. In either case, the maximum point can be ex-pressed as $\hat{\theta} = \dfrac{\sum_{i=1}^{n} x_i}{mn}$. Furthermore, $\hat{\theta}$ falls between 0 and 1, since $0 \le x_i \le m$ for each observation. So we can conclude that $\hat{\theta}$ is the MLE of the parameter based on these observations. For $m = 1$, all x_i's are either 0 or 1 and $\hat{\theta}$ becomes the sample mean or sample proportion.

Note that the logarithm of the likelihood function is called the *log-likelihood function* and is denoted by $l(\theta|x_1, x_2, \cdots, x_n)$. For convenience, we may also write the likelihood function and the log-likelihood function as $L(\theta)$ and $l(\theta)$ respectively.

Although it is both common and convenient to find the MLE of a pa-rameter θ by locating the zero or zeros of the first derivative of the (log-) likeli-hood function, it is important to note that the zero or zeros can only provide us candidates for MLE. Even if we verify that the second derivative is negative, the zero of the first derivative is only the local maximum point of the likelihood func-tion on the interior. If the MLE occurs at the boundary, the first derivative method can not locate it. We should therefore check the boundary before the result calcu-lated from the first derivative method is adopted. Furthermore, if a likelihood function is not continuous on θ or the domain of the parameter to be estimated is discrete, the first derivative method should be avoided.

Other algebraic methods such as complete squares, Cauchy inequality or Lagrange Multipliers, can be used for finding the global maximum point of the (log-) likelihood function.

Example 8.6 Assume that $x_1, x_2, \cdots, x_n$ are the realization of the random samples $X_1, X_2, \cdots, X_n$ distributed as $U[0,\theta]$. Recall that the pdf for $U[0,\theta]$ is $f(x|\theta) = \begin{cases} \frac{1}{\theta} & \text{if } x \le \theta \\ 0 & \text{if } x > \theta \end{cases}$. The likelihood function can be written as

$$L(\theta|x_1, x_2, \cdots, x_n) = \begin{cases} \dfrac{1}{\theta^n} & \text{if all } x_i \le \theta \\ 0 & \text{otherwise} \end{cases}.$$

Note that "all $x_i \leq \theta$" is equivalent to saying $x_{(n)} \leq \theta$, where $x_{(n)}$ is the maximum of $x_1, x_2, \cdots, x_n$. For this likelihood function, the first derivative method cannot help us to find the maximum point for θ. However, a simple graph of the likelihood function can easily determine the MLE. Figure 8.1 shows that the likelihood functional values were 0 when $\theta < x_{(n)}$ and decreases after it reaches $x_{(n)}$.

Therefore, it is easy to conclude that the MLE $\hat{\theta} = x_{(n)}$.

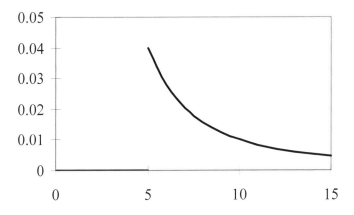

Figure 8.1 The likelihood function of the uniform distribution when $x_{(n)} = 5$

Example 8.7 Let $x_1, x_2, \cdots, x_n$ be a sequence of random observations from a Poisson distribution with parameter λ, where $\lambda > 0$. The likelihood function is given by (1) $L(\lambda) = \prod_{i=1}^{n} \dfrac{\lambda^{x_i} e^{-\lambda}}{x_i!}$ if $\bar{x} \neq 0$ and (2) $L(\lambda) = e^{-n\lambda}$, if $\bar{x} = 0$ (implying $x_i = 0, i = 1, 2, \ldots, n$). For case (2), the likelihood function is monotone decreasing in $(0, \infty)$ and thus has no maximum point. For case (1), the log-likelihood function can be expressed as $l(\lambda) = \log(\lambda)(\sum_{i=1}^{n} x_i) - n\lambda - \sum_{i=1}^{n} \log(x_i!)$. By differentiating $l(\lambda)$ with respect to λ and setting the first derivative function equal to zero, we obtain $\hat{\lambda} = \dfrac{\sum_{i=1}^{n} x_i}{n} = \bar{x}$ for $\bar{x} \neq 0$. Using L'Hôpital's rule to check the limit of the likelihood function as $\lambda \to \infty$ or as $\lambda \to 0$, we conclude that, in both cases, the limit approaches 0. Therefore, $\hat{\lambda} = \bar{x}$ maximizes the likelihood function on $(0, \infty)$ and the MLE for λ can be verified as $\hat{\lambda} = \bar{x}$ for $\bar{x} \neq 0$.

Example 8.8 The yearly number of deaths due to diseases of the gastrointestinal tract in the United States, in 1989 to 1996, is assumed to follow a Poisson distribution with an unknown parameter λ. Assume that numbers of deaths in these years are independent of each other. Peterson and Calderon (2003) reported these numbers as 1319, 1463, 1757, 2114, 2801, 3375, 3870 and 4308 in chronological order from 1989 to 1996. Using the formula derived in Example 8.7, we have $\hat{\lambda} = \bar{x} = 2625.88$.

Example 8.9 Assume that $x_1, x_2, \cdots, x_n$ is an observed random sample from a population that has a $N(\mu, \sigma^2)$ distribution, where μ is unknown. The likelihood function for μ can be written as

$$L(\mu) = (2\pi\sigma^2)^{-n/2} \exp(-\sum_{i=1}^{n} \frac{(x_i - \mu)^2}{2\sigma^2}).$$

Consequently, the log-likelihood function is $l(\mu) = -\frac{n}{2} \log(2\pi\sigma^2) - \sum_{i=1}^{n} \frac{(x_i - \mu)^2}{2\sigma^2}$ and its first derivative function can be simplified as $l'(\mu) = \sum_{i=1}^{n} \frac{(x_i - \mu)}{\sigma^2}$. Solving the equation $l'(\mu) = 0$ gives us the MLE $\hat{\mu} = \bar{x}$, since the second derivative is clearly negative and, when $\mu \to \pm\infty$, the value of the likelihood function goes to 0. Note that the MLE for this example can also be obtained by rewriting the exponent of the likelihood function as

$$-\sum_{i=1}^{n} \frac{(x_i - \mu)^2}{\sigma^2} = \frac{-1}{\sigma^2} \left[\sum_{i=1}^{n} (x_i - \bar{x})^2 + n(\bar{x} - \mu)^2 \right].$$

This term will be maximized when $\hat{\mu} = \bar{x}$ by completing the squares.

Example 8.10 In Example 8.9, if σ^2 is also unknown and needs to be estimated, our log-likelihood function will be bivariate with the same form

$$l(\mu, \sigma^2) = -\frac{n}{2} \log(2\pi\sigma^2) - \sum_{i=1}^{n} \frac{(x_i - \mu)^2}{2\sigma^2}.$$

However, the procedures for finding MLE require computation of the partial derivative with respect to each parameter:

$$\frac{\partial l(\mu,\sigma^2)}{\partial \mu} = \sum_{i=1}^{n} \frac{(x_i - \mu)}{\sigma^2} \text{ and } \frac{\partial l(\mu,\sigma^2)}{\partial \sigma^2} = -\frac{n}{2\sigma^2} + \sum_{i=1}^{n} \frac{(x_i - \mu)^2}{2(\sigma^2)^2}.$$

Setting $\dfrac{\partial l(\mu,\sigma^2)}{\partial \mu} = 0$ and $\dfrac{\partial l(\mu,\sigma^2)}{\partial \sigma^2} = 0$, and solving the equations simultaneously we obtain $\hat{\mu} = \bar{x}$ and $\hat{\sigma}^2 = \dfrac{1}{n}\sum_{i=1}^{n}(x_i - \bar{x})^2$. Note that the solution $\hat{\mu} = \bar{x}$ is indeed a global maximum point for any value of σ^2 (Here the global maximum point is the point that maximizes the function on the entire parameter space). This can be verified by the method described at the end of Example 8.9. To verify that $\hat{\sigma}^2$ achieves its global maximum, we need to check the cases when $\sigma^2 \to 0$ or when $\sigma^2 \to \infty$. In either case, the likelihood function approaches 0. Therefore, we can conclude that $\hat{\mu} = \bar{x}$ and $\hat{\sigma}^2 = \dfrac{1}{n}\sum_{i=1}^{n}(x_i - \bar{x})^2$ are indeed the MLE.

Strictly speaking, in order to demonstrate that a pair of solutions for the first partial derivative equations are local maximum points for a (log-)likelihood function, we have to show that the second order partial derivative matrix is negative definite (see, for example Khuri(2003)). Specifically, we have to show that one of two second order (non-mixed) partial derivatives is negative at the roots of the first derivative of (log-)likelihood function and the determinant of the second order derivative matrix, referred to as the Jacobian, is positive at that point.

Example 8.11 It is assumed that the systolic blood pressure (SBP) and diastolic blood pressure (DBP) follow a bivariate normal distribution with parameters $\theta = (\mu_1, \mu_2, \sigma_1^2, \sigma_2^2, \rho)^T$, where

      μ_1 represents the mean of the SBP,

      μ_2 represents the mean of the DBP,

      σ_1^2 represents the variance of the SBP,

      σ_2^2 represents the variance of the DBP, and

      ρ represents the correlation between the SBP and the DBP in the population.

In the data collected in the Heartfelt study (see Meininger and et. al.(1999)), there were 391 adolescents with valid resting blood pressure, both SBP and DBP. The MLEs for the above parameters will be found using these observations.

Let y_{1i} and y_{2i}, $i = 1, 2, \cdots, n$ denote the SBP and the DBP (in mmHg) respectively for the ith subject. Recall from Chapter 5 that the bivariate normal probability density is

$$f(y_1, y_2) = \left(2\pi\sigma_1\sigma_2\sqrt{1-\rho^2} \right)^{-1}$$

$$\exp\left\{ \frac{-1}{2(1-\rho^2)} \left[\left(\frac{y_1 - \mu_1}{\sigma_1} \right)^2 - 2\rho\left(\frac{y_1 - \mu_1}{\sigma_1} \right)\left(\frac{y_2 - \mu_2}{\sigma_2} \right) + \left(\frac{y_2 - \mu_2}{\sigma_2} \right)^2 \right] \right\}$$

The log-likelihood for θ is

$$l(\theta) = -n\log(2\pi) - \frac{n}{2}\log(\sigma_1^2\sigma_2^2) - \frac{n}{2}\log(1-\rho^2)$$

$$- \frac{1}{2(1-\rho^2)} \left[\sum_{i=1}^{n} \left(\frac{y_{1i} - \mu_1}{\sigma_1} \right)^2 - 2\rho\sum_{i=1}^{n} \left(\frac{y_{1i} - \mu_1}{\sigma_1} \right)\left(\frac{y_{2i} - \mu_2}{\sigma_2} \right) + \sum_{i=1}^{n} \left(\frac{y_{2i} - \mu_2}{\sigma_2} \right)^2 \right].$$

The first step is to solve the log-likelihood equations $\dfrac{\partial l}{\partial \mu_1} = 0$, $\dfrac{\partial l}{\partial \mu_2} = 0$,

$\dfrac{\partial l}{\partial \sigma_1^2} = 0$, $\dfrac{\partial l}{\partial \sigma_2^2} = 0$ and $\dfrac{\partial l}{\partial \rho} = 0$. This is equivalent to solving the following set of equations simultaneously.

$$\sum_{i=1}^{n} \frac{(y_{1i} - \mu_1)}{\sigma_1} - \rho\sum_{i=1}^{n} \frac{(y_{2i} - \mu_2)}{\sigma_2} = 0 \tag{8.2}$$

$$\sum_{i=1}^{n} \frac{(y_{2i} - \mu_2)}{\sigma_2} - \rho\sum_{i=1}^{n} \frac{(y_{1i} - \mu_1)}{\sigma_1} = 0 \tag{8.3}$$

$$-n + \frac{1}{(1-\rho^2)} \frac{\sum_{i=1}^{n}(y_{1i} - \mu_1)^2}{\sigma_1^2} - \frac{\rho}{(1-\rho^2)} \frac{\sum_{i=1}^{n}(y_{1i} - \mu_1)(y_{2i} - \mu_2)}{\sigma_1\sigma_2} = 0 \tag{8.4}$$

$$-n + \frac{1}{(1-\rho^2)} \frac{\sum_{i=1}^{n}(y_{2i} - \mu_2)^2}{\sigma_2^2} - \frac{\rho}{(1-\rho^2)} \frac{\sum_{i=1}^{n}(y_{1i} - \mu_1)(y_{2i} - \mu_2)}{\sigma_1\sigma_2} = 0 \tag{8.5}$$

$$
n\rho(1-\rho^2) - \rho \left(\frac{\sum_{i=1}^{n}(y_{1i}-\mu_1)^2}{\sigma_1^2} + \frac{\sum_{i=1}^{n}(y_{2i}-\mu_2)^2}{\sigma_2^2} \right)
$$

(8.6)

$$
+(1+\rho^2)\frac{\sum_{i=1}^{n}(y_{1i}-\mu_1)(y_{2i}-\mu_2)}{\sigma_1\sigma_2} = 0
$$

Note that Equations (8.2) through (8.6) are obtained by multiplying each

of $\dfrac{\partial l}{\partial \mu_1}=0$, $\dfrac{\partial l}{\partial \mu_2}=0$, $\dfrac{\partial l}{\partial \sigma_1^2}=0$, $\dfrac{\partial l}{\partial \sigma_2^2}=0$ and $\dfrac{\partial l}{\partial \rho}=0$ by $\sigma_1(1-\rho^2)$,

$\sigma_2(1-\rho^2)$, $2\sigma_1^2$, $2\sigma_2^2$ and $(1-\rho^2)^2$ respectively.

Multiplying Equation (8.3) by ρ and adding to Equation (8.2), we ob-

tain $\hat{\mu}_1 = \dfrac{\sum_{i=1}^{n} y_{1i}}{n} = \overline{y}_1$ by simple algebra. Using similar algebraic operations, we

obtain $\hat{\mu}_2 = \dfrac{\sum_{i=1}^{n} y_{2i}}{n} = \overline{y}_2$. Substituting μ_1 and μ_2 by $\hat{\mu}_1$ and $\hat{\mu}_2$ respectively into

Equations (8.4) and (8.5) and using simple algebraic manipulations, we obtain

$\dfrac{S_1}{\sigma_1^2} = \dfrac{S_2}{\sigma_2^2}$ implying $\hat{\sigma}_1^2 = cS_1$ and $\hat{\sigma}_2^2 = cS_2$ for some $c > 0$, where

$S_1 = \sum_{i=1}^{n}(y_{1i}-\overline{y}_1)^2$ and $S_2 = \sum_{i=1}^{n}(y_{2i}-\overline{y}_2)^2$. Substituting $\hat{\mu}_1$, $\hat{\mu}_2$, $\hat{\sigma}_1^2$, and $\hat{\sigma}_2^2$ into

Equations (8.4) and (8.6), and adding them together, we can show that, after a little

algebraic manipulations, $\hat{\rho} = \dfrac{S_{12}}{\sqrt{S_1 S_2}}$, where $S_{12} = \sum_{i=1}^{n}(y_{1i}-\overline{y}_1)(y_{2i}-\overline{y}_2)$. Replac-

ing all the parameters by their estimates in Equation (8.4), we obtain $c = \dfrac{1}{n}$. Thus,

$\hat{\sigma}_1^2 = \dfrac{S_1}{n}$ and $\hat{\sigma}_2^2 = \dfrac{S_2}{n}$. It remains for the readers to check that the matrix of the

second derivative is negative definite and that the maximal point falls into the interior of the parameter space. In the above algebraic operations, we assume that
$S_1 \neq 0$, $S_2 \neq 0$, and $S_{12}^2 \neq S_1 S_2$. If these conditions are not met, one of the parameters may not achieve the MLE.

Applying the above formulas for calculating the estimated parameters for the Heartfelt study data, we obtain $\hat{\mu}_1 = 106.85$, $\hat{\mu}_2 = 61.96$, $\hat{\sigma}_1^2 = 96.35$, $\hat{\sigma}_2^2 = 91.37$ and $\hat{\rho} = 0.43$.

Invariance Property

Maximum likelihood estimators possess a very useful property known as the *invariance property*. It allows us to find the MLE for a function of a parameter or parameters. This function does not have to be one-to-one. This extension broadens its practical application. For example, it makes it easy for us to find the MLE for θ^2 when the MLE for θ has been found. This property will be stated below. The readers are referred to Zehna (1966) or Pal and Berry (1992) for the proof.

This property states that if $\hat{\theta}$ is the MLE for θ, then $\tau(\hat{\theta})$ is the MLE for $\tau(\theta)$ for any function $\tau()$.

The invariance property can be applied to Example 8.11, if our aim is to find the MLE for the covariance. Since $\sigma_{12} = Cov(Y_1, Y_2) = \rho\sqrt{\sigma_1^2\sigma_2^2}$ (from Chapter 3), we have $\hat{\sigma}_{12} = \hat{\rho}\sqrt{\hat{\sigma}_1^2\hat{\sigma}_2^2}$.

Up to now, we have identified MLE using analytical methods and we have expressed all MLE in explicit form. However, not all MLE can be found algebraically. Computational methods discussed in Chapter 11 can also be used to find MLE numerically. These methods become increasingly important in a real-life data analysis when very often the analytic solution for a MLE can not be found.

Example 8.12   A health promotion study was designed to estimate the weekly mean number of unprotected sexual encounters in a group of 300 adolescents following a sex education program. These adolescents were randomly assigned to three groups. 100 each were interviewed at the end of the 4th week, 8th week and 12th week after a common safe sex campaign. While the response regarding the number of encounters may not be reliable, it may be used to indicate if these subjects ever indulged in an unprotected sex from the time of the safe sex campaign to the time that the subject was interviewed. The result showed that the first group had 97 subjects who had no unprotected sex in 4 weeks; the second group had 91 in 8 weeks and the third group had 83 in 12 weeks.

Assume that the number of unprotected sexual encounters for this group of teens follows a Poisson distribution with mean μt, where t is the number of weeks after the campaign. Since we can only identify the number of teens with or without unprotected sex, the observation becomes a binomial distribution with probability of "success" equal to $e^{-\mu t}$. Note that $e^{-\mu t}$ is the probability of an adolescent who had zero unprotected sexual encounters in a period of t weeks. The likelihood function can then be written as

$$L(\mu) = (e^{-4\mu})^{97}(1-e^{-4\mu})^3(e^{-8\mu})^{91}(1-e^{-8\mu})^9(e^{-12\mu})^{83}(1-e^{-12\mu})^{17}.$$

The log-likelihood function is

$$l(\mu) = -2112\mu + 3\log(1-e^{-4\mu}) + 9\log(1-e^{-8\mu}) + 17\log(1-e^{-12\mu}).$$

The first derivative function of $l(\mu)$ can be calculated as

$$l'(\mu) = -2112 + 3\frac{4e^{-4\mu}}{1-e^{-4\mu}} + 9\frac{8e^{-8\mu}}{1-e^{-8\mu}} + 17\frac{12e^{-12\mu}}{1-e^{-12\mu}}.$$

An algebraic solution for solving the above equation for μ is not available. A numerical approach will be attempted. In this case, we choose Newton-Raphson's method (see Chapter 11) to find the solution. The MLE can be found by iteratively evaluating $\mu_{n+1} = \mu_n - \dfrac{l'(\mu_n)}{l''(\mu_n)}$ until the absolute difference of the $(n+1)^{th}$ iteration value μ_{n+1} and the nth iteration value μ_n is less than a pre-determined small positive number, where

$$l''(\mu) = 12\frac{(1-e^{-4\mu})(-4e^{-4\mu}) + 4(e^{-4\mu})^2}{(1-e^{-4\mu})^2}$$

$$+72\frac{(1-e^{-8\mu})(-8e^{-8\mu}) + 8(e^{-8\mu})^2}{(1-e^{-8\mu})^2} + 204\frac{(1-e^{-12\mu})(-12e^{-12\mu}) + 12(e^{-12\mu})^2}{(1-e^{-12\mu})^2}.$$

A SAS program is provided for performing the numerical optimization. The MLE is found to be $\hat{\mu} = 0.0129$. That means there were on average 0.0129 unprotected sexual encounters per week for each adolescent in this group.

Note that if the log likelihood function is concave and unimodal, then the Newton-Raphson's method guarantees the iteration will converge to the MLE.

Example 8.13 Suppose that the probability that a family owns a house can be represented by a logistic function of the form $\dfrac{e^{a+bx}}{1+e^{a+bx}}$, where x represents the family income (in thousands of dollars). Clearly, the probability that a family does not own a house is $\dfrac{1}{1+e^{a+bx}} = 1 - \dfrac{e^{a+bx}}{1+e^{a+bx}}$. The hypothetical data in Table 8.1 are on home ownership, where $y = 1$ if the family owns their own home and 0 otherwise. The researcher's goal is to determine a and b.

Table 8.1 Home Ownership and Family Income (in thousands).

y	1	1	0	1	1	0	1	1	0	0	1	0	1	0
x	75	50	61	48	90	73	48	58	40	35	38	28	51	23

Note that the likelihood function in this example is

$$L(a,b) = \left(\prod_{i=1}^{8} \frac{e^{a+bx_i}}{1+e^{a+bx_i}} \right) \left(\prod_{i=1}^{6} \frac{1}{1+e^{a+bx_i}} \right),$$

where the first term is the product of probabilities owning a home and the second term is that for not owning a home.

It is apparent that the MLE for a and b can not have an analytical solution. The Newton-Raphson method can be used for finding a numerical solution. The detailed descriptions of Newton-Raphson method will be discussed in Chapter 11. We first calculate the score statistic, defined as the gradient vector of the log likelihood function $S\left(\begin{pmatrix} a \\ b \end{pmatrix} \right) = \begin{pmatrix} \dfrac{\partial \log L(a,b)}{\partial a} \\ \dfrac{\partial \log L(a,b)}{\partial b} \end{pmatrix}$. The MLE can then be found iteratively as

$$\begin{pmatrix} a \\ b \end{pmatrix}^{(k+1)} = \begin{pmatrix} a \\ b \end{pmatrix}^{(k)} + I^{-1} \left(\begin{pmatrix} a \\ b \end{pmatrix}^{(k)} \right) S\left(\begin{pmatrix} a \\ b \end{pmatrix}^{(k)} \right),$$

where $I^{-1}\left(\begin{pmatrix} a \\ b \end{pmatrix}^{(k)} \right)$ is the inverse of the information matrix

$$I\left(\begin{pmatrix} a \\ b \end{pmatrix}^{(k)} \right) = (-1) \begin{pmatrix} \dfrac{\partial^2 \log L(a,b)}{\partial a^2} & \dfrac{\partial^2 \log L(a,b)}{\partial a \partial b} \\ \dfrac{\partial^2 \log L(a,b)}{\partial a \partial b} & \dfrac{\partial^2 \log L(a,b)}{\partial b^2} \end{pmatrix}_{\begin{pmatrix} a \\ b \end{pmatrix} = \begin{pmatrix} a \\ b \end{pmatrix}^{(k)}},$$

where $\begin{pmatrix} a \\ b \end{pmatrix}^{(k)}$ are the values estimated at the kth iteration.

The detailed description of the information matrix will be discussed later.

The numerical estimation gives us $\hat{a} = 2.2170$ and $\hat{b} = -0.0503$. The SAS program is provided in an appendix.

Note that the Newton-Raphson method is a common procedure for finding a numerical solution for the MLE. However, this method frequently encounters computational difficulties when calculating the information matrix at each iteration. There have been several suggestions on replacing $I^{-1}\left(\begin{pmatrix} a \\ b \end{pmatrix}^{(k)}\right)$ by a simpler matrix, called the Quasi-Newton methods. Detailed introduction of the Newton-type methods can be found in section 1.3 of McLachlan and Krishnnan (1997). Other optimization procedures can also be applied for the calculation of MLE, if the Newton type method does not work.

8.3.2 MLE with Incomplete Observations and the EM Algorithm*

In the last subsection, we discuss methods of finding MLE when the data are completely observed. In practice, there are situations in which observations are missing or some of the data are not observable. The EM algorithm is an appropriate method that provides an iterative procedure for computing MLEs in these situations. The EM algorithm first computes the expected log likelihood of the complete data given the available observations (called E-step). It then maximizes this conditionally expected log likelihood over the parameters (called M-step). These procedures repeat alternatively until the change in difference of log likelihood is very small.

Example 8.14 In Example 8.11, if 3 measurements of SBP and 5 measurements of DBP were accidentally deleted, do we have to remove all 8 subjects from the analysis or can we use all available data to find these MLE?

In order to present this problem in a general data framework that is applicable for any pair of observations distributed as a bivariate normal random vector, we let $Y = (Y_1, Y_2)^T$ denote the bivariate normal random vector with parameters specified in Example 8.11. Furthermore, we assume m_1 subjects had only the first component (SBP) recorded and missed the measurement of the second component (DBP), and m_2 subjects had only the second component (DBP) recorded and missed the measurement of the first component (SBP). Suppose that the "missingness" does not depend on the observed values or the missing values. This missing mechanism is called *Missing Completely at Random* (MCAR). Detailed discussions on missing mechanism can be found in Little and Rubin (1987).

Let $y_i = (y_{1i}, y_{2i})^T$ $i = 1, 2, \cdots, n$ be the data that is observed or would have been observed. Here, we assume that there are $m = n - m_1 - m_2$ subjects with complete observations and $y_{2,m+1}, \cdots, y_{2,m+m_1}$ and $y_{1,m+m_1+1}, \cdots, y_{1,n}$ are missing or unobserved.

The log likelihood of the complete data for $\theta = (\mu_1, \mu_2, \sigma_1^2, \sigma_2^2, \rho)$ is

$$l_C(\theta) = -n\log(2\pi) - n\log(\sigma_1\sigma_2) - \frac{n}{2}\log(1-\rho^2)$$

$$-\frac{1}{2(1-\rho^2)}\left\{\frac{T_{11}}{\sigma_1^2} + \frac{T_{22}}{\sigma_2^2} + T_1\left(\frac{-2\mu_1}{\sigma_1^2} + \frac{2\rho\mu_2}{\sigma_1\sigma_2}\right)\right.$$

$$+T_2\left(\frac{-2\mu_2}{\sigma_2^2} + \frac{2\rho\mu_1}{\sigma_1\sigma_2}\right) + T_{12}\left(-\frac{2\rho}{\sigma_1\sigma_2}\right)$$

$$\left. + \left(\frac{n\mu_1^2}{\sigma_1^2} + \frac{n\mu_2^2}{\sigma_2^2} - \frac{2n\rho\mu_1\mu_2}{\sigma_1\sigma_2}\right)\right\},$$

where

$$T_1 = \sum_{j=1}^{n} y_{1j}, \quad T_2 = \sum_{j=1}^{n} y_{2j}, \quad T_{11} = \sum_{j=1}^{n} y_{1j}^2, \quad T_{22} = \sum_{j=1}^{n} y_{2j}^2, \text{ and } T_{12} = \sum_{j=1}^{n} y_{1j}y_{2j}.$$

It is required to calculate the conditional expectation of the complete-data log likelihood at each step,

$$Q(\theta, \theta^{(k)}) = E_{\theta^{(k)}}(l_c(\theta)|y_1, \cdots, y_m, u, v),$$

where $u = (y_{1,m+1}, \cdots, y_{1,m+m_1})^T$ and $v = (y_{2,m+m_1+1}, \cdots, y_{2,n})^T$ are the observed portion of the incompletely-recorded subjects and $\theta^{(k)} = (\mu_1^{(k)}, \mu_2^{(k)}, \sigma_1^{2(k)}, \sigma_2^{2(k)}, \rho^{(k)})$ is the estimated parameter vector at the kth iteration.

For the bivariate normal random vector, the conditional distribution of Y_{2i} given $Y_{1i} = y_{1i}$ follows a normal distribution with mean $\mu_2 + \rho\frac{\sigma_2}{\sigma_1}(y_{1i} - \mu_1)$ and variance $\sigma_2^2(1-\rho^2)$. Similarly, the conditional distribution of Y_{1i} given $Y_{2i} = y_{2i}$ follows a normal distribution with mean $\mu_1 + \rho\frac{\sigma_1}{\sigma_2}(y_{2i} - \mu_2)$ and variance $\sigma_1^2(1-\rho^2)$. Thus,

$$E_{\theta^{(k)}}(Y_{2i}|y_{1i}) = \mu_2^{(k)} + \rho^{(k)}\frac{\sigma_2^{(k)}}{\sigma_1^{(k)}}(y_{1i} - \mu_1^{(k)}),$$

$$E_{\theta^{(k)}}(Y_{1i}|y_{2i}) = \mu_1^{(k)} + \rho^{(k)}\frac{\sigma_1^{(k)}}{\sigma_2^{(k)}}(y_{2i} - \mu_2^{(k)}),$$

$$E_{\theta^{(k)}}(Y_{2i}^2|y_{1i}) = [E_{\theta^{(k)}}(Y_{2i}|y_{1i})]^2 + \sigma_2^{2(k)}(1 - \rho^{2(k)})$$

and

$$E_{\theta^{(k)}}(Y_{1i}^2|y_{2i}) = [E_{\theta^{(k)}}(Y_{1i}|y_{2i})]^2 + \sigma_1^{2(k)}(1 - \rho^{2(k)}).$$

It is now easy to calculate the conditional expectation of T_1, T_2, T_{11}, T_{22} and T_{12}, given the observed data when the parameters $\theta^{(k)}$ are estimated at the kth iteration:

$$T_1^{(k)} = E_{\theta^{(k)}}(T_1|y_1,\cdots,y_m,u,v)$$
$$= \sum_{j=1}^{m+m_1} y_{1j} + m_2\mu_1^{(k)} + \frac{\sigma_1^{(k)}}{\sigma_2^{(k)}}\rho^{(k)}\sum_{j=m+m_1+1}^{n}(y_{2j} - \mu_2^{(k)}),$$

$$T_2^{(k)} = E_{\theta^{(k)}}(T_2|y_1,\cdots,y_m,u,v)$$
$$= \sum_{j=1}^{m} y_{2j} + \sum_{j=m+m_1+1}^{n} y_{2j} + m_1\mu_2^{(k)} + \frac{\sigma_2^{(k)}}{\sigma_1^{(k)}}\rho^{(k)}\sum_{j=m+1}^{m+m_1}(y_{1j} - \mu_1^{(k)}),$$

$$T_{11}^{(k)} = E_{\theta^{(k)}}(T_{11}|y_1,\cdots,y_m,u,v)$$
$$= \sum_{j=1}^{m+m_1} y_{1j}^2 + \sum_{j=m+m_1+1}^{n} [E_{\theta^{(k)}}(Y_{1i}|y_{2i})]^2 + m_2\sigma_1^{2(k)}(1 - \rho^{2(k)}),$$

$$T_{22}^{(k)} = E_{\theta^{(k)}}(T_{22}|y_1,\cdots,y_m,u,v)$$
$$= \sum_{j=1}^{m} y_{2j}^2 + \sum_{j=m+m_1+1}^{n} y_{2j}^2 + \sum_{j=m+1}^{m+m_1} [E_{\theta^{(k)}}(Y_{2i}|y_{1i})]^2 + m_1\sigma_2^{2(k)}(1 - \rho^{2(k)}),$$

$$T_{12}^{(k)} = E_{\theta^{(k)}}(T_{12}|y_1,\cdots,y_m,u,v)$$
$$= \sum_{j=1}^{m} y_{1j}y_{2j} + \sum_{j=m+1}^{m+m_1} y_{1j}E_{\theta^{(k)}}(Y_{2i}|y_{1i}) + \sum_{j=m+m_1+1}^{n} y_{2j}E_{\theta^{(k)}}(Y_{1i}|y_{2i}),$$

and $Q(\theta, \theta^{(k)})$ can be expressed as a function of $\theta, \theta^{(k)}$, $T_1^{(k)}$, $T_2^{(k)}$, $T_{11}^{(k)}$, $T_{22}^{(k)}$, $T_{12}^{(k)}$, and other observed values. This completes the E-step. The M-step maximizes $Q(\theta, \theta^{(k)})$ over all θ and the vector that attains the maximum is $\theta^{(k+1)}$. From the derivations presented in Example 8.11, the estimated parameters $\theta^{(k+1)} = (\hat{\mu}_1^{(k+1)}, \hat{\mu}_2^{(k+1)}, \hat{\sigma}_1^{2(k+1)}, \hat{\sigma}_2^{2(k+1)}, \hat{\rho}^{(k+1)})^T$ at $(k+1)^{\text{th}}$ iteration can be written as

$$\hat{\mu}_1^{(k+1)} = \frac{T_1^{(k)}}{n}, \quad \hat{\mu}_2^{(k+1)} = \frac{T_2^{(k)}}{n}, \quad \hat{\sigma}_1^{2(k+1)} = \frac{T_{11}^{(k)} - n(\hat{\mu}_1^{(k+1)})^2}{n},$$

$$\hat{\sigma}_2^{2(k+1)} = \frac{T_{22}^{(k)} - n(\hat{\mu}_2^{(k+1)})^2}{n} \quad \text{and} \quad \hat{\rho}^{(k+1)} = \frac{T_{12}^{(k)} - n\hat{\mu}_1^{(k+1)}\hat{\mu}_2^{(k+1)}}{n\sqrt{\hat{\sigma}_1^{2(k+1)}\hat{\sigma}_2^{2(k+1)}}}.$$

The E-step and M-step are alternated repeatedly until the difference of the (log)-likelihood values of two consecutive iterations is less than a preassigned small positive number.

For the data in Example 8.11, if we delete the SBP for the first 3 subjects and the DBP for the last 5 subject, the MLE using EM algorithm will give us $\hat{\mu}_1 = 106.75$, $\hat{\mu}_2 = 61.99$, $\hat{\sigma}_1^2 = 98.45$, $\hat{\sigma}_2^2 = 93.38$ and $\rho = 0.39$. A SAS program is available.

Note that in this example, if both SBP and DBP of a subject are not recorded, we will exclude this individual in the estimation.

Example 8.15 It is known that a person's blood type can be classified as A, B, O and AB. These blood types are determined by three allelomorphic genes O, A and B and that O is recessive to A and B. If we assume that the probability of having n allelomorphic gene O, A and B is r, p and q (where $r + p + q = 1$), then the expected probabilities of genotypes AA, AO, BB, BO, OO and AB are p^2, $2pr$, q^2, $2qr$, r^2 and $2pq$ respectively. Hence, the probabilities of the blood types of A, B, O and AB are $p^2 + 2pr$, $q^2 + 2qr$, r^2 and $2pq$ respectively, since both genotypes AA and AO have a blood type A and both genotypes BB and BO have a blood type B. A data set in Cavalli-Sforza (1971) that recorded the blood types of 166 individuals in African Pygmy population is used for this example. Among them, 88 (n_O) have blood type O, 44 (n_A) have blood type A, 27 (n_B) have blood type B and 7 (n_{AB}) have blood type AB. If the objective is to find the MLE for $\theta = (p, q)^T$, how should we approach this problem? We first formulate the log likelihood function for θ, that is

$$l(\theta) = C + 2n_O \log\left(1 - p - q\right) + n_A \log\left(p^2 + 2p(1 - p - q)\right)$$
$$+ n_B \log\left(q^2 + 2q(1 - p - q)\right) + n_{AB} \log\left(pq\right),$$

where C represents the appropriate constant term. There is no closed form for solving the gradient of $l(\theta)$. The Newton-Raphson method may be used to determine a numeric solution.

To apply the EM algorithm, we choose the natural complete data vector $y = (n_O, n_{AA}, n_{AO}, n_{BB}, n_{BO}, n_{AB})^T$, where $u = (n_{AA}, n_{AO}, n_{BB}, n_{BO})^T$ is the unobservable vector of frequencies for genotypes AA, AO, BB and BO respectively and the sum of all components in y is equal to n. Hence, the complete log likelihood function is

$$l_C(\theta) = C + 2n_{AA} \log p + n_{AO} \log p + n_{AO} \log(1 - p - q) + 2n_{BB} \log q$$
$$+ n_{BO} \log q + n_{BO} \log(1 - p - q) + 2n_O \log(1 - p - q) + n_{AB} \log p$$
$$+ n_{AB} \log q$$
$$= C + \left(2n_{AA} + n_{AO} + n_{AB}\right) \log p + \left(2n_{BB} + n_{BO} + n_{AB}\right) \log q$$
$$+ \left(n_{AO} + n_{BO} + 2n_O\right) \log(1 - p - q).$$

From Problem 8.13, the complete log likelihood function is the same as what we have in multinomial distribution with 3 possible outcomes, where $N_1 = 2n_{AA} + n_{AO} + n_{AB}$, $N_2 = 2n_{BB} + n_{BO} + n_{AB}$, and $N_3 = n_{AO} + n_{BO} + 2n_O$ are the observed counts for categories 1, 2 and 3 respectively. Clearly, $N_1 + N_2 + N_3 = 2n$. Therefore, in the M-step, we can derive $\hat{p} = \dfrac{m_A}{n}$ and $\hat{q} = \dfrac{m_B}{n}$, where $m_A = n_{AA} + \dfrac{1}{2}n_{AO} + \dfrac{1}{2}n_{AB}$ and $m_B = n_{BB} + \dfrac{1}{2}n_{BO} + \dfrac{1}{2}n_{AB}$. The E-step requires the calculations of $E(n_{AA}|y)$, $E(n_{AO}|y)$, $E(n_{BB}|y)$ and $E(n_{BO}|y)$. These can be obtained by the property that the conditional distribution of one component in a multinomial random vector, given the sum of several components, is binomial. This property provides the following computations for the E-step:

$$n_{AA}^{(k)} = n_A \frac{(p^{(k)})^2}{(p^{(k)})^2 + 2p^{(k)}r^{(k)}}, \quad n_{AO}^{(k)} = n_A \frac{2p^{(k)}r^{(k)}}{(p^{(k)})^2 + 2p^{(k)}r^{(k)}},$$

$$n_{BB}^{(k)} = n_B \frac{(q^{(k)})^2}{(q^{(k)})^2 + 2q^{(k)}r^{(k)}} \quad \text{and} \quad n_{BO}^{(k)} = n_B \frac{2q^{(k)}r^{(k)}}{(q^{(k)})^2 + 2q^{(k)}r^{(k)}}.$$

The M-step can now be implemented by $p^{(k+1)} = \dfrac{m_A^{(k)}}{n}$ and $q^{(k+1)} = \dfrac{m_B^{(k)}}{n}$, where

$$m_A^{(k)} = n_{AA}^{(k)} + \frac{1}{2}n_{AO}^{(k)} + \frac{1}{2}n_{AB} \text{ and } m_B^{(k)} = n_{BB}^{(k)} + \frac{1}{2}n_{BO}^{(k)} + \frac{1}{2}n_{AB}.$$

Repeating the E and M steps alternatively until the change of the log-likelihood is small, the resulting estimates of r, p and q will be the MLE. For this data set, it takes 5 iterations to obtain the MLE as $(\hat{p}, \hat{q}, \hat{r}) = (0.16733, 0.10805, 0.72462)$.

A similar data set was analyzed by McLachlan and Krishnan (1997) using the EM algorithm.

8.4 Bayes Estimators

For the estimation methods we have discussed thus far, the targeted parameters have been treated as unknown but fixed constants. No prior information on these parameters was assumed for estimation. As long as our distributional form is correct, the estimate of the parameter can be any value in the parameter space that is inferred by the observed data. There are some situations in practice where we have prior knowledge on the parameter, particularly, the random behavior of the parameter. We will use Example 8.16 below to illustrate the importance of using prior information, if it is available on the parameter in estimation. This is the Bayesian approach. For this approach, the parameter is assumed to be a random quantity that can be described by a probability distribution, called *the prior distribution*. A prior distribution not only can represent prior knowledge of the parameter as in Example 8.16; it can also be interpreted as the personal belief about the parameter before the data are collected. The description of this personal belief is similar to the subjective distribution discussed in Chapter 2. The Bayesian approach uses the observed data to update the prior distribution by calculating the conditional distribution of the parameter given the data. This updated prior distribution is called the posterior distribution. A *Bayes estimate* is the mean of the posterior distribution.

Example 8.16 It is well known that a couple, in which one individual has blood type A and the other has blood type B, can have children with blood type A, B and AB. Our interest is in estimating the probability θ that a child born to such a couple will have blood type AB. Assume that one such couple has n children and all of them have been tested to have blood type AB. Under the assumption of independence, we can use the binomial distribution $B(n, \theta)$ to describe the number X of these n children who have blood type AB. In addition, using basic genetics, we can easily calculate that θ can be 1, ½ and ¼ with probability ¼, ½ and ¼ respectively, depending on the couple's genotypes as explained in Example 8.15. For example, if the couple's genotypes are AA and BB, then the child's blood type will be AB with probability 1. If we ignore this information on θ and use either the Method of Moment Estimate or MLE, we may result in estimating θ as 1.

If we use the knowledge on θ, i.e. the prior distribution of θ, and apply Bayes theorem discussed in Chapter 2, we can calculate the posterior distribution of θ as follows:

$$P(\theta = 1 | X = n) = \frac{P(\theta = 1, X = n)}{P(X = n)}$$

$$= \frac{P(X = n | \theta = 1) P(\theta = 1)}{P(X = n | \theta = 1) P(\theta = 1) + P\left(X = n \middle| \theta = \frac{1}{2}\right) P\left(\theta = \frac{1}{2}\right) + P\left(X = n \middle| \theta = \frac{1}{4}\right) P\left(\theta = \frac{1}{4}\right)}$$

$$= \frac{\dfrac{1}{4}}{\dfrac{1}{4} + \left(\dfrac{1}{2}\right)^{n+1} + \left(\dfrac{1}{4}\right)^{n+1}},$$

since X is distributed as a $B(n, \theta)$ random variable, therefore $P(X = n | \theta) = \theta^n$. Similarly,

$$P\left(\theta = \frac{1}{2} \middle| X = n\right) = \frac{\left(\dfrac{1}{2}\right)^{n+1}}{\dfrac{1}{4} + \left(\dfrac{1}{2}\right)^{n+1} + \left(\dfrac{1}{4}\right)^{n+1}} \quad \text{and} \quad P\left(\theta = \frac{1}{4} \middle| X = n\right) = \frac{\left(\dfrac{1}{4}\right)^{n+1}}{\dfrac{1}{4} + \left(\dfrac{1}{2}\right)^{n+1} + \left(\dfrac{1}{4}\right)^{n+1}}.$$

From the above calculations, we conclude that θ may not always be equal to 1, even when it is given that $X = n$. In fact, θ can be 1, ½ or ¼ and each value associates a probability that θ can happen. It is reasonable to say that an overall estimate of θ should be the weighted sum of these values. Bayes estimate chooses the mean of the posterior probability mass function on 1, ½ and ¼ as its estimate that is equal to

$$E(\theta | X = n) = \frac{1 \times \dfrac{1}{4} + \dfrac{1}{2} \times \left(\dfrac{1}{2}\right)^{n+1} + \dfrac{1}{4} \times \left(\dfrac{1}{4}\right)^{n+1}}{\dfrac{1}{4} + \left(\dfrac{1}{2}\right)^{n+1} + \left(\dfrac{1}{4}\right)^{n+1}} = 1 - \frac{\left(\dfrac{1}{2}\right)^{n+2} + \dfrac{3}{4}\left(\dfrac{1}{4}\right)^{n+1}}{\dfrac{1}{4} + \left(\dfrac{1}{2}\right)^{n+1} + \left(\dfrac{1}{4}\right)^{n+1}}.$$

Note that, when n is large, the difference between this estimate and the other two estimates is small. When n is small, the difference is large. For example, if the couple has only two children (i.e. $n=2$), Bayes estimate above is 0.81, compared with the estimates derived by MLE procedures and the Method of Moments, each gives the estimate 1 for θ.

Definition 8.3 Let $X_1, X_2, \cdots, X_n$ be a random sample obtained from a population with either pdf or pmf $f(x|\theta_1, \theta_2, \cdots, \theta_k)$, where $\boldsymbol{\theta} = (\theta_1, \theta_2, \cdots, \theta_k) \in \boldsymbol{\Theta}$. If we denote the prior distribution by $\pi(\boldsymbol{\theta})$, then the posterior distribution, given the sample $(X_1, \ldots, X_n) = \boldsymbol{x}$, can be defined as $\pi(\boldsymbol{\theta}|\boldsymbol{x}) = \dfrac{f(\boldsymbol{x}|\boldsymbol{\theta})\pi(\boldsymbol{\theta})}{m(\boldsymbol{x})}$, where $m(\boldsymbol{x})$ is the marginal distribution of the random vector $\boldsymbol{X} = (X_1, \ldots, X_n)$ and can be calculated as $m(\boldsymbol{x}) = \int_{\boldsymbol{\Theta}} f(\boldsymbol{x}|\boldsymbol{\theta})\pi(\boldsymbol{\theta})d\boldsymbol{\theta}$. The Bayes estimate based on the sample $\boldsymbol{x}$, is defined as the mean of the posterior distribution $\pi(\boldsymbol{\theta}|\boldsymbol{x})$.

Note that when the population distribution is discrete, the above integration should be replaced by a summation.

Example 8.17 Let $X_1, \ldots, X_n$ represent the SAT-Math scores of n college freshmen randomly chosen from a large state university. Assume that these scores follow a normal distribution $N(\theta, \sigma^2)$, where σ^2 is known and θ, from previous information, is believed to follow a normal distribution with known mean μ and known variance τ^2, If we wish to find the Bayes estimator, we have to calculate the posterior distribution given the samples.

$$\pi(\theta \mid x_1, \ldots, x_n) = \frac{\left[\prod_{i=1}^{n} f(x_i|\theta)\right]\pi(\theta)}{m(\boldsymbol{x})}$$

$$= \frac{\left[\prod_{i=1}^{n} \frac{1}{\sqrt{2\pi}} \exp\left(-\frac{(x_i - \theta)^2}{2\sigma^2}\right)\right] \frac{1}{\sqrt{2\pi}} \exp\left(-\frac{(\theta - \mu)^2}{2\tau^2}\right)}{m(\boldsymbol{x})} \qquad (8.7)$$

$$\propto \exp\left[-\frac{1}{2}\left(\frac{(\theta - \mu)^2}{\tau^2} + \sum_{i=1}^{n} \frac{(x_i - \theta)^2}{\sigma^2}\right)\right]$$

where $m(\boldsymbol{x}) = \int \prod_{i=1}^{n} f(x_i \mid \theta)\pi(\theta)d\theta$.

Since $\sum_{i=1}^{n} \frac{(x_i - \theta)^2}{\sigma^2} = \sum_{i=1}^{n} \frac{(x_i - \bar{x} + \bar{x} - \theta)^2}{\sigma^2} = \sum_{i=1}^{n} \frac{(x_i - \bar{x})^2}{\sigma^2} + \frac{n(\bar{x} - \theta)^2}{\sigma^2}$ and the term

$\sum_{i=1}^{n} \frac{(x_i - \bar{x})^2}{\sigma^2}$ does not depend on the parameter θ, Equation (8.7) can be written

as $\pi(\theta \mid x_1, \ldots, x_n) \propto \exp\left[-\frac{1}{2}\left(\frac{(\theta - \mu)^2}{\tau^2} + \frac{(\bar{x} - \theta)^2}{\sigma^2 / n} \right) \right]$. Furthermore, we can re-

write the exponent of this equation as

$$\frac{(\theta - \mu)^2}{\tau^2} + \frac{(\bar{x} - \theta)^2}{\sigma^2 / n} = \theta^2\left(\frac{n}{\sigma^2} + \frac{1}{\tau^2} \right) - 2\theta\left(\frac{n\bar{x}}{\sigma^2} + \frac{\mu}{\tau^2} \right) + \left(\frac{n\bar{x}^2}{\sigma^2} + \frac{\mu^2}{\tau^2} \right)$$

$$= \varphi\left[\theta - \frac{1}{\varphi}\left(\frac{n\bar{x}}{\sigma^2} + \frac{\mu}{\tau^2} \right) \right]^2 - \frac{1}{\varphi}\left(\frac{n\bar{x}}{\sigma^2} + \frac{\mu}{\tau^2} \right)^2 + \left(\frac{n\bar{x}^2}{\sigma^2} + \frac{\mu^2}{\tau^2} \right),$$

where $\varphi = \dfrac{n}{\sigma^2} + \dfrac{1}{\tau^2}$.

Therefore, $\pi(\theta \mid x_1, \ldots, x_n) \propto \exp\left\{ -\frac{1}{2}\varphi\left[\theta - \frac{1}{\varphi}\left(\frac{n\bar{x}}{\sigma^2} + \frac{\mu}{\tau^2} \right) \right]^2 \right\}$ and we can con-

clude that $\pi(\theta \mid x_1, \ldots, x_n)$ is normally distributed with mean $\dfrac{1}{\varphi}\left(\dfrac{n\bar{x}}{\sigma^2} + \dfrac{\mu}{\tau^2} \right)$ and

variance $\dfrac{1}{\varphi}$. The Bayes estimator for this case is

$$\hat{\theta}_B = \frac{1}{\varphi}\left(\frac{n\bar{x}}{\sigma^2} + \frac{\mu}{\tau^2} \right) = \frac{\tau^2\bar{x} + \dfrac{\sigma^2}{n}\mu}{\dfrac{\sigma^2}{n} + \tau^2}\left(\frac{\tau^2\bar{x} + \dfrac{\sigma^2}{n}\mu}{\tau^2\dfrac{\sigma^2}{n}} \right) = \frac{\tau^2\bar{x} + \dfrac{\sigma^2}{n}\mu}{\dfrac{\sigma^2}{n} + \tau^2}$$

that is a weighted sum of $\bar{x}$ and μ.

Example 8.18 A public health study has found that 6 of 100 randomly chosen children develop chronic bronchitis in the first year of their life. Assume that the probability θ that a child in the first year of life will develop chronic bronchitis, follows the beta distribution with known parameters $\alpha = 1.5$ and $\beta = 3$. Suppose that the number X of infants under study develop chronic bronchitis follows a binomial distribution $B(n, \theta)$ and the parameter θ has a prior distribution

$$\pi(\theta) = \frac{\Gamma(\alpha+\beta)}{\Gamma(\alpha)\Gamma(\beta)} \theta^{\alpha-1}(1-\theta)^{\beta-1} I_{[0,1]}(\theta).$$

Given $X = x$, the posterior distribution can be derived as

$$\pi(\theta|x) = \frac{1}{m(x)} \binom{n}{x} \theta^x (1-\theta)^{n-x} \frac{\Gamma(\alpha+\beta)}{\Gamma(\alpha)\Gamma(\beta)} \theta^{\alpha-1}(1-\theta)^{\beta-1} \qquad (8.8)$$

$$= C\theta^{x+\alpha-1}(1-\theta)^{n-x+\beta-1}, \text{ for } 0 \le \theta \le 1$$

where C is a constant and $m(x)$ is the marginal distribution of X and is found by integrating the joint distribution of X and θ. Equation (8.8) indicates that the posterior distribution follows a beta distribution $Beta(\alpha+x, \beta+n-x)$, since C is the integrating factor that makes the integral of $\pi(\theta|x)$ equal 1. From a property of the beta distribution discussed in chapter 5, we can calculate the posterior mean as $\frac{\alpha+x}{(\alpha+x)+(\beta+n-x)} = \frac{\alpha+x}{\alpha+\beta+n}$. In this study, the Bayes estimator for the rate at which children develop chronic bronchitis in the first year of life is approximately equal to 7% by substituting $n=100$, $x=6$, $\alpha=1.5$ and $\beta=3$.

It looks like we use only one observation to perform this point estimation. In fact, each of 100 children randomly chosen can be viewed as one observation and whether the infant developed chronic bronchitis can be viewed as an observation from a Bernoulli random variable. We indeed have 100 observations and only their sum was actually observed. If we were to use all the observations, instead of their sum, the result will be identical. Further discussion of this issue appears section 8.5.

Example 8.19 The number of customers shopping at a particular store on the Friday night is assumed to follow a Poisson distribution with parameter θ. Suppose the distribution of θ follows a $gamma(\alpha, \beta)$ distribution. The store manager observed n consecutive Friday nights and recorded the number of customers in this store. These observations were denoted as $x_1, x_2, \ldots, x_n$. The posterior distribution of θ, given these observations is

$$\pi(\theta | x_1, x_2, \ldots, x_n) = C e^{-n\theta} \theta^{\sum_{i=1}^{n} x_i} \theta^{\alpha-1} e^{-\beta\theta} = C e^{-(n+\beta)\theta} \theta^{\alpha-1+\sum_{i=1}^{n} x_i} \qquad \text{for } \theta > 0,$$

where C is the constant that makes the integral of this posterior distribution equal 1. Therefore, the posterior distribution is also a gamma distribution with parameters $\left(\alpha + \sum_{i=1}^{n} x_i, n + \beta \right)$ and the Bayes estimator is $\dfrac{\alpha + \sum_{i=1}^{n} x_i}{n + \beta}$.

In Examples 8.17−8.19, the prior distribution and its corresponding posterior distribution belong to the same class of distributions. This class of distributions is called a *conjugate family* of the class of the distributions for the observed samples. In summary, the normal family is conjugate for itself, the beta family is conjugate for the binomial family and the gamma family is conjugate for the Poisson family. Detailed discussions on the conjugate family can be found in any Bayesian statistics book, for example Berger (1985).

Although the conjugate prior may offer its simplicity, computational convenience and attractive mathematical properties, the readers should be aware that there are many other prior distributions that may be chosen and the choice of the prior should not depend on conjugacy. For example, the prior distribution in Example 8.16 can not be substituted by the conjugate prior, since the original one represents the biological understanding of the parameter.

Remark 8.1 The mode of the posterior distribution is sometimes referred to as the *generalized maximum likelihood estimate* for the parameter θ.

Example 8.20 The generalized maximum likelihood estimate (GMLE) for example 8.17 is the same as the Bayes estimate, since the posterior distribution is normal and, in a normal distribution, the mean and the mode are identical. In general, if a posterior distribution is symmetric and unimodal, these two estimates are identical.

Example 8.21 To find the GMLE for Example 8.18, we calculate the derivative of the logarithmic transformation of the posterior distribution in Equation (8.8). By equating this derivative to 0, the GMLE is the solution of the following equation

$$\frac{\partial}{\partial \theta} \log \left[\pi(\theta | x) \right] = (x + \alpha - 1) \frac{1}{\theta} - (n - x + \beta - 1) \frac{1}{1 - \theta} = 0.$$

Since $\dfrac{\partial^2}{\partial \theta^2} \log \left[\pi(\theta | x) \right] \Big|_{\theta = \hat{\theta}} < 0$ for $\alpha > 1$ and $\beta > 1$, the GMLE is the the soluteion of this equation and can be expressed as $\hat{\theta} = \dfrac{\alpha + x - 1}{\alpha + \beta + n - 2}$.

Up to now, we have successfully used analytical methods to find Bayesian estimates and GMLE. In practice, we will frequently encounter problems that are difficult to find an algebraic form of the posterior mean or posterior mode (GMLE). In particular, when the number of parameters is large or, equivalently the dimension of the parameter space is high, algebraic solution for the Bayes estimator is uncommon. In this situation, we may apply different computational methods to find the posterior mean or posterior mode. In Example 8.22 below, we will use a Monte Carlo method to find the Bayes estimate. In Example 8.23, the Bayes estimate will be carried out using Markov Chain Monte Carlo (MCMC) method. The readers are referred to Carlin and Louis (2000) for more detailed discussions on Bayesian computational methods.

Example 8.22 Let $X_1,\ldots,X_n$ represent a random sample from a normal distribution $N(\mu,\sigma^2)$, where the unknown parameters μ and σ^2 are assumed to follow a prior distribution $\pi(\mu,\sigma^2) = (\sigma^2)^{-1}$. The joint posterior distribution is

$$\pi(\mu,\sigma^2 \,|\, X_1,\ldots,X_n) = C_0(\sigma^2)^{-1}(\sigma^2)^{-n/2}\exp\left(-\frac{1}{2\sigma^2}\sum_{i=1}^n(X_i-\mu)^2\right)$$

$$= C_0(\sigma^2)^{-(n+2)/2}\exp\left(-\frac{1}{2\sigma^2}(\sum_{i=1}^n(X_i-\bar{X})^2 + n(\bar{X}-\mu)^2)\right),$$

where C_0 is the appropriate constant that makes the integral of $\pi(\mu,\sigma^2)$ to equal 1.

The conditional distribution of μ, given σ^2 and $X_1,\ldots,X_n$, can be found to be $N(\bar{X},\dfrac{\sigma^2}{n})$ by calculating $\dfrac{\pi(\mu,\sigma^2 \,|\, X_1,\ldots,X_n)}{\pi(\sigma^2 \,|\, X_1,\ldots,X_n)}$. The conditional distribution of σ^2, given $X_1,\ldots,X_n$ is then equal to

$$\pi(\sigma^2 \,|\, X_1,\ldots,X_n) = \int \pi(\mu,\sigma^2 \,|\, X_1,\ldots,X_n)d\mu$$

$$= C_0(\sigma^2)^{-(n+2)/2}\exp\left[-\frac{1}{2\sigma^2}\sum_{i=1}^n(X_i-\bar{X})^2\right]\int\exp\left[\frac{(\bar{X}-\mu)^2}{2\sigma^2/n}\right]d\mu$$

$$= C_1(\sigma^2)^{-(n+1)/2}\exp\left(-\frac{1}{2\sigma^2}\sum_{i=1}^n(X_i-\bar{X})^2\right).$$

From Problem 33, Chapter 5, $\sigma^2 | X_1, \ldots, X_n$ has the same distribution (called the

inverse gamma distribution) as $\dfrac{\sum_{i=1}^{n}(X_i - \bar{X})^2}{W}$, where W is distributed as χ^2_{n-1}

and $\sum_{i=1}^{n}(X_i - \bar{X})^2$ is known.

To find the Bayes estimate, we can now generate random samples from $\sigma^2 | X_1, \ldots, X_n$ by first generating samples from χ^2_{n-1} and multiplying its recipro-

cal by $\sum_{i=1}^{n}(X_i - \bar{X})^2$. For each sample σ_j^2, we generate random samples for μ_j

from $N(\bar{X}, \dfrac{\sigma_j^2}{n})$. The generated data set $\{(\mu_j, \sigma_j^2): j = 1, 2, \cdots, N\}$ will be distrib-

uted as $\pi(\mu, \sigma^2 | X_1, \ldots, X_n)$ and can be used for calculating empirical Bayes estimates.

Example 8.23 Pulse (heartbeat) is a measurement of artery contraction that blows blood through the blood vessels. It is frequently used in the study of cardiovascular disease and its prevention. In the data collected by the Healthy Growth project (see Taylor and et. al. [2002]), there were 83 African American girls participating in this study on the relationship between physical activity and cardiovascular health. Besides data on physical activity and diet were gathered by questionnaires, and anthropometric and cardiovascular (including pulse) measurements were collected on each subject. There were two trained research assistants measuring the pulse of 72 subjects. One of the statistical questions was to estimate the common mean and the variances of the pulse measured by these two research assistants (RA).

Let x_{ij} denote the pulse of jth subject taken by the ith RA, $i = 1, 2$, and $j = 1, 2, \ldots, n_i$. Assume that x_{ij} are normally distributed with common mean θ

and variance $\dfrac{1}{\tau_i}$. The prior distribution for the parameters θ, τ_1, τ_2 is believed to

be

$$\pi(\theta, \tau_1, \tau_2) = \frac{1}{\tau_1 \tau_2} \text{ for } \tau_i > 0.$$ The joint posterior distribution for θ, τ_1, τ_2, given

the observations can be obtained as

$$\pi(\theta, \tau_1, \tau_2 | x_{11}, \ldots, x_{1n_1}, x_{21}, \ldots, x_{2n_1})$$

$$= C_0 \prod_{i=1}^{2} \tau_i^{-1} \tau_i^{n_i/2} \exp\left(-\frac{\tau_i}{2} \sum_{j=1}^{n_i} (x_{ij} - \theta)^2\right)$$

$$= C_0 \prod_{i=1}^{2} \tau_i^{(n_i-2)/2} \exp\left(-\frac{\tau_i}{2} (\sum_{j=1}^{n_i} (x_{ij} - \bar{x}_i)^2 + n_i(\bar{x}_i - \theta)^2)\right),$$

where again C_0 is the appropriate constant that makes the integral of $\pi(\theta, \tau_1, \tau_2)$ equal to 1. The conditional posterior distribution of the parameter τ_1, given θ and τ_2 is a gamma density with parameters $\frac{n_1}{2}$ and $\frac{1}{2}(\sum_{j=1}^{n_1} (x_{1j} - \bar{x}_1)^2 + n_1(\bar{x}_1 - \theta)^2)$. Similarly, the conditional posterior distribution of the parameter τ_2, given θ and τ_1 is a gamma density with parameters $\frac{n_2}{2}$ and $\frac{1}{2}(\sum_{j=1}^{n_2} (x_{2j} - \bar{x}_2)^2 + n_2(\bar{x}_2 - \theta)^2)$. Clearly, these two conditional posterior distributions are independent. Thus $\tau_1 | \theta, \tau_2$ has the same distribution as $\tau_1 | \theta$ and similarly for $\tau_2 | \theta, \tau_1$. The conditional posterior distribution of the parameter θ, given τ_1 and τ_2 is

$$\pi(\theta | \tau_1, \tau_2, x_{11}, \ldots, x_{1n_1}, x_{21}, \ldots, x_{2n_1})$$

$$= \frac{\pi(\theta, \tau_1, \tau_2 | x_{11}, \ldots, x_{1n_1}, x_{21}, \ldots, x_{2n_1})}{\pi(\tau_1, \tau_2 | x_{11}, \ldots, x_{1n_1}, x_{21}, \ldots, x_{2n_1})}$$

$$= C_1 \exp\left[-\frac{1}{2}\left(\tau_1 n_1 (\bar{x}_1 - \theta)^2 + \tau_2 n_2 (\bar{x}_2 - \theta)^2\right)\right] \quad (8.9)$$

$$= C_2 \exp\left[-\frac{1}{2}\left((\tau_1 n_1 + \tau_2 n_2)\theta^2 - 2(\tau_1 n_1 \bar{x}_1 + \tau_2 n_2 \bar{x}_2)\theta\right)\right]$$

$$= C_3 \exp\left[-\frac{(\tau_1 n_1 + \tau_2 n_2)}{2}\left(\theta - \frac{\tau_1 n_1 \bar{x}_1 + \tau_2 n_2 \bar{x}_2}{\tau_1 n_1 + \tau_2 n_2}\right)^2\right].$$

where C_1, C_2 and C_3 are appropriate constants. From Equation (8.9), the conditional posterior distribution of the parameter θ, given τ_1 and τ_2 can be identified as a normal distribution with mean $\dfrac{\tau_1 n_1 \bar{x}_1 + \tau_2 n_2 \bar{x}_2}{\tau_1 n_1 + \tau_2 n_2}$ and variance $\dfrac{1}{\tau_1 n_1 + \tau_2 n_2}$.

We can now apply Gibbs sampler algorithm, as discussed in Chapter 11, to simulate the posterior samples and use the Monte Carlo method to find the associated Bayes estimates.

Choosing $\theta^{(0)} = \dfrac{n_1 \bar{x}_1 + n_2 \bar{x}_2}{n_1 + n_2}$, $\tau_1^{(0)} = \dfrac{1}{\sum\limits_{j=1}^{n_1} (x_{1j} - \bar{x}_1)^2}$ and $\tau_2^{(0)} = \dfrac{1}{\sum\limits_{j=1}^{n_2} (x_{2j} - \bar{x}_2)^2}$, as

the initial values of θ, τ_1 and τ_2, we first determine $\theta^{(1)}$ by drawing from the conditional normal distribution using Equation (8.9) with $\tau_1 = \tau_1^{(0)}$ and $\tau_2 = \tau_2^{(0)}$. Then we determine $\tau_1^{(1)}$ by drawing from conditional gamma distribution, given

$\theta = \theta^{(1)}$, with parameters $\dfrac{n_1}{2}$ and $\dfrac{1}{2}(\sum\limits_{j=1}^{n_1} (x_{1j} - \bar{x}_1)^2 + n_1(\bar{x}_1 - \theta^{(1)})^2)$. Lastly, we

determine $\tau_2^{(1)}$ by drawing from conditional gamma distribution, given $\theta = \theta^{(1)}$. Then we generate $\theta^{(2)}$ from Equation (8.9) with $\tau_1 = \tau_1^{(1)}$ and $\tau_2 = \tau_2^{(1)}$, and generate $\tau_1^{(2)}$ and $\tau_2^{(2)}$ from gamma distribution with $\theta = \theta^{(2)}$. We continue this process until there are m iterations $(\theta^{(m)}, \tau_1^{(m)}, \tau_2^{(m)})$, for a large m. Then $\theta^{(m)}$, $\tau_1^{(m)}$ and $\tau_2^{(m)}$ would be a sample from $\pi(\theta, \tau_1, \tau_2 | x_{11}, \ldots, x_{1n_1}, x_{21}, \ldots, x_{2n_2})$.

Independently repeating these Gibbs processes M times will produce 3-tuple posterior samples $(\theta_j^{(m)}, \tau_{1j}^{(m)}, \tau_{2j}^{(m)})$, $j = 1, 2, \ldots, M$ for θ, τ_1, τ_2. The Bayes estimates and all other statistical inferences based on the posterior distribution can be obtained from the M sample values generated by the Gibbs sampler. Gelfand et al (1990) recommends that m be approximated 50 iterations and M be less than or equal to 1000 samples.

For our data set, the common mean is 78.33, $\sum\limits_{j=1}^{n_1} (x_{1j} - \bar{x}_1)^2 = 2240.8$,

$n_1 = 30$, $\sum\limits_{j=1}^{n_2} (x_{2j} - \bar{x}_2)^2 = 9122.0$ and $n_2 = 42$. We use $m = 50$ and $M = 1000$

to generate the posterior samples. The results are $\hat{\theta}_B = 78.20$, $\hat{\tau}_{1B} = 17087.84$ and $\hat{\tau}_{1B} = 33976.51$.

Note that the methodology used in this example was first published by Gregurich and Broemeling (1997).

8.5 Sufficient Statistics

As we can see in the previous two sections, we need to use the information in a sample in order to calculate maximum likelihood and Bayes estimates and make statistical inferences. However, it is very cumbersome to write down the likelihood function or the posterior distribution when the sample size is large. Fortunately, most of the likelihood functions and posterior distributions can be described in terms of some summary statistics such as the sample sum, the sample mean, the sample variance and the largest sample value or the smallest sample value. These phenomena lead us to think about two questions: (1) Is there a legitimate way we can reduce the data presented by the sample and preserve the important information for calculating a point estimate or conducting statistical inferences? and (2) Will data reduction affect our point estimation?

Broadly, any *statistic* $T(X_1,\ldots,X_n)$ defined as a function of the sample $(X_1,\ldots,X_n)$, presents a form of data reduction or a type of data summary. However, we would like to see our statistic contains the same information about the parameter or parameters as the original sample. Specifically, if two sample points $\mathbf{x}=(x_1,\ldots,x_n)$ and $\mathbf{y}=(y_1,\ldots,y_n)$ have the same value of the statistic, i.e. $T(\mathbf{x})=T(\mathbf{y})$, we hope the inference about θ will be the same whether $\mathbf{x}$ or $\mathbf{y}$ is observed.

Definition 8.4 A statistic $T(X_1,\ldots,X_n)$ is said to be *sufficient* for θ, if the conditional distribution of the vector $(X_1,\ldots,X_n)$ given $T(X_1,\ldots,X_n)=t$ does not depend on θ for any value t.

Example 8.24 Let $X_1,\ldots,X_n$ be the outcomes of n independent Bernoulli trials with probability of success θ. Consider the statistic $T(X_1,\ldots,X_n)=\sum_{i=1}^{n}X_i$. Since $\{X_1=x_1,\ldots,X_n=x_n\}\subseteq\{T=t\}$, we have

$$P(X_1=x_1,\ldots,X_n=x_n,T=t)=P(X_1=x_1,\ldots,X_n=x_n)\quad\text{for }\sum_{i=1}^{n}x_i=t.\text{ From the}$$

property of independent Bernoulli trials, we have

$$P(X_1=x_1,\ldots,X_n=x_n)=\theta^{\sum_{i=1}^{n}x_i}\left(1-\theta\right)^{n-\sum_{i=1}^{n}x_i}=\theta^{t}\left(1-\theta\right)^{n-t}.\text{ As we discussed in}$$

Chapter 4, T has a binomial distribution with pmf $P(T=t)=\binom{n}{t}\theta^{t}\left(1-\theta\right)^{n-t}$.

Thus, the conditional distribution of $X_1,\ldots,X_n$ given T is

$$P(X_1 = x_1, \ldots, X_n = x_n | T = t) = \frac{P(X_1 = x_1, \ldots, X_n = x_n, T = t)}{P(T = t)} = \frac{1}{\binom{n}{t}},$$

if $\sum_{i=1}^{n} x_i = t$, and $P(X_1 = x_1, \ldots, X_n = x_n | T = t) = 0$, if $\sum_{i=1}^{n} x_i \neq t$. Clearly, the conditional distribution of the sample given T does not depend on θ and hence T is a sufficient statistic.

This example demonstrates that, for a sequence of Bernoulli trials, the entire set of sample observations does not provide us more knowledge about the success rate than their sum does, since the conditional distribution of the sample given the sum is independent of the success rate.

Example 8.25 Guillain-Barré syndrome (GBS) is a rare neurologic disease often resulting in paralysis. Kinnunen et al. (1998) reported the cases of GBS that were admitted to the neurological units of hospitals in Finland. There were 28 cases reported in the 9 month period from April to December, 1984 and 52 cases reported in the 10 month period from January to October, 1985.

Suppose that the number of GBS cases in 1984, denoted by X, is distributed as a Poisson random variable with mean 9λ and that number in 1985, denoted by Y, also follows the Poisson distribution with mean 10λ. Let $T(X, Y) = X + Y$ and we will show that T is a sufficient statistic, assuming X and Y are independent. For $k = 1, 2, \ldots, n$, we have

$$P(X = k, Y = n - k | T = n) = \frac{P(X = k)P(Y = n - k)}{P(T = n)}$$

$$= \frac{\frac{(9\lambda)^k}{k!} e^{-9\lambda} \frac{(10\lambda)^{n-k}}{(n-k)!} e^{-10\lambda}}{\frac{(19\lambda)^n}{n!} e^{-19\lambda}} = \binom{n}{k} \left(\frac{9}{19}\right)^k \left(\frac{10}{19}\right)^{n-k}.$$

This conditional distribution does not depend on the parameter λ (the monthly mean number of cases), hence the sum of two sample points T is a sufficient statistic for λ. Thus, the total number of cases in both years will provide us as much information on estimating λ as both case numbers recorded for each year.

In general, the calculation of the conditional distribution of the sample given the value of statistic T is cumbersome, especially for the continuous case. Fortunately, another criterion for checking sufficiency is available. This method was presented in various forms by Fisher (1922), Neyman (1935) and Halmos and Savage (1949).

Theorem 8.1 (Factorization Theorem) Let $X_1, \ldots, X_n$ be a random sample from the distribution (pdf or pmf) $f(x|\theta)$, with $\theta \in \Theta$. A statistic $T(X_1, \ldots, X_n)$ is a sufficient statistic for θ if and only if for any $\mathbf{x} = (x_1, \ldots, x_n)$ and any $\theta \in \Theta$, the joint distribution $f(\mathbf{x}|\theta) = f(x_1|\theta) \cdots f(x_n|\theta)$ can be written as

$$f(\mathbf{x}|\theta) = g(T(\mathbf{x})|\theta) h(\mathbf{x}), \tag{8.10}$$

where $h(\mathbf{x})$ does not depend on θ and $g(T(\mathbf{x})|\theta)$ depends on $\mathbf{x}$ only through $T(\mathbf{x})$.

Proof. We shall present the proof only for the discrete case. The proof for the continuous case can be found in, for example, Lehmann and Casella (1998).

Suppose $T(\mathbf{x})$ is a sufficient statistic and let

$$h(\mathbf{x}) = P(\mathbf{X} = \mathbf{x}|T(\mathbf{X}) = T(\mathbf{x})) \quad \text{and} \quad g(T(\mathbf{x}) = t|\theta) = P(T(\mathbf{X}) = t|\theta).$$

Thus $h(\mathbf{x})$ does not depend on θ and $g(T(\mathbf{x}) = t|\theta)$ is the pmf for $T(\mathbf{x})$. Now, observe that

$$\begin{aligned} f(\mathbf{x}|\theta) &= P(\mathbf{X} = \mathbf{x}|\theta) \\ &= P(\mathbf{X} = \mathbf{x}, T(\mathbf{X}) = T(\mathbf{x})|\theta) \\ &= P(T(\mathbf{X}) = T(\mathbf{x})|\theta) P(\mathbf{X} = \mathbf{x}|T(\mathbf{X}) = T(\mathbf{x})) \\ &= g(T(\mathbf{x})|\theta) h(\mathbf{x}). \end{aligned}$$

This has shown Equation (8.10) by assuming $T(\mathbf{x})$ is sufficient.

Now suppose Equation (8.10) holds.

Let $A(t)$ be the set of all $\mathbf{x}$ such that $T(\mathbf{x}) = t$. Then $P(T = t|\theta) = \sum_{\mathbf{y} \in A(t)} f(\mathbf{y}|\theta)$. If $T(\mathbf{x}) = t$, we have

$$P(\mathbf{X} = \mathbf{x}|T = t) = \frac{P(\mathbf{X} = \mathbf{x}, T = t)}{P(T = t)} = \frac{f(\mathbf{x}|\theta)}{\sum_{\mathbf{y} \in A(t)} f(\mathbf{y}|\theta)} = \frac{g(T(\mathbf{x})|\theta) h(\mathbf{x})}{\sum_{\mathbf{y} \in A(t)} g(T(\mathbf{y})|\theta) h(\mathbf{y})}$$

$$= \frac{h(\mathbf{x})}{\sum_{\mathbf{y} \in A(t)} h(\mathbf{y})}.$$

Note that the first equality is the result of $\{\mathbf{X} = \mathbf{x}\} \subseteq \{T = t\}$, the second equality follows the definition of $A(t)$ as explained above, the third equality comes from Equation (8.10) and the last equality is due to $T(\mathbf{x}) = T(\mathbf{y})$ and the fact that $g(T(\mathbf{y})|\theta)$ depends on $\mathbf{y}$ only through $T(\mathbf{y})$. Since $\dfrac{h(\mathbf{x})}{\displaystyle\sum_{\mathbf{y} \in A(t)} h(\mathbf{y})}$ does not depend on θ, this shows that the conditional distribution of $\mathbf{x}$ given $T(\mathbf{x}) = t$ does not depend on θ and hence $T(\mathbf{x})$ is a sufficient statistic.

Example 8.26 Let $X_1, \ldots, X_n$ be a random sample from a normal distribution with unknown mean θ and a known variance σ^2. The joint pdf of the sample can be written as

$$f(x_1, \cdots, x_n | \theta) = \frac{1}{(2\pi\sigma^2)^{n/2}} \exp\left[-\frac{1}{2\sigma^2} \sum_{i=1}^n (x_i - \theta)^2 \right]$$

$$= \frac{1}{(2\pi\sigma^2)^{n/2}} \exp\left[-\frac{1}{2\sigma^2} \left(\sum_{i=1}^n x_i^2 - 2\theta \sum_{i=1}^n x_i + n\theta^2 \right) \right]$$

$$= \left\{ \exp\left(\frac{\theta}{\sigma^2} \sum_{i=1}^n x_i - \frac{n\theta^2}{2\sigma^2} \right) \right\} \left\{ \frac{1}{(2\pi\sigma^2)^{n/2}} \exp\left(-\frac{1}{2\sigma^2} \sum_{i=1}^n x_i^2 \right) \right\}$$

$$= g(T(\mathbf{x})|\theta) h(\mathbf{x}).$$

Hence $T(x_1, \ldots, x_n) = \displaystyle\sum_{i=1}^n x_i$ is a sufficient statistic. Therefore, for estimating the mean in a normal population with known variance, the sample sum will provide enough information.

Example 8.27 Mr. Smith catches the 7 o'clock bus every weekday morning. It is known that he arrives at the bus station before 7 am and his arrival times, recorded (with a negative sign) as minutes before 7 am are uniformly distributed between θ and 0, where $\theta < 0$. Let $x_1, \ldots, x_n$ represent his arrival times on n days. Then the joint density of the sample is

$$f(x_1, \cdots, x_n | \theta) = \frac{1}{(-\theta)^n} I_{(\theta,0)}(x_1) \cdots I_{(\theta,0)}(x_n).$$

Since all the observations are negative, $I_{(\theta,0)}(x_1)\cdots I_{(\theta,0)}(x_n)=1$ is equivalent to $\theta < \min(x_1,\cdots,x_n)$. Therefore, we can rewrite $I_{(\theta,0)}(x_1)\cdots I_{(\theta,0)}(x_n)$ in the above equation as $I_{(\theta,0)}(\min(x_1,\cdots,x_n))$. The joint density can be written as

$$f(x_1,\cdots,x_n|\theta) = \frac{1}{(-\theta)^n}I_{(\theta,0)}(\min(x_1,\cdots,x_n)) = g(T(\mathbf{x})|\theta)h(\mathbf{x}),$$

where $h(\mathbf{x})=1$ and $g(T(\mathbf{x})|\theta) = \frac{1}{(-\theta)^n}I_{(\theta,0)}(\min(x_1,\cdots,x_n))$ that depends on the sample on through $T(\mathbf{x}) = \min(x_1,\ldots,x_n)$. Hence $T(\mathbf{x}) = \min(x_1,\ldots,x_n)$ is a sufficient statistic for estimating θ.

Example 8.28 Consider $x_1,\ldots,x_n$ as a random sample taken from a Poisson population with mean λ. Letting $T(x_1,\ldots,x_n) = \sum_{i=1}^{n}x_i$, we have

$$f(x_1,\cdots,x_n|\lambda) = \prod_{i=1}^{n}\frac{\lambda^{x_i}}{x_i!}e^{-\lambda} = \lambda^{\sum_{i=1}^{n}x_i}e^{-n\lambda}\prod_{i=1}^{n}\frac{1}{x_i!} = g(T(\mathbf{x})|\lambda)h(\mathbf{x}),$$

where $g(t|\lambda) = \lambda^t e^{-n\lambda}$ and $h(\mathbf{x}) = \prod_{i=1}^{n}\frac{1}{x_i!}$. Therefore, $T(x_1,\ldots,x_n) = \sum_{i=1}^{n}x_i$ is a sufficient statistic.

Definition 8.5 A k-dimensional statistic $\mathbf{T}(\mathbf{x}) = (T_1(\mathbf{x}),\cdots,T_k(\mathbf{x}))$ is jointly sufficient for $\boldsymbol{\theta}$ if the joint density of the random sample satisfies Equation (8.10). Specifically, there exists a function of k-dimensional vector $g(\mathbf{t}|\boldsymbol{\theta})$ and a function of n-dimensional vector $h(\mathbf{x})$ such that

$$f(\mathbf{x}|\boldsymbol{\theta}) = g(\mathbf{T}(\mathbf{x})|\boldsymbol{\theta})h(\mathbf{x}).$$

From the expression above, it is clear that the function $g(\mathbf{T}(\mathbf{x})|\boldsymbol{\theta})$ depends on $\mathbf{x}$ only through the vector $\mathbf{T}(\mathbf{x})$ and the function $h(\mathbf{x})$ does not depend on $\boldsymbol{\theta}$.

Example 8.29 Let us revisit Example 8.26 for a sample of size n, taken from a normal population with both mean and variance unknown. Let $\boldsymbol{\theta} = (\mu,\sigma^2)$. We can write the joint density function of the sample as

$$f(\mathbf{x}|\mathbf{\theta}) = \frac{1}{(2\pi\sigma^2)^{n/2}} \exp\left(-\sum_{i=1}^{n} \frac{(x_i - \mu)^2}{2\sigma^2}\right)$$

$$= \frac{1}{(2\pi)^{n/2}} (\sigma^2)^{-\frac{n}{2}} \exp\left(-\frac{1}{2\sigma^2}\sum_{i=1}^{n} x_i^2 + \frac{\mu}{\sigma^2}\sum_{i=1}^{n} x_i - \frac{n\mu^2}{2\sigma^2}\right)$$

$$= g(\mathbf{T}(\mathbf{x})|\mathbf{\theta}) h(\mathbf{x}),$$

where $\mathbf{T}(\mathbf{x}) = (T_1(\mathbf{x}), T_2(\mathbf{x}))$ with $T_1(\mathbf{x}) = \sum_{i=1}^{n} x_i$ and $T_2(\mathbf{x}) = \sum_{i=1}^{n} x_i^2$. If we let $h(\mathbf{x}) = 1$ and set $g(\mathbf{T}(\mathbf{x})|\mathbf{\theta})$ equal to the second to the last expression, $\mathbf{T}(\mathbf{x})$ is jointly sufficient for $\mathbf{\theta} = (\mu, \sigma^2)$.

It is easy to see that, if we make a one-to-one transformation $\mathbf{u}(\mathbf{t})$ on the $\mathbf{T}(\mathbf{x})$ such that $\mathbf{u}(\mathbf{T}(\mathbf{x})) = \mathbf{T}^*(\mathbf{x})$ and $\mathbf{u}^{-1}(\mathbf{T}^*(\mathbf{x})) = \mathbf{T}(\mathbf{x})$, then we can rewrite the density function $f(\mathbf{x}|\theta)$ as

$$f(\mathbf{x}|\theta) = g(\mathbf{T}(\mathbf{x})|\theta) h(\mathbf{x}) = g\left[\mathbf{u}^{-1}(\mathbf{T}^*(\mathbf{x}))|\theta\right] h(\mathbf{x}) = g^*(\mathbf{T}^*(\mathbf{x})|\theta) h(\mathbf{x}),$$

where $g^*(\mathbf{t}|\theta) = g(\mathbf{u}^{-1}(\mathbf{t})|\theta)$. Hence $\mathbf{T}^*(\mathbf{x})$ is another sufficient statistic.

For Example 8.29, we can manipulate the algebra and find that the sample mean $T_1(\mathbf{x}) = \frac{1}{n}\sum_{i=1}^{n} x_i$ and the sample variance $T_2(\mathbf{x}) = \frac{1}{n-1}\sum_{i=1}^{n}(x_i - \overline{x})^2$ are also jointly sufficient statistics for $\mathbf{\theta} = (\mu, \sigma^2)$.

There may be more than one sufficient statistic for a parameter or a set of parameters of a population distribution from which the sample is taken. The entire sample and the order statistics are two obvious sufficient statistics for any sample. The latter is left as an exercise for the reader to prove. However these two sufficient statistics do not reduce the data.

It is natural for us to find a sufficient statistic that can perform maximal data reduction and preserve the whole information for statistical inferences. This sufficient statistic is called a *minimal sufficient statistic*. The readers are referred to other books (for example, Dudewicz and Mishra (1988)) for details.

Let us now examine how the MLE is determined by the sufficient statistic. Rewrite the likelihood function as

$$L(\theta|x_1,\ldots,x_n) = \prod_{i=1}^{n} f(x_i|\theta) = g(T(\mathbf{x})|\theta) h(\mathbf{x}).$$

We can easily see that the maximization of $L(\theta|x_1,\ldots,x_n)$ on θ, is equivalent to the maximization of $g(T(\mathbf{x})|\theta)$ on θ, since $h(\mathbf{x})$ is positive and does not depend on θ. We can see that the MLE is a function of $T(\mathbf{x})$ only. Similarly, for the Bayes estimator, we can rewrite the posterior density as

$$\pi(\theta|\mathbf{x}) = \frac{f(\mathbf{x}|\theta)\pi(\theta)}{\int f(\mathbf{x}|\tau)\pi(\tau)d\tau} = \frac{g(T(\mathbf{x})|\theta)h(\mathbf{x})\pi(\theta)}{\int g(T(\mathbf{x})|\tau)h(\mathbf{x})\pi(\tau)d\tau} = \frac{g(T(\mathbf{x})|\theta)\pi(\theta)}{\int g(T(\mathbf{x})|\tau)\pi(\tau)d\tau},$$

that depends on $T(\mathbf{x})$ only. Therefore, the Bayes estimate calculated from posterior mean, also depends only on $T(\mathbf{x})$, not the entire sample.

8.6 Exponential Families

There is a class of distributions that have a natural sufficient statistic whose dimension is independent of the sample size. These distributions are said to belong to *the exponential family*. Due to the recent development of generalized linear models, the exponential family has become particular important.

Definition 8.6 A distribution is said to be a member of the one-parameter exponential family, if its pdf or pmf can be written as

$$f(x|\theta) = \left\{\exp\left(c(\theta)T(x) + b(\theta) + S(x)\right)\right\}I_A(x) \tag{8.11}$$

for some real-valued functions $c(\theta)$ and $b(\theta)$ on Θ, real-valued functions $T(x)$ and $S(x)$ on $\square$, and a set A of real values that does not depend on θ.

Although (8.11) describes the distributional form for a univariate random variable, it can be applied to a multivariate random variable and to the joint distribution of iid random variables.

Note that, if $X_1,\ldots,X_n$ is a random sample from a distribution belonging to a one-parameter exponential family, $\sum_{i=1}^{n}T(X_i)$ is a sufficient statistic. This can be seen by identifying $\exp\left(c(\theta)\sum_{i=1}^{n}T(X_i) + nb(\theta)\right)$ as $g(T(X_1,\ldots,X_n)|\theta)$ and $\exp\left(\sum_{i=1}^{n}S(X_i)\right)$ as $h(X_1,\ldots,X_n)$. It is left as an exercise to show that the joint distribution of the sample is a member of a one-parameter exponential family.

Example 8.30 The normal distribution with unknown mean μ and known variance σ^2 is a member of a one-parameter exponential family. The density can be written as

$$f(x|\mu) = (2\pi\sigma^2)^{-1/2} \exp\left(-\frac{(x-\mu)^2}{2\sigma^2}\right) I_{(-\infty,\infty)}(x)$$

$$= \exp\left(\frac{\mu x}{\sigma^2} - \frac{\mu^2}{2\sigma^2} - (\frac{x^2}{2\sigma^2} + \frac{1}{2}\log(2\pi\sigma^2))\right) I_{(-\infty,\infty)}(x).$$

By choosing

$$c(\mu) = \frac{\mu}{\sigma^2}, \quad b(\mu) = -\frac{\mu^2}{2\sigma^2}, \quad T(x) = x, \quad S(x) = -\left(\frac{x^2}{2\sigma^2} + \frac{1}{2}\log(2\pi\sigma^2)\right) \quad \text{and}$$

$A = (-\infty, \infty)$, we can easily verify it.

Example 8.31 Suppose X has a binomial distribution with parameters n and θ, $0 < \theta < 1$. Letting $A = \{0, 1, \cdots, n\}$, we can rewrite the pmf as

$$p(x|\theta) = \binom{n}{x} \theta^x (1-\theta)^{n-x} I_A(x)$$

$$= \left\{\exp\left[x\log\left(\frac{\theta}{1-\theta}\right) + n\log(1-\theta) + \log\binom{n}{x}\right]\right\} I_A(x).$$

Therefore, the distribution of X is a member of one-parameter exponential family with $c(\theta) = \log\left(\frac{\theta}{1-\theta}\right)$, $b(\theta) = n\log(1-\theta)$, $T(x) = x$, and $S(x) = \log\binom{n}{x}$.

Example 8.32 Consider a random variable with pdf

$$f(x|\theta) = \begin{cases} \theta^{-1}\exp\left(1 - \frac{x}{\theta}\right) & \text{for } 0 < \theta < x < \infty \\ 0 & \text{otherwise} \end{cases}.$$

Although we can write $c(\theta) = \dfrac{1}{\theta}$, $b(\theta) = -\log(\theta)$, $T(x) = x$, and $S(x) = 1$, the indicator function $I_A(x) = I_{(\theta,\infty)}(x)$ does depend on the parameter. This distribution is not a member of the one-parameter exponential family.

Example 8.33 The pmf of the Poisson distribution can be written as

$$p(x|\lambda) = \frac{\lambda^x e^{-\lambda}}{x!} I_A(x) = \left\{\exp(x\log(\lambda) - \lambda - \log(x!))\right\} I_A(x),$$

where $A = \{0, 1, 2, \cdots, n, \cdots\}$. Note that, in addition to the normal distribution (Example 8.30), the binomial distribution (Example 8.31) and the Poisson distribution (Example 8.33), other commonly known distributions that are members of the exponential family, include the gamma distribution and the inverse Gaussian distribution. These five distributions as outcome variables are the major focus of generalized linear models. Interested readers are referred to McCullagh and Nelder (1989).

By letting $\xi = c(\theta)$, a distribution in an exponential family can be reparameterized as

$$f(x|\xi) = \left\{\exp\left(\xi T(x) + b^*(\xi) + S(x)\right)\right\} I_A(x), \tag{8.12}$$

where $b^*(\xi) = -\log \displaystyle\int_A \exp\left(\xi T(x) + S(x)\right) dx$ in the continuous case, and the integral is replaced by a sum over all values of x in A for the discrete case. In fact, the term $b^*(\xi)$ is a constant term that makes the pdf integral equal to 1. Letting $\Psi = \left\{\xi : \xi = c(\theta), \theta \in \Theta, b^*(\xi) < \infty\right\}$, the distribution given by Equation (8.12) belongs to the *one-parameter exponential family in natural form*.

Definition 8.7 A distribution is said to be a member of the k-parameter exponential family, if its pdf or pmf can be written as

$$f(x|\theta) = \left\{\exp\left(\sum_{i=1}^{k} c_i(\theta) T_i(x) + b(\theta) + S(x)\right)\right\} I_A(x) \tag{8.13}$$

for some real-valued functions $c_1(\theta), c_2(\theta), \cdots, c_k(\theta)$ and $b(\theta)$ on Θ, real-valued functions $T(x)$ and $S(x)$ on $\square$, and a set A of real values that does not depend on θ.

Similar to the case of the one-parameter exponential family, the vector

$(\sum_{j=1}^{n} T_1(X_j), \cdots, \sum_{j=1}^{n} T_k(X_j))$ is a natural sufficient statistic of the k-parameter expo-

nential family.

Example 8.34 Let $X_1, \ldots, X_n$ be a random sample from a member of the one-parameter exponential family. The log-likelihood function becomes

$l(\theta) = c(\theta) \sum_{i=1}^{n} T(X_i) + nb(\theta) + \sum_{i=1}^{n} S(X_i)$ and the likelihood equation can be writ-

ten as $l'(\theta) = c'(\theta) \sum_{i=1}^{n} T(X_i) + nb'(\theta) = 0$. When $c(\theta) = \theta$, this equation is

equivalent to the equation $\sum_{i=1}^{n} T(X_i) + nb'(\theta) = 0$. From Problem 33, Chapter 5,

that $b'(\theta) = -E(T(X))$, we can rewrite the likelihood equation as

$$nE(T(X)|\theta) = \sum_{i=1}^{n} T(X_i). \tag{8.14}$$

Since $b''(\theta) = -Var(T(X))$, the second derivative of the log-likelihood function

is $l''(\theta) = nb''(\theta) = -\text{var}(T(X)) < 0$. The solution $\hat{\theta}$ of Equation (8.14) is the

unique MLE. This is also true, if $c(\theta) \neq \theta$, but $c(\hat{\theta})$ is an interior point of the

range of $c(\theta)$. This result can be applied to Examples 8.30, 8.31 and 8.33 for

calculating the MLE. The results match with the direct calculations presented in

Examples 8.9, 8.5 and 8.7 respectively.

     Note that a similar result for the random sample taken from population

distributed as a *k*-parameter exponential family is available. In this case, the solu-

tion for the equations

$$nE(T_j(X)|\mathbf{\theta}) = \sum_{i=1}^{n} T_j(X_i), \text{ for } j = 1, 2, \cdots, k$$

is the unique MLE $\hat{\mathbf{\theta}}$ for $\mathbf{\theta}$ if $(c_1(\hat{\mathbf{\theta}}), c_2(\hat{\mathbf{\theta}}), \cdots, c_k(\hat{\mathbf{\theta}}))$ is an interior point of the

range of $(c_1(\mathbf{\theta}), c_2(\mathbf{\theta}), \cdots, c_k(\mathbf{\theta}))$.

Example 8.35 Consider the normal distribution with both parameters unknown.
We can easily show that it is a two-parameter exponential family by identifying

the corresponding functions $T_1(x) = x$, $T_2(x) = x^2$, $c_1(\theta) = \dfrac{\mu}{\sigma^2}$, $c_2(\theta) = -\dfrac{1}{2\sigma^2}$,

$$b(\theta) = -\frac{1}{2}\left(\frac{\mu}{\sigma^2} + \log(2\pi\sigma^2)\right), \quad S(x) = 0 \text{ and } A = \square.$$

If we observe a random sample $x_1, \ldots, x_n$ from this distribution, we can clearly see that $(\sum_{i=1}^{n} T_1(x_i), \sum_{i=1}^{n} T_2(x_i)) = (\sum_{i=1}^{n} x_i, \sum_{i=1}^{n} x_i^2)$ is a sufficient statistic for $\theta = (\mu, \sigma^2)$ as we have shown in Example 8.29.

Furthermore, the solution for the equations $\sum_{i=1}^{n} X_i = nE(X|\boldsymbol{\theta}) = n\mu$ and

$\sum_{i=1}^{n} X_i^2 = nE(X^2|\boldsymbol{\theta}) = n(\mu^2 + \sigma^2)$ gives $(\bar{x}, \frac{1}{n}\sum_{i=1}^{n}(x_i - \bar{x})^2)$ as the MLE of (μ, σ^2) as is shown in Example 8.10.

Distributions in the exponential family enjoy having a natural sufficient statistic and a natural unique MLE. The following example indicates that it also has a conjugate family under Bayesian estimation.

Example 8.36 Consider a random sample $X_1, \ldots, X_n$ taken from a distribution belonging to a k-parameter exponential family. Its joint pdf can then be written as

$$f(x_1, \ldots, x_n|\theta)$$
$$= \left\{ \exp\left(\sum_{i=1}^{k} c_i(\theta)\left(\sum_{j=1}^{n} T_i(x_j) \right) + nb(\theta) + \sum_{j=1}^{n} S(x_j) \right) \right\} I_A(\mathbf{x}). \tag{8.15}$$

If we choose a member of the $(k+1)$-parameter exponential family

$\pi(\theta) = C_1 \left\{ \exp\left(\sum_{i=1}^{k} \omega_i c_i(\theta) + \omega_{k+1} b(\theta) \right) \right\} I_\Theta(\theta)$ as the prior density, then the posterior distribution will be proportional to

$$\pi(\theta|x_1, \ldots, x_n) \propto \left\{ \exp\left(\sum_{i=1}^{k} c_i(\theta)\left(\omega_i + \sum_{j=1}^{n} T_i(x_j) \right) + (\omega_{k+1} + n)b(\theta) \right) \right\} I_\Theta(\theta),$$

which belongs to $(k+1)$-parameter exponential family, with $(c_1(\theta), \cdots, c_k(\theta), b(\theta))$ as the natural sufficient statistic. The corresponding parameter functions are

$$\left(\omega_1 + \sum_{j=1}^{n} T_1(x_j), \cdots, \omega_k + \sum_{j=1}^{n} T_k(x_j), \omega_{k+1} + n \right).$$

Therefore, this class of (k+1)-parameter exponential family and the k-parameter exponential family described in Equation (8.15) form a conjugate family.

8.7 Other Estimators*

In addition to the three estimators discussed in sections 8.2−8.4, there are many other estimators. In this section, we will first introduce estimators using the re-sampling method: the bootstrap estimator and the jackknife estimator. We will also briefly describe some non-parametric estimators: the least squares estimator, the L^I estimator, the M-estimator and the L-estimator.

The method of moments estimator, the MLE and the Bayes estimator require the specification of the functional form. The non-parametric estimator does not require a known functional form. Although the MLE is a most commonly-used method, it may not be appropriate for every parameter estimation, since it is not necessary that every random sample has a distribution with a known functional form. Among those that have a known distributional form and are suitable for the likelihood approach, there are several different modifications of the likelihood-based estimator such as the quasi-likelihood estimator, profile likelihood estimator and partial likelihood estimator (see Murphy and van der Vaart [2000]).

For Bayesian method, in addition to the Bayes estimator or generalized maximum likelihood estimator discussed in section 8.3, there are several other estimators such as the robust Bayes estimator, empirical Bayes estimator and Hierarchical Bayes estimator. Interested readers are referred to books such as Carlin and Louis (2000).

The *bootstrap estimator* depends on replications of the bootstrap sample defined below. Given a random sample $x_1, \ldots, x_n$, a *bootstrap sample* is defined to be a random sample of size n taken from a population distributed as the empirical distribution F_n formed by the given sample. By the definition in Chapter 5, F_n assigns probability $\dfrac{1}{n}$ on each observed value x_i. In other words, a bootstrap sample $\{x_1^*, x_2^*, \cdots, x_n^*\}$ is a sample taken with replacement from the population of n objects $\{x_1, \ldots, x_n\}$.

Using the bootstrap sample, we can create the plug-in estimate of a parameter θ. For example, if θ is the mean, $\theta = \int x \, dF(x)$ and the plug-in estimate is $\hat{\theta}^* = \int x \, dF_n(x) = \dfrac{1}{n}\sum_{i=1}^{n} x_i^*$, where $\theta = \int x \, dF(x)$ and $\int x \, dF_n(x)$ are Riemann-Stieltjes integrals. If we repeat the bootstrap sample B times, the bootstrap estimate for the variance can be defined by

$$\frac{1}{B-1}\sum_{i=1}^{B}(\hat{\theta}_i^* - \hat{\theta}_{\cdot}^*)^2, \text{ where } \hat{\theta}_{\cdot}^* = \frac{1}{B}\sum_{i=1}^{B}\hat{\theta}_i^*.$$

The *jackknife estimate* of the variance is defined by

$$\frac{n-1}{n}\sum_{i=1}^{n}(\hat{\theta}_{(i)} - \hat{\theta}_{(\cdot)})^2, \text{ where } \hat{\theta}_{(\cdot)} = \frac{1}{n}\sum_{i=1}^{n}\hat{\theta}_{(i)} \text{ and } \hat{\theta}_{(i)} = \frac{1}{n-1}\sum_{k=1,k\neq i}^{n}x_k^*.$$

The *least squares estimator* minimizes the sum of squares of the deviations of the observed values from the parameter $\sum_{i=1}^{n}(X_i - \theta)^2$. It is often used to estimate the parameters in regression analysis. If $U_1,\ldots,U_k$, called *independent variables*, are believed or are suspected of having a certain relationship with X, called the *dependent variable*, the least squares method is usually applied to estimate parameters by minimizing $\sum_{i=1}^{n}(x_i - h(u_{i1},\ldots,u_{ik};\boldsymbol{\theta}))^2$, where u_{ij}, $i = 1,\cdots,n$ and $j = 1,\cdots,k$, are observations from U_j, and x_i are observations from X.

The L^1 *estimator* minimizes the sum of the absolute deviation between the observed values and the parameter $\sum_{i=1}^{n}|X_i - \theta|$. Like the least squares estimator, the L^1 estimator is also applied to find the relationship between the dependent variable and independent variables. Although it is more intuitive, L^1 estimator does not enjoy as many good mathematical properties as does the least squares estimator. For example, we can not find the derivative of $\sum_{i=1}^{n}|X_i - \theta|$ with respect to θ at every point of the parameter space.

A way to generalize the least squares estimator and the L^1 estimator is to create a general "distance function" for minimization. An estimator, called the *M-estimator* is the value that minimizes $\sum_{i=1}^{n}Q(x_i, h(u_{i1},\ldots,u_{ik};\boldsymbol{\theta}))$, for some function of two arguments $Q(x,v)$. It is equivalent to solving the equation

$$\sum_{i=1}^{n}\frac{\partial}{\partial v}Q(x_i, h(u_{i1},\ldots,u_{ik};\boldsymbol{\theta})) = 0 \qquad (8.16)$$

For the general case, we define the M-estimator as the solution to the equation

$$\sum_{i=1}^{n} \psi(x_i, h(u_{i1}, \ldots, u_{ik}; \boldsymbol{\theta})) = 0.$$

If, in Equation (8.16), we let $Q(x, v)$ be the square of the difference between x and v, then this M-estimator is the same as the least square estimator. If we let $Q(x, v)$ be the absolute difference between x and v, then this M-estimator is the same as the L^1 estimator. If the distribution of the sample is known as $f(x, u)$ and we choose $Q(x, v) = -\log(f(x, u))$, then the M-estimator is the same as the MLE. The well known Huber estimator choosing $Q(x, v)$ as

$$Q(x, v) = \begin{cases} \dfrac{1}{2}(x - v)^2 & if \ |x - v| \le K \\ K|x - v| - \dfrac{1}{2}K^2 & if \ |x - v| > K \end{cases}$$

is a special case of the M-estimator.

An *L-estimator* is a linear combination of order statistics $\breve{\theta} = \sum_{i=1}^{n} c_i X_{(i)}$. It is most useful for the estimation of the location parameter θ through it the expression of the pdf of the observation can be written as a function of $x - \theta$, i.e. $f(x|\theta) = g(x - \theta)$. By properly choosing c_i, $\breve{\theta}$ can be the sample mean $(c_i = \dfrac{1}{n})$, the minimum $(c_1 = 1$ and $c_i = 0, i \ne 0)$, the maximum $(c_n = 1$ and $c_i = 0, i \ne n)$ or any other sample quantiles. It is very useful in estimation when the samples include outlying or inferential observations.

8.8 Criteria of a Good Point Estimator

In previous sections, we have introduced some techniques for finding point estimators of parameters. We have also demonstrated that we can usually apply more than one of these methods in a given scenario. Sometimes, different estimation methods may result in the same estimate. For example, both the MLE and the moment estimate for the mean of a normal population are the sample average (Examples 8.1 & 8.9). Most of the times, point estimates obtained from different methods will not be identical. For example, the MLE and the Bayes estimator of the mean (with a known variance) of a normal population are different (see Examples 8.9 and 8.17). In this situation, we may face a challenge of choosing a good estimator. The first task for us is to determine the criteria of a good estimator. Although an ideal point estimate is one which is perfectly accurate (i.e.

$\Pr(T(X_1,\ldots,X_n)=\theta)=1$) and absolutely precise (i.e. $Var(T(X_1,\ldots,X_n))=0$), in practice it is impossible to find such an estimate unless it is in a trivial case. In this section, we will present some criteria for comparing two estimators and examine several estimators against these criteria.

8.8.1 The Bias and the Mean Squared Error

A common measurement of the accuracy of an estimator is the *bias*, defined as the expected difference between the estimator and the parameter it targets. Mathematically, the bias of $T(X_1,\ldots,X_n)$ can be described as

$$b_\theta(T) = E(T(X_1,\ldots,X_n)|\theta) - \theta.$$

The bias measures the magnitude and direction of the expected location of the point estimator away from its target. An estimator whose bias is identically equal to 0, is called *unbiased*. Therefore, if $T(X_1,\ldots,X_n)$ is an unbiased estimator, we have $E(T(X_1,\ldots,X_n)|\theta) = \theta$, for all θ. In estimating the population mean μ, the sample mean $\bar{X}$ is an unbiased estimator since $E(\bar{X}) = \mu$. Similarly, a single sample observation X_1 is also an unbiased estimator since $E(X_1) = \mu$. If we are faced with a choice of two estimators that have the same accuracy using the bias criterion, we may compare the precision, for example,

$$\frac{\sigma^2}{n} = Var(\bar{X}) < Var(X_1) = \sigma^2, \text{ for } n > 1.$$

A criterion for comparing estimators called the *mean squared error* takes into consideration both accuracy and precision.

Definition 8.8 The *mean squared error* (MSE) of an estimator $T(X_1,\ldots,X_n)$ of a parameter θ is defined as $E_\theta[T(X_1,\ldots,X_n)-\theta]^2$.

It is noted that MSE is a function of θ and will be denoted as $R(\theta,T)$, as defined by decision theorists. $R(\theta,T)$ can be expressed as

$$R(\theta,T) = Var_\theta[T(X_1,\ldots,X_n)] + [E_\theta T(X_1,\ldots,X_n)-\theta]^2, \tag{8.17}$$

where $Var_\theta[T(X_1,\ldots,X_n)]$ measures precision of the estimator and $[E_\theta T(X_1,\ldots,X_n)-\theta]^2$ is the square of the bias that is a measure of the accuracy.

If two estimators $S(X_1,\ldots,X_n)$ and $T(X_1,\ldots,X_n)$ are such that $R(\theta,S) \le R(\theta,T)$ for all θ with strict inequality holding for some θ, then the estimator $T(X_1,\ldots,X_n)$ will not be reasonably acceptable and is called *inadmissible*.

Example 8.37 Let $X_1,\ldots,X_n$ be a random sample taken from a population with a distribution with unknown mean μ and unknown variance σ^2. Assume that the distributional form of X_i is also not known. Then the sample mean $\bar{X}$ is an unbiased estimator of μ and its MSE is $E(\bar{X}-\mu)^2 = \text{var}(\bar{X})+0 = \dfrac{\sigma^2}{n}$. The sample variance $S^2 = \dfrac{1}{n-1}\sum_{i=1}^{n}(X_i-\bar{X})^2$ is also an unbiased estimator of σ^2, since

$$E\left(\sum_{i=1}^{n}(X_i-\bar{X})^2\right) = E\left(\sum_{i=1}^{n}(X_i-\mu+\mu-\bar{X})^2\right) = E\left(\sum_{i=1}^{n}(X_i-\mu)^2\right) - nE(\mu-\bar{X})^2$$
$$= (n-1)\sigma^2.$$

The MSE of S^2 is then equal to its variance that was shown in section 3.8.1 as $Var(S^2) = \dfrac{1}{n}\left(\kappa - \dfrac{n-3}{n-1}(\sigma^2)^2\right)$, where $\kappa = E(X-\mu)^4$ is the kurtosis of the distribution.

If we assume the population is normally distributed, the MLE $\hat{\sigma}^2$ (see Example 8.10) for σ^2 is not unbiased. Its bias is equal to $-\dfrac{\sigma^2}{n}$. The MSE of $\hat{\sigma}^2$ can be calculated as $\dfrac{(\sigma^2)^2(2n-1)}{n^2}$ using the following calculations:

1. $\dfrac{n\hat{\sigma}^2}{\sigma^2}$ has a chi-squared distribution with $n-1$ degrees of freedom.

2. The variance of a chi-square random variable χ_p^2 is $2p$.

3. $Var(\hat{\sigma}^2) = Var(\dfrac{\sigma^2}{n}\dfrac{n\hat{\sigma}^2}{\sigma^2}) = \left(\dfrac{\sigma^2}{n}\right)^2 2(n-1)$.

4. Equation (8.17) and $(bias)^2 = \left(\dfrac{-\sigma^2}{n}\right)^2$.

A question raised is "is the MSE a good criterion for comparing estimators?" Although the MSE takes into consideration both accuracy and precision, it depends on the parameter or parameters. To compare two estimators using the MSE is essentially comparing two functions of the parameter. It may happen that one estimator is smaller than the other in the MSE for some portion of the range of the parameter and is larger than the same estimator for the other portion of the range. Consider an estimator $\frac{1}{4}\bar{X}$ of the population mean μ. The MSE is

$$E(\frac{1}{4}\bar{X} - \mu)^2 = \frac{1}{16}\frac{\sigma^2}{n} + \frac{9}{16}\mu^2.$$

It is clear that, when μ is small ($\mu^2 < \frac{15\sigma^2}{9n}$), $\frac{1}{4}\bar{X}$ is better than $\bar{X}$ and, when μ is large, the relationship reverses. Figure 8.2 illustrates the comparison of the MSE between $\bar{X}$ and $\frac{1}{4}\bar{X}$ when $\sigma^2 = 1$ and $n = 20$. We can even consider a more extreme estimator $T(X_1, \ldots, X_n) = 7$ that has a perfect precision. This estimator is better than $\bar{X}$ or any other estimators when $\mu = 7$, but performs poorly in general. Up to here, we agree that the MSE is a reasonable criterion for comparing estimators, but it is not always satisfactory. The problem for the MSE may be that there are too many types of estimators. The MSE may not be good for comparing them. However, if we focus on the estimators that are unbiased, the MSE may be a more practical criterion.

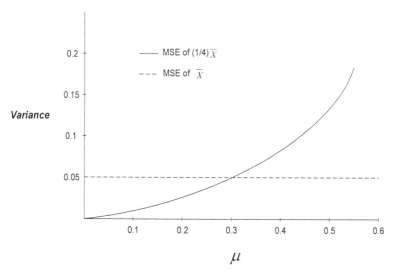

Figure 8.2 Comparison of MSE between $\bar{X}$ and $\frac{1}{4}\bar{X}$ when $\sigma^2 = 1$ and $n = 20$

Definition 8.9 An unbiased estimator $T^*(X_1,\ldots,X_n)$ of θ is called a uniformly minimum variance unbiased estimator (UMVUE), if, for any other unbiased estimator $T(X_1,\ldots,X_n)$ of θ, $Var(T^*|\theta) \le Var(T|\theta)$ for all θ.

The definitions of the bias, the unbiased estimator, the MSE and the UMVUE can be modified if the parameter to be estimated is a function of θ, say $\tau(\theta)$. For example,

$$\text{bias} = E\big(T(X_1,\ldots,X_n)\big) - \tau(\theta).$$

By focusing on the unbiased estimators, we may be able to find the best unbiased estimator using the notion of the UMVUE. However, there are still some problems. First, an unbiased estimator may not always exist. In the case that an UMVUE exists, it may not be admissible. That means we may find an estimator T_0 with the mean squared error smaller than or equal to this UMVUE at any point of the parameter with strict inequality for some points. In this case, T_0 is not unbiased. This dilemma may lead us to face a difficult choice for the criterion of a good estimator: UMVUE or a biased estimator with a smaller MSE. Moreover, to find an UMVUE or even just to verify if an unbiased estimator is an UMVUE, is usually strenuous and sometimes impossible. It involves calculating the variances of the candidate estimator and an arbitrary estimator without a specific form. This can be complex as shown in the calculation of the MSE of the sample variance in Example 8.37.

If we are to avoid calculating and comparing the variance of an arbitrary estimator, there are two ways we can find an UMVUE. One is to find the lower bound of the variances of all unbiased estimators and then show that a candidate unbiased estimator has the variance equal to this lower bound. The other is to find a way that can create a variance-reduced estimator and show that, under some conditions, it can reach an UMVUE.

8.8.2 Variance Reduction for an Estimator

To find an estimator that can reduce the variance and keep the same bias, let us go back to the general case. Consider an estimator $T(X_1,\ldots,X_n)$ for $\tau(\theta)$ and a sufficient statistic $S(X_1,\ldots,X_n)$ for θ (of course, S is also a sufficient statistic for $\tau(\theta)$). It is obvious from the tower property, that $E\big(E(T|S)\big) = E(T)$ (Chapter 3). Furthermore, $Var\big(E(T|S)\big) \le Var(T)$ when $E\big(|T(X_1,\ldots,X_n)|\big) < \infty$. It can also be proved that $E(T|S)$ is indeed an estimator for $\tau(\theta)$, that means $E(T|S)$ is not a function of θ. We can conclude that the MSE for $E(T|S)$ is less than the MSE for T.

We may now state the Rao-Blackwell theorem.

Theorem 8.2 (Rao-Blackwell Theorem) Suppose $S(X_1,\ldots,X_n)$ is a sufficient statistic for θ and $T(X_1,\ldots,X_n)$ is an estimator of $\tau(\theta)$ such that

$$E\left(\left|T(X_1,\ldots,X_n)\right|\right) < \infty \text{ for all } \theta.$$

Then $S^*(X_1,\ldots,X_n) = E\left(T(X_1,\ldots,X_n)\big|S(X_1,\ldots,X_n)\right)$ is an estimator of $\tau(\theta)$ and $E\left(S^*(X_1,\ldots,X_n)-\tau(\theta)\right)^2 \le E\left(T(X_1,\ldots,X_n)-\tau(\theta)\right)^2$, for all θ.

Moreover, the inequality is strict for some θ unless T is an unbiased estimator of $\tau(\theta)$ and is also a function of the sufficient statistic S.

The Rao-Blackwell theorem suggests that an estimator that is not a function of a sufficient statistic is inadmissible. We can always construct a better estimator by taking expectation of this estimator conditioning it on a sufficient statistic. If $T(X_1,\ldots,X_n)$ is an unbiased estimator, the constructed estimator $S^*(X_1,\ldots,X_n)$ is also unbiased. It may well be a candidate for the UMVUE. However, the Rao-Blackwell theorem does not tell us which one of the constructed $S^*(X_1,\ldots,X_n)$ is better or if they are the same, when we start with two different unbiased estimators. Lehmann and Scheffé applied and extended this result by showing that their variances are identical, under some conditions. Their theorem can be used for finding an UMVUE.

This theorem requires the definition of a complete family of distributions.

Definition 8.10 A family of distributions $f(x|\theta)$ (pmf or pdf) is called *complete* if, $E\left(U(X)|\theta\right)=0$ for all θ implies $U(X)=0$ with probability 1 for all θ. A statistic whose distribution is a member of a complete family is called a complete statistic.

Example 8.38 Let $X_1,\ldots,X_n$ be a random sample from a *Poisson*(λ) population. From Example 8.33, we know that $\sum_{i=1}^{n} X_i$ is a sufficient statistic. From a result in Chapter 4, $T(X_1,\ldots,X_n) = \sum_{i=1}^{n} X_i$ has a Poisson distribution with mean $n\lambda$. If there is a function $g(\ )$ such that $E\left(g(T)|\lambda\right)=0$ for all $\lambda > 0$, then

$$e^{-n\lambda} \sum_{k=0}^{\infty} \frac{g(k)(n\lambda)^k}{k!} = 0 \text{ for all } \lambda > 0. \text{ Since } e^{-n\lambda} \ne 0 \text{ for any } \lambda, \text{ we can obtain}$$

$\sum_{k=0}^{\infty} \dfrac{g(k)(n\lambda)^k}{k!} = 0$ for all $\lambda > 0$. Letting $h(\lambda) = \sum_{k=0}^{\infty} \dfrac{g(k)(n\lambda)^k}{k!}$, the Taylor series

expansion gives $h(\lambda) = \sum_{k=0}^{\infty} \dfrac{h^{(k)}(0)\lambda^k}{k!}$ that implies $h^{(k)}(0) = g(k)n^k$ for all k.

Since $h(\lambda) \equiv 0$, we can conclude $h^{(k)}(0) = 0$ for all k, and hence $g(k) = 0$ for all non-negative integer k. This implies $g(X) = 0$ with probability 1, since the support of the Poisson distribution is the set of all non-negative integers. Therefore, $T(X_1, \ldots, X_n)$ is a complete sufficient statistic.

Using the notion of completeness, we revisit the Rao-Blackwell theorem by assuming S is a complete sufficient statistic. If there are two unbiased estimators U_1 and U_2 of $\tau(\theta)$, then $g_1(S) = E(U_1|S)$ and $g_2(S) = E(U_2|S)$ are both unbiased. Furthermore, $E\big((g_1 - g_2)(S)|\theta\big) = E\big(g_1(S)|\theta\big) - E\big(g_2(S)|\theta\big) = 0$ for all θ. The property of completeness implies that $(g_1 - g_2)(S) = 0$ or equivalently, $g_1(S) = g_2(S)$ (with probability 1). This means that the unbiased estimator created by taking the expectation of an unbiased estimator U conditioning on a complete sufficient statistic is independent of the choice of U.

We now state the Lehmann-Scheffé theorem.

Theorem 8.3 (Lehmann-Scheffé Theorem) Suppose $S(X_1, \ldots, X_n)$ is a complete sufficient statistic for θ and $U(X_1, \ldots, X_n)$ is an unbiased estimator of $\tau(\theta)$, then $S^*(X_1, \ldots, X_n) = E\big(U(X_1, \ldots, X_n)|S(X_1, \ldots, X_n)\big)$ is a UMVUE of $\tau(\theta)$. If $Var\big(S^*(X_1, \ldots, X_n)\big) < \infty$ for all θ, then $S^*(X_1, \ldots, X_n)$ is the unique UMVUE.

Therefore, if we can find an unbiased estimator that is a function of a complete sufficient statistic, then this unbiased estimator is an UMVUE, since $E\big(g(T(X_1, \ldots, X_n))|T(X_1, \ldots, X_n)\big) = g(T(X_1, \ldots, X_n))$. If we have an unbiased estimator and a complete sufficient statistic, we can construct a UMVUE by taking the expectation of the unbiased estimator conditioning on this complete sufficient statistic.

Example 8.39 Let $X_1, \ldots, X_n$ be a random sample from a population having a Poisson distribution with mean λ. From Example 8.38, we know that $\sum_{i=1}^{n} X_i$ is complete sufficient statistic. From Chapter 3, we also know that $I_{\{0\}}(X_1)$ is an unbiased estimator of $P(X = 0) = e^{-\lambda}$, where $I_{\{0\}}(X_1)$ defined as

$$I_{\{0\}}(X_1) = \begin{cases} 1 & if\ X_1 = 0 \\ 0 & if\ X_1 \neq 0 \end{cases}.$$

Using the Lehmann-Scheffé theorem, we can conclude that $E\left(I_{\{0\}}(X_1)\Big|\sum_{i=1}^{n}X_i\right)$ is the UMVUE. From Problem 4.25, we know that $X_1\Big|\sum_{i=1}^{n}X_i = m$ has a binomial distribution $B\left(m, \dfrac{1}{n}\right)$. Now, we examine this conditional expectation

$$E\left(I_{\{0\}}(X_1)\Big|\sum_{i=1}^{n}X_i = m\right) = P\left(X_1 = 0\Big|\sum_{i=1}^{n}X_i = m\right) = \left(1 - \frac{1}{n}\right)^m.$$

Therefore, we conclude that the UMVUE of $e^{-\lambda}$ is $\left(1 - \dfrac{1}{n}\right)^{\sum_{i=1}^{n}X_i}$.

Example 8.40 Let $X_1, \ldots, X_n$ be n independent Bernoulli trials with probability of success θ, where $0 < \theta < 1$. From Example 8.24, $S(X_1, \ldots, X_n) = \sum_{i=1}^{n}X_i$ is a sufficient statistic. If $E(g(S)|\theta) = 0$ for all θ, then $\sum_{k=0}^{n}g(k)\binom{n}{k}\theta^k(1-\theta)^{n-k} = 0$, which implies $(1-\theta)^n\sum_{k=0}^{n}g(k)\binom{n}{k}\rho^k = 0$, where $\rho = \dfrac{\theta}{1-\theta}$. Since $(1-\theta)^n \neq 0$, the polynomial $\sum_{k=0}^{n}g(k)\binom{n}{k}\rho^k$ of ρ equals to 0 and thus $g(k)\binom{n}{k} = 0$ for all k. That means $g(k) = 0$ for all k. $S(X_1, \ldots, X_n)$ is now a complete sufficient statistic. By the Lehmann-Scheffé theorem, $\overline{X}$ that is a function of $S(X_1, \ldots, X_n)$ is the unique UMVUE since it is unbiased and has a finite variance.

Example 8.41 Let $X_1, \ldots, X_n$ be a random sample taken from a population having a uniform distribution with support in $(0, \theta)$. Similar to the result derived in Example 8.27, $S(X_1, \ldots, X_n) = X_{(n)}$, the nth order statistic, is a sufficient statistic. From a result in Chapter 6, the pdf of $X_{(n)}$ is $f(x) = \dfrac{n}{\theta}\left(\dfrac{x}{\theta}\right)^{n-1}$ for $0 < x < \theta$.

Therefore, if we assume $g(\ )$ is a continuous function, we can calculate

$$E\left(g(X_{(n)})\middle|\theta\right) = \frac{n}{\theta^n}\int_0^\theta g(x)x^{n-1}dx = 0, \text{ for all } \theta > 0 \text{ that implies } g(X_{(n)}) = 0. \text{ This}$$

can be shown by differentiating both sides of $\int_0^\theta g(x)x^{n-1}dx = 0$ with respect to θ,

and getting $g(\theta)\theta^{n-1} = 0$ for all $\theta > 0$. For the case when g is not continuous, the property that $E\left(g(X_{(n)})\middle|\theta\right) = 0$ implies $g(X_{(n)}) = 0$ still holds. However, its derivation requires knowledge of Lebesque integration theory which is beyond the scope of this book.

Since $E\left(X_{(n)}\middle|\theta\right) = \int_0^\theta n\left(\frac{x}{\theta}\right)^n dx = \frac{n\theta}{n+1}$, the estimator $\frac{n+1}{n}X_{(n)}$ is an

unbiased estimator that is also a function of a complete sufficient statistic. By the

Lehmann-Scheffé theorem, $\frac{n+1}{n}X_{(n)}$ is UMVUE.

From the above example, we can sense that completeness is not easy to prove. However, there is a natural complete statistic in the distribution that is a member of the exponential family. In a random sample taken from a distribution in a k-parameter exponential family described in Equation (8.13), the statistic

$$(\sum_{j=1}^n T_1(X_j), \cdots, \sum_{j=1}^n T_k(X_j)) \text{ is complete and sufficient, under some regularity con-}$$

ditions. This condition actually states that the range of $c_1(\theta), c_2(\theta), \cdots, c_k(\theta)$ contains a k-dimensional open interval. This condition is usually satisfied in distributions with continuous parameter space and would not be emphasized in the examples shown below.

Example 8.42 Consider the random sample taken from a normal population with both mean μ and variance σ^2 unknown. Example 8.35 and the accompanying

remark show that $(\sum_{i=1}^n T_1(X_i), \sum_{i=1}^n T_2(X_i)) = (\sum_{i=1}^n X_i, \sum_{i=1}^n X_i^2)$ in this 2-parameter ex-

ponential family is a complete sufficient statistic for (μ, σ^2). Example 8.37

shows that the sample mean $\bar{X}$ and the sample variance $S^2 = \frac{1}{n-1}\sum_{i=1}^n (X_i - \bar{X})^2$

are unbiased estimators of μ and σ^2 respectively. Since both are also functions

of the complete sufficient statistic $(\sum_{i=1}^n X_i, \sum_{i=1}^n X_i^2)$, $(\bar{X}, S^2)$ are the UMVUE of

(μ, σ^2).

Example 8.43 Let $X_1, \ldots, X_n$ be a random sample taken from a $N(\mu, 1)$ population. From Example 8.30, we know that $N(\mu, 1)$ is a member of one parameter exponential family with $T(X_1, \ldots, X_n) = \sum_{i=1}^{n} X_i$. Following the remark shown after Example 8.41, we can also conclude that $\sum_{i=1}^{n} X_i$ is a complete sufficient statistic. If we are interested in estimating $\Phi(\mu) = P(X \geq 0 | \mu)$, we first identify a simple unbiased estimator

$$I_{(0,\infty)}(X_1) = \begin{cases} 1 & if \ X_1 \geq 0 \\ 0 & if \ X_1 < 0 \end{cases}.$$

From the Lehmann-Scheffé theorem, we can show that $E\left(I_{[0,\infty)}(X_1) \bigg| \sum_{i=1}^{n} X_i\right)$ is the UMVUE.

Note that, from Problem 5.34, the conditional distribution of X_1, given $\sum_{i=1}^{n} X_i = t$, is distributed as $N\left(\dfrac{t}{n}, \sigma^2(1 - \dfrac{1}{n})\right)$. Therefore, since $\sigma^2 = 1$,

$$E\left(I_{(0,\infty)}(X_1) \bigg| \sum_{i=1}^{n} X_i = t\right) = P\left[N\left(\frac{t}{n}, (1 - \frac{1}{n})\right) \geq 0\right]$$

$$= P\left(Z \geq \frac{-\dfrac{t}{n}}{\sqrt{(1 - \dfrac{1}{n})}}\right) = \Phi\left(\frac{\dfrac{t}{n}}{\sqrt{(1 - \dfrac{1}{n})}}\right),$$

using the property of the standard normal cdf.

In summary, our UMVUE of $\Phi(\mu)$ is $\Phi\left(\dfrac{\bar{X}}{\sqrt{(1 - \dfrac{1}{n})}}\right)$.

Example 8.44 Let $X_1, \ldots, X_n$ be random sample taken from a $gamma(\alpha, \beta)$ distribution. The joint pdf can be written as

$$f_{X_1,\ldots,X_n}(x_1,\ldots,x_n) = \exp\left((\alpha-1)\log(\prod_{i=1}^n x_i) - \beta\sum_{i=1}^n x_i - n\log\Gamma(\alpha) + \alpha n\log(\beta)\right)$$

.

It is clear that this distribution is a member of the exponential family and the statistic $\left(\log(\prod_{i=1}^n x_i), \sum_{i=1}^n x_i\right)$ is a complete sufficient statistic for (α,β). Furthermore, $\bar{X}$ is an unbiased estimator of $\dfrac{\alpha}{\beta}$ since $E(\bar{X}) = \dfrac{\alpha}{\beta}$ (see Chapter 5).

Therefore, $\bar{X}$ is an UMVUE of $\dfrac{\alpha}{\beta}$, by the Lehmann-Scheffé theorem .

8.8.3 The Lower Bound of the Variance of an Estimator

At the end of subsection 8.8.1, we suggest two approaches for finding an UM-VUE that will not require calculating and comparing the variance of an arbitrary estimator. In subsection 8.8.2, we have presented methods that can create a variance-reduced estimator from an arbitrary estimator. These methods will help find an UMVUE from an unbiased estimator. In this subsection, we will derive the lower bound of the variance of an estimator. If an unbiased estimator has the variance that attains this lower bound, then this unbiased estimator will be the UMVUE.

For a given random sample $X_1,\ldots,X_n$ with pdf or pmf $f(x|\theta)$, consider the statistic $\dfrac{\partial}{\partial\theta}\log\left(\prod_{i=1}^n f(X_i|\theta)\right)$ that is known as the *score statistic* and is denoted by $S(X_1,\ldots,X_n|\theta)$. When the parameter is a vector, the score statistic is defined as the gradient vector of the log-likelihood with respect to $\boldsymbol{\theta}$ and is referred to as the *score vector* (see also Example 8.13). If we assume that the score statistic is a smooth function of θ and all the interchanges of expectations, summations and differentiations do not cause any mathematical problem, then the mean of the score statistic can be calculated as

$$E\left[S(X_1,\ldots,X_n|\theta)\right] = E\left(\sum_{i=1}^n \frac{\partial}{\partial\theta}\log\left(f(X_i|\theta)\right)\right) = E\left(\sum_{i=1}^n \frac{\frac{\partial}{\partial\theta}\left(f(X_i|\theta)\right)}{f(X_i|\theta)}\right).$$

Since the derivatives in the above equation are finite, Fubini's theorem allows us to exchange the expectation and the summation. We can therefore obtain

$$E\left[\sum_{i=1}^{n}\frac{\frac{\partial}{\partial\theta}\bigl(f(X_i|\theta)\bigr)}{f(X_i|\theta)}\right]=\sum_{i=1}^{n}E\left[\frac{\frac{\partial}{\partial\theta}\bigl(f(X_i|\theta)\bigr)}{f(X_i|\theta)}\right].$$

Since

$$E\left[\frac{\frac{\partial}{\partial\theta}\bigl(f(X_i|\theta)\bigr)}{f(X_i|\theta)}\right]=\int\frac{\frac{\partial}{\partial\theta}\bigl(f(x_i|\theta)\bigr)}{f(x_i|\theta)}f(x_i|\theta)\,dx_i=\int\frac{\partial}{\partial\theta}\bigl(f(x_i|\theta)\bigr)dx_i=0\qquad(8.18)$$

the mean of the score statistic is $E\bigl[S(X_1,\dots,X_n|\theta)\bigr]=0.$

The validity of the last two equalities in (8.18) is due to the fact that $\int f(x_i|\theta)dx_i=1$ and hence $\dfrac{\partial}{\partial\theta}\int f(x_i|\theta)dx_i=0.$

The variance of the score statistic is called the *Fisher information* and can be expressed as

$$Var\bigl(S(X_1,\dots,X_n|\theta)\bigr)=\sum_{i=1}^{n}Var\left[\frac{\partial}{\partial\theta}\log\bigl(f(X_i|\theta)\bigr)\right]=nE\left[\frac{\partial}{\partial\theta}\log\bigl(f(X_1|\theta)\bigr)\right]^2,$$

since $X_1,\dots,X_n$ are independent and the covariance of $\dfrac{\partial}{\partial\theta}\log\bigl(f(X_i|\theta)\bigr)$ and $\dfrac{\partial}{\partial\theta}\log\bigl(f(X_j|\theta)\bigr)$ is equal to 0 for $i\neq j.$

Note that the Fisher information is a function of the parameter θ and sample size n and is usually denoted by $I_n(\theta)$. For $n=1$, we denote $I(\theta)=I_1(\theta)$. The variance-covariance matrix of the score vector is referred to as the *information matrix*.

Consider the covariance of the score statistic and another statistic $T(X_1,\dots,X_n)$. Since $E\bigl(S(X_1,\dots,X_n|\theta)\bigr)=0,$ the product of the expectations of $T(X_1,\dots,X_n)$ and $S(X_1,\dots,X_n)$ is 0. We can calculate their covariance directly as the expectation of the product:

$$Cov\Big(T(X_1,\ldots,X_n),S(X_1,\ldots,X_n|\theta)\Big)$$

$$= E\left[T(X_1,\ldots,X_n)\frac{\partial}{\partial\theta}\log\left(\prod_{i=1}^{n}f(X_i|\theta)\right)\right]$$

$$= \int T(x_1,\ldots,x_n)\frac{\dfrac{\partial}{\partial\theta}\prod_{i=1}^{n}f(x_i|\theta)}{\prod_{i=1}^{n}f(x_i|\theta)}\prod_{i=1}^{n}f(x_i|\theta)dx_1,\ldots,x_n$$

$$= \frac{\partial}{\partial\theta}\int T(x_1,\ldots,x_n)\prod_{i=1}^{n}f(x_i|\theta)dx_1,\ldots,x_n$$

$$= \frac{\partial}{\partial\theta}E\Big(T(X_1,\ldots,X_n)|\theta\Big).$$

Recall the Cauchy-Schwarz Inequality in Chapter 3 states that, for any two random variables V and W,

$$(VarV)(VarW) \geq \big(Cov(V,W)\big)^2$$

or equivalently, $VarV \geq \dfrac{\big(Cov(V,W)\big)^2}{VarW}.$

Letting $V = T(X_1,\ldots,X_n)$ and $W = S(X_1,\ldots,X_n|\theta)$, we have

$$Var\big(T(X_1,\ldots,X_n)\big) \geq \frac{\left(\dfrac{\partial}{\partial\theta}E\Big(T(X_1,\ldots,X_n)|\theta\Big)\right)^2}{nE\left(\dfrac{\partial}{\partial\theta}\log\big(f(X_1|\theta)\big)\right)^2}. \tag{8.19}$$

Although the derivation of Inequality (8.19) assumes that the distribution is continuous, the result is also true for the discrete case when we substitute the integration by the summation. We can now state the Cramér-Rao Inequality.

Theorem 8.4 (Cramér-Rao Inequality) Let $X_1,\ldots,X_n$ be iid with pdf or pmf $f(x|\theta)$ and $T(X_1,\ldots,X_n)$ be an estimator. Assume that the support of the distribution does not depend on the parameter θ. If the variance of T is finite and the operations of expectation and differentiation by θ can be interchanged in

$$\frac{\partial}{\partial \theta} E\left(T(X_1,\ldots,X_n)\middle|\theta\right), \text{ then inequality (8.19) holds.}$$

The right hand side of Inequality (8.19) is called the *Cramér-Rao lower bound*.

Corollary 8.5 If $T(X_1,\ldots,X_n)$ is an unbiased estimator of θ in the Cramér-Rao Inequality, then the variance of $T(X_1,\ldots,X_n)$ attains the Cramér-Rao lower bound if and only if

$$T(X_1,\ldots,X_n) = a(\theta)S(X_1,\ldots,X_n\middle|\theta) + b(\theta) \text{ with probability 1,}$$

for some functions $a(\theta)$ and $b(\theta)$.

The proof of Corollary 8.5 will be left as an exercise.

Note that the left hand side of the above equality does not depend on θ, so that the right hand side should not depend on θ either, i.e. the functions $a(\theta)$ and $b(\theta)$ cancel the dependence of $S(X_1,\ldots,X_n\middle|\theta)$ on the parameter θ.

If $T(X_1,\ldots,X_n)$ is an unbiased estimator of θ, then the right-handed side of inequality (8.19) becomes $\dfrac{1}{nE\left(\dfrac{\partial}{\partial \theta}\log\left(f(X_1\middle|\theta)\right)\right)^2}$ which is equal to

$\dfrac{1}{nI(\theta)}$. If $\mathbf{X} = (X_1,\ldots,X_n)$ are not independent, then Inequality (8.19) can be extended to

$$Var\left(T(\mathbf{X})\right) \geq \frac{\left(\dfrac{\partial}{\partial \theta} E\left(T(\mathbf{X})\middle|\theta\right)\right)^2}{E\left(\dfrac{\partial}{\partial \theta}\log\left(f(\mathbf{X}\middle|\theta)\right)\right)^2}.$$

The equality

$$E\left(\frac{\partial^2}{\partial \theta^2}\left(\log\left(f(X_1\middle|\theta)\right)\right)\right) = -E\left(\frac{\partial}{\partial \theta}\log\left(f(X_1\middle|\theta)\right)\right)^2 \tag{8.20}$$

is useful in numerical computation of the Cramér-Rao lower bound and will be derived below. From the differentiation

$$\frac{\partial^2}{\partial\theta^2}\big(\log\big(f(x|\theta)\big)\big) = \frac{\partial}{\partial\theta}\left[\frac{\frac{\partial}{\partial\theta}f(x|\theta)}{f(x|\theta)}\right] = \frac{\left[\frac{\partial^2}{\partial\theta^2}f(x|\theta)\right]f(x|\theta) - \left(\frac{\partial}{\partial\theta}f(x|\theta)\right)^2}{\big(f(x|\theta)\big)^2}$$

$$= \frac{\left[\frac{\partial^2}{\partial\theta^2}f(x|\theta)\right]}{f(x|\theta)} - \left(\frac{\frac{\partial}{\partial\theta}f(x|\theta)}{f(x|\theta)}\right)^2,$$

we can compute

$$E\left(\frac{\partial^2}{\partial\theta^2}\big(\log\big(f(X_1|\theta)\big)\big)\right) = \int \frac{\partial^2}{\partial\theta^2}f(x|\theta)dx - E\left(\frac{\frac{\partial}{\partial\theta}f(X_1|\theta)}{f(X_1|\theta)}\right)^2$$

$$= -E\left(\frac{\partial}{\partial\theta}\log\big(f(X_1|\theta)\big)\right)^2.$$

The last equality of the above equation is due to the facts that

$$\int \frac{\partial^2}{\partial\theta^2}f(x|\theta)dx = \frac{\partial^2}{\partial\theta^2}\int f(x|\theta)dx = \frac{\partial^2}{\partial\theta^2}(1) = 0 \text{ and}$$

$$E\left(\frac{\frac{\partial}{\partial\theta}f(X_1|\theta)}{f(X_1|\theta)}\right) = 0.$$

Hence Equality (8.20) is verified.

Example 8.45 Consider the sample mean $\bar{X}$ of n observations randomly taken from a $N(\mu,\sigma^2)$ population. Then

$$I(\theta) = E\left(\frac{\partial}{\partial\mu}\log\big(f(X|\mu)\big)\right)^2 = E\left(\frac{X-\mu}{\sigma^2}\right)^2 = \frac{\sigma^2}{\sigma^4} = \frac{1}{\sigma^2}.$$

By the Cramér-Rao Inequality, the lower bound of the variance for an unbiased estimator is $\frac{1}{nI(\theta)} = \frac{\sigma^2}{n}$. Since $\bar{X}$ is unbiased and $Var(\bar{X}) = \frac{\sigma^2}{n}$, we conclude that $\bar{X}$ is the UMVUE. This confirms the result in Example 8.42.

Example 8.46 Consider the sample proportion $\bar{X}$ of size n in a Bernoulli distribution with the rate of success θ. Then

$$
\begin{aligned}
I(\theta) &= E\left(\frac{\partial}{\partial \theta} \log\left(\theta^X (1-\theta)^{1-X}\right)\right)^2 \\
&= E\left(\frac{\partial}{\partial \theta}\left(X \log(\theta) + (1-X)\log(1-\theta)\right)\right)^2 \\
&= E\left(\frac{X}{\theta} - \frac{1-X}{(1-\theta)}\right)^2 \\
&= P(X=1)\frac{1}{\theta^2} + P(X=0)\frac{1}{(1-\theta)^2} \\
&= \frac{1}{\theta(1-\theta)}.
\end{aligned}
$$

By the Cramér-Rao Inequality, the lower bound of the variance for an unbiased estimator is $\dfrac{1}{nI(\theta)} = \dfrac{\theta(1-\theta)}{n}$. Since $\bar{X}$ is an unbiased estimator of θ and $Var(\bar{X}) = \dfrac{\theta(1-\theta)}{n}$, $\bar{X}$ is the UMVUE.

Example 8.47 Consider a random sample $X_1, \ldots, X_n$ taken from a distribution belonging to the one-parameter exponential family as described in (8.11). If $c(\theta)$ is continuous and $c'(\theta) \neq 0$ for all θ, then the variance of $\sum_{i=1}^{n} T(X_i)$ achieves the Cramér-Rao lower bound and is a UMVUE of $E\left(\sum_{i=1}^{n} T(X_i)\middle| \theta\right)$.

The proof of this example will be left as an exercise.

The Cramér-Rao inequality leads to a standardized comparison of the estimators.

Definition 8.11 An unbiased estimator $T(X_1, \ldots, X_n)$ that satisfies the condition in Theorem 8.4 and has the variance equal to the Cramér-Rao lower bound is called *efficient*. The ratio $\dfrac{\left.1\middle/nI(\theta)\right.}{Var(T(X_1,\ldots,X_n)|\theta)}$ of the Cramér-Rao lower bound

and the variance of $T(X_1,\ldots,X_n)$ is called the *efficiency* of $T(X_1,\ldots,X_n)$.

When we compare two unbiased estimators T_1 and T_2, the ratio of their

variances $\dfrac{Var(T_1|\theta)}{Var(T_2|\theta)}$ is called the *relative efficiency* of T_2 with respect to T_1.

Example 8.48 Let $X_1,\ldots,X_{2k}$ be a random sample taken from $N(\mu,\sigma^2)$ population. If we denote $\overline{X}_m, m=1,2,\cdots,2k,$ as the average of the first m samples,

then $Var(\overline{X}_m)=\dfrac{\sigma^2}{m}$ and $E(\overline{X}_m|\theta)=\theta.$ The relative efficiency of $\overline{X}_k$ with re-

spect to $\overline{X}_{2k}$ is $\dfrac{1}{2}$ and the efficiency of $\overline{X}_k$ is also $\dfrac{1}{2}$, since $\overline{X}_{2k}$ achieves the Cramér-Rao lower bound.

8.8.4 Consistency and Large Sample Comparisons

In the previous three subsections we have presented some criteria to evaluate point estimators and discussed some methods on finding the best unbiased estimator. All of the criteria we have discussed have been based on finite samples and finite-sample properties. In this subsection, the asymptotic properties that were discussed in Chapter 7 will be used to provide us with another viewpoint for comparing estimators. These criteria that allow the sample size to become infinite, can be useful when finite-sample comparisons fail or encounter some difficulty. However, the theoretical foundation for deriving these properties may involve some advanced mathematics that is beyond the scope of this text. We plan to present the definitions related to these large-sample criteria along with some basic properties. The proof and detailed derivations will be skipped.

In the study of the large sample property, although we may still refer to an estimator, we realistically consider a sequence of estimators of the same form. For example, when we examine the property of $\overline{X}$, we really investigate the asymptotic behavior of the sequence $X_1, \dfrac{X_1+X_2}{2}, \dfrac{X_1+X_2+X_3}{3},\cdots.$ From this prospective, one important property we expect an estimator to have is that, when the sample size becomes infinite the estimator converges to the parameter it targets.

Mathematically, an estimator $T_n=T(X_1,\ldots,X_n)$ of θ is called a *consistent* estimator, if T_n converges in probability to θ. In addition to checking the definition, some properties that we have known, can be used for showing consistency. Chebychev's Inequality is a common tool for examining consistency. We can also write down the MSE of the estimator as a sum of the variance and the square of the bias [see Equation (8.17)] and show both terms converge to 0.

Example 8.49 Consider the sample mean $\overline{X}$ in a population with finite mean μ and variance σ^2. Since $\overline{X}$ is an unbiased estimator of μ and $Var(\overline{X}) = \dfrac{\sigma^2}{n}$ converges to 0, $\overline{X}$ is a consistent estimator.

Certainly, a distribution of an estimator is informative. For the large sample case, under some conditions, an estimator can be approximated by a normal distribution. By treating an estimator as a sequence of random variables, the concepts of distributional approximation are almost identical to what we discussed in chapter 7.

For an estimator $T_n = T(X_1, \ldots, X_n)$ of θ, if there exist two sequences $\mu_n(\theta)$ and $\sigma_n^2(\theta)$ such that $\dfrac{T_n(X_1, \ldots, X_n) - \mu_n(\theta)}{\sigma_n(\theta)}$ converges to $N(0,1)$ in distribution, then T_n is called *asymptotic normal*. The sequences $\mu_n(\theta)$ and $\sigma_n^2(\theta)$ are not unique. If $n\sigma_n^2(\theta) \to \sigma^2(\theta) > 0$, we refer to $\sigma^2(\theta)$ as the *asymptotic variance* of T_n. If $\sqrt{n}\left(\mu_n(\theta) - \theta\right)$ converges to 0, then T_n is called *asymptotic unbiased*. If the asymptotic variance of T_n is equal to $\dfrac{1}{I(\theta)}$, then T_n is called *asymptotic efficient*.

It is known that the MLE is consistent, asymptotic normally distributed, asymptotic unbiased and asymptotic efficient.

Problems

1. Two observations are taken on a discrete random variable with pmf $p(x|\theta)$, where $\theta = 1$ or 2. (a) Find the MLE of θ on all possible samples, (b) Find the Bayes estimate when the prior of θ is defined as $\pi(1) = \dfrac{1}{3}$ and $\pi(2) = \dfrac{2}{3}$, (c) Is either one of them an unbiased estimator? and (d) Find the MSE for both estimators.

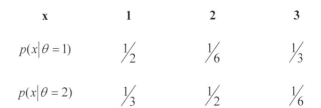

x	1	2	3
$p(x\vert\theta=1)$	$\frac{1}{2}$	$\frac{1}{6}$	$\frac{1}{3}$
$p(x\vert\theta=2)$	$\frac{1}{3}$	$\frac{1}{2}$	$\frac{1}{6}$

2. Let $X_1,\ldots,X_n$ be a random sample taken from $gamma(\alpha,\beta)$ distribution. Find the method of moment estimator for the unknown α and β and the MLE for β when α is known.

3. Assume that the occurrence of at least one near accident within three months after getting the first driver's license, has a Bernoulli distribution with parameter p. Twenty of 80 new drivers interviewed admitted at least one near accident within three months. (a) What is the MLE for p? (b) If $0.05 \le p \le 0.15$ is known, what is the MLE? (c) If p is uniformly distributed on $[0.05,0.15]$, what is the Bayes estimate for p?

4. Assume $X_1,\ldots,X_n$ is a random sample from the geometric distribution $p(x\vert\theta)=(1-\theta)^{x-1}\theta$ for $x=1,2,\ldots$ and $(=0$ otherwise) , where $\theta>0$. Find the method of moment estimator of θ.

5. Assume that the number of fertility cycles to conception follows a geometric distribution defined in problem 4. A partial data provided and analyzed by Morgan(2000), is listed below. Find the MLE of θ. (Hint: find the probability that conception takes more than 12 cycles).

Cycle	1	2	3	4	5	6	7	8	9	10	11	12	>12
Number of women	29	16	17	4	3	9	4	5	1	1	1	3	7

6. Let $X_1,\ldots,X_n$ be a random sample taken from a discrete unifom distribution $p(x\vert\theta)=\frac{1}{\theta}$ for $x=1,2,\ldots\theta$. Find the the method of moment estimator and the MLE for the unkown θ.

7. Let $X_1,\ldots,X_n$ be a random sample taken from a population with pdf defined as $f(x\vert\theta)=\frac{\theta}{x^2}\mathrm{I}_{[\theta,\infty)}(x)$ for some $\theta>0$. (a) Find the MLE for θ. (b) Find the bias for the MLE.

8. Let $x_1, \ldots, x_n$ be a random sample from a population with pdf $f(x|\theta)$. Show that the order statistics is a joint sufficient statistics for θ.

9. Let $X_1, \ldots, X_n$ be a random sample taken from a Poisson distribution with parameter λ. Find an unbiased estimator.

10. Let $X_1, \ldots, X_n$ be a random sample taken from a Bernoulli distribution with rate of success θ. (a) Find the MLE for $\theta(1-\theta)$ (b) Is the MLE unbiased?

11. In problem 10, show that $\dfrac{\left(\sum_{i=1}^{n} X_i\right)\left(n - \sum_{i=1}^{n} X_i\right)}{n(n-1)}$ is an unbiased estimator for $\theta(1-\theta)$.

12. (a) Find a UMVUE for $\theta(1-\theta)$ in problem 10. (b) Is this UMVUE achieve the Cramér-Rao lower bound?

13. Let N_1, N_2, N_3 be the observed counts from a multinomial distribution with 3 possible outcomes, where $N_1 + N_2 + N_3 = N$. Assume that p, q, and $1 - p - q$ are the occurrence probabilities on the event 1, event 2 and event 3, respectively. (a) Show that the likelihood function for the parameters (p,q) is $l(p,q) = C + N_1 \log p + N_2 \log q + N_3 \log(1 - p - q)$, where C is a constant. (b) Show that the MLE of (p,q) is $(\dfrac{N_1}{N}, \dfrac{N_2}{N})$.

14. Suppose that $X_1, \ldots, X_n$ form a random sample from a distribution for which the pdf is defined as follow: $f(x|\theta) = \dfrac{1}{2}\exp(-|x-\theta|)$ for $-\infty < x < \infty$. Find the MLE of θ.

15. A group of 25 rats under some medication in a lab is expected to survive for a few more days. Assume that their remaing life time (a continuous variable) follows a exponential distribution with parameter λ. The daily numbers of deaths were recorded up to 10 days. (a) Let N_g denote the number of deaths between the beginning of $(k-1)^{th}$ and the end of kth day (i.e. the time interval $(k-1,k]$), $k = 1,2,\ldots,10$ and N_{11} denote the number of rats surviving

more than 10 days. Show that $(N_1, N_2, \ldots, N_{10}, N_{11})$ is a sufficient statistic for λ, and (b) Find the MLE of λ based on the following data.

k	1	2	3	4	5	6	7	8	9	10	11
N_k	0	0	1	4	2	2	3	6	5	0	2

16. Let $X_1, \ldots, X_n$ be a random sample taken from a population with pdf defined as $f(x|\theta) = c\theta^c x^{-(c+1)}$ for $x \geq \theta$, where c is a known positive constant and $\theta > 0$. Find the MLE of θ.

17. Suppose that $X_1, \ldots, X_n$ is a random sample taken from a population that has pdf defined as $f(x|\theta) = \dfrac{1}{\theta}\exp(-\dfrac{x}{\theta})I_{(0,\infty)}(x)$. Assume that the prior distribution of θ is $\pi(\theta|\alpha,\beta) = \dfrac{1}{\Gamma(\alpha)\beta^\alpha \theta^{(\alpha+1)}}\exp(-\dfrac{1}{\beta\theta})I_{(0,\infty)}(\theta)$. Find the posterior distribution.

18. Let $X_1, \ldots, X_n$ be a random sample taken from a population with pdf defined as $f(x|\theta) = \dfrac{3x^2}{\theta^3}I_{[0,\theta^3]}(x)$. (a) Find the MLE of θ and show that this estimator is a sufficient statistic. (b) Calculate the MSE for the MLE. (c) Find the Cramér-Rao lower bound for the MLE.

19. The following are 15 observations from a uniform distribution on $[\theta-1, \theta+1]$. Show that any estimate T such that $5.01 \leq T \leq 6.83$ is a MLE of θ.

    5.01 5.10 5.14 5.16 5.47 5.57 5.74 6.03 6.29 6.37 6.45 6.71 6.77
    6.78 6.82.

20. Forced expiratory volume (FEV) is a standard measure of pulmonary function. The distribution of a boy's FEV is assumed to follow a normal distribution with mean $\alpha + \beta * (\text{height})$ and variance σ^2. The FEV is measured in liters and height is measured in centimeters. The following is the hypothetical data with 2 missing observations on FEV.

FEV	2.09	2.52	2.70	3.26	2.12	2.04	2.34	1.83	miss	miss
Height	146	157	161	171	148	147	152	143	168	157

    Use EM algorithm to find the MLE for α and β.

21. Let $X_1, \ldots, X_n$ be a random sample from a uniform distribution on $[\alpha, \beta]$. Find the sufficient statistic for the unknown α and β.

22. In problem 8.21, find the MLE for α and β.

23. Let $(X_1, Y_1), (X_2, Y_2), \cdots, (X_n, Y_n)$ be a random sample taken from a bivariate normal distribution with parameters $\mu_1, \mu_2, \sigma_1^2, \sigma_1^2, \rho$. Show that $\sum_{i=1}^{n} X_i$,

$\sum_{i=1}^{n} Y_i$, $\sum_{i=1}^{n} X_i^2$, $\sum_{i=1}^{n} Y_i^2$ and $\sum_{i=1}^{n} X_i Y_i$ are sufficient statistic for the parameters.

24. Suppose that $X_1, \ldots, X_n$ is a random sample taken from a distribution with the pdf defined as $f(x|\theta) = \beta \exp(-\beta(x-\theta)) I_{[\theta,\infty)}(x)$. Find the joint sufficient statistics for the unknown β and θ.

25. Suppose that $X_1, \ldots, X_n$ is a random sample taken from a population with pdf $f(x|\theta) = \theta a x^{a-1} \exp(-\theta x^a)$, for $x > 0$, $\theta > 0$ and $a > 0$. (a) Find a sufficent statistic for θ and (b) Find the MLE for θ.

26. Suppose $X_1, \ldots, X_n$ is a random sample taken from a uniform distribution on $[-\theta, \theta]$. Show that $T(X_1, \ldots, X_n) = \max(|X_{(1)}|, |X_{(n)}|)$ is a sufficient statistic for θ.

27. Show that the beta distribution is a member of the exponential family.

28. Show that the gamma distribution is a member of the exponential family.

29. Suppose that $X_1, \ldots, X_n$ is a random sample taken from a uniform distribution on $(0, \theta)$. If the prior distribution for θ is

$$\pi(\theta|\theta_0, \alpha) = \frac{\alpha}{\theta_0} \left(\frac{\theta_0}{\theta}\right)^{\alpha+1} I_{(\theta_0, \infty)}(\theta).$$ Find the posterior distribution of θ.

30. If $X_1, \ldots, X_n$ is a random sample from a distribution belonging to a one-parameter exponential family, show that the joint distribution of the sample is also a member of one-parameter exponential family.

31. Let $X_1, \ldots, X_n$ be iid observations from a Bernoulli trials. (a) Find the Cramér-Rao lower bound for $\bar{X}$ and (b) Show that $\bar{X}$ is UMVUE.

32. Suppose that $X_1, \ldots, X_n$ is a random sample taken from $Beta(\theta, 1)$. Show that $T(X_1, \ldots, X_n) = -\dfrac{1}{n} \sum_{i=1}^{n} \log(X_i)$ a UMVUE of $\tau(\theta) = \dfrac{1}{\theta}$ and find its MSE.

33. Calculate the MSE for the UMVUE proposed in Example 8.39.

34. Prove Example 8.47.

35. Let $T_1(X_1, \ldots, X_n)$ and $T_2(X_1, \ldots, X_n)$ be two independent unbiased estimators of θ. Find the value a such that $aT_1 + (1-a)T_2$ is a UMVUE of θ.

36. Let $X_1, \ldots, X_n$ be a random sample from a population with pdf $f(x, \theta) = \theta(\theta+1)x^{\theta-1}(1-x), 0 < x < 1, \theta > 1$. (a) Find the method of moment estimator of θ. (b) Show that this estimator is an asymptotic unbiased estimator of θ with its asymptotic variance equal to $\theta(\theta+2)^2 / 2(\theta+3)$ in distribution.

37. In problem 8.10, show that the MLE is asymptotic efficient if $\theta \neq \dfrac{1}{2}$.

38. Prove Corollary 8.5 (Hint: use the equality statement of Cauchy-Schwartz inequality).

References

Berger, J.O. (1985). *Statistical Decision Theory and Bayesian analysis*. 2ⁿᵈ ed. Springer-Verlag. New York.

Carlin, B.P. and Louis, T.A. (2000). *Bayes and Empirical Bayes Methods for Data Analysis*. 2ⁿᵈ ed. Chapman & Hall/CRC.

Dudewicz, E.J. and Mishra, S.N. (1988). *Modern Mathematical Statistics*. Wiley, New York.

Fisher, R.A. (1922), "On the mathematical foundations of theoretical statistics," reprinted in *Contributions to Mathematical Statistics* (by R. A. Fisher) (1950), J Wiley & Sons, New York.

Gelfand, A.E., Hills, S.E., Racine-Poon, A. and Smith, A.F.M. (1990). Illustration of Bayesian inference in normal data models using Gibbs sampling. *Journal of the American Statistical Association*. 85:972-985.

Gregurich, M.A. and Lyle, D.B. (1997). A Bayesian analysis for estimating the common mean of independent normal populations using the Gibbs sampler. *Communications in Statistics- Simulation.* 26:35-51.

Halmos, P.R. and Savage, L.J. (1949). Applications of the Radon-Nikodym Theorem to the Theory of Sufficient Statistics. *Annals of Mathematical Statistics.* 20: 225-241.

Khuri, A.I. (2003). *Advanced Calculus with Applications in Statistics.* 2nd ed. Wiley-Interscience.

Kinnunen, E., Junttila, O., Haukka, J., and Hovi, T. (1998). Nationwise oral poliovirus vaccination campaign and the incidence of Guillian-Barre' syndrome. *American Journal of Epidemiology.* 147(1):69-73.

Lehmann, E.L. and Casella, G. (1998.) *Theory of Point Estimation,* 2nd edition. Springer-Verlag. New York.

Luepker, R.V. and et al. (2000). Effect of a community intervention on patient delay and emergency medical service use in acute coronary heart disease: The Rapid Early Action for Coronary Treatment (REACT) Trial. *Journal of American Medical Association.* 284:60-67.

McCullagh. P. and Nelder. J.A. (1989). *Generalized Linear Models,* 2nd edition. Chapman & Hall. New York.

Meininger, J.C, Liehr, P., Mueller, W.H., Chan, W., Smith, G.L., and Portman, R.L. (1999). Stress-induced alternations of blood pressure and 24 h ambulatory blood pressure in adolescents. *Blood Pressure Monitoring.* 4(3/4), 115-120.

Morgan, B.J.T. (2000). *Applied Stochastic Modelling.* Arnold. London.

Murphy, S.A. and van der Vaart, A.W. (2000). On profile likelihood. *Journal of the American Statistical Association.* 95:449-465.

Neyman, J. (1935). Su un Theorema Concernente le Cosiddette Statistiche Sufficienti. *Inst. Ital. Atti. Giorn.* 6:320-334.

Pal N. and Berry, J. (1992). On invariance and maximum likelihood estimation. *The American Statistician.* 46:209-212.

Peterson, C.A. and Calderon, R.L. (2003). Trends in enteric disease as a cause of death in the United States, 1989-1996. *American Journal of Epidemiology.* 157:58-65.

Taylor, W., Chan, W., Cummings, S., Simmons-Morton, B., McPherson, S., Sangi-Haghpeykar, H., Pivarnik, J., Mueller, W., Detry, M., Huang, W., Johnson-Masotti, A. and Hsu, H. (2002), Health growth: project description and baseline findings. *Ethnicity and Disease.* 12:567-577.

Zehna, P.W. (1966). Invariance of maximum likelihood estimators. *Annals of Mathematical Statistics.* 37:744.

9

Hypothesis Testing

9.1 Statistical Reasoning and Hypothesis Testing

Applied statisticians are integral members of the research team. These productive statisticians focus on the design of the data collection activity, the actual data collection effort with its resultant quality control, and the analysis of the collected data. This final procedure concentrates on two activities — estimation and inference. The estimation process was discussed in Chapter Eight. Inference is the process by which an observer deduces conclusions about a population based on the results of a sample. Although statistical theory has made critically important contributions to this thought process, its involvement has received consistent criticism. This chapter will focus on these inference activities, commonly referred to as hypothesis testing.

9.1.1 Hypothesis Testing and the Scientific Method

A singularly remarkable trait of human beings is their drive to shape and control their surroundings. This ability is presaged by the need to understand their environment; the motivation to learn is common to all human cultures. Roman scholars, Indian shamans, nano-technologists, and space station mission specialists have the same characteristics. As learned men and women, they have sought an understanding of their world. In each culture, understanding is first and foremost based on an honest representation of the facts as best as they can be observed. This natural learning process has itself undergone remarkable growth, and there is no sign that the learning process will decelerate – in fact there is every indication that this learning process will continue its spiraling upward growth.

In Europe, learning was primarily the province of nobility and religious officials until the 16th century. The substitution of machine activity for human exertion was the direct effect of the Industrial Revolution. However, its indirect effects reverberated across the European culture. One of these was the creation of leisure time for the common laborer. People, in that time as well as now, used this new time to travel more, learn more, and, in general, question the circumstances of their environment. The construction of secure housing, the development of safe sanitation, and consecutive years of peace combined to shift the cultural paradigm. Families replaced the incessant question "Will we survive?" with "How can we prosper?" They began to ask hard questions about the rules that governed their environment, and, no longer satisfied with the untested answers from kings and religious officials, they turned to men who based their theories on observation. Men such as Copernicus, Kepler, Galileo, Newton, and Bacon all argued that natural law should be based on observation and deduction, not from the solitary contemplation of religious minds. The scientific method was forged in this new, intellectual cauldron. This method of creating a theory or hypothesis, collecting evidence in an objective way, and finally drawing a conclusion on the theory's tenability based on that collected evidence continues to be the bedrock of the scientific learning process to this day.

9.1.2 The Role of Statistics in Early Inference Testing

An important root of modern Western probability theory has been the development of probability theory from games of chance. Mathematicians of that day played an early role in the development of estimation theory. However, their attempts to generalize the results of their sample-based results to the population at large were tightly and politically circumscribed

Four hundred years ago, the monarchies of France, England, Prussia, and Italy required taxes in order to prosecute their costly wars. However, since tax rates are related to the size of the population, and there were no good estimates of population size, these monarchies decided to begin a systematic count of their population. The regals employed census takers in order to provide an accurate count of their populations — populations that were reluctant to participate in this activity. These first census takers had to resort to extraordinary lengths to count the population of London, and their estimates of the population size were among the first that were based on a systematic plan to count city populations.

The estimation procedures utilized by these early statisticians were innovative. However, this newly developed opportunity to examine the population provided a new view of the people on whose allegiance the nobility relied. The population count was gradually transformed into a survey instrument, in which important, and sometimes leading questions were inserted. While the census takers had been first asked to merely count the number of men in a household, they were now told to inquire about the behavior of men. Questions that examined how many men were married, how many were fighting men, how many men ignored their civic duties, and how many men attended church found their way into the new surveys. In an environment in which religion dominated the political, economic and social worlds, and where Catholics, Protestants, and Puritans vied for domination, answers to these questions were potentially explosive. It was felt that the interpretations of responses to these questions were too important to be left to those who collected the data. In fact, the entire process of tabulating and drawing conclusions from data was not called statistics at all, but instead was referred to as "political arithmetic" (Pearson). The isolation of statisticians from the inference issue was further solidified when this principle was accepted by the Statistical Society of London (which would later become the Royal Statistical Society). The first emblem that this group chose to represent their organization was a sheaf of wheat, neatly and tightly bound, representing the data that was carefully collected and accurately tabulated. Beneath this picture appeared the inscription "Allis exterendum", meaning "Let Others Thrash it Out" (Pearson).

This point of view has changed with the increasing complexity of the researcher's observations and experiments. While continuing to be resisted by some, the need for the incorporation of mathematical and statistical theory into the methodology by which scientists draw conclusions from their evidence has increased. Dissemination of the works of Albert Einstein and the Russian mathematician Kolmogorov revealed that naïve observations about nature can be sharpened, and other times shown to be incorrect, by a disciplined mathematical treatment. This quantitative development can in turn reveal a striking and powerful new "observation" that was missed by the traditional scientists in the field.[*] And, to the consternation of many both inside and outside the world of statistics, these efforts revealed that an unusual and, to some, a theoretical approach to a problem that is commonly viewed as solved can provide useful insight and illumination.

9.2   Discovery,   the   Scientific   Method,   and   Statistical Hypothesis Testing

The basis of the scientific method is hypothesis testing. The researcher formulates a hypothesis, then conducts an experiment that she believes will put this hypothesis to the test. If the experiment is methodologically sound and is executed according to its plan[†], then the experiment will be a true test of that hypothesis. This process, although time consuming, is the most reliable way to gain new knowledge and insight. It is in this context of a well-developed, prospectively stated research plan that statistical hypothesis testing makes its greatest contribution.

---

[*] Exploration of the "abnormal" orbit of the planet Mercury in our solar system is a fine demonstration of a revelation about the limitations of seasoned astronomical observers. See Ronald Clark. *Einstein: The Life and Times*. 1984. New York City. Avon Books, Inc.

[†] This plan is often called a protocol.

9.2.1 The Promise and Peril of "Discovery"

It is critical to recognize the importance of the central hypothesis to this scientific thought process. It is the central hypothesis that leads to the design of the research that itself produces data that tests that hypothesis. In this circumstance, statistical hypothesis testing provides the necessary quantitative tool that permits a defensible test of the central scientific hypothesis.

This deliberative procedure stands in contradistinction to another learning process known as "discovery" or "exploration". The discovery process involves the recognition of a finding that was not anticipated, and occurs as a surprise. Such findings can have important implications. The arrival of Christopher Columbus at the island of San Salvador led to the unanticipated "discovery" of the New World. Madam Curie "discovered" radiation. These researchers did not anticipate and were not looking for their discoveries. They stumbled upon them precisely because these discoveries were not in their view.

Discovery is an important process, and, as the examples of the previous section have demonstrated, can lead to new knowledge that reshapes our world. However, a "discovery" by itself must be confirmed before it can be accepted as a trustworthy, new illumination of the scientific issues at hand. Eccentricities of the discovery process e.g., faulty instrumentation, sample to sample variability, and outright mistakes in measurements and observations can each mislead the honest researcher and her audience. Yet it is important to include this unplanned, haphazard but nevertheless revealing discovery tool into the scientific thought process. This integration can be achieved by allowing the discovery to become the central *a priori* hypothesis for a new experiment that seeks to verify the discovery. For example, while it is true that Columbus discovered the new world, he had to return three additional times before he was given credit for his discovery. Columbus was essentially forced to prove that he could find the New World when he was actually looking for it. On the other hand, an important new discovery for the treatment of patients with congestive heart failure (Pitt et al. 1997) had to be reversed when a confirmatory finding could not repeat the original, spectacular surprise result (Pitt, 2000). Given that claims of discovery can commonly misdirect us, it is important for us to distinguish between an evaluation that uses a research effort to confirm a prospectively stated hypothesis, a process that we will define as *confirmatory analysis* (truly "*re-searching*") versus the identification of a finding for which the research effort was not specifically designed to reliably detect ("searching"). This latter category, into which "discoveries" and other surprise results fall we will term "*hypothesis generating analyses*" or "*exploratory analyses*". While hypothesis generating analyses can provide important new insight into our understanding, their analyses and interpretation must be handled differently than the results of confirmatory analyses. We will therefore set exploratory analyses aside, and now focus on the incorporation of statistical hypothesis testing in its most productive, confirmatory analysis setting.

9.2.2 Good Experimental Design and Execution

The growth and development of mathematical and computational analysis in the late 20th century has contributed immensely to our ability to conduct complicated research evaluations. However, mathematics cannot guarantee good research design, and without a well designed experiment, even the most advanced mathematics and statistics would be of limited utility. A solid statistical analysis is necessary, but not sufficient for a good research product.

The importance of good research design and execution has been recognized for generations, but one of the earliest and most forceful articulations of the importance of these principles emerged from the agricultural literature. After inheriting a farm in 1763, Arthur Young published a three volume book entitled *A Course of Experimental Agriculture* (Young, 1771). In it, he described ideas that we now clearly recognize as essential to good experimental methodology. One issue that Young identified was the problem introduced by biased investigator perspective. Each of Young's volumes begins with examples of how some authors slant the presented data to support their favored conclusion.

In addition, Young was well aware of the perils associated with drawing conclusions that extend far beyond the experimental setting to an inappropriately broad set of circumstances. Young warned that his own conclusions about the influences on crop development and growth could not be trusted to apply to a different farm with different soil and land management practices. By carefully noting that his results would not apply as a guide to long-term agricultural policy, he stressed the pitfalls of what we now call extrapolation.

Young also stressed that experiments must be comparative, and he insisted that, when comparing a new method and standard method, both must be present in the experiment.[*] He was able to recognize that even in comparative experiments, many factors other than the intervention being tested (e.g., soil fertility, drainage, and insects) contribute to increasing or decreasing crop yields in each of the experimental plots. In addition, because the sum effect of these influences can affect the plot yield in different ways and directions, the results of a single experiment in one year could not be completely trusted. Young therefore deduced that replication would be valuable in producing accurate estimates of an effect of an intervention on crop yield. Finally, Young stressed careful measurements, especially of the yield. When it was time to determine the experiment's outcome, all his expenses that could be traced to the intervention being tested were recorded in pounds, shillings, pence, halfpennies, and

---

[*] This was an important, early admonition against the use of historical controls.

farthings. These are important principles of clarity of observation and the ability to attribute an effect to an intervention.

Seventy-eight years later, James Johnson followed with the book *Experimental Agriculture* (1849), a treatise devoted to practical advice on experimental design (Owen). A major contribution of Johnson was the emphasis he gave to the supremacy of good experimental execution, and his observations are remarkably prescient for today's turbulent scientific environment. Johnson noted that a badly conceived or badly executed experiment was more than a mere waste of time and money — like a bad compass, it provided misdirection. The incorporation of the misleading results of a poor experiment into standard textbooks halts what would have been promising lines of research. This is because these new avenues of research would appear to be a waste of effort in the face of the findings that, unknown to the research community, are spurious. In addition, as other farmers attempted to use these false research results, the crop yields would not be as predicted, with a resultant waste of resources and money.

It is important to note that the aforementioned issues, while containing mathematics, are not entirely mathematical. Good research efforts require careful observation, clarity of thought, a clear view of the research question, a well designed and well executed experiment, and honesty. Without these integral components, the research effort will ultimately fail. Statistics in and of itself cannot transform a poorly conceived and designed research effort to an acceptable one.

9.2.3 Scientific vs. Statistical Hypotheses

There are of course important similarities between scientific and statistical hypothesis evaluations that allow statistical hypothesis testing to be very naturally and easily folded into the scientific method. Each procedure begins with a hypothesis, the veracity of which the investigator seeks to determine. In each case, the data must be the product of a well designed and well conducted research effort. Since the data are most commonly quantitative, both the scientific method and statistical hypothesis testing involve important mathematical computations. However, there is an important difference in the approach that is used by each that we must carefully examine.

The scientific method begins with a hypothesis or initial idea that the researcher hopes to prove. Statistical hypothesis testing commonly begins with a hypothesis that the researcher hopes to disprove or nullify.

For example, a researcher whose job it is to increase customer satisfaction at a bank would like to reduce the time customers are required to wait in line before they can see a bank teller. She may believe that having one long line of customers serviced by multiple bank tellers is more efficient than having individual lines, one line for each bank teller (Figure 9.1). The researcher starts with the idea that the single line operation will produce a shorter average total time for the bank customer, thereby increasing customer satisfaction with their banking experience. In order to demonstrate this, she designs a research effort that will provide affirmative evidence that her underlying thesis is correct. This is natural and appropriate approach to this research question.

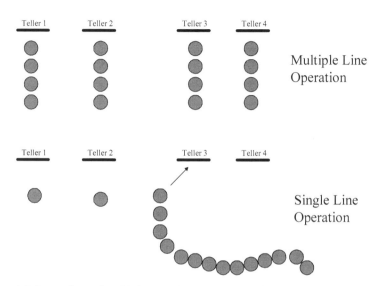

Figure 9.1 Comparison of MultiLine vs. Single Line Operation.

However, the application of statistical hypothesis testing in this setting would examine this problem from a different perspective. Statistical hypothesis testing starts with the assumption that the average total time for the single line approach is not shorter, but is the same as the average customer total time for the multiple line setup. It would then attempt to identify evidence that produces a contradiction of this underlying hypothesis of total time equality. This contradiction would lead to the rejection of the equality hypothesis in favor of a hypothesis of total time inequality.

In this setting, if the researcher plans to use statistical hypothesis testing to examine her hypothesis, she would first state a hypothesis (called the null hypothesis) that the average mean total time for the two different operations (multiple line vs. single line) is the same. Assuming that the distribution of total times follows an exponential distribution[*], the assumptions of equal total times

---

[*] The distribution $f_X(x) = \lambda e^{-\lambda x}$ for non-negative values of x is referred to as the exponential distribution here. In the field of operations research, this distribution is called the negative exponential distribution.

implies that the parameters of the two exponential distributions are equal. She would then carry out a research effort that would allow her to compute estimates of the parameters of the two distributions. If the original hypothesis of equality of total times were true, then it would be very likely that the sample estimates of parameter from the multi-line operation would be very close to the sample estimate of the parameter from the single line operation. In this case the null hypothesis would not be rejected.

However, the occurrence of two sample estimates that are very different requires the researcher to consider other possibilities. The first is that a sampling error has occurred. A sampling error in this case means that in the population of service systems, there is no difference in the mean total time for the multiple line vs. the single line operation. However, in this circumstance, the researcher happened to draw a sample of banks and participants that did not reflect the true state of nature in the population but instead provided a misleading result that the mean total times were different. The second explanation is that the estimates of the parameters from the two distributions are different because in fact the parameters themselves are different*. This latter result is the conclusion that the researcher would like to draw; however she cannot be assured that this result is due to a true difference in the operational characteristics of the two bank operations and not due to sampling error. Recognizing the potential difficulty in distinguishing between these two possible explanations for the difference in waiting times during the planning phase of her work, she designs the study so that it is very unlikely that a result will occur due to sampling error. She can never exclude the possibility that sampling error has produced the result of different parameter estimates. However, if the researcher can make the occurrence of a sampling error very unlikely, then the only conclusion that can be drawn from well executed research producing different parameter estimates is that the total times are different.

When left to the imagination, there are many functions of the sample data that one might be tempted to use to evaluate the effect of the different queue structure on the customer burden. However, there are properties of test statistics that we will see at once are desirable. Two of them are the property of invariance and the property of unbiasedness.

9.2.4 Invariance Testing

Statistical decision rule invariance is a property of the uniformity of the result of the hypothesis test. It would be meaningless to conclude that the effect of queue structure led to one conclusion if the customer total time was measured in seconds, another conclusion if the total time was measured in minutes, and yet a

---

* We set aside the possibility that the research effort was itself flawed for this first example. However, a poorly conceived, or well conceived but poorly executed experiment would shed no useful light on different total times for the multi-channel and single channel operation.

third conclusion when the customer total time was measured in terms of fractions of hours. Invariance under a suitable group of transformations means the decision rule changes under the transformation such that the final decision is independent of the transformation. From Lehman (Lehman, 1986), a group G of transformations g of the sample space leaves a statistical decision problem invariant if it satisfies the following conditions:

(1)   It leaves the invariant the family of distributions which describe the newly transformed random variable. For example, if the original random variable X has a cumulative distribution function $F_X(x)$, with unknown parameter θ, then the distribution of the new random variable $Y = g(X)$ with cumulative distribution $F_Y(y)$ is also a function of the unknown parameter θ.

(2)   Consequences of a decision rule of the original random variable remain the same under any transformation g in G. [*]

When these two properties are met, the decision rule is independent of the particular coordinate system that is adopted by the researcher. Invariance is a very useful property of a test statistic.

9.2.5 Unbiasedness
An alternative impartiality restriction for a hypothesis test, also from Lehman, is the property of unbiasedness. An *unbiased test* for the unknown parameter of a probability distribution is a statistical test that is likely to come closer to the correct decision about θ than any wrong decision in the long term. [†] This property is analogous to the unbiasedness property of estimators. As in that case, we cannot be assured that our decision rule will come to the correct conclusion in this one sample. However, we can say in the long run, the less plausible the null hypothesis is, the greater the probability that the unbiased test will reject the null hypothesis.

9.2.6 The Null Hypothesis
This indirect approach to statistical hypothesis testing (i.e., assuming an hypothesis that one does not believe in and therefore hopes to disprove) is not new, but has nevertheless received a great deal of criticism. Mathematicians will recognize this idea as that of a "proof by contradiction", in which the researcher assumes a statement is true, yet clear, directed reasoning from this premise

---

[*] Lehman discusses this property in terms of loss functions and homomorphisms which are beyond the scope of this chapter's discussion.
[†] This is commonly expressed as the property of *L-unbiased tests*, and is a property of the loss function that is associated with a decision.

produces a contradiction that proves that the original statement was false.[*] Ronald Fisher termed the original hypothesis that the researcher believes is false as the "null hypothesis", a term that is used to the present day. Fisher called this hypothesis the null hypothesis because it was the hypothesis that was likely to be nullified by the data.

However, it is the observation that the occurrence of naturally occurring events with their sampling variability that could be used to disprove the null hypothesis that was a major contribution to statistical inference testing. In 1710, the physician Dr. John Arbuthnot was concerned about the tendency for there to be more male births than female births registered in London in each of 82 consecutive years (Moyé, 2000). He understood that this tendency might be due to chance alone, and not due to some other influence. However, he believed that, if it was due to chance alone, the probability of this event should be relatively high. He correctly computed that the probability that in each of 82 successive years there were more male than female births was $\frac{1}{2}^{82}$. Since this probability was very small, he interpreted this result as strong evidence against the conclusion that 82 consecutive years in which male births predominate was likely to occur by chance alone. This was perhaps the earliest use of a formal probability calculation for a purpose in statistical inference.

Here, the contradiction in this "proof by contradiction" was the exceedingly small probability that the observed event would occur by chance alone. Since this probability was small, Dr. Arbuthnot believed that the thesis or hypothesis that the excessive male births occurred by chance alone should be rejected and replaced by the idea that there was a nonrandom, systematic explanation for the birth pattern. Using probability to reject the explanation that an event occurred randomly or "due to chance" is the hallmark of statistical hypothesis testing. In the 20th century Ronald Fisher developed this concept into what we recognize as modern statistical hypothesis testing.

Born in East Finchley, London in 1890, Ronald Aylmer Fisher grew to be one of the giants in the mid-twentieth century application of statistics to the inference process. Fisher's earliest writing on the general strategy in field experimentation appeared in the first edition of his book *Statistical Methods for Research Workers* (Fisher, 1925), and in a short paper that appeared in 1926, entitled "The arrangement of field experiments" (Fisher 1926). This brief paper contained many of the principal ideas on the planning of experiments, including the idea of significance testing.

9.3 Simple Hypothesis Testing

Recall the banking example from section 9.2 in which the researcher is interested in demonstrating that the average total time for the single line is less

---

[*] Among the most famous of the "proofs by contradiction" is the proof that the $\sqrt{2}$ is irrational.

than the ten minute average wait for the current operation. In this case, she assumes that the average total time that any customer spends in the system follows an exponential probability distribution with parameter λ. We now need to state her research concept in terms of a null hypothesis, i.e., the hypothesis to be nullified by the data.

9.3.1 Null and Alternative Hypotheses and the Test Statistic

Recall that the mean value for a random variable that follows an exponential distribution with parameter λ is λ^{-1}. Therefore, the average total time can be translated into a statement about λ, and the null hypothesis becomes $\lambda = \frac{1}{10} = 0.10$. We write this as H_0: $\lambda = 0.10$. Therefore, if the bank researcher believes that the mean total time for the single line operation is less than ten minutes, then she believes that $\lambda_s > 0.10$. We will call this assumption the *alternative hypothesis*, H_a. Thus the statement of this problem in terms of a statistical hypothesis is

$$H_0: \lambda = 0.10 \quad \text{vs.} \quad H_a: \lambda > 0.10$$

The researcher must now decide how to use the data to distinguish between these two possibilities. It stands to reason that, under the appropriate conditions discussed in section 9.2, if the researcher determines that the average total time for the single line operation is substantially less than ten minutes, she will have rejected the null hypothesis. Therefore, we need to find a *statistic*, or a function of the data, that contains information about the parameters λ_s. This actually just requires a little familiarity with the exponential distribution.

We note that if the total time associated with the ith individual, x_i follow a exponential distribution with parameter λ, then $\sum_{i=1}^{n} x_i$ follow a gamma distribution with parameters λ and n. This is easily demonstrated through the use of moment generating functions (Chapter 5). We quickly find the moment generating function of an exponential random variable X from Chapter 5 as

$$M_X(t) = \frac{\lambda}{\lambda - t}. \tag{9.1}$$

Now, we know from Chapter 3 that if $Y = \sum_{i=1}^{n} X_i$, then we can find the moment generating function of Y, Thus

$$M_Y(t) = \left(\frac{\lambda}{\lambda - t}\right)^n.$$

Recall from Chapter 8 that the minimal sufficient statistic for a parameter is that function of the data that contains all of the sample information in a condensed form about the value of a parameter. The identification of this minimal sufficient statistic is therefore a natural goal of this researcher, since it stands to reason that she will want to utilize all of the information contained in her sample to estimate the parameter on which her research hypothesis rests.

In order to identify the sufficient statistic, also remember that the gamma distribution is a member of the exponential family of distributions and that the natural, sufficient statistic for λ is the sum of the observations. The researcher has the resources to accurately measure the total times of 25 bank customers. Let $S_{25} = \sum_{i=1}^{25} X_i$. Thus, the sum of the total times for these 25 customers is the natural sufficient statistic for λ and will play a pivotal role in the execution of the statistical hypothesis test for λ. We can examine the probability distribution for S_{25}, which is gamma with parameters $\lambda = 0.10$ and $n = 25$ (Figure 9.2). We see that under the null hypothesis, we would expect the mean of the total times to be $\dfrac{n}{\lambda} = \dfrac{25}{0.1} = 250$. If the researcher's hypothesis is correct, she would expect this distribution to be shifted to the left. Therefore, the smaller the sum of the observations, the more likely it is that the null hypothesis is wrong, and the more likely it is that her hypothesis of the shorter total times under the single line operation is correct.

Since working with areas under a gamma distribution can be somewhat awkward, we invoke the following lemma in order to compute probabilities for a more familiar distribution.

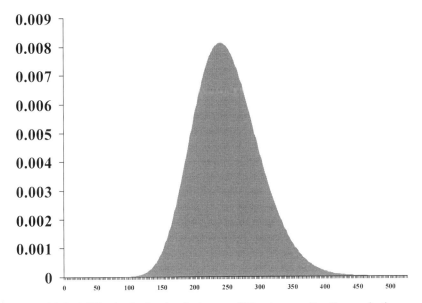

Figure 9.2. Probability density function for the sum of 25 customer waiting times under the null hypothesis (Gamma, λ=0.10, n = 25).

Lemma 9.1: If X_i, i = 1, 2, 3, …, n are independent and identically distributed random variables that follow an exponential distribution with parameter λ, and $S_n = \sum_{i=1}^{n} X_i$, then the quantity $2\lambda S_n$ follows a χ^2 distribution with $2n$ degrees of freedom.

This fact is actually very simple to demonstrate.

Let Y be a random variable that follows a gamma distribution with parameters λ and n, we know that if $W = 2\lambda Y$, then the probability density function of the random variable W is,

$$f_W(w) = \frac{\left(\dfrac{1}{2}\right)^n}{\Gamma(n)} w^{n-1} e^{-\frac{1}{2}w} dw$$

which is the probability density function for the chi-square distribution with 2n degrees of freedom.

We now apply this result to our hypothesis test. Recall that the researcher plans to obtain the total times of $n = 25$ bank customers. If the null hypothesis is true, then the researcher could construct a test statistic by multiplying the sum of the total times from her sample of n customers by 2λ. Thus under the null hypothesis the total time is ten minutes, $\lambda = 10^{-1}$, and her test statistic $T = 2\lambda S_n = 0.20 S_n$ would be expected to follow a χ^2 distribution with 50 degrees of freedom. Note that the test statistic is constructed under the assumption of the null hypothesis. Only under the null hypothesis would the test statistic $T = 0.20 S_n$ follow this distribution.

What value would we expect this test statistic to assume? Since the χ^2 distribution has been carefully studied, we are familiar with its shape and properties (Figure 9.3). For example we know that the expected value of a χ^2 distribution is the number of degrees of freedom i.e., $E[T] = n$ (Chapter 5). However, although this is the anticipated mean of the chi square distribution, the consideration of sample-to-sample variability would lead us to expect that the test statistic would in general not be equal to the mean of its distribution. However, it is likely that if the null hypothesis were true, the test statistic would be centrally located, and appear in the regions of highest probability for this χ^2 distribution. Therefore, it is more plausible to discuss the range of likely values of the test statistic in terms of extremes. Keeping in mind that, while any value between zero and infinity is permissible for a random variable obtained from a χ^2 distribution, if the null hypothesis is true, we would not anticipate that the test statistic would be very small. The more likely values of the test statistic would cluster toward the center of the distribution and not be close to zero.

The researcher applies this principle to the banking problem by saying that if she gets a value of a test statistic that is less than or equal to the 5th percentile of a χ^2 distribution with $2n$ degrees of freedom, she will reject the null hypothesis that the mean total time is 10 minutes in favor of the alternative hypothesis that the mean total time is less than ten minutes, and accept the concept that the single line customer service operation has reduced the total time. The 5th percentile value for the chi-square distribution with 50 degrees of freedom is 34.8. Thus any value of the test statistic < 34.8 while still possible under the null hypothesis of the mean waiting time = 10 would be considered so unlikely that the null hypothesis itself would not be plausible. This range of values in which the test statistic would lead to rejection of the null hypothesis is known as the *critical region*. If the test statistic falls in the critical region, the researcher would conclude that the null hypothesis was rejected and that the level of significance was < 0.05 since the probability that this result (or a result more extreme than this) that appears in the left tail of the distribution was less than 0.05.

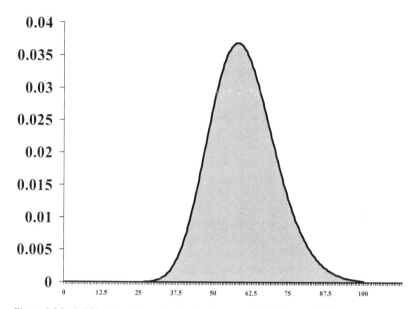

Figure 9.3 Probability density function for the chi square distribution with 50 degrees of freedom.

After all of this preliminary work, the investigator is ready to proceed with data collection and analysis. She selects her customers and measures the total time of each of the 25 selected individuals. The sum of their total times is determined to be 125 minutes. The test statistic is $(0.20)(125) = 25$. The test statistic falls in the critical region (Figure 9.4). The researcher states that, since the test statistic occurred in the critical region, the results of her research are *statistically significant.* It is important to note that the occurrence of this extreme event (i.e., a small test statistic) is still possible under the null hypothesis. However, since the likelihood of this event is so small, its possibility is excluded. In this case, the investigator would reject the assumption of the null hypothesis that the average total time under the single line operation was 10 minutes. This leaves her the only option of assuming that the alternative hypothesis is true, i.e., that the total time was less then ten minutes. By rejecting the hypothesis that she did not believe, she demonstrated the plausibility of the hypothesis that she did believe.

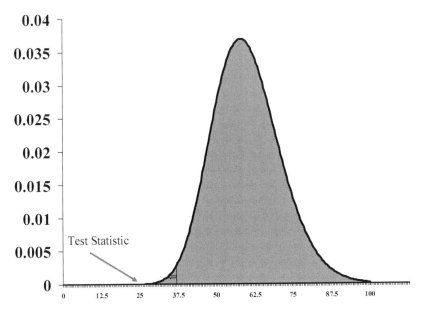

Figure 9.4 The test statistic value of 25 falls in the critical region of the chi-square distribution with 30 degrees of freedom.

We must keep in mind that this researcher's conclusions about the average total time in the single line operation are tightly tethered to the circumstances under which her research sample was obtained. If the methodology that is implemented to generate the data from the sample is poor, then any conclusion drawn from the statistical analysis is likely to be wrong. For example, the use of more bank tellers than are customarily available, the choice of a day on which customer activity is low (i.e., prolonged stormy weather), inaccurate measures of the total time for individual customers, the exclusion of some customers' data from the sample, and overtly fraudulent behavior are all occurrences that would work to undermine the ability of the researcher to draw reliable conclusions from her research effort. Poor sampling methodology is not overcome by statistical analysis. Instead, poor sampling methodology is transmitted to and ultimately corrupts the statistical evaluation.

9.4 Statistical Significance

The concept of statistical significance is well accepted in many scientific circles, but fiercely contested in others. This has been the case since its inception by Fisher in 1925. at which time the concept of hypothesis testing was a new and radical concept. Published papers by Berkson (1942) and Fisher (1942) provide some feel for the repartee exchanged between these strongly opinionated researchers as they struggled to understand all of the implications of the new hypothesis testing scenario. In general, there were two major areas of controversy. The first was that scientists were in general reluctant to work to reject a hypothesis that they did not believe, preferring instead to accept a hypothesis that they did believe. To many, significance testing appeared to be a very indirect, and to incorporate an uncomfortable, reverse logic approach to hypothesis testing than that prescribed by the scientific method.

The second major problem introduced by statistical hypothesis testing was the interpretation of the data based on an extreme value of the test statistic. The difficulty was that many workers were looking for arguments of certainty (i.e., a statement that the null hypothesis was wrong), but all that hypothesis testing could provide was an assessment of the likelihood of this event.

Gradually, however, the notion of statistical hypothesis testing came to be accepted in agronomy and other scientific disciplines. There have been adaptations of these procedures to deal with specific problems that arise, e.g. the propagation of type I error in the scenario of multiple testing. In addition, there remain important criticisms of the underlying theory (Lang), as well as the imperfect implementation of hypothesis testing (Nester, 1996). Finally, alternatives to hypothesis testing continue to appear (Birnbaum, 1962). However, statistical hypothesis testing, when implemented correctly and jointly with a well designed and well executed research effort serves as a useful tool that can shed important new life on the plausibility of scientific hypotheses (Moyé, 2000).

9.4.1 Type I and Type II Errors

A closer inspection of the decision process used by the bank researcher reveals a potential problem. The decision to reject the null hypothesis was based on the conclusion that the value of test statistic ($T = 25$) was implausible under the null hypothesis assumption that $\lambda = 0.10$. However, this was not an impossible result under the null hypothesis; in fact, this result will sometimes be produced by the null hypothesis. A test statistic < 34.8, under the null hypothesis, implies that the sum of the total times $S_{25} < \dfrac{34.8}{2\lambda} = \dfrac{34.8}{0.20} = 174$. We can depict the critical region in terms of the original measures, and not just the test statistic (Figure 9.5).

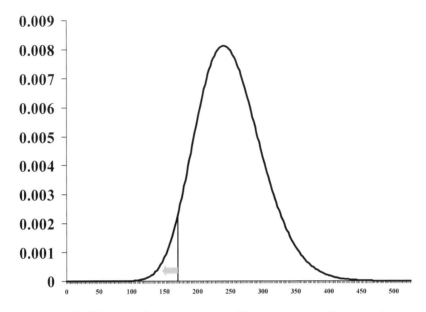

Figure 9.5 Probability density function for the sum of 25 customer waiting times under the null hypothesis and the critical region.

If the null hypothesis did produce this extremely low value of the test statistic, then the bank researcher's conclusion to reject the null hypothesis would have been the wrong one. The decision to reject the null hypothesis when the null hypothesis is in fact true is defined as the *type I error*. We can easily compute the probability of this error as

$$P\left[Type\ I\ error\right] = \int_{0}^{25} \frac{\left(0.5\right)^{n}}{\Gamma(n)} w^{n-1} e^{-\frac{w}{2}} dw \ < \ 0.005$$

Type I errors in statistical hypothesis testing have important consequences. In this banking example, the occurrence of a type I error means that the researcher draws the conclusion that the single line queue offers an advantage when it actually does not. The bank, embracing this conclusion, would restructure its handling of customers, a process that would incur both financial expense, and logistical adjustments. If a type I error occurred, these costs and adjustments would be for no tangible gain, because in reality, the single line operation does not decrease the total time.

Consideration of the implications of this error would lead the bank researcher to minimize the possibility of its occurrence. For example, she might prospectively decide to reject the null hypothesis only if the test statistic was less than 20, rather than less than 34.8. This prospective decision would require that a profound decrease in the total time produced by the single line operation be observed before the null hypothesis would be rejected. This change in the size of the critical region would certainly produce less false positive results.

However, there is an additional difficulty that is produced by the requirement of a more extreme test statistic to reject the null hypothesis. Assume that the bank researcher is correct, and the average total time is less than 10 minutes. As an illustration, assume the average total time under the single line operation is 5 minutes. In this case $\lambda = 0.20$, and the sum of the sample of 25 total times follows a gamma distribution with parameter $\lambda = 0.20$ and $n = 25$. The density of the total times under the alternative hypothesis is shifted to the left of that for the null hypothesis (Figure 9.6). If the decision rule is to reject the null hypothesis when $S_{25} < 174$, then we see that there is some probability that we will inappropriately retain the null hypothesis when the alternative hypothesis is true. This is known as a *type II error* or a *beta (β) error*. In this circumstance, we can compute whether the probability of a type II error for the critical region, reject the null hypothesis when $S_{25} < 174$. The probability of a type II error is the probability that the test statistic does not fall in the critical region when the alternative hypothesis is true. Under the alternative hypothesis, we know that S_{25} follows a gamma distribution with $\lambda = 0.20$ and $n = 25$. Using lemma 9.1 we may compute

$$P\left[S_{25} > 174\right] = P\left[2\lambda S_{25} > (2\lambda)174\right] = P\left[2\lambda S_{25} > (2)(0.20)174\right]$$
$$= P\left[\chi^2_{50} > 69.6\right] = 0.04.$$

Thus, the decision rule to reject the null hypothesis of $\lambda = 0.10$ vs. the specific alternative solution of $\lambda = 0.20$ has type I error of < 0.005 and a type II error of 0.04.

Examination of Figure 9.6 reveals the implications of decreasing the type I error. By decreasing the type I error, the critical region is moved farther to the left. However, this leftward movement increases the area to the right of the critical region under the probability distribution of the alterative hypothesis. Therefore, the likelihood of a type II error has increased. In general, if everything else about the research remains the same, decreasing the type I error rate of a statistical test increases the type II error rate of the same test.[*]

---

[*] We will see later in this chapter that increasing the sample size will allow one to maintain low type I and type II error rates.

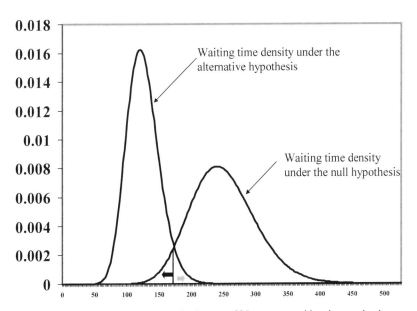

Figure 9.6 Probability density function for the sum of 25 customer waiting times under the alternative and null hypotheses.

9.3.4. Choosing the Type I Error.

Typically in hypothesis testing, the type I error rate for a research effort is set at the 0.05 level of significance. The motivation for this level is completely historical and traditional.

Ronald Fisher's earliest writing on the general strategy in field experimentation appeared in the first edition of his book *Statistical Methods for Research Workers* (1925) and in a short paper in 1926 entitled "The arrangement of field experiments". This brief paper was the source of many of the principal ideas of significance testing, and it is in this paper that the notion of a 5 percent level of significance first appears[*].

The example Fisher used to demonstrate his reasoning for significance testing was an assessment of the influence of manure on crop yield. In this circumstance, two neighboring acres of land were used in a study. The first was treated with manure; the second was left without manure but was otherwise treated the same. Crops were planted on both, and, upon harvest, the results

---

[*] This account is taken from Moyé, "Statistical Reasoning in Medicine – *The Intuitive P value Primer*, Springer 2000).

were evaluated. The yield from the manure-treated acre was 10 percent higher than that of the non-manure acre. Although this suggested a benefit attributable to the manure, Fisher grappled with how one could decide that the greater yield was due to the manure and not just to the random play of chance. He focused on the likelihood that the yield would be 10 percent greater with no manure (i.e., , without any differences between the plots). The differences between the yields of plots which were themselves treated the same was termed sampling variability.

Fisher began by noting that if the sampling variability did not produce a 10 percent difference in crop yield, then

"…the evidence would have reached a point which may be called the verge of significance; for it is convenient to draw the line at about the level at which we can say 'Either there is something in the treatment or a coincidence has occurred such as does not occur more than once in twenty trials.' This level, which we may call the 5 per cent level point, would be indicated, though very roughly, by the greatest chance deviation observed in twenty successive trials."

This is the first mention of the 5 percent level of significance. Note that the underlying philosophy is rooted in the notion of what might have been expected to occur naturally, i.e., , produced by random variation. Fisher went on to add

"If one in twenty does not seem high enough odds, we may, if we prefer it, draw the line at one in fifty (the 2 per cent point) or one in a hundred (the 1 per cent point). Personally, the writer prefers to set the low standard of significance at the 5 per cent point, and ignore entirely all results which fail to reach this level) ".(Fisher 1925)

The significance level of 0.05 was born from these rather casual remarks. It cannot be emphasized enough that there is no deep mathematical theory that points to 0.05 as the optimum type I error level — only tradition.

9.5 The Two Sample Test
9.5.1 Motivation
The example that we have developed in this chapter thus far is the statistical examination of likelihood that the parameter of a probability distribution has a specified value. While the utility of this level of examination is useful, it has the important drawback that one must know, or at least suspect the value of the parameter that is the basis of the comparison. Such certainty is commonly not the case in application.

We assume that the parameter has a value under the null hypothesis. However, the confidence in this value may not be justified. Returning to the banking problem of the previous section, it is reasonable to question the source of the assumption under the null hypothesis that $\lambda = 10$. This assumption was based on the belief that the total time for the multiple line operation was known to be ten minutes with certainty. However, this may not be a realistic assumption. The total time for the multiple line operation will itself be a random variable, with its own standard error. Certainly that natural variability should be considered in any statistical evaluation of the total time. If, for example, the total time was not ten minutes but seven minutes, the critical region that the researcher created for the evaluation of the null hypothesis would be wrong, and the researcher would run the risk of drawing misleading conclusions from her research effort.

Finally, there may be circumstances under which the average total time for the multiple line operation may have changed over time. Perhaps the average number of tellers the bank uses has decreased. This would serve to increase the total time, as fewer servers would be available to assist the customers. If the total time for the single line operation was computed based on a new system in which there were 25% fewer tellers than when the total time for the multiple line operation was calculated, the total time for the single line operation would be greater, and a comparison between the single line and multiple line operation would be inappropriate.[*]

A change in the conditions over time can represent an important research shortcoming. One must carefully scrutinize the information used to establish the value of the parameter in the null hypothesis. A more useful, appropriate, but expensive procedure is to collect information about the value of the parameter under the multiple line and single line operation simultaneously. A research effort designed in this fashion could be constructed so that extraneous factors that influence total time (i.e., , number of customers and the weather) affect the total times of both systems equally.

9.5.2. The New Null Hypothesis

This alteration in the research design that collects information about the total time for each of the multiple line and single line operations produces a fundamental alteration in the null and alternative hypotheses. We will continue to assume that the total time (total time + service time) follows an exponential distribution with parameter λ. However, customers in the single line operation will experience total times from a exponential distribution from which $\lambda = \lambda_s$. Let us also assume that, for customers in the multiple line operation $\lambda = \lambda_m$. When the problem is formulated in this manner, the null hypothesis is that the total times under the two operations are equivalent, or $\lambda_s = \lambda_m$. We write as H_0:

---

[*] It is important to note that there is nothing in the mechanical execution of the hypothesis test that would warn the researcher of this problem. Statistical hypothesis testing is no substitute for a clear and objective appraisal of the research design's ability to provide important illumination about the scientific question at hand.

$\lambda_s = \lambda_m$. Recall that the mean value for a random variable that follows an exponential distribution with parameter λ is λ^{-1}. Therefore, if the bank researcher believes that the mean waiting time for the single line operation is less than that of the multiple line operation, she believes that $\lambda_s^{-1} < \lambda_m^{-1}$, or $\lambda_s > \lambda_m$. We will call this assumption the *alternative hypothesis*, H_a. Thus the statement of this problem is

$$H_0: \lambda_s = \lambda_m \quad \text{vs.} \quad H_a: \lambda_s > \lambda_m$$

There remains the possibility that the single line operation could produce not a shorter but a longer average total time, in which case $\lambda_s < \lambda_m$. Since a longer total time for the single line operation has the same implications as an equivalent total time, there is no need to differentiate the finding that $\lambda_s = \lambda_m$ from the finding that $\lambda_s < \lambda_m$*. This observation will be included in the null hypothesis

$$H_0: \lambda_s \leq \lambda_m \quad \text{vs.} \quad H_a: \lambda_s > \lambda_m$$

This is the statement of the null and alternative hypothesis in this banking problem. We now must decide how to use the data to distinguish between these two possibilities. It stands to reason that, under the appropriate conditions discussed in section 9.2, if the researcher determines that the average total time for the single line operation is substantially less than the average total time for the multiple line operation, she will have rejected the null hypothesis. So we need to find a *statistic*, or a function of the data, that contains information about the parameters λ_s and λ_m. In order to do this, we will build on our understanding of the test statistic.

Recall from lemma 9.1 that, if X_i, $i = 1,2,3,\ldots,n$ are i.i.d. random variables that follow an exponential distribution with parameter λ, and we define $S_n = \sum_{i=1}^{n} X_i$, then $2\lambda S_n$ follows a χ^2 distribution with $2n$ degrees of freedom. In this two sample problem if we let the collection of x_i, $i = 1,2,3,\ldots,n$ be the collection of total times for the n customers identified from the single line operation, then $2\lambda_s S_n(s)$ follows a χ^2 distribution with $2n$ degrees of freedom. Similarly we can denote the collection of total times from the m customers evaluated in the multilane operation as Y_i, $i = 1,2,3,\ldots,m$ and define $T_m = \sum_{i=1}^{m} Y_i$. We would then also conclude that $2\lambda_m T_m$ follows a χ^2 distribution with $2m$ degrees of freedom. Since the customers' total times are independent of one

---

* We will see in other examples that this assumption is not valid and that any important deviation from parameter equality must be included in the alternative hypothesis.

another, we have two independent χ^2 random variables. Recall from Chapter 5 that the ratio of two independent χ^2 distributions, each divided by its degrees of freedom, follows an F distribution. Thus we may define our test statistic as

$$\frac{2\lambda_m T_m / 2m}{2\lambda_s S_n / 2n} \text{ follows an } F_{2m,2n} \text{ distribution.} \qquad (9.2)$$

Under the null hypothesis that $\lambda_m = \lambda_n$, the test statistic reduces to the ratio of the sum of the total times for the multiple line and single line operations for $m = n$.

The investigator can measure the total time of 30 customers in each of the multiple line and single line operation. Under the alternative hypothesis that the average customer total time for the single line operation is less than that of the multiple line operation, the test statistic would be large. The researcher makes the *a priori* decision to reject the null hypothesis if the test statistic $TS = \dfrac{T_m}{S_n}$ falls in the upper 1% of the F distribution with $F_{60,60}$ distribution. Thus, she would reject the null hypothesis when $TS > 1.84$. From her data, she finds $T_m = 305$, and $S_n = 165$. In this case the test statistic $TS = \dfrac{305}{165} = 1.85$. The test statistic falls in the critical region and the null hypothesis is rejected in favor of the alternative hypothesis. Although a ratio of 1.85 is possible under the null hypothesis, the likelihood that it would occur by change alone is less than 1 in 100.

9.6 Two-Sided vs. One-Sided Testing

Our discussion thus far has been focused on research efforts where the scientist's focus (i.e., the alternative hypothesis) has been on one tail of the distribution. However, it is commonly the case that extreme values of the test statistic that fall in either tail of the probability density function have important implications and must therefore be separated from the null hypothesis.

In measurement of heart function, the output of the heart is an important measurement which conveys critical information about the health of this vital organ. Cardiac output that is too low can be a sign of congestive heart failure, a chronic disease that often, despite the use of therapy, leads to death. In addition, abnormally high cardiac outputs are also associated with diseases, such as thyroid disease, anemia, and some vitamin deficiencies. The evaluation of the cardiac output can therefore contribute a critical piece of the clinical assessment.

Since the heart is a pump, the output of this pump is a function of how rapidly it pumps (the heart rate, R), and how much blood is ejected from the heart with each beat (stroke volume, or V). Therefore, the cardiac output W may be computed as $W = VR$. An investigator is interested in assessing the change in cardiac output from time 1 to time 2, or $W_2 - W_1$. Specifically, he wishes to know the probability distribution of $W_2 - W_1$

In order to address this question, we must identify the distribution of cardiac output. Let us assume that the heart rate and stroke volume are independent of one another and normally distributed. We may standardize each of them so that they have a mean value of zero and a variance of one. Then the distribution of the cardiac output computed from the standardized heart rate and stroke volume is the probability distribution of the product of two standard normal random variables.

We may apply the double expectation argument developed in chapter three to find the moment generating function of this random variable W.

$$M_W(t) = E_W\left[e^{Wt}\right] = E_W\left[e^{VRt}\right] = E_R\left[E_V\left[e^{VRt} \mid R\right]\right].$$

Since V is a standard normal random variable, we know from a moment generating function argument that $E_V\left[e^{RVt} \mid R\right] = e^{\frac{R^2 t^2}{2}}$. To find the moment generating function of W, we must identify $E_R\left[e^{\frac{R^2 t^2}{2}}\right]$. Since the random variable R is also follows a standard normal probability distribution, we can compute

$$E_R\left[e^{\frac{R^2 t^2}{2}}\right] = \int_{-\infty}^{\infty} e^{\frac{r^2 t^2}{2}} f_R(r) = \int_{-\infty}^{\infty} e^{\frac{r^2 t^2}{2}} \frac{1}{\sqrt{2\pi}} e^{-\frac{r^2}{2}} dr = \int_{-\infty}^{\infty} \frac{1}{\sqrt{2\pi}} e^{-\frac{r^2(1-t^2)}{2}} dr \tag{9.3}$$

The integrand of the last integral on the right in equation (9.3) is almost the density of a normal random variable with mean $\mu = 0$ and variance $\sigma^2 = \frac{1}{1-t^2}$. We multiply and divide by $\frac{1}{\sqrt{1-t^2}}$ to compute

$$M_W(t) = \frac{1}{\sqrt{1-t^2}} \int_{-\infty}^{\infty} \frac{1}{\sqrt{2\pi \frac{1}{1-t^2}}} e^{-\frac{r^2}{2\left(\frac{1}{1-t^2}\right)}} dr = \frac{1}{\sqrt{1-t^2}} \tag{9.4}$$

The investigator is interested in measuring the difference in cardiac output over time $S = W_1 - W_2$. The moment generating function of this difference may be written as $M_S(t) = M_W(t)M_W(-t) = \dfrac{1}{1-t^2}$. However, the moment generating function $M_S(t)$ turns out to be the moment generating function of the double exponential distribution $f_X(x) = \dfrac{1}{2}e^{-|x|}1_{(-\infty,\infty)}(x)$. To see this, for $|t| < 1$,

$$M_X(t) = E\left[e^{tX}\right] = \int_{\Omega_X} e^{tx} f_X(x)dx = \int_{-\infty}^{\infty} e^{tx}\frac{1}{2}e^{-|x|}dx = \frac{1}{2}\left[\int_{-\infty}^{0} e^{tx}e^{-|x|}dx + \int_{0}^{\infty}e^{tx}e^{-|x|}dx\right]$$

Each of these evaluations is straightforward

$$\int_{-\infty}^{0} e^{tx}e^{-|x|}dx = \int_{-\infty}^{0} e^{tx}e^{x}dx = \int_{-\infty}^{0} e^{(1+t)x}dx = \frac{1}{1+t}\int_{-\infty}^{0}(1+t)e^{(1+t)x}dx = \frac{1}{1+t}$$

$$\int_{0}^{\infty} e^{tx}e^{-|x|}dx = \int_{0}^{\infty} e^{tx}e^{-x}dx = \int_{0}^{\infty} e^{-(1-t)x}dx = \frac{1}{1-t}\int_{0}^{\infty}(1-t)e^{-(1-t)x}dx = \frac{1}{1-t}$$

And the moment generating function is simply

$$M_X(t) = \frac{1}{2}\left[\frac{1}{1+t} + \frac{1}{1-t}\right] = \frac{1}{2}\frac{2}{\left(1-t^2\right)} = \frac{1}{1-t^2}$$

We may easily graph this symmetric distribution (Figure 9.7). If the researcher is interested in examining the change in the normed cardiac output, then we state the null hypothesis as H_0: $S = 0$. However, how should extreme values be judged? Since decreases in cardiac output can be associated with heart failure, the researcher would certainly be interested in distinguishing low cardiac outputs from the null hypothesis, suggesting that the alternative hypothesis should be H_a: $S < 0$. However, increases in cardiac outputs can also be signs of serious illness. This suggests that both abnormally low and abnormally high cardiac outputs should be separated from the null hypothesis. We portray this circumstance as H_a: $S \neq 0$. Thus the null and alternative hypotheses are

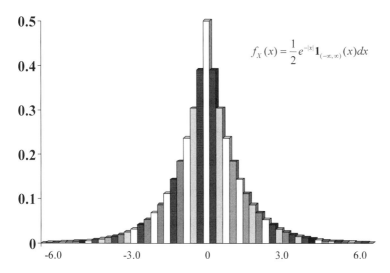

$$f_X(x) = \frac{1}{2}e^{-|x|}\mathbf{1}_{(-\infty,\infty)}(x)dx$$

Figure 9.7 The probability density function of the double exponential distribution.

$$H_0: S = 0 \qquad H_a: S \neq 0$$

The clinical observation that both extremely low and extremely high cardiac outputs must be distinguished from the null hypothesis has led to an alternative hypothesis that has two directions (commonly referred to as *two-sided*, or *two-tailed*). This is in contradistinction to our hypothesis testing that the banking researcher engaged in earlier in this chapter in which interest rested exclusively with determining whether the average total time for the single line operation was less than that from the multiple line operation. In the banking circumstance, there was no difference in the consequence of the conclusion that the total times were equal to the multiple line average, from the conclusion that the single line average total time was greater than the multiple line total time. In this public health paradigm of cardiac output assessments in which low measurements and high measurements lead to conclusions with implications that are different than those of the null hypothesis, statisticians are best served by carrying out two-sided testing, although there continues to be important discussion concerning this point in the literature (Fisher, 1991; Knottnerus, 2001; Moyé and Tita, 2001).

Since both low and high cardiac outputs must be distinguished from the null hypothesis, we will need to have the critical region divided into two regions. Most commonly, this is carried out by dividing the critical region, placing half of it in the lower tail of the distribution and the remaining half in the upper tail. In this case if we choose a type I error rate of 0.05, then 0.025 would be allocated for extremely low values of normed cardiac output, and 0.025 would be set aside for the upper tail of the distribution. A quick computation shows how to compute the location of the critical region for this symmetric distribution.

$$\alpha = P\big[\,|\,X\,|>b\big] = 2P\big[\,X>b\big] = 2\int_b^\infty f_X(x)dx = 2\left(\frac{1}{2}\right)\int_b^\infty e^{-|x|}dx$$

$$= \int_b^\infty e^{-x}dx = e^{-b}$$

Therefore we need to solve the equation $\alpha = e^{-b}$, or $b = -\ln\alpha$ in order to find the lower bound of the critical region on the right side of this symmetric probability density function. For example, if $\alpha = 0.05$, then $b = -\ln(0.05) = 3.00$. We would reject the null hypothesis if $|x| \geq 3$ (Figure 9.8).

While we chose to divide the critical region into two equal components in this example, this is not the only defensible allocation. An alternative is mathematically valid, although the specific allocation the researcher should receive requires the approval of the research community. For example, suppose the researcher chooses to allocate a total type I error of 0.05, with 25% apportioned to the upper tail, and 75% apportioned to the lower tail. The critical regions are easily identified. For the upper tail,

$$0.0125 = P\big[\,X>b\big] = \int_b^\infty f_X(x)dx = \left(\frac{1}{2}\right)\int_b^\infty e^{-|x|}dx$$

$$= \int_b^\infty e^{-x}dx = \frac{e^{-b}}{2}$$

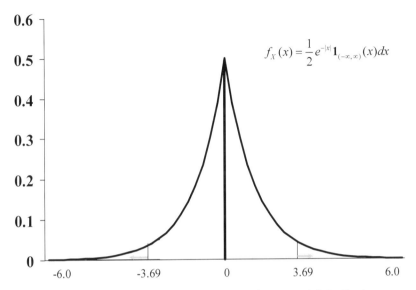

$$f_X(x) = \frac{1}{2} e^{-|x|} 1_{(-\infty, \infty)}(x) dx$$

Figure 9.8 Symmetric critical regions and the double exponential distribution

or $b = -\ln((2)(0.0125)) = 3.69$. Similarly for the lower tail of the probability density function, we have

$$0.0375 = P[X < b] = \int_{-\infty}^{b} f_X(x) dx = \left(\frac{1}{2}\right) \int_{-\infty}^{b} e^{-|x|} dx$$

$$= \left(\frac{1}{2}\right) \int_{-\infty}^{b} e^x dx = \frac{e^b}{2}$$

or
$b = \ln((2)(0.0375)) = -2.59$. The total type I error $= 0.05$, but now it has been distributed in unequal portions, in accordance with the researcher's plan (Figure 9.9)

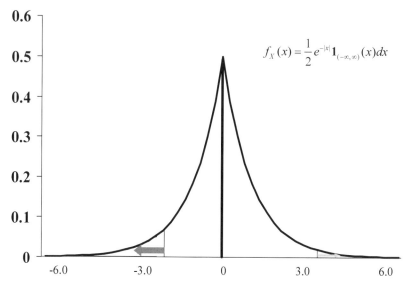

Figure 9.9 Asymmetric critical regions and the double exponential distribution.

9.7 Likelihood Ratios and the Neyman-Pearson Lemma

The tests developed thus far in this chapter permit an examination of the validity of a predefined scientific idea based on the underlying distribution of the dataset. The approaches were intuitive based largely on the demonstration of the underlying probability distribution of the mathematical function of the observation that we were computing. However, it must be admitted, that there are many functions of the data that we could compute. The recognition of this begs the question as to which function produces the best test.

A *best test* is defined as the test which, of all tests that are constructed with a specific type I error level, has the lowest type II error level (Hoel, 1971). The procedure by which a best test is constructed is guided by a theorem that was first proved by two statisticians, Jerzy Neyman and Egdon Pearson. The Neyman-Pearson lemma is based on the concept of a likelihood function, that was introduced in the chapter on estimation theory.

Neyman-Pearson Lemma
Consider a sample of observations collected to test a hypothesis

$$H_0: \theta = \theta_0 \text{ versus } H_a: \theta = \theta_a$$

Then if we can find a critical region of size α and a constant k such that

$$\frac{L\left(x_1, x_2, x_3, \ldots x_n \mid H_a\right)}{L\left(x_1, x_2, x_3, \ldots x_n \mid H_0\right)} \geq k \text{ inside A}$$

and

$$\frac{L\left(x_1, x_2, x_3, \ldots x_n \mid H_a\right)}{L\left(x_1, x_2, x_3, \ldots x_n \mid H_0\right)} < k \text{ outside A}$$

then the region A is the best critical region of size α. An easily understood proof of this important lemma is provided by (Hoel, 1971).

The Neyman-Pearson lemma provides important assurance for the scientist who is engaged in hypothesis testing. We pointed out earlier that there are many functions of the data that can naively be created to "test" the tenability of the null hypothesis. The Neyman-Pearson lemma directs us to one function of the data that will yield a critical region the probability of which is α, the size of the type I error. While other functions of the data can also provide critical regions of size α, these latter regions will be inferior to the region derived using the Neyman Pearson lemma for the hypothesis test. Specifically, the other non Neyman-Pearson derived region will lead to a test that is less powerful than that which is based on the Neyman Pearson lemma. In order to develop the best test for a statistical hypothesis test, we need only identify the underlying probability distribution function of the data (that will be a function of the unknown parameter θ). We can then work to identify the region A, a task that focuses on finding the value of k that bounds the region.

The Neyman-Pearson lemma was derived and is directly applicable to simple hypothesis testing, i.e., , H_0: $\theta = \theta_0$ vs. H_a: $\theta = \theta_a$. However, the principle of the Neyman-Pearson lemma can sometimes be extended to hypotheses where the value of θ under the null hypothesis is known simply not to be equal to θ_0. In this circumstance, our focus in developing useful hypothesis tests will remain on the likelihood ratio test. However, we will use maximum likelihood techniques that received attention in the estimation chapter of this text to allow us to continue to develop hypothesis testing techniques in this useful setting of composite hypotheses. The integration of these two topics is commonly denoted as *likelihood ratio testing*.

The previous examples used several interesting examples of hypothesis testing. These illustrations have hopefully produced some appreciation of the flexibility of well planned hypothesis testing to provide specific and direct information about the research question that is under investigation. At this point we will examine the development of hypothesis tests from yet another perspective, that of the likelihood function.

The common implementation of the normal distribution in hypothesis testing will hopefully come as no surprise to the reader. Our work in Chapter Seven demonstrated that, though there is a collection of random variables that are normally distributed, many more are not intrinsically normal. Instead, through the central limit theorem, these random variables have arithmetic means whose probability distribution can be approximated by the normal distribution. Also, in Chapter 9 we demonstrated that estimates of the parameters μ and σ^2 from a random sample of observations are readily available. These estimates are unbiased and have the smallest possible variance of any unbiased estimates. In addition, these estimates are easily and inexpensively computed, not an unimportant consideration in research endeavors that have limited financial resources. Thus a distribution which is so ubiquitous and facile is correctly anticipated to be a natural nidus around which hypothesis testing theory is developed.

Rao has investigated the likelihood ratio tests. These tests have the pleasing property of being a function of the minimal sufficient statistic when the probability density function can be factored using the factorization theorem. In addition, the likelihood ratio test is unbiased.

9.7.1 Uniformly Most Powerful Tests

In estimation theory, we saw that there are commonly several estimators that one may find themselves considering. The analogous problem exists in hypothesis testing. In this setting, there are several (perhaps many) test statistics that the investigator could choose to carry out her hypothesis test. In order to guide the selection of a best test (when one is available), statisticians have studied useful properties of competing test statistics. In the case of estimation theory, we have already seen some properties of these properties, such as the properties of unbiasedness, precision, and consistency.

There are analogous properties that have been developed for test statistics as well. We have already seen two of these properties — unbiasedness and invariance. A third property is the power of the test. The power of a test is defined in terms of the null (H_0) and alternative (H_a) hypotheses and the probability of a type I error, α. Consider two test statistics that are competing for consideration in testing H_0 versus H_a. For each test statistic we can identify a critical region, such that if the test statistic falls in that critical region, the null hypothesis will be rejected. The power of the test is the likelihood that the test statistic will fall in the critical region given that the alternative hypothesis is true. The researcher wishes to maximize the probability of this event, i.e., wishes to maximize the power of the hypothesis test. If these tests have different power for the same, null hypothesis, alternative hypothesis, and type I error level α, then the researcher will choose the test statistic with the greatest power (everything else, e.g., the cost of generating the test statistic being equal).

In some circumstances a test statistic will be demonstrated to have superior power to all other test statistics for a given type of alternative hypothesis. Such a test is described as a *uniformly most powerful test* (UMP). We will point out these UMP tests as we examine different hypothesis testing scenarios. In some cases, we will only be able to identify a test that is locally most powerful. These locally powerful tests are tests that maximize power for a small number of alternative hypotheses, and the worker hopes that the property of high power will extend to more distant alternatives as well. Lehman (1986) provides a thorough discussion of the notion of UMP testing in exponential families.

9.8 One-Sample Testing and the Normal Distribution

Assume that we have a random sample of n observations, which are i.i.d. normal with mean μ and variance σ^2. The scientific question of interest is focused on the value of μ, a value that we do not know. Assume that we know the value of σ^2, so we do not have to expend resources on its estimation. The null hypothesis is that $\mu = \mu_0$, where μ_0 is known. The alternative hypothesis is that $\mu = \mu_a$. Without any loss of generality we will assume that $\mu_a > \mu_0$. How would we decide which of these two hypotheses, $H_0: \mu = \mu_a$ vs. $H_a: \mu = \mu_a$ is best, based on n observations?

Simple reflection on this suggests the following argument. We know that in a sample of n observations, the quantity $\overline{X}$ incorporates all of the information that the sample contains about μ, and that $\overline{X}$ is a very good estimate of μ. We would therefore expect that $\overline{X}$ will be close to μ. Therefore, if the true value of μ is in fact μ_0 then $\overline{X}$ will be close to this quantity. If on the other hand, the true mean of the population from which our random sample was drawn is not μ_0 but μ_a, then we would expect that $\overline{X}$ to be larger than μ_0 and closer to μ_a. Thus, the proximity of our estimate $\overline{X}$ to μ_0 will provide important information about the relative tenability of the null and the alternative hypotheses. For us "relative tenability" will be translated into "relative likelihood". We will compute the likelihood of the data under each of the null and alternative hypotheses and compare them.

We can easily write the probability density function of X_i given μ and

σ^2 as $f_X(x_i) = \dfrac{1}{\sqrt{2\pi\sigma^2}} e^{\frac{-(x_i - \mu)^2}{2\sigma^2}} \mathbf{1}_{(-\infty, \infty)} x_i \, dx_i$. We can therefore write the likelihood

function* as

$$L(x_1, x_2, x_3, \ldots x_n) = \left(2\pi\sigma^2\right)^{-\frac{n}{2}} e^{\frac{-\sum_{i=1}^{n}(x_i - \mu)^2}{2\sigma^2}} \tag{9.5}$$

Under the null hypothesis, equation (9.5) becomes

$$L(x_1, x_2, x_3, \ldots x_n \mid H_0) = \left(2\pi\sigma^2\right)^{-\frac{n}{2}} e^{\frac{-\sum_{i=1}^{n}(x_i - \mu_0)^2}{2\sigma^2}} \tag{9.6}$$

Under the alternative hypothesis, equation (9.5) becomes

$$L(x_1, x_2, x_3, \ldots x_n \mid H_a) = \left(2\pi\sigma^2\right)^{-\frac{n}{2}} e^{\frac{-\sum_{i=1}^{n}(x_i - \mu_a)^2}{2\sigma^2}} \tag{9.7}$$

Let's now examine the ratio of these likelihoods

$$LR = \frac{L(x_1, x_2, x_3, \ldots x_n \mid H_a)}{L(x_1, x_2, x_3, \ldots x_n \mid H_0)} \tag{9.8}$$

We should pause now to consider the implication of this ratio. Since equation (9.8) is the ratio of the likelihood of the data under the alternative to the likelihood of the data under the null hypothesis, larger values of the ratio mean that the data are relatively more likely under the alternative hypothesis and therefore more supportive of the alternative hypothesis than of the null hypothesis. Thus, we would have more evidence for the alternative hypothesis, and reject the null hypothesis for large values of this likelihood ratio. This will be our basic decision rule.

In this particular circumstance, the likelihood ratio is

---

* The indicator function does not make any contribution to our argument so we will not explicitly write it for the duration of this chapter's discussion of hypothesis testing and the normal distribution.

$$\frac{L\left(x_1, x_2, x_3, \ldots x_n \mid H_a\right)}{L\left(x_1, x_2, x_3, \ldots x_n \mid H_0\right)} = \frac{\left(2\pi\sigma^2\right)^{-\frac{n}{2}} e^{\frac{-\sum\limits_{i=1}^{n}(x_i - \mu_a)^2}{2\sigma^2}}}{\left(2\pi\sigma^2\right)^{-\frac{n}{2}} e^{\frac{-\sum\limits_{i=1}^{n}(x_i - \mu_0)^2}{2\sigma^2}}} = e^{-\frac{1}{2\sigma^2}\left[\sum\limits_{i=1}^{n}(x_i - \mu_a)^2 - \sum\limits_{i=1}^{n}(x_i - \mu_0)^2\right]} \quad (9.9)$$

Since the natural log is a monotonically increasing function, taking the log of the likelihood ratio does not reverse or change our decision rule. We will define $l\left(x_1, x_2, x_3, \ldots x_n\right)$ as the log of the likelihood ratio. Therefore

$$l\left(x_1, x_2, x_3, \ldots x_n\right) = -\frac{1}{2\sigma^2}\left[\sum_{i=1}^{n}(x_i - \mu_a)^2 - \sum_{i=1}^{n}(x_i - \mu_0)^2\right]$$

$$-2\sigma^2 l\left(x_1, x_2, x_3, \ldots x_n\right) = \sum_{i=1}^{n}(x_i - \mu_a)^2 - \sum_{i=1}^{n}(x_i - \mu_0)^2 \quad (9.10)$$

We can easily see that

$$\sum_{i=1}^{n}(x_i - \mu_a)^2 = \sum_{i=1}^{n}(x_i - \bar{x} + \bar{x} - \mu_a)^2 = \sum_{i=1}^{n}\left([x_i - \bar{x}] + [\bar{x} - \mu_a]\right)^2$$

$$= \sum_{i=1}^{n}(x_i - \bar{x})^2 + 2\sum_{i=1}^{n}(x_i - \bar{x})(\bar{x} - \mu_a) + \sum_{i=1}^{n}(\bar{x} - \mu_a)^2$$

$$= \sum_{i=1}^{n}(x_i - \bar{x})^2 + 2\sum_{i=1}^{n}(x_i - \bar{x})(\bar{x} - \mu_a) + n(\bar{x} - \mu_a)^2$$

Since the expression $\left(\bar{x} - \mu_a\right)$ is a constant with respect to the summation, then

$$2\sum_{i=1}^{n}(x_i - \bar{x})(\bar{x} - \mu_a) = 2(\bar{x} - \mu_a)\sum_{i=1}^{n}(x_i - \bar{x}) = 0$$

Therefore $\sum\limits_{i=1}^{n}(x_i - \mu_a)^2 = \sum\limits_{i=1}^{n}(x_i - \bar{x})^2 + n(\bar{x} - \mu_a)^2$. Similarly, we have

$\sum\limits_{i=1}^{n}(x_i - \mu_0)^2 = \sum\limits_{i=1}^{n}(x_i - \bar{x})^2 + n(\bar{x} - \mu_0)^2$. Placing these results into the right hand side of the last equation in expression (9.10), we find

$$-2\sigma^2 l\left(x_1,\ x_2, x_3, \ldots x_n\right) = \sum_{i=1}^{n}\left(x_i - \overline{x}\right)^2 + n\left(\overline{x} - \mu_a\right)^2$$

$$-\left[\sum_{i=1}^{n}\left(x_i - \overline{x}\right)^2 + n\left(\overline{x} - \mu_0\right)^2\right] \tag{9.11}$$

$$= n\left(\overline{x} - \mu_a\right)^2 - n\left(\overline{x} - \mu_0\right)^2$$

Further simplification reveals

$$-\frac{2\sigma^2}{n} l\left(x_1,\ x_2, x_3, \ldots x_n\right) = \left(\overline{x} - \mu_a\right)^2 - \left(\overline{x} - \mu_0\right)^2$$

$$= \overline{x}^2 - 2\overline{x}\mu_a + \mu_a^2 - \left(\overline{x}^2 - 2\overline{x}\mu_0 + \mu_0^2\right)$$

$$= -2\overline{x}\left(\mu_a - \mu_0\right) + \left(\mu_a^2 - \mu_0^2\right)$$

We may now write

$$-\frac{2\sigma^2}{n} l\left(x_1,\ x_2, x_3, \ldots x_n\right) = -2\overline{x}\left(\mu_a - \mu_0\right) + \left(\mu_a^2 - \mu_0^2\right)$$

or

$$\frac{\sigma^2 l\left(x_1,\ x_2, x_3, \ldots x_n\right) + \dfrac{n}{2}\left(\mu_a^2 - \mu_0^2\right)}{n\left(\mu_a - \mu_0\right)} = \overline{x} \tag{9.12}$$

Our conclusion from this exercise is that the likelihood ratio for the hypothesis test H_0: $\mu = \mu_0$ vs. H_a: $\mu = \mu_a$ is proportional to the sample mean. Thus the distribution of the sample mean will provide useful information about the magnitude of the likelihood ratio, and therefore the plausibility of the null hypothesis.

We know that, under the null hypothesis, the quantity $\overline{X}$ follows a normal distribution with mean μ_0 and variance $\dfrac{\sigma^2}{n}$. Therefore, under the null hypothesis, the quantity $TS = \dfrac{\overline{X} - \mu_0}{\sqrt{\sigma^2/n}}$ will follow a standard normal distribution. If the alternative hypothesis is true, then we expect that this test statistic will be large since we assume $\mu_a > \mu_0$. Therefore our critical region will be in the upper, or right hand tail of the standard normal distribution (Figure 9.10).

If the type I error rate is prospectively set at 0.05, then the test critical region will include all values of the test statistic that are at or greater than the

95th percentile value of 1.645. Thus the procedure to carry out this test is as follows

1) State the null and alternative hypothesis
2) Identify the formula for the test statistic
3) Find the critical region for the test statistic under the null hypothesis
4) Collect the data in a well designed and well executed research endeavor
5) Compute the test statistic
6) If the test statistic falls in the critical region, reject the null hypothesis in favor of the alternative.

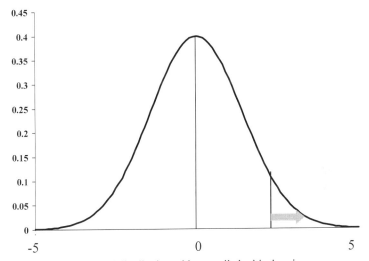

Figure 9.10 Standard normal distribution with one tailed critical region.

This case of statistical testing is termed *simple hypothesis testing*. It is simple because the two hypotheses involve choosing one value of μ from the two offered. In simple hypothesis testing, the test statistic $TS = \dfrac{\overline{X} - \mu_0}{\sqrt{\sigma^2/n}}$ is uniformly most powerful.

9.8.1 One-Sample Composite Hypothesis Testing

In the previous example of statistical hypothesis testing for the mean value of a normal distribution, the researcher knew the value of that mean under each of the null hypothesis and under the alternative hypothesis. Typically, however, the researcher knows the value under the null, but not under the alternative hypothesis. The statement of the null and alternative hypothesis in this scenario is

$$H_0: \mu = \mu_0: \qquad H_a: \mu \neq \mu_0$$

As we think through the development of the likelihood ratio in this circumstance, we can anticipate that we will have to estimate the value of the mean under the alternative hypothesis. This is the principal difference between simple hypothesis testing, in which the value of the parameter under both the null and alternative hypothesis test is known, and composite hypothesis testing, when the value is unknown and must be estimated from the data.

Proceeding as we did in the previous section, we can write the probability density function of x_i given μ and σ^2 (again we assume that the value of σ^2 is known) as $f_X(x) = \dfrac{1}{\sqrt{2\pi\sigma^2}} e^{\dfrac{-(x_i-\mu)^2}{2\sigma^2}}$ for $-\infty \leq x \leq \infty$. We can therefore write the likelihood function as

$$L(x_1, x_2, x_3, \ldots x_n) = \left(2\pi\sigma^2\right)^{-\frac{n}{2}} e^{\dfrac{-\sum_{i=1}^{n}(x_i-\mu)^2}{2\sigma^2}} \qquad (9.13)$$

Under the null hypothesis, equation (9.13) as before becomes

$$L(x_1, x_2, x_3, \ldots x_n \mid H_0) = \left(2\pi\sigma^2\right)^{-\frac{n}{2}} e^{\dfrac{-\sum_{i=1}^{n}(x_i-\mu_0)^2}{2\sigma^2}} \qquad (9.14)$$

Under the alternative hypothesis, equation (9.13) becomes

$$L(x_1, x_2, x_3, \ldots x_n \mid H_a) = \left(2\pi\sigma^2\right)^{-\frac{n}{2}} e^{\dfrac{-\sum_{i=1}^{n}(x_i-\mu_a)^2}{2\sigma^2}} \qquad (9.15)$$

However, the value of μ_a is unknown. What we must do now is identify the estimate of μ_a that maximizes the likelihood function. Fortunately, from Chapter 8, we have the maximum likelihood estimate available, a quantity that meets our exact requirement. Thus we can now write

$$L(x_1, x_2, x_3, \ldots x_n \mid H_a) = \left(2\pi\sigma^2\right)^{-\frac{n}{2}} e^{\dfrac{-\sum_{i=1}^{n}(x_i-\bar{x})^2}{2\sigma^2}}$$

Proceeding, we see that the likelihood ratio is

$$LR = \frac{L\left(x_1, x_2, x_3, \ldots x_n \mid H_a\right)}{L\left(x_1, x_2, x_3, \ldots x_n \mid H_0\right)} = \frac{\left(2\pi\sigma^2\right)^{-\frac{n}{2}} e^{-\frac{\sum_{i=1}^{n}(x_i-\bar{x})^2}{2\sigma^2}}}{\left(2\pi\sigma^2\right)^{-\frac{n}{2}} e^{-\frac{\sum_{i=1}^{n}(x_i-\mu_0)^2}{2\sigma^2}}} = e^{-\frac{1}{2\sigma^2}\left[\sum_{i=1}^{n}(x_i-\bar{x})^2 - \sum_{i=1}^{n}(x_i-\mu_0)^2\right]}$$

Setting $l = \log(LR)$, we can compute

$$-2\sigma^2 l = \left[\sum_{i=1}^{n}(x_i - \bar{x})^2 - \sum_{i=1}^{n}(x_i - \mu_0)^2\right] \tag{9.16}$$

Writing

$$\sum_{i=1}^{n}(x_i - \mu_0)^2 = \sum_{i=1}^{n}(x_i - \bar{x} + \bar{x} - \mu_0)^2 = \sum_{i=1}^{n}(x_i - \bar{x})^2 + n(\bar{x} - \mu_0)^2$$

And equation (9.16) becomes

$$-2\sigma^2 l = -n(\bar{x} - \mu_0)^2$$

Which is equivalent to

$$2l = \frac{(\bar{x} - \mu_0)^2}{\sigma^2/n} \tag{9.17}$$

Under the null hypothesis, the quantity $\dfrac{(\bar{X} - \mu_0)}{\sqrt{\sigma^2/n}}$ follows a standard normal distribution. This is the key statistic when the variance σ^2 is known. Therefore, the square of a standard normal random variable is proportional to the likelihood ratio in this one sample composite hypothesis testing circumstance. It is important to keep in mind the square function from equation (9.17). This exponent focused our attention on the fact that values of $\bar{X}$ that are both much smaller than or much larger than μ_0 will lead to rejection of the null hypothesis. Thus, the composite hypothesis test will have a two tails, with a divided critical region as discussed in section 9.5.

One-sided composite hypothesis tests are also available. For example the statement H0: $\mu = \mu_0$ vs. H_a: $\mu > \mu_0$, focuses attention on values of $\bar{x}$ that are greater than μ_0. In this circumstance, the null hypothesis is rejected only for large values of $\dfrac{(\bar{x} - \mu_0)}{\sqrt{\sigma^2/n}}$.

In this circumstance, the null and alternative hypotheses span many values of μ. This is termed *composite hypothesis testing*.

9.8.2 One-sample Test of the Means with Unknown Variance

The hypothesis tests that we have carried out so far on the normal distribution have assumed that the only unknown distributional parameter was μ. With the value of σ^2 available, no resources had to be expended on its evaluation. However, this ideal circumstance is commonly not the case. In many hypothesis testing circumstances, neither the mean nor the variance of the population is known. Therefore, the sample which is used to carry out the hypothesis test on the mean must also be used to estimate the population variance σ^2.

In this circumstance, we might expect our ability to adequately carry out the hypothesis test on the mean of the distribution to be reduced. Assume that we have a random sample $x_1, x_2, x_3, \ldots, x_n$ that is selected from a normal distribution. In order to carry out the statistical hypothesis test on the mean denoted by $H_0: \mu = \mu_0$ vs. $H_a: \mu \neq \mu_a$ in the case where the researcher knows σ^2, the researcher uses the statistic

$$z = \frac{\bar{X} - \mu}{\sqrt{\sigma^2/n}}$$

In the case where σ^2 is unknown, a reasonable decision would be to replace σ^2 by the maximum likelihood estimator for σ^2, s^2, producing a test statistic

$$t = \frac{\bar{X} - \mu}{\sqrt{s^2/n}}$$

where $s^2 = \dfrac{\sum_{i=1}^{n}(x_i - \bar{x})^2}{n-1}$ The desirable properties of maximum likelihood statistics have been discussed in Chapter 8, and we would expect that our estimate s^2 would be close to σ^2. However, although this may be the case, the distribution of the statistic t is not the same as that of z. The distribution of t has been derived, revealing that the additional requirement of estimating σ^2 introduces additional variability into t that is not present in z.

In section 9.8.3 we will use the likelihood ratio method to demonstrate that the statistic t is the appropriate test statistic for the statistical hypothesis test on the mean of a normal distribution when the variance is unknown.

9.8.3 The Likelihood Ratio Test

We begin this development of the one-sample test on the mean of the normal distribution with a restatement of the problem. We let X_1, X_2, X_3, …, X_n be an i.i.d. sample from a normal distribution with unknown mean μ and unknown variance σ^2. The hypothesis test of interest is the composite hypothesis test

$$H_0: \mu = \mu_0 \text{ vs. } H_a: \mu \neq \mu_a$$

We begin as before by writing $f_X(x) = \dfrac{1}{\sqrt{2\pi\sigma^2}} e^{\frac{-(x_i - \mu)^2}{2\sigma^2}}$. We can therefore write the likelihood function as

$$L(x_1, x_2, x_3, \ldots x_n) = (2\pi\sigma^2)^{-\frac{n}{2}} e^{\frac{-\sum_{i=1}^{n}(x_i - \mu)^2}{2\sigma^2}} \tag{9.18}$$

Under the null hypothesis, equation (9.18) as before becomes

$$L(x_1, x_2, x_3, \ldots x_n \mid H_0) = (2\pi\sigma^2)^{-\frac{n}{2}} e^{\frac{-\sum_{i=1}^{n}(x_i - \mu_0)^2}{2\sigma^2}} \tag{9.19}$$

However, now we must estimate the unknown parameter σ^2 in equation (9.19). Since this likelihood ratio approach requires that we maximize the likelihood function of the data under each of the null and alternative hypotheses, it follows that we should use parameter estimates that produce this maximization. We therefore use the maximum likelihood estimator

$$s^2 = \frac{\sum_{i=1}^{n}(x_i - \overline{x})^2}{n}$$

to estimate σ^2. In this circumstance, we now write

$$\max L(x_1, x_2, x_3, \ldots x_n \mid H_0) = (2\pi s^2)^{-\frac{n}{2}} e^{\frac{-\sum_{i=1}^{n}(x_i - \mu_0)^2}{2s^2}}$$

Under the alternative hypothesis, we must estimate both the mean and the variance of the normal distribution. Thus, we have

$$\max L\left(x_1, x_2, x_3, \ldots x_n \mid H_a\right) = \left(2\pi s^2\right)^{-\frac{n}{2}} e^{\dfrac{-\sum\limits_{i=1}^{n}(x_i-\bar{x})^2}{2s^2}} \tag{9.20}$$

We can now write the likelihood ratio LR as

$$LR = \frac{\max L\left(x_1, x_2, x_3, \ldots x_n \mid H_a\right)}{\max L\left(x_1, x_2, x_3, \ldots x_n \mid H_0\right)} = \frac{\left(2\pi s^2\right)^{-\frac{n}{2}} e^{\dfrac{-\sum\limits_{i=1}^{n}(x_i-\bar{x})^2}{2s^2}}}{\left(2\pi s^2\right)^{-\frac{n}{2}} e^{\dfrac{-\sum\limits_{i=1}^{n}(x_i-\mu_0)^2}{2s^2}}}$$

which upon simplification becomes

$$LR = \frac{e^{\dfrac{-\sum\limits_{i=1}^{n}(x_i-\bar{x})^2}{2s^2}}}{e^{\dfrac{-\sum\limits_{i=1}^{n}(x_i-\mu_0)^2}{2s^2}}}$$

The remainder of our work is familiar to us. We next compute the natural log of the likelihood ratio, l and multiply by $-2S^2$ to find

$$-2s^2 l = \sum_{i=1}^{n}\left(x_i - \bar{x}\right)^2 - \sum_{i=1}^{n}\left(x_i - \mu_0\right)^2 \tag{9.21}$$

We saw before that $\sum\limits_{i=1}^{n}\left(x_i - \mu_0\right)^2 = \sum\limits_{i=1}^{n}\left(x_i - \bar{x}\right)^2 - n\left(\bar{x} - \mu_0\right)^2$. Substituting this result into equation (9.21) and dividing both sides by S^2 reveals

$$-2l = \left[\frac{\left(\bar{x} - \mu_0\right)}{\sqrt{S^2 / n}}\right]^2$$

Thus, the likelihood ratio test suggests that we would reject the null hypothesis for extreme values of the statistic t where

$$t = \frac{\left(\overline{X} - \mu_0\right)}{\sqrt{S^2/n}}$$
(9.22)

which is our desired result. The distribution of the test statistic in (9.22) has the *t* distribution with $n - 1$ degrees of freedom when $\mu = \mu_0$ as discussed in Chapter 5 (Figure 9.11).

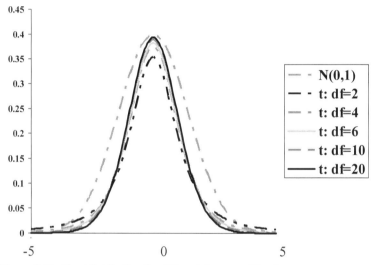

Figure 9.11 The *t* distribution for different degrees of freedom.

9.8.4 One-sample Testing on the Variance

We have used the likelihood ratio test thus far to construct formal statistical hypothesis tests on the mean of a normal distribution. However, this procedure can also be used to construct a hypothesis test for the population variance from the normal distribution.

Consider a sample of i.i.d. observations $X_1, X_2, X_3, \ldots, X_n$ from a $N(\mu, \sigma^2)$ distribution. Our goal is to compute a likelihood ratio based on a statistical test for the hypothesis

$$H_0: \sigma^2 = \sigma_0^2 \quad \text{vs. } H_a: \sigma^2 \neq \sigma_0^2$$

where σ_0^2 is known. In this case, we will assume that the population mean μ is known, and all effort is concentrated on estimating the variance of the distribution based on the sample of n observations. We begin by writing

$$L(x_1, x_2, x_3 ... x_n) = \left(2\pi\sigma^2\right)^{-\frac{n}{2}} e^{-\frac{\sum_{i=1}^{n}(x_i-\mu)^2}{2\sigma^2}}$$

We can now write

$$L(x_1, x_2, x_3 ... x_n \mid H_0) = \left(2\pi\sigma_0^2\right)^{-\frac{n}{2}}. \qquad (9.23)$$

Similarly, we may write

$$L(x_1, x_2, x_3 ... x_n \mid H_a) = \left(2\pi\sigma_a^2\right)^{-\frac{n}{2}} e^{-\frac{\sum_{i=1}^{n}(x_i-\mu)^2}{2\sigma_a^2}} \qquad (9.24)$$

Now, substituting the maximum likelihood estimator for σ^2, $s^2 = \dfrac{\sum_{i=1}^{n}(x_i - u)^2}{n}$ as σ_a^2 in equation (9.24) we may write

$$L(x_1, x_2, x_3 ... x_n \mid H_a) = \left(2\pi s_0^2\right)^{-\frac{n}{2}} e^{-\frac{\sum_{i=1}^{n}(x_i-\mu)^2}{2s_0^2}} = \left(2\pi s_0^2\right)^{-\frac{n}{2}} e^{-\frac{n}{2}} \qquad (9.25)$$

$$LR = \frac{L(x_1, x_2, x_3 ... x_n \mid H_a)}{L(x_1, x_2, x_3 ... x_n \mid H_0)} = \frac{\left(2\pi s^2\right)^{-\frac{n}{2}}}{\left(2\pi\sigma_0^2\right)^{-\frac{n}{2}}} = \left[\frac{S^2}{\sigma_0^2}\right]^{-\frac{n}{2}} \qquad (9.26)$$

We now identify the quantity -2 multiplied by the natural log of the likelihood ratio as

$$2l = -n \ln\left[\frac{s_0^2}{\sigma_0^2}\right] \qquad (9.27)$$

We don't have to identify the distribution of the right hand side of equation (9.27); we can recall at once from our earlier discussions that the statistic

$\dfrac{ns_0^2}{\sigma_0^2} = \dfrac{(n-1)S_0^2}{\sigma_0^2}$ follows a χ^2 distribution with $n - 1$ degrees of freedom (Chapter 5). Thus, using this distribution, we can compute critical regions for a hypothesis test for the variance of a normal distribution.

9.8.5 Example: Examination of Temperature Variation

Use of the χ^2 distribution can lead to remarkable skewed sampling distributions when the data set is small. As an example, consider a meteorologist who is interested in examining environmental temperatures. Specifically, she is interested in determining if there has been a change in the variability of morning temperatures in a major US metropolitan area. Typically, morning temperatures in June are very stable, exhibiting little day–to–day variability. The current understanding is that the variance of these temperatures is 1 degree Fahrenheit during this month of the year.

The researcher is interested in examining whether this sample of temperatures produces a test statistic that is consistent with the null hypothesis that the variance is 1. She plans to observe these temperatures for a week. The sampling distribution of the researcher's test statistic is the χ^2 with 6 degrees of freedom. An examination of this probability distribution demonstrates substantial skewness (Figure 9.12), leading the researcher to suspect that larger values of the sample variance may not be inconsistent with the null hypothesis that $\sigma^2 = 1$. The alternative hypothesis for this investigator is that $\sigma^2 \neq 1$, suggesting a two-sided hypothesis test. The critical region that this researcher has selected a priori is a symmetric one with the total type I error rate of 0.10. The 5 percentile value for this distribution is 1.64; the 95 percentile value is 12.59. Thus a test statistic that falls between 1.64 and 12.59 is consistent with the null hypothesis. The seven temperatures that the researcher has collected are 77, 79, 80, 81, 81, 80, 76. The variance of these temperatures is 1.95, suggesting that the variability of these temperatures may be greater than the expected value of $\sigma^2 = 1$.

However, the test statistic in this example is $6(1.95) = 11.7$ a value that does not fall in either the lower or the upper bound of this critical region. Thus, the null hypothesis is not rejected, and the data are consistent with a variance of 1 degree Fahrenheit.

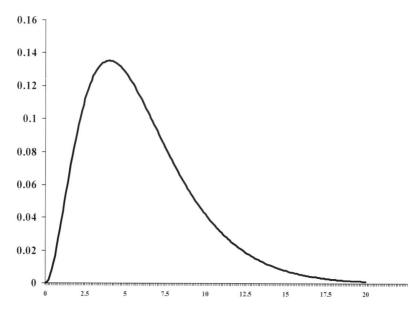

Figure 9.12 Probability density function for the chi square distribution with 6 degrees of freedom.

9.9. Two-Sample Testing for the Normal Distribution

9.9.1. Testing for Equality of Means: Variance Known

The development work that we have carried out so far for executing statistical hypothesis tests for the mean of a normal distribution when we have one-sample generalizes directly to hypothesis testing in the two sample case. In this circumstance, the researcher samples from each of two populations, in the hope of learning whether the means of the two populations are the same. Let X_1, X_2, X_3, ... , X_n be i.i.d. $N(\mu, \sigma^2)$. Analogously, allow Y_1, Y_2, Y_3, ..., Y_m be i.i.d. $N(\tau, \sigma^2)$. Then we may state the statistical hypothesis test as

$$H_0: \mu = \tau \text{ vs. } H_a: \mu \neq \tau$$

Note that in this circumstance, we assume the variances of the two populations are equal and that its value is known. Therefore, all resources and effort are exerted on estimating the means of these two distributions. We will proceed to find a likelihood ratio test statistic for this hypothesis test. We begin by writing

$$L(x_1, x_2, x_3 \ldots x_n, y_1, y_2, y_3, \ldots y_m) = L(x_1, x_2, x_3 \ldots x_n) L(y_1, y_2, y_3, \ldots y_m)$$

$$= \left(2\pi\sigma^2\right)^{-\frac{n}{2}} e^{-\frac{\sum_{i=1}^{n}(x_i - \mu)^2}{2\sigma^2}} \left(2\pi\sigma^2\right)^{-\frac{n_y}{2}} e^{-\frac{\sum_{i=1}^{m}(y_i - \tau)^2}{2\sigma^2}} \tag{9.28}$$

$$= \left(2\pi\sigma^2\right)^{-\frac{(n+m)}{2}} e^{-\frac{\sum_{i=1}^{n}(x_i - \mu)^2 + \sum_{i=1}^{m}(y_i - \tau)^2}{2\sigma^2}}$$

The last line of expression (9.28) is the likelihood function under the null hypothesis, $L(H_0)$. We now compute the likelihood ratio under the alternative hypothesis, using the maximum likelihood estimator for the two parameters μ and τ that we must estimate.

$$L(H_a) = \left(2\pi\sigma^2\right)^{-\frac{(n+m)}{2}} e^{-\frac{\sum_{i=1}^{n}(x_i - \bar{x})^2 + \sum_{i=1}^{m}(y_i - \bar{y})^2}{2\sigma^2}}$$

We now compute the likelihood ratio

$$LR = \frac{\left(2\pi\sigma^2\right)^{-\frac{(n+m)}{2}} e^{-\frac{\sum_{i=1}^{n}(x_i - \bar{x})^2 + \sum_{i=1}^{m}(y_i - \bar{y})^2}{2\sigma^2}}}{\left(2\pi\sigma^2\right)^{-\frac{(n+m)}{2}} e^{-\frac{\sum_{i=1}^{n}(x_i - \mu)^2 + \sum_{i=1}^{m}(y_i - \mu)^2}{2\sigma^2}}} = e^{-\left[\frac{\sum_{i=1}^{n}(x_i - \bar{x})^2 + \sum_{i=1}^{m}(y_i - \bar{y})^2}{2\sigma^2} - \frac{\sum_{i=1}^{m}(y_i - \bar{y})^2 + \sum_{i=1}^{m}(y_i - \mu)^2}{2\sigma^2}\right]}$$

and we can immediately compute $-2\sigma^2 l$

$$-2\sigma^2 l = \sum_{i=1}^{n}\left(x_i - \bar{x}\right)^2 + \sum_{i=1}^{m}\left(y_i - \bar{y}\right)^2 - \left[\sum_{i=1}^{n}(x_i - \mu)^2 + \sum_{i=1}^{m}(y_i - u)^2\right] \tag{9.29}$$

Recall that

$$\sum_{i=1}^{n}(x_i - u)^2 = \sum_{i=1}^{n}\left(x_i - \bar{x}\right)^2 + n\left(\bar{x} - \mu\right)^2$$

$$\sum_{i=1}^{m}(v_i - u)^2 = \sum_{i=1}^{m}\left(v_i - \bar{y}\right)^2 + m\left(\bar{y} - u\right)^2$$

Substituting these results into expression (9.29) and simplifying, we obtain

$$-2\sigma^2 l = n\left(\overline{x}-\mu\right)^2 - m\left(\overline{y}-u\right)^2$$

This is a function of $\overline{X}$ and $\overline{Y}$, the natural sufficient statistics for μ from the two samples respectively. Specifically, the likelihood function is related to the distance between the sample and population means of the two groups respectively. We know that if $\overline{X}$ is normally distributed with mean μ and variance σ^2/n independent of $\overline{Y}$ which itself follows a normal distribution with mean τ and variance σ^2/m, then the statistic

$$\frac{\left(\overline{X}-\mu\right)-\left(\overline{Y}-\mu\right)}{\sqrt{\sigma^2\left(\dfrac{1}{n}+\dfrac{1}{m}\right)}} \tag{9.30}$$

follows a standard normal distribution under H_0. Expression (9.30) reduces to

$$\frac{\overline{X}-\overline{Y}}{\sqrt{\sigma^2\left(\dfrac{1}{n}+\dfrac{1}{m}\right)}}$$

Note that this statistic is proportional to the difference between the sample means and is not related to the ordering of $\overline{x}$ and $\overline{y}$. Thus, for the statistical hypothesis test of the means, we can set up a two-sided critical region, rejecting the null hypothesis if either $\overline{x}$ is much larger or much smaller than $\overline{y}$.

In the circumstance where the variance is unknown but presumed to be equal, we may utilize much of the preceding development. However, we will have to estimate the variance by an estimate that is based on each of the samples of the n x's and the m y's. This pooled estimate S_p is

$$s_p^2 = \frac{\displaystyle\sum_{i=1}^{n}\left(x_i-\overline{x}\right)^2 + \sum_{i=1}^{m}\left(y_i-\overline{y}\right)^2}{n+m-2}$$

We may find the test statistic by noting that the statistic $\left(n+m-2\right)s_p^2/\sigma^2$ follows a χ^2 distribution with $n-m-2$ degrees of freedom. Thus we may create a *t*-statistic by taking the ratio of a $N(0,1)$ random variable divided by the square root of an independent χ^2 divided by its degrees of freedom. Thus

$$\frac{N(0,1)}{\sqrt{\chi^2_\upsilon/\upsilon}} = \frac{\dfrac{\overline{X}-\overline{Y}}{\sqrt{\sigma^2\left(\dfrac{1}{n}+\dfrac{1}{m}\right)}}}{\sqrt{\dfrac{(n+m-2)S_p^2}{\sigma^2}\bigg/ n+m-2}} = \frac{\overline{X}-\overline{Y}}{\sqrt{S_p^2\left(\dfrac{1}{n}+\dfrac{1}{m}\right)}}$$

which follows a t distribution with $n + m - 2$ degrees of freedom.

For the case when the variances are known and unequal (i.e., $\sigma_x^2 \neq \sigma_y^2$), we only need write

$$L(x_1,x_2,x_3...x_n,y_1,y_2,y_3,...y_m) = L(x_1,x_2,x_3...x_n)L(y_1,y_2,y_3,...y_m)$$

$$= \left(2\pi\sigma^2\right)^{-\frac{n}{2}} e^{-\frac{\sum_{i=1}^{n}(x_i-\mu)^2}{2\sigma_x^2}} \left(2\pi\sigma^2\right)^{-\frac{n_y}{2}} e^{-\frac{\sum_{i=1}^{m}(y_i-\tau)^2}{2\sigma_y^2}}$$

$$= \left(2\pi\sigma^2\right)^{-\frac{(n+m)}{2}} e^{-\left[\frac{\sum_{i=1}^{n}(x_i-\mu)^2}{2\sigma_x^2}+\frac{\sum_{i=1}^{m}(y_i-\tau)^2}{2\sigma_y^2}\right]}$$

We now compute the likelihood ratio

$$LR = \frac{\left(2\pi\sigma_x\sigma_y\right)^{-\frac{(n+m)}{2}} e^{-\left[\frac{\sum_{i=1}^{n}(x_i-\overline{x})^2}{2\sigma_x^2}+\frac{\sum_{i=1}^{m}(y_i-\overline{y})^2}{2\sigma_y^2}\right]}}{\left(2\pi\sigma_x\sigma_y\right)^{-\frac{(n+m)}{2}} e^{-\left[\frac{\sum_{i=1}^{n}(x_i-\mu)^2}{2\sigma_x^2}+\frac{\sum_{i=1}^{m}(y_i-\mu)^2}{2\sigma_y^2}\right]}} = e^{-\left[\frac{\sum_{i=1}^{n}(x_i-\overline{x})^2+\sum_{i=1}^{n}(x_i-\mu)^2}{2\sigma_x^2}-\frac{\sum_{i=1}^{m}(y_i-\overline{y})^2+\sum_{i=1}^{m}(y_i-\mu)^2}{2\sigma_y^2}\right]}$$

and we can immediately compute $-2\sigma_x\sigma_y l$

$$-2\sigma_x\sigma_y l \;=\; \sum_{i=1}^{n}\left(x_i-\overline{x}\right)^2+\sum_{i=1}^{m}\left(y_i-\overline{y}\right)^2 \;-\; \left[\sum_{i=1}^{n}\left(x_i-\mu\right)^2+\sum_{i=1}^{m}\left(y_i-u\right)^2\right] \quad (9.31)$$

Proceeding as before, we find

$$-2\sigma_x\sigma_y l \;=\; n\left(\overline{X}-\mu\right)^2-m\left(\overline{Y}-u\right)^2$$

This is a function of $\overline{X}$ and $\overline{Y}$, the natural sufficient statistics for μ from the two samples respectively. Specifically, the likelihood function is related to the distance between the sample and population means of the two groups respectively. We know that if $\overline{X}$ is normally distributed with mean μ and variance $\sigma_x^2\big/ n$ independent of $\overline{Y}$ which itself follows a normal distribution with mean τ and variance $\sigma_y^2\big/ m$, then the statistic

$$\frac{\left(\overline{X}-\mu\right)-\left(\overline{Y}-\mu\right)}{\sqrt{\dfrac{\sigma_x^2}{n}+\dfrac{\sigma_y^2}{m}}} \quad (9.32)$$

follows a standard normal distribution. Expression (9.30) reduces to

$$\frac{\overline{X}-\overline{Y}}{\sqrt{\dfrac{\sigma_x^2}{n}+\dfrac{\sigma_y^2}{m}}}$$

9.9.2 Two-Sample Testing on the Variance

We can move easily to the two sample variance testing issue. In this circumstance, the researcher samples from each of two populations, in the hope of learning whether the variances of the two populations are the same. Let X_1, X_2, X_3, ... , X_n be i.i.d. $N(\mu, \sigma_x^2)$. Analogously, allow Y_1, Y_2, $Y_3,$ ..., Y_m be i.i.d. $N(\tau, \sigma_y^2)$. Then we may state the statistical hypothesis test as

$$H_0: \sigma_x^2 = \sigma_y^2 \text{ vs. } H_a: \sigma_x^2 \neq \sigma_y^2$$

Note that in this case, we assume that the means of the two populations are unknown, and the entire mathematical effort is devoted to estimating the variances of these two distributions. We will proceed to find a likelihood ratio test statistic for this hypothesis test. We begin by writing

$$L(x_1, x_2, x_3 \ldots x_n, y_1, y_2, y_3, \ldots y_m) = L(x_1, x_2, x_3 \ldots x_n) L(y_1, y_2, y_3, \ldots y_m)$$

$$= \left(2\pi\sigma_x^2\right)^{-\frac{n}{2}} e^{-\frac{\sum_{i=1}^{n}(x_i - \mu)^2}{2\sigma_x^2}} \left(2\pi\sigma_y^2\right)^{-\frac{m}{2}} e^{-\frac{\sum_{i=1}^{m}(y_i - \tau)^2}{2\sigma_y^2}} \tag{9.33}$$

$$= \left(2\pi\right)^{-\frac{(n+m)}{2}} \left(\sigma_x^2\right)^{-\frac{n}{2}} \left(\sigma_y^2\right)^{-\frac{m}{2}} e^{-\left(\frac{\sum_{i=1}^{n}(x_i - \mu)^2}{2\sigma_x^2} + \frac{\sum_{i=1}^{m}(y_i - \tau)^2}{2\sigma_y^2}\right)}$$

We now write the likelihood function for this two sample problem under the null hypothesis as

$$L(x_1, x_2, x_3 \ldots x_n, y_1, y_2, y_3, \ldots y_m) = \left(2\pi\right)^{-\frac{(n+m)}{2}} \left(\sigma^2\right)^{-\left(\frac{n}{2} + \frac{m}{2}\right)} e^{-\left(\frac{\sum_{i=1}^{n}(x_i - \mu)^2}{2\sigma^2} + \frac{\sum_{i=1}^{m}(y_i - \tau)^2}{2\sigma^2}\right)} \tag{9.34}$$

It remains for us to identify the maximum value of this likelihood function under the null hypothesis. $L(x_1, x_2, x_3 \ldots x_n, y_1, y_2, y_3 \ldots y_m \mid H_0)$ is maximized when we substitute S_p^2 for σ^2 where S_p^2 is the estimate for σ^2 that is based on both the sample of x's and the sample of y's. Specifically,

$$S_p^2 = \frac{\sum_{i=1}^{n}\left(x_i - \overline{x}\right)^2 + \sum_{i=1}^{m}\left(y_i - \overline{y}\right)^2}{n+m}$$

and thus,

$$L(x_1, x_2, x_3 \ldots x_n, y_1, y_2, y_3, \ldots y_m) = \left(2\pi s_p^2\right)^{-\frac{(n+m)}{2}} e^{-\frac{\sum_{i=1}^{n}(x_i - \overline{x})^2}{2s_p^2} + \frac{\sum_{i=1}^{m}(y_i - \overline{y})^2}{2s_p^2}} \tag{9.35}$$

$$= \left(2\pi s_p^2\right)^{-\frac{(n+m)}{2}} e^{-\frac{\sum_{i=1}^{n}(x_i - \overline{x})^2 + \sum_{i=1}^{m}(y_i - \overline{y})^2}{2s_p^2}}$$

We can now write

$$L(x_1, x_2, x_3 \ldots x_n, y_1, y_2, y_3 \ldots y_m \mid H_0) = \left(2\pi s_p^2\right)^{-\frac{(n+m)}{2}} e^{-\frac{(n+m)}{2}}.$$

This follows since the exponent in the last line of expression (9.35) can be rewritten as

$$\frac{\sum_{i=1}^{n}\left(x_i-\overline{x}\right)^2+\sum_{i=1}^{m}\left(y_i-\overline{y}\right)^2}{2s_p^2}=\frac{(n+m)s_p^2}{2s_p^2}=\frac{(n+m)}{2}$$

We may proceed in the same fashion as we did earlier in order to write the likelihood function under the alternative hypothesis as

$$L(x_1,x_2,x_3...x_n,y_1,y_2,y_3...y_m\mid H_a)$$

$$=\left(2\pi\right)^{-\frac{(n+m)}{2}}\left(\sigma_x^2\right)^{-\frac{n}{2}}\left(\sigma_y^2\right)^{-\frac{m}{2}}e^{-\frac{\sum_{i=1}^{n}\left(x_i-\overline{x}\right)^2}{2\sigma_x^2}+\frac{\sum_{i=1}^{m}\left(y_i-\overline{y}\right)^2}{2\sigma_y^2}} \qquad (9.36)$$

$$=\left(2\pi\right)^{-\frac{(n+m)}{2}}\left(\sigma_x^2\right)^{-\frac{n}{2}}\left(\sigma_y^2\right)^{-\frac{m}{2}}e^{-\frac{n+m}{2}}.$$

Now, substituting the maximum likelihood estimator s_x^2 for σ_x^2 and s_y^2 for σ_y^2 in equation (9.24) we may write

$$L(x_1,x_2,x_3...x_n,y_1,y_2,y_3...y_m\mid H_a)=\left(2\pi\right)^{-\frac{(n+m)}{2}}\left(s_x^2\right)^{-\frac{n}{2}}\left(s_y^2\right)^{-\frac{m}{2}}e^{-\frac{n+m}{2}} \qquad (9.37)$$

And the likelihood ratio for equality of variances becomes

$$LR=\frac{L(x_1,x_2,x_3...x_n,y_1,y_2,y_3...y_m\mid H_a)}{L(x_1,x_2,x_3...x_n,y_1,y_2,y_3...y_m\mid H_0)}$$

$$=\frac{\left(2\pi\right)^{-\frac{(n+m)}{2}}\left(s_x^2\right)^{-\frac{n}{2}}\left(s_y^2\right)^{-\frac{m}{2}}e^{-\frac{n+m}{2}}}{\left(2\pi s_p^2\right)^{-\frac{(n+m)}{2}}e^{-\frac{n+m}{2}}} \qquad (9.38)$$

$$=\frac{\left(s_p^2\right)^{\frac{(n+m)}{2}}}{\left(s_x^2\right)^{\frac{n}{2}}\left(s_y^2\right)^{\frac{m}{2}}}=\frac{\left(s_p^2\right)^{\frac{n}{2}}\left(s_p^2\right)^{\frac{m}{2}}}{\left(s_x^2\right)^{\frac{n}{2}}\left(s_y^2\right)^{\frac{m}{2}}}=\left(\frac{s_p^2}{s_x^2}\right)^{\frac{n}{2}}\left(\frac{s_p^2}{s_y^2}\right)^{\frac{m}{2}}$$

We proceed by writing

$$\left(\frac{s_p^2}{s_x^2}\right) = \frac{\dfrac{\sum\limits_{i=1}^{n}\left(x_i-\overline{x}\right)^2 + \sum\limits_{i=1}^{m}\left(y_i-\overline{y}\right)^2}{n_x+n_y}}{\dfrac{\sum\limits_{i=1}^{n_x}\left(x_i-\overline{x}\right)^2}{n_x}} = \frac{n}{n+m}\cdot\frac{\sum\limits_{i=1}^{n_x}\left(x_i-\overline{x}\right)^2 + \sum\limits_{i=1}^{n_y}\left(y_i-\overline{y}\right)^2}{\sum\limits_{i=1}^{n_x}\left(x_i-\overline{x}\right)^2}$$

$$= \left(1+\frac{m}{n}\right)^{-1}\left[1+\frac{\sum\limits_{i=1}^{m}\left(y_i-\overline{y}\right)^2}{\sum\limits_{i=1}^{n}\left(x_i-\overline{x}\right)^2}\right] = \left(1+\frac{m}{n}\right)^{-1}\left(1+\frac{ms_y^2}{ns_x^2}\right)$$

Analogously, $\left(\dfrac{s_p^2}{s_y^2}\right) = \left(1+\dfrac{n}{m}\right)^{-1}\left(1+\dfrac{ns_x^2}{ms_y^2}\right)$. We may now write the likelihood ratio as

$$LR = \left[\left(1+\frac{m}{n}\right)^{-1}\left(1+\frac{ms_y^2}{ns_x^2}\right)\right]^{\frac{n}{2}}\left[\left(1+\frac{n}{m}\right)^{-1}\left(1+\frac{ms_y^2}{ns_x^2}\right)\right]^{\frac{m}{2}}.$$

We need proceed no further to observe that the value of the likelihood ratio is directly related to the ratio of the variance estimates S_x^2 and S_y^2. The task before us is to find a function of the ratio *of* these two variances whose probability distribution is easily identified. Recall that the statistic $\dfrac{(n-1)S_x^2}{\sigma_x^2}$ follows a χ^2 distribution with $n-1$ degrees of freedom, and similarly $\dfrac{(m-1)S_y^2}{\sigma_y^2}$ follows a χ^2 distribution with $m-1$ degrees of freedom. Since the two-samples are independent, the ratio of these independent χ^2 statistics follows an F distribution with n_x-1 and n_y-1 degrees of freedom. However, under the condition of the null hypothesis $\sigma_x^2=\sigma_y^2$, this ratio becomes $\dfrac{(n-1)S_x^2}{(m-1)S_y^2}$. Therefore, for the hypothesis test of

$$H_0\text{: } \sigma_x^2 = \sigma_y^2 \quad \text{vs. } H_a\text{: } \sigma_x^2 \neq \sigma_y^2$$

we may use the test statistic

$$\frac{(n-1)S_x^2}{(m-1)S_y^2} = \frac{nS_x^2}{mS_y^2} \approx F_{n-1,\,m-1} \tag{9.39}$$

and we will reject the null hypothesis if this test statistic is either too small or too large.

9.9.3 The Behrens-Fisher Problem

Hypothesis testing for the equality of the variance from two populations is often carried out as a preamble to hypothesis testing on the means of two populations. The motivation for this approach is that the test statistic that is used for carrying out hypothesis testing on the means depends on the conclusion of the hypothesis testing that is carried out on the variances.

As before, let $X_1, X_2, X_3, \ldots, X_n$ be i.i.d. $N(\mu, \sigma_x^2)$. Analogously, allow $Y_1, Y_2, Y_3, \ldots, Y_m$ be i.i.d. $N(\tau, \sigma_y^2)$. Then may state the statistical hypothesis test as

$$H_0: \sigma_x^2 = \sigma_y^2 \text{ vs. } H_a: \sigma_x^2 \neq \sigma_y^2$$

If the null hypothesis is not rejected, then it is reasonable to conclude that the variances of the two populations are the same, and can be estimated by combining the sample estimates. If we assume for example that

$$S_x^2 = \frac{\sum_{i=1}^{n}(x_i - \bar{x})^2}{n-1} \text{ and } S_y^2 = \frac{\sum_{i=1}^{m}(y_i - \bar{y})^2}{m-1}, \text{ then a common pooled estimator } S_p^2$$

of σ^2 is

$$S_p^2 = \frac{(n-1)S_x^2 + (m-1)S_y^2}{n+m-2}, \tag{9.40}$$

and the t test for an evaluation of the equality of means is

$$\frac{\bar{X} - \bar{Y}}{\sqrt{\left(\frac{1}{n} + \frac{1}{m}\right)S_p^2}} \text{ follows a } t \text{ distribution} \tag{9.41}$$

with $n_x + n_y - 2$ degrees of freedom.

When the hypothesis of equality (or homogeneity) of variance is rejected, then a reasonable test statistic is

$$\frac{\bar{x} - \bar{y}}{\sqrt{\dfrac{s_x^2}{n} + \dfrac{s_y^2}{m}}}. \tag{9.42}$$

The distribution of expression (9.42) depends on the circumstances. If the sample size is large, then it is reasonable to assume that under the null hypothesis, expression (9.42) follows a standard normal distribution. However, when the sample size is small the distribution of (9.42) is problematic. Most commonly a t distribution is assumed, but the degrees of freedom v must be computed as:

$$v = \frac{\left(\dfrac{s_x^2}{n} + \dfrac{s_y^2}{m}\right)^2}{\dfrac{\left(\dfrac{s_x^2}{n}\right)^2}{n-1} + \dfrac{\left(\dfrac{s_y^2}{m}\right)^2}{m-1}} \tag{9.43}$$

There are several algorithms that have been developed that guide the user as to how to carry out hypothesis testing on the means of samples from two populations (Miller, 1981). However, the user must also be cognizant that with every hypothesis test, the type I error increases. Therefore the researcher's best tactic is to review the literature and the experience of others in drawing conclusions about the distributional assumptions of the data and the likelihood that the variances are equal before the data for her research effort are collected. In this way, the researcher can select the correct sample size and carry out one hypothesis test that will answer the important scientific question.

9.9.4 Paired Testing

The two-sample testing that we have studied thus far in this section has focused on the collection of two independent samples. Examples of experiments that are appropriate for this setting are weights from a sample of males and a sample of females, duration of hospitalizations due to onsite accidents from construction workers in two different cities, and measures of frozen CO_2 levels on the north poles of Venus and Mars. The critical assumption about independence is that information about the values of observations in one sample provides no information about the value of that information in another sample.

However, there is a class of samples that violates this independence assumption. Consider for example an investigator who is interested in comparing the change in ozone levels during a day in a major urban center. In this research endeavor, ozone levels are collected in the morning and again in the evening on n days randomly selected between May 1 and Oct 1. In this setting, we can reasonably expect that the assumption of independence of the ozone samples at the two time points would be violated. Since days on which the ozone levels are higher than average in the morning are likely to produce evening ozone levels that are higher than the average evening ozone level, the "within day" ozone levels are related. The presence of this dependence does not invalidate the experiment. It merely suggests that a modification of the analysis is in order.

We will now parameterize this model. Let $X_{i,j}$ be the jth ozone level on the ith day where $j = 1, 2$ and $i = 1, 2, 3, \ldots , n$. Let $X_{i,j}$ follow a normal distribution with mean μ_j $j = 1, 2$, and variance σ^2. This data set represents n pairs of data points $(x_{11}, x_{1,2})$, $(x_{21}, x_{2,2})$, $(x_{31}, x_{3,2})$, $\ldots$, $(x_{n,1}, x_{n,2})$ (Table 9.1). The statistical hypothesis is

$$H_0: \mu_1 = \mu_2 \quad \text{versus} \quad H_a: \mu_1 \neq \mu_2$$

If we let $\bar{x}_1 = \dfrac{\displaystyle\sum_{i=1}^{n} x_{i,1}}{n}$ and $\bar{x}_2 = \dfrac{\displaystyle\sum_{i=1}^{n} x_{i,2}}{n}$, then from the previous section on the two-sample mean test based on samples from a normal distribution, we may write the test statistic as

$$\frac{\overline{X}_1 - \overline{X}_2}{\sqrt{Var\left[\overline{X}_1 - \overline{X}_2\right]}} \text{ follows a } N(0,1)$$

and it remains for us to identify the $Var\left[\bar{x}_1 - \bar{x}_2\right]$. Begin by considering $Var\left[x_{i,1} - x_{i,2}\right]$. We can write at once that

$$Var\left[X_{i,1} - X_{i,2}\right] = Var\left[X_{i,1}\right] + Var\left[X_{i,2}\right] - 2Cov\left[X_{i,1}, X_{i,2}\right]$$
$$= 2\sigma^2 - 2Cov\left[X_{i,1}, X_{i,2}\right].$$

(9.44)

Table 9.1. Data Layout for Paired Analysis

Obs	First Obs	Second	Difference
1	x_{11}	x_{12}	d_1
2	x_{21}	x_{22}	d_2
3	x_{31}	x_{32}	d_3
.	.	.	.
.	.	.	.
.	.	.	.
n	x_{n1}	x_{n2}	d_n

Recall that the correlation coefficient $\rho = \dfrac{Cov\left[X_{i,1}, X_{i,2}\right]}{\sqrt{\sigma_{x_{i,1}}^2 \sigma_{x_{i,2}}^2}} = \dfrac{Cov\left[X_{i,1}, X_{i,2}\right]}{\sigma^2}$, or

$Cov\left[X_{i,1}, X_{i,2}\right] = \rho\sigma^2$. Substituting this result into expression (9.44), we may write $Var\left[X_{i,1} - X_{i,2}\right] = 2\sigma^2 - 2\rho\sigma^2 = 2\sigma^2\left(1-\rho\right)$. Finally, we can write

$$Var\left[\bar{x}_1 - \bar{x}_2\right] = 2\frac{\sigma^2}{n}\left(1-\rho\right),$$ and complete the test statistic for the paired test.

$$\frac{\bar{x}_1 - \bar{x}_2}{\sqrt{2\dfrac{\sigma^2}{n}\left(1-\rho\right)}} \text{ follows a } N(0,1).$$

This research setting commonly occurs when the value of σ^2 is unknown. We may address this issue by recalling that a t statistic is defined as the ratio of a standard normal random variable and the square root of a χ_v^2

divided by its degrees of freedom v. Recall that $\dfrac{(n-1)S^2}{\sigma^2}$ is a χ^2_{n-1} that is

independent of $\dfrac{\overline{X}_1 - \overline{X}_2}{\sqrt{2\dfrac{\sigma^2}{n}(1-\rho)}}$. Thus we may construct

$$t_v = \dfrac{\dfrac{\overline{X}_1 - \overline{X}_2}{\sqrt{2\dfrac{\sigma^2}{n}(1-\rho)}}}{\sqrt{\dfrac{(n-1)S^2}{\sigma^2}}} = \dfrac{\overline{X}_1 - \overline{X}_2}{\sqrt{2\dfrac{S^2}{n}(1-\rho)}}$$

This test statistic, commonly used to compare the means of paired data, is the *paired t-test*. The computation of this test statistic is simplified by recognizing that the estimate of $Var\left[\overline{X}_1 - \overline{X}_2\right]$ is simply the sample variance of the n differences d_l where $d_l = X_{i,1} - X_{i,2}$.

9.10 Likelihood Ratio Test and the Binomial Distribution

The binomial distribution is one of the simplest distributions to understand and derive and utilize as the basis of hypothesis testing. Events that (1) are dichotomous in nature and (2) occur frequently but should not be considered rare, are typically the data that contribute to a statistical hypothesis. In this section we will explore the use of one-sample and two-sample hypothesis testing that is based on the binomial distribution.

9.10.1 Binomial Distribution: One-sample Testing

Let X be a random variable that follows a Bernoulli distribution with parameters p for small samples. Then we will write

$$P[X = 1] = p \qquad (9.45)$$

We are interested in carrying out a hypothesis test for value of p based on the collection of a sample of n i.i.d. random variables that follow a Bernoulli distribution. We can therefore write the null hypothesis as

$$H_0: p = p_0 \quad \text{versus} \quad H_a: p \neq p_0$$

We will construct a likelihood ratio test to examine and compare the evidence for the null and alternative hypotheses. The likelihood function for n random variables $x_1, x_2, x_3, \ldots, x_n$ may be written as

$$L(x_1, x_2, x_3, \ldots, x_n) = p^{\sum_{i=1}^{n} x_i} (1-p)^{n - \sum_{i=1}^{n} x_i}$$

Under H_0, $p = p_0$, allowing us to write the value of the likelihood function under the null hypothesis as

$$L(x_1, x_2, x_3, \ldots, x_n \mid H_0) = p_0^{\sum_{i=1}^{n} x_i} (1-p_0)^{n - \sum_{i=1}^{n} x_i} \tag{9.46}$$

Under the alternative hypothesis, we replace the unknown parameter p with the best estimate of this parameter $\overline{X}_n$. This estimator maximizes the likelihood of the data under the alternative hypothesis. We now write

$$L(x_1, x_2, x_3, \ldots, x_n \mid H_a) = \overline{x}^{\sum_{i=1}^{n} x_i} (1-\overline{x})^{n - \sum_{i=1}^{n} x_i}. \tag{9.47}$$

We can now write the likelihood ratio for the statistical hypothesis on the value of p.

$$LR = \frac{L(x_1, x_2, x_3, \ldots, x_n \mid H_a)}{L(x_1, x_2, x_3, \ldots, x_n \mid H_0)} = \frac{\overline{x}^{\sum_{i=1}^{n} x_i} (1-\overline{x})^{n - \sum_{i=1}^{n} x_i}}{p_0^{\sum_{i=1}^{n} x_i} (1-p_0)^{n - \sum_{i=1}^{n} x_i}} \tag{9.48}$$

We observe that the likelihood ratio is a function of the separation of $\overline{x}$ and p_0.

Taking logs we see
$$\log(LR) =$$
$$\sum_{i=1}^{n} x_i \left[\log(\overline{x}) - \log(p_0) \right] + \left(n - \sum_{i=1}^{n} x_i \right) \left[\log(1-\overline{x}) - \log(1-p_0) \right]$$

Thus we will reject the null hypothesis for values of $\overline{x}$ that are observed to be "far" from p_0 the value of the parameter under the null hypothesis.

9.10.2 Example

A particular baseball batter historically gets on base 20% of the time. In his last eleven at bats, he has gained base 8 times. Is it likely that his on-base percentage has increased? We will use statistical hypothesis testing here to illustrate the application of the binomial distribution in this setting. We may write the null and alternative hypothesis as

$$H_0: p = 0.20 \text{ versus } H_a: p > 0.20.$$

We have to ask how much evidence of 8 successful attempts to get on base out of 11 attempts contradict the null hypothesis. Under the null hypothesis the probability of an event at this level or more extreme is

$$\sum_{k=8}^{11} \binom{11}{k} 0.20^k 0.80^{11-k} \approx 0.00021$$

We would conclude that the data are unsupportive of the null hypothesis.

9.10.3 Large Sample Testing for the Binomial Distribution

When there is a large number of subjects in the sample, we might avail ourselves of large sample techniques. From Chapter 7, recall that the sampling distribution of $S_n = \sum_{i=1}^{n} X_i$ is normally distributed with mean np and variance $np(1 - p)$. Since the test statistic is computed under the null hypothesis, we might write that, under H_0,

$$\frac{\overline{X} - p_0}{\sqrt{p_0(1 - p_0)/n}} \sim N(0,1) \tag{9.49}$$

9.10.4 Example in Politics

A politician in an election anticipates that she will win the election with 55% of the voting electorate of over 1 million voters casting a ballot for her. Out of a sample of 500 votes, only 200 have voted for her. Although the pattern of voting may be more revealing (i.e., which precincts have responded in these early tabulations) we can address the sampling probabilities. In this case, $p_0 = 0.55$, and we may write the test statistic as

$$\frac{\bar{x}-p_0}{\sqrt{p_0\left(1-p_0\right)/n}}=\frac{0.40-0.55}{\sqrt{(0.55)(0.45)/500}}=-6.7$$

suggesting that, everything else being equal, the early returns are not promising for the inquiring candidate.

■

9.10.5 Binomial Distribution: Two-Sample Testing

The development of the two-sample binomial hypothesis follows the development of the one-sample test of the previous subsection. In this setting, the statistician has two samples of observations. The first sample contains n observations x_1, x_2, x_3, ..., x_n from a Bernoulli (p_x) probability distribution. The second sample of m observations, y_1, y_2, y_3, ..., y_m observations is from a Bernoulli (p_y) distribution. The statistical hypothesis to be evaluated is

$$H_0: p_x = p_y \quad \text{versus} \quad H_a: p_x \neq p_y$$

As before, we construct the likelihood function of the data under the null and alternative hypothesis test. Beginning with the alternative hypothesis, the likelihood function is easily written as

$$L\left(x_1,x_2,x_3,...,x_n,y_1,y_2,y_3,...,y_m \mid H_a\right) = p_x^{\sum_{i=1}^{n} x_i}\left(1-p_x\right)^{n-\sum_{i=1}^{n} x_i} p_y^{\sum_{i=1}^{m} y_i}\left(1-p_y\right)^{m-\sum_{i=1}^{m} y_i}$$

Under the alternative hypothesis, we replace the unknown parameter p_x with the best estimator of this parameter $\bar{x}$ and replace p_y with the estimator of this quantity based on the data from the second sample $\bar{y}$. These estimators maximize the likelihood of the data under the alternative hypothesis. We can now write

$$L\left(x_1,x_2,x_3,...,x_n,y_1,y_2,y_3,...,y_m \mid H_a\right) = \left(\bar{x}\right)^{\sum_{i=1}^{n} x_i}\left(1-\bar{x}\right)^{n-\sum_{i=1}^{n} x_i}\left(\bar{y}\right)^{\sum_{i=1}^{m} y_i}\left(1-\bar{y}\right)^{m-\sum_{i=1}^{m} y_i}$$

Under the null hypothesis, we must find a common estimator for the parameter p based on a combined sample estimate. The assertion that $p_x = p_y = p$ implies that each of the two samples of observations are taken from the same Bernoulli (p) distribution. In this circumstance, the maximum likelihood estimator T based on $n + m$ observations is

$$T = \frac{\sum_{i=1}^{n} x_i + \sum_{i=1}^{m} y_i}{n+m}$$

The likelihood function under the null hypothesis as

$$L\left(x_1, x_2, x_3, \ldots, x_n, y_1, y_2, y_3, \ldots, y_m \mid H_0\right) = T^{\sum_{i=1}^{n} x_i + \sum_{i=1}^{m} y_i} \left(1 - T\right)^{n+m-\sum_{i=1}^{n} x_i - \sum_{i=1}^{m} y_i}$$

We can therefore write the likelihood ratio as

$$LR = \frac{\left(\overline{x}\right)^{\sum_{i=1}^{n} x_i} \left(1 - \overline{x}\right)^{n - \sum_{i=1}^{n} x_i} \left(\overline{y}\right)^{\sum_{i=1}^{m} y_i} \left(1 - \overline{y}\right)^{m - \sum_{i=1}^{m} y_i}}{T^{\sum_{i=1}^{n} x_i + \sum_{i=1}^{m} y_i} \left(1 - T\right)^{n+m-\sum_{i=1}^{n} x_i - \sum_{i=1}^{m} y_i}}$$

$$= \left[\frac{\left(\overline{x}\right)^{\sum_{i=1}^{n} x_i} \left(1 - \overline{x}\right)^{n - \sum_{i=1}^{n} x_i}}{\left(T\right)^{\sum_{i=1}^{n} x_i} \left(1 - T\right)^{n - \sum_{i=1}^{n} x_i}}\right]\left[\frac{\left(\overline{y}\right)^{\sum_{i=1}^{m} y_i} \left(1 - \overline{y}\right)^{m - \sum_{i=1}^{m} y_i}}{\left(T\right)^{\sum_{i=1}^{m} y_i} \left(1 - T\right)^{m - \sum_{i=1}^{m} y_i}}\right]$$

This likelihood ratio is a function of how different the sample estimates $\overline{x}$ and $\overline{y}$ are from the combined estimator T. The greater the difference $\left|\overline{x} - \overline{y}\right|$, the larger the value of the likelihood ratio and the greater the plausibility of the alternative hypothesis becomes.

Following the development of the large sample test statistic for the one-sample binomial test, we can write

$$\frac{\overline{X} - \overline{Y}}{\sqrt{Var\left[\overline{X} - \overline{Y}\right]}} \sim N(0,1) \tag{9.50}$$

It remains for us to compute the denominator of the test statistic in expression (9.50). As before we use the large sample estimator T based on $n + m$ observations to write the test statistic as

$$\frac{\overline{X} - \overline{Y}}{\sqrt{T\left(1 - T\right)/(n + m)}} \sim N(0,1) \tag{9.51}$$

9.10.6 Clinical Example

A clinical study is testing the effectiveness of a new therapy that may help to reduce the probability that high risk patients will have a cerebrovascular accident (also known as a stroke). Two hundred patients receive standard medical care, and three hundred additional patients receive standard medical care plus the new therapy. After a year of observations, 50 of the patients who received standard medical care had a stroke, while 63 of the patients who received the new therapy had this event occur. In this example,

$$\bar{x} = \frac{50}{200} = 0.25 : \bar{y} = \frac{63}{300} = 0.21 \; : T = \frac{50+63}{200+300} = 0.226 \qquad (9.52)$$

and the test statistic is

$$\frac{0.25-0.21}{\sqrt{0.226(1-0.226)/(500)}} = 2.14$$

The two-sided p-value associated with this z-score is $2(1 - F_z(2.14)) = 0.032$, suggesting that, if the research is well-designed and well-executed, then the type I error is low enough to suggest that it is likely that the findings in the study of a reduced stroke rate associated with the new therapy is not a sampling error, but reflects a true population finding. ∎

.

9.11 Likelihood Ratio Test and the Poisson Distribution

The Poisson probability distribution is most useful for the statistical evaluation of the occurrence of rare events over the course of time. Like the binomial distribution, the Poisson distribution lends itself to statistical hypothesis testing. We will examine the use of this distribution in one-sample and two-sample testing.

9.11.1 Poisson Distribution: One-Sample Testing

In one-sample testing that employs the Poisson distribution, the hypothesis test of interest is one that is based on a decision about the value of the parameter λ. Specifically, the statistical hypothesis is

$$H_0: \lambda = \lambda_0 \qquad \text{versus} \qquad H_a: \lambda \neq \lambda_0$$

As before, we will construct the likelihood ratio L such that large values of L provide evidence against the null and for the alternative hypothesis. We begin by assuming that we have a collection of n independent and identically distributed random variables $X_1, X_2, X_3, \ldots, X_n$ that follow a Poisson distribution with parameter λ. The probability mass function for x_i is

$$P[X = x_i] = \frac{\lambda^{x_i}}{x_i!} e^{-\lambda}$$

We start by identifying the form of the likelihood function $L(x_1, x_2, x_3, \ldots, x_n)$ as

$$L(x_1, x_2, x_3, \ldots, x_n) = \frac{\lambda^{\sum_{i=1}^{n} x_i}}{\prod_{i=1}^{n} x_i!} e^{-n\lambda}$$

We can quickly write the value of the likelihood function under the null hypothesis

$$L(x_1, x_2, x_3, \ldots, x_n \mid H_0) = \frac{\lambda_0^{\sum_{i=1}^{n} x_i}}{\prod_{i=1}^{n} x_i!} e^{-n\lambda_0} \qquad (9.53)$$

In order to maximize the likelihood of the data under the alternative hypothesis we write the likelihood function, substituting the maximum likelihood estimator $\overline{X}_n$ for the parameter λ as follows:

$$L(x_1, x_2, x_3, \ldots, x_n \mid H_a) = \frac{\overline{X}^{\sum_{i=1}^{n} x_i}}{\prod_{i=1}^{n} x_i!} e^{-n\overline{X}} \qquad (9.54)$$

The likelihood ratio L can now be constructed

$$L = \frac{L(x_1, x_2, x_3, \ldots, x_n \mid H_a)}{L(x_1, x_2, x_3, \ldots, x_n \mid H_0)}$$

$$= \frac{\dfrac{\overline{X}_n^{\sum_{i=1}^{n} x_i} e^{-n\overline{x}_n}}{\prod_{i=1}^{n} x_i!}}{\dfrac{\lambda_0^{\sum_{i=1}^{n} x_i} e^{-n\lambda_0}}{\prod_{i=1}^{n} x_i!}} = \frac{\overline{X}_n^{\sum_{i=1}^{n} x_i} e^{-n\overline{x}_n}}{\lambda_0^{\sum_{i=1}^{n} x_i} e^{-n\lambda_0}} \qquad (9.55)$$

Thus the likelihood ratio becomes

$$L = \left[\frac{\overline{x}_n}{\lambda_0} \right]^{\sum_{i=1}^{n} x_i} e^{-n(\overline{x}_n - \lambda_0)},$$

and we see that the tenability of the null hypothesis depends on a comparison of the values of $\overline{x}_n$ and λ_0. Thus, the farther $\overline{x}_n$ is from λ_0, the greater the evidence the sample of data contains against the null hypothesis and for the alternative hypothesis.

9.11.2 Example

A neighborhood expresses frequent concerns about the dangers of a traffic intersection. Over the past two years, there have been 16 accidents at this uncontrolled intersection. The municipality has finally responded to this concern and has converted the uncontrolled intersection to a controlled one by installing a stop sign on each of the four roads that enter the problematic intersection. This change was made one year ago and, at the time, the stop signs were put in place, a prospective decision was made to evaluate the effect the stop signs had on the subsequent accident rate. Subsequent to that time, there have been three accidents. How likely is it that the change in accident rate is due solely to chance alone and not the stop sign?

Assuming that accidents occur independently of one another, we can use the Poisson distribution as a basis of the hypothesis test on the parameter λ. Over the past two years, there have been eight accidents per year. Specifically, the hypothesis test to be evaluated is whether over the past year, the average number of accidents is different than that observed historically, or

$$H_0: \lambda = 8 \quad \text{versus} \quad H_a: \lambda < 8.$$

We can use the Poisson distribution with parameter $\lambda = 8$ to compute the probability that three or fewer accidents have to occur over the year.

$$\sum_{k=0}^{3} \frac{8^k}{k!} e^{-8} = e^{-8} \left[1 + \frac{8^1}{1!} + \frac{8^2}{2!} + \frac{8^3}{3!} \right] = 0.042$$

It is unlikely that three accidents would have occurred in a year by chance alone under the null hypothesis. The engineers conclude that the traffic light, while not completely solving the traffic problem, has had a beneficial impact.

From a large sample perspective, we can construct a hypothesis test based on the observation that $S_n = \sum_{i=1}^{n} x_i$ has an asymptotic normal distribution with mean $n\lambda_0$ and variance $n\lambda_0$. We can then write that a large sample test statistic would be

$$\frac{S_n - n\lambda_0}{\sqrt{n\lambda_0}} \text{ follows a } N(0,1) \tag{9.56}$$

9.11.3 Poisson Distribution: Two-Sample Testing

Developing the two-sample hypothesis test that is based on the Poisson distribution will follow the development of the one-sample testing circumstance. In this case, we will assume that we have two independent collections of observations. The first is a collection of n independent and identically distributed random variables $X_1, X_2, X_3, \ldots, X_n$ that follow a Poisson distribution with parameter λ_x. A second sample of m observations denoted by $Y_1, Y_2, Y_3, \ldots,$ Y_m follows a Poisson distribution with parameter λ_y. The statistical hypothesis test is

$$H_0: \lambda_x = \lambda_y \qquad \text{vs.} \qquad H_a: \lambda_x \neq \lambda_y$$

We start by identifying the form of the likelihood function $L\left(x_1, x_2, x_3, \ldots, x_n, y_1, y_2, y_3, \ldots y_n\right)$ as

$$L\left(x_1, x_2, x_3, \ldots, x_n, y_1, y_2, y_3, \ldots y_n\right) = \frac{\lambda_x^{\sum_{i=1}^{n} x_i}}{\prod_{i=1}^{n} x_i!} e^{-n\lambda_x} \frac{\lambda_y^{\sum_{i=1}^{m} y_i}}{\prod_{i=1}^{m} y_i!} e^{-n\lambda_y}$$

We need to compare the ratio of the likelihood functions under the null and alternative hypotheses is order to visualize the test statistic. The likelihood function under the alternative hypothesis will have the maximum likelihood estimator $\bar{x}$ substituted for λ_x, and similarly λ_y will be replaced by $\bar{y}$. We can therefore write

$$L\left(x_1, x_2, x_3, \ldots, x_n, y_1, y_2, y_3, \ldots y_n \mid H_a\right) = \frac{\bar{x}^{\sum_{i=1}^{n} x_i}}{\prod_{i=1}^{n} x_i!} e^{-n\bar{x}} \frac{\bar{y}^{\sum_{i=1}^{m} y_i}}{\prod_{i=1}^{m} y_i!} e^{-m\bar{y}} \tag{9.57}$$

In order to write the likelihood function of the collected data under the null hypothesis where $\lambda_x = \lambda_y = \lambda$, we must construct a combined estimator of λ from the two sample sets. In this case, the estimator is the sample mean based on all n + m observations

$$W = \frac{\sum_{i=1}^{n} x_i + \sum_{i=1}^{m} y_i}{n+m} = \frac{n\overline{x} + m\overline{y}}{n+m}$$

The likelihood function under the null hypothesis may now be written as

$$L\left(x_1, x_2, x_3, \ldots, x_n, y_1, y_2, y_3, \ldots y_n \mid H_0\right) = \frac{W^{\sum_{i=1}^{n} x_i + \sum_{i=1}^{m} y_i}}{\prod_{i=1}^{n} x_i! \prod_{i=1}^{m} y_i!} e^{-(n+m)W}$$

The likelihood ratio L can now be constructed

$$L = \frac{L\left(x_1, x_2, x_3, \ldots, x_n, y_1, y_2, y_3, \ldots y_n \mid H_a\right)}{L\left(x_1, x_2, x_3, \ldots, x_n, y_1, y_2, y_3, \ldots y_n \mid H_0\right)} = \frac{\dfrac{\overline{x}^{\sum_{i=1}^{n} x_i}}{\prod_{i=1}^{n} x_i!} e^{-n\overline{x}} \dfrac{\overline{y}^{\sum_{i=1}^{m} y_i}}{\prod_{i=1}^{m} y_i!} e^{-m\overline{y}}}{\dfrac{W^{\sum_{i=1}^{n} x_i + \sum_{i=1}^{m} y_i}}{\prod_{i=1}^{n} x_i! \prod_{i=1}^{m} y_i!} e^{-(n+m)W}}$$

$$\quad (9.58)$$

$$= \left[\frac{\overline{x}}{W}\right]^{\sum_{i=1}^{n} x_i} \left[\frac{\overline{y}}{W}\right]^{\sum_{i=1}^{m} y_i}$$

From equation (9.58), we see that if the null hypothesis is true, then we would expect (within a small degree of sampling error) that $\overline{x} = \overline{y}$. Thus $\overline{x} = \overline{y} = W$, and the likelihood ratio would be one. The farther apart $\overline{x}$ and $\overline{y}$ are, then the less tenable the null hypothesis becomes. Thus, the likelihood ratio test for comparing the parameters of two Poisson distributions reduces to an examination of the differences between the means $\overline{x}$ and $\overline{y}$.

From a large sample perspective, we can construct a hypothesis test based on the observations from a suitable large sample. We know that

$S_n = \sum_{i=1}^{n} X_i$ has an asymptotic normal distribution. Similarly, we observe that

$S_m = \sum_{i=1}^{m} Y_i$ follow independent normal distributions with mean and variance $m\lambda$.

Using the combined sample mean W as the best estimate for λ, we may write the large sample test statistic as

$$\frac{\sum_{i=1}^{n} X_i - \sum_{i=1}^{m} Y_i}{\sqrt{(n+m)W}} \sim N(0,1)$$

or, basing the hypothesis test on the sample means, we may write

$$\frac{\overline{X} - \overline{Y}}{\sqrt{\left(\dfrac{1}{n} + \dfrac{1}{m}\right)W}} \sim N(0,1)$$

An interesting small sample test statistic follows from Chapter 4. Assume that $S_n = \sum_{i=1}^{n} x_i$ and $T_m = \sum_{i=1}^{m} y_i$. If we can determine the probability distribution of S_n given the sum $S_n + T_m$, we will have a useful exact text for the hypothesis test that examines the equality of the Poisson parameters λ_x and λ_y. We seek $P[S_n = s \mid S_n + T_m = s + t]$. We can write this as

$$
\begin{aligned}
P[S_n = s \mid S_n + T_m = s+t] &= \frac{P[S_n = s \cap T_m = t]}{P[S_n + T_m = s+t]} \\
&= \frac{P[S_n = s]P[T_m = t]}{P[S_n + T_m = s+t]}
\end{aligned}
\tag{9.59}
$$

Each of the probabilities in the last line of expression (9.59) may be expressed using the Poisson distribution function.

$$P[S_n = s \mid S_n + T_m = s+t] = \frac{P[S_n = s\,]P[T_m = t]}{P[S_n + T_m = s+t]} = \frac{\dfrac{\lambda_x^s}{s!}e^{-\lambda_x}\dfrac{\lambda_y^t}{t!}e^{-\lambda_y}}{\dfrac{(\lambda_x + \lambda_y)^{s+t}}{(s+t)!}e^{-\lambda_x+\lambda_y}}$$

This last term can be easily written as

$$\frac{(s+t)!}{s!t!}\left(\frac{\lambda_x}{\lambda_x + \lambda_y}\right)^s \left(\frac{\lambda_y}{\lambda_x + \lambda_y}\right)^t = \binom{s+t}{s}\left(\frac{\lambda_x}{\lambda_x + \lambda_y}\right)^s \left(\frac{\lambda_y}{\lambda_x + \lambda_y}\right)^t \qquad (9.60)$$

which we recognize at once as being a binomial probability. Under the null hypothesis of $\lambda_x = \lambda_y = \lambda$, we can write this binomial probability as

$$P[S_n = s \mid S_n + T_m = s+t] = \binom{s+t}{s}\left(\frac{1}{2}\right)^{s+t}$$

9.12 The Multiple Testing Issue

Executing several hypothesis tests in a single research effort is a commonplace occurrence. These *multiple analyses* are a natural byproduct of the complexity of experiments, and are carried out for either reasons of logistical efficiency, to provide additional information about the link between the effect being evaluated and its impact on the endpoint measure, or to explore new ideas and establish new relationships between the intervention or exposure and the effect. These well-motivated concerns for efficiency, the requirement for good evidence necessary to solidify the causal relationship between the intervention and the observed effect, and the need to explore together demand that multiple analyses remain a common occurrence in research. The continued incorporation of multiple analyses in clinical experiments has led to an increased interest in issues surrounding their use. Since each of these analyses involves a statistical hypothesis test, and each hypothesis test produces a *p*-value, a relevant question is how should these *p*-values be interpreted?

Some have argued in articles (Dowdy) and editorials (Rothman) that these additional *p*-values should be ignored. Others have argued that they should be interpreted as though the value of 0.05 is the cutoff point for statistical significance, regardless of how many *p*-values have been produced by the study. This is called using "nominal significance testing" or "marginal significance."

Others have debated whether investigators should be able to analyze all of the data, and then choose the results they want to disseminate (Fisher and Moyé 1999; Fisher, 1999; Moyé, 1999).

The early discussion in this chapter forces us to reject the results of the investigator who, after inspecting the magnitudes of each of the p-values, makes an after the fact, or *post hoc* choice, from among them. This "wait and see what analysis is positive" approach violates the underlying critical assumption of the p-value construction (i.e., , that the data with its embedded sampling error should not choose either the endpoint or the analysis). This violation invalidates the interpretation of the p-value.

9.12.1 Nominal Significance Testing

The strategy of interpreting each of several p-values from a single experiment, one at a time, based on whether they are greater or less than the traditional threshold of 0.05 may seem like a natural alternative to the *post hoc* decision structure that we just rejected. In fact, the nominal p-value approach is very alluring at first glance. The rule to use nominal p-values is easily stated prospectively at the beginning of the trial, and is easy to apply at that trial's end.

However, the consequences of this approach must be given careful attention. Consider, for example, two analyses from a randomized clinical experiment that is designed to measure the effect of therapy of an intervention in patients with atherosclerotic cardiovascular disease. Let us assume in this hypothetical example that the research was well-designed with two endpoints in mind: (1) the cumulative total mortality rate and (2) the cumulative incidence of fatal/nonfatal heart attacks. We will also assume that the analysis has been carried out concordantly (i.e., , according to the experiment's prospectively written protocol). The first analysis reveals that the intervention reduces the cumulative incidence rate of total mortality by a clinically meaningful magnitude, producing a p-value of 0.045. The second analysis reveals that the intervention reduces the cumulative incidence of fatal and nonfatal stroke in a clinically meaningful way, again with a p-value of 0.045. The investigators have clearly met the clinical, statistical, and traditional threshold of results with p-values of less than 0.05. Should each of these be accepted nominally, i.e., , should the investigator conclude that the study produced evidence that the intervention (when used in the population from which the research sample was obtained) will reduce total mortality and will also reduce the fatal and nonfatal stroke rate?

In this specific example in which there are two analyses, one on the effect of the intervention on the total mortality rate and the second on the intervention's impact of the fatal and nonfatal stroke rate, a type I error means that the population has produced by chance alone a sample that either (1) gives a false and misleading signal that the intervention reduced the cumulative total mortality incidence rate, (2) gives a false and misleading signal that the intervention reduced the fatal and nonfatal heart attack rate, or (3) gives a false and misleading signal suggesting that the intervention reduced both the cumulative mortality rate and the fatal and nonfatal stroke rate. There are three errors of which we must now keep track when there are two endpoint analyses, and the misleading events of interest can occur in combination. Therefore, a more complicated measure of type I error rate is required. This complex measure of type I error rate might be described as an overall research α, but has previously been termed the familywise (type I) error probability (or error level), or FWER (Westfall and Young, 1993) and will be designated as ξ.

There is a critical difference between the standard type I error level for a single endpoint and ξ. The type I error probability for a single, individual endpoint focuses on the occurrence of a misleading positive result for a single analysis. This is the single test error level, or test-specific error level. The familywise error level focuses on the occurrence of at least one type I error in the entire collection of analyses. Thus, ξ incorporates the test-specific type I error levels for each of the analyses taken one at a time, and in addition, considers the combinations of type I errors when the statistical hypothesis tests are considered jointly. In the preceding example where there were two endpoints, total mortality and fatal/nonfatal stroke, the occurrence of a familywise error is a measure of the likelihood that we have drawn the wrong positive conclusion about the benefit of therapy for total mortality alone, fatal/nonfatal stroke alone, or have made an error about both.

9.12.2 Initial Computations for ξ

The familywise error level can be easily computed if we assume that the result of one hypothesis test provides no information about the result of the other hypothesis test.[*] Recall that the probability of a type I error for the hypothesis test examining the effect of therapy on the cumulative incidence of the total mortality rate is 0.045, and that the same rate has been chosen for the stroke endpoint. First, compute the probability that there is no type I error for the total mortality rate effect and no type I error for the fatal/nonfatal stroke rate effect using the α error levels for each as follows:

P[no type I error for the total mortality effect and no type I error for the fatal/nonfatal heart attack effect]

---

[*] The performance of this computation when the result of a hypothesis test for one endpoint provides information about the result of another hypothesis test is the specific subject of Chapters 5–6.

= P[no type I error for total mortality effect] * P [no type I error for the fatal/nonfatal stroke rate effect]

=(1 − P[a type I error occurred for the total mortality effect]) *(1 − P [no type I error occurred for the fatal/nonfatal stroke rate])

$$= (1 - 0.045)(1 - 0.045) = (0.955)^2 = 0.9120.$$

Thus, 0.9120 is the probability that there is no false signal from the sample about the beneficial effect of the intervention on each of the total mortality rate and the fatal/nonfatal stroke rate. This event is the best of all possible worlds. The familywise error level in which we are interested is the reverse of this, i.e., , that there is at a type I error for either the total mortality finding, the fatal/nonfatal stroke finding, or both. Thus, we easily compute ξ as

$$\xi = 1 - 0.9120 = 0.088. \tag{9.61}$$

This value of 0.088 is greater than 0.045 (the test-specific p-value for each of the two analyses) and presents the results of this experiment in a very different light. By accepting a less than one in twenty chance of a type I error for either the effect of the intervention on either (1) the cumulative mortality rate or (2) the cumulative incidence rate for fatal/nonfatal strokes, we accept almost a one in ten chance of falsely concluding that the intervention will be effective in the population when it is not. Recognition of this error level inflation is the heart of the problem with accepting nominal significance for multiple analyses. For any realistic collection of single test error levels, the greater the number of endpoints, the larger the familywise error level ξ becomes. For example, if two analyses each have an α level value of 0.05, $\xi = 0.098$.

The familywise error level increases with the number of statistical hypothesis tests. However, this relationship can be used in another manner. Note that if a researcher wishes to keep ξ at less than 0.05, the number of analyses whose results can be controlled (i.e., , the number of analyses that can be carried out and still keep the familywise error level ≤ 0.05) depends on the significance level at which the individual analyses are to be evaluated. For example, if each of the individual analyses is to be judged at the 0.05 level (i.e., , the p-value resulting from the analyses must be less than 0.05 in order to claim the result is statistically significant), then only one analysis can be controlled, since the familywise error level for two analyses exceeds the 0.05 threshold. The researcher can control the familywise error level for two analyses if each is judged at the 0.025 level. If each test is evaluated at the 0.01 level, then five independent hypothesis tests can be carried out.

9.12.3 The Bonferroni Inequality

The previous section's discussion provides important motivation to control the type I error level in clinical trial hypothesis testing. One of the most important, easily used methods to accomplish this prospective control over type I error rate is through the use of the Bonferroni procedure (Miller, 1981). This procedure is developed here.

Assume in a research effort that there are K analyses, each analysis consisting of a hypothesis test. Assume also that each hypothesis test is to be carried out with a prospectively defined type I error probability of α; this is the test-specific type I error level or the test-specific α level. We will also make the simplifying assumption that the result of each of the hypothesis tests is independent of the others. This last assumption allows us to multiply type I error rates for the statistical hypothesis tests when we consider their possible joint results.

Our goal in this evaluation is to easily compute the familywise type I error level, ξ. This is simply the probability that there is a least one type I error among each of the K statistical hypothesis tests. An exact computation for the familywise type I error rate is readily available. Let α be the test-specific alpha error probability for each of K tests. Note that the type I error rate is the same for each hypothesis test. We need to find the probability that there is not a single type I error among these K statistical hypothesis tests. Under our assumption of independence, this probability is simply the product of the probabilities that there is no type I error for each of the K statistical tests. Write

$$(1-\alpha)(1-\alpha)(1-\alpha)\dots(1-\alpha) = \prod_{j=1}^{K}(1-\alpha). \tag{9.62}$$

Therefore ξ, the probability of the occurrence of at least one type I error, is one minus the probability of no type I error among any of the K tests, or

$$\xi = 1 - \prod_{j=1}^{K}(1-\alpha) = 1 - (1-\alpha)^{K}. \tag{9.63}$$

Finding the value of ξ exactly requires some computation. Bonferroni simplified this using Boole's inequality (Chapter 2) which states that the probability of the occurrence of at least one of a collection of events is less than or equal to the sum of the probabilities of these events. This is all that we need to know to write

$$\xi = P[\textit{at least one type I error}] \leq \sum_{i=1}^{K}\alpha_{i}. \tag{9.64}$$

If each of the test-specific type I error levels is the same value α, (9.64) reduces to

$$\xi \leq K\alpha. \tag{9.65}$$

The correspondence between the Bonferroni and the exact ξ is closest when the type I error for each individual test is low. A greater divergence between the two measures is seen as the type I error rate for each individual test increases to 0.05.

Equation (9.65) of course can be rewritten as

$$\alpha \le \frac{\xi}{K}, \tag{9.66}$$

expressing the fact that a reasonable approximation for the α level for each of K hypothesis test can be computed by dividing the familywise error level by the number of statistical hypothesis tests to be carried out. This is the most common method of applying the Bonferroni approach.

As an example, consider an clinical investigator who wishes to carry out a study to test the effect of a new medication on each of three endpoints. The trial has three different treatment arms: intervention dose 1, intervention dose 2, and placebo. Each of the three treatment comparisons must be made against each of the three endpoints, producing a total of nine analyses to be executed. The Bonferroni adjustment for multiple analysis testing provided in (9.66) demonstrates that if the familywise type I error rate is to be maintained at the 0.05 level, then each hypothesis test will need to be evaluated at the $0.05/9 = 0.0056$ level of significance. Alternative approaches include sequentially rejective procedures and resampling procedures (Westfall and Young, 1993; Westfall, Young and Wright, 1993; Westfall, Young).

9.13 Nonparametric Testing

Carrying out hypothesis testing for mean and the variance of a distribution have been relatively straightforward tasks for us up to this point. The mathematics have proceeded smoothly primarily because we were able to identify the underlying probability distribution of the random variable (e.g., the sample mean) in question. This accomplishment allowed us to find exact test statistics and to identify (again, exactly) the probability distribution of these test statistics, thus paving the way for significance testing.

Unfortunately, hypothesis testing does not commonly permit the approaches that we have developed thus far in this chapter. Many times, the distribution of the underlying random variable is not normal. This non-normal distribution may be identifiable but might not be tractable, thereby presenting important difficulties in its mathematical manipulation. If the mean of the distribution is the parameter of interest, then we may invoke the central limit theorem to construct a hypothesis test on the estimate of this parameter that will be based on the normal distribution. However, the invocation of the central limit theorem may be disallowed because only a small number of observations is available for analysis.

In order to effectively deal with this combination of circumstances when one is faced with a small number of observations, we turn to a class of tests that are not based on strong distributional assumptions. These tests have come to be known as *nonparametric tests,* and will be the topic of this section.

9.13.1 The Mann-Whitney (Rank-Sum) Test for Two Unpaired Samples

As we have seen, a common question that the researcher must address is an inquiry into the populations from which two samples are drawn. This can commonly be reduced to the examination of the medians of the two samples. The *Mann-Whitney* test (also commonly referred to as the *rank sum* test), makes no distributional assumptions about the underlying probability distribution of two samples of random variables.

The test is actually very easy to construct. We presume that a sample of n observations x_1, x_2, x_3, ... , x_n from one population and a sample of m observations y_1, y_2, y_3, ... , y_m are available. In order to execute the rank sum test, these two samples are combined into one large sample. In this single, combined sample, the individual observations are ordered from smallest to largest. In doing so, each observation is assigned a rank (the minimum observation is assigned rank 1, the largest is assigned rank $n + m$). Once this ranking has been accomplished, the observations are then placed back in their original samples and the sum of the ranks in each of the two samples is then computed and compared. We would expect that, if the two samples were in fact drawn from the same population, then the ranks would be equivalent in the two samples with equal sample sizes.

The test statistic is the sum of the ranks in the group that has the smallest sum. This rank sum is then compared to percentile values from tables that have been tabulated.

9.13.2 The Signed Rank Test

The signed rank test is a nonparametric approach that is commonly used for the examination of paired data. It is a useful alternative to the paired t test. Like the Mann-Whitney test the signed rank test is easily executed. When provided with n pairs of observations, one first computes the n differences, identifies the absolute values of these differences and then ranks them (the smallest difference is assigned a rank of 1, the next largest difference is assigned a rank of 2, etc.). After this computation, the signs of these difference are reassigned to the individual ranks. One then can compute the sum of the positive ranks and the sum of the negative ranks. The smaller of the absolute values of these sums is compared to the percentile of the distribution of signed ranks (Mann, 1947). It is the percentile values of this distribution that was identified by Wilcoxon and hence the signed rank test is attributed to Wilcoxon. Suggestions for procedures to be followed when the number of positive signed ranks is equal to the number of negative signed ranks are detailed in Snedecor and Cochran.

9.14 Goodness of Fit Testing

The focus of our hypothesis testing efforts have concentrated on the parameters of probability distributions. However, another useful approach is to examine not just the parameters of a probability distribution through the parameter estimators but to instead examine the actual relative frequencies of data in the strata of the distribution. This perspective compares the observed probability distribution to the expected probability distribution. Thus the null hypothesis is not simply a statement about the location of the parameter of the distribution, but is instead a more global statement about the distribution of the sample. This emphasis produces an examination of observed vs. expected and has produced an entire evaluation described as contingency tables. Essentially, these contingency tables, and goodness of fit examinations themselves, compare probability distribution by examining the relative frequencies of events.

9.14.1 Notation

In the simplest case, goodness of fit testing examines the relative frequency of dichotomous variables. Consider the comparison of the frequency of disease among patients unexposed to an environmental agent to the disease's frequency among individuals who are exposed to the agent. If the agent is in fact a risk factor of the disease, then we would expect that the disease is more likely among the exposed than among the unexposed. Thus we would expect the distribution of the disease to be different among those patients exposed to the agent then in those who are unexposed. The data for such a research effort are easily displayed In this setting $n_{i,j}$ is the number of patients in the ith row and j^{th} column of the data table. We can also denote the total number of the observations in the ith row as n_i and the total number of observations in the jth column as n_j. The total number of observations in the entire data table is $n..$ Figure 9.13).

9.14.2 Chi Square Testing

From Figure 9.13, we can easily compute the proportion of patients in each of the cells, representing the observed data. However, a comparison of this data requires us to compute the expected distribution of the data. The direct calculation of the expected frequency distribution is directly related to the null hypothesis. In this circumstance, the null hypothesis is that the distribution of disease among the exposed and unexposed is the same. Thus, the expected distribution would be based on a combined estimate of disease from both the exposed and unexposed populations.

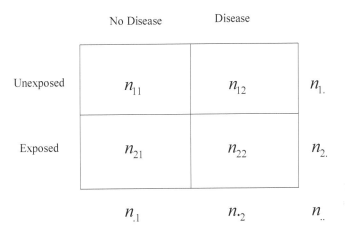

Figure 9.13 Layout and notation for a 2 by 2 table.

We begin with the expected proportion of patients with disease. This is simply the total number of patients with the disease divided by the total number of patients or $\frac{n_{12} + n_{22}}{n_{..}} = \frac{n_{.2}}{n_{..}}$. The expected number of patients with disease in the unexposed group is simply the proportion of patients with disease multiplied by the number of unexposed patients, or $\frac{n_{1.}n_{.2}}{n_{..}}$. Following this procedure for each of the four cells of this data cell we can calculate the expected number of patients in each of the four cells (Figure 9.14) using the expression that for the ith row and j^{th} column the expected number of observations e_{ij} is

$$e_{ij} = \frac{n_{i.}n_{.j}}{n_{..}}. \tag{9.67}$$

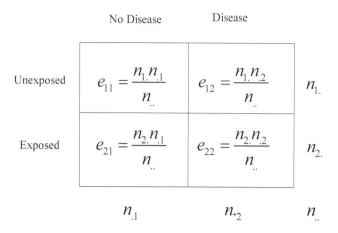

Figure 9.14   Expected number of subjects in each cell of a 2 x 2 layout.

This procedure permits us to compute a test statistic which at its heart is simply a cell by cell comparison of observed vs. expected number of subjects in each cell. We may write this as

$$\sum_{i=1}^{2}\sum_{j=1}^{2}\frac{\left(n_{ij}-e_{ij}\right)^2}{e_{ij}} \tag{9.68}$$

This statistic has a χ^2 distribution with 1 degree of freedom. Its square root follows a standard $N(0,1)$ distribution.

This procedure can be followed for any table with r rows and c columns. We may write that, for suitably large n, the test statistic

$$\sum_{i=1}^{r}\sum_{j=1}^{c}\frac{\left(n_{ij}-e_{ij}\right)^2}{e_{ij}} \sim \chi^2_{(r-1)(c-1)} \tag{9.69}$$

9.14.3 Example
Consider a researcher interested in determining if there is a relationship between exposure of Native Americans to toxins and liver damage. In order to identify and evaluate this relationship, the researcher first identifies Native Americans based on their exposure. These subjects are classified as either being exposed or unexposed to the toxin. After making this assessment, on all subjects, she then

has another individual determine the disease status of all patients.[*] Subjects can either not have the disease, have signs of the disease revealed by a clinical test but not have the disease (characterized as dormant disease), or have the active disease (Table 9.1) manifested by clinical symptoms. The question is whether exposure to the toxin is related to the occurrence of the disease.

Table 9.2. Relationship Between Disease and Exposure

	No Disease	Dormant Disease	Active Disease	Total
Unexposed	275	15	10	300
Exposed	5	40	5	50
Total	280	55	15	350

From the data in Table 9.2, we can use expression (9.67) to compute the expected number of observations in each cell. For example, the expected number of observations in the unexposed–no disease cell e_{11} is computed as

$$e_{11} = \frac{(280)(300)}{350} = 240.$$

The expected number of observations is computed for each of the six cells of Table 9.2 (Table 9.3).

Expression (9.69) is most accurate when the expected number of observations in each cell is five or more. Since this is not the case for the exposed–active disease cell, its data may be combined with that of the exposed–dormant disease cell.

The χ^2 test statistic is easily computed to be

$$\frac{(275-240)^2}{240} + \frac{(15-47.1)^2}{47.1} + \frac{(10-12.9)^2}{12.9} + \frac{(40-5)^2}{40} + \frac{(45-10)^2}{10} = 180.7.$$

---

[*] A second individual carries out the disease assessment without knowledge of the exposure state of the subject to insure that disease ascertainment is not influenced by knowledge of the subject's exposure state.

Table 9.3. Comparison of Observed versus Expected

		No Disease	Dormant Disease	Active Disease	Total
Unexposed	Observed	275	15	10	300
	Expected	*240*	*47.1*	*12.9*	
Exposed	Observed	5	40	5	50
	Expected	*40*	*7.9*	*2.1*	
Total		280	55	15	350

The value of this test statistic is in the extreme tail of the χ^2 distribution with $(2-1)(3-1) = 2$ degrees of freedom. Close inspection of Table 9.3 reveals a large number of excess cases of dormant disease among the exposed Native Americans.

9.14.4 Maentel-Haenszel Procedures

The *Maentel-Haenszel test* was developed in 1959, and is a stalwart statistical testing procedure in epidemiology, or the study of the determinants and distribution of disease in populations. After examining the data from the previous example, the researcher may be interested in determining the relationship between toxin exposure and disease separately for several different demographic or risk factor groups. Applying the methodology of Section 9.13.3 would require that several hypothesis tests be executed, one for each strata. However, increasing the number of hypothesis tests increases the likelihood of the occurrence of at least one type I error. The Maental-Haenszel test is commonly used in this circumstance in order to avoid an unnecessary increase in the number of hypothesis tests that must be executed.

Consider the analysis of data from four 2 x 2 tables (Figure 9.15).

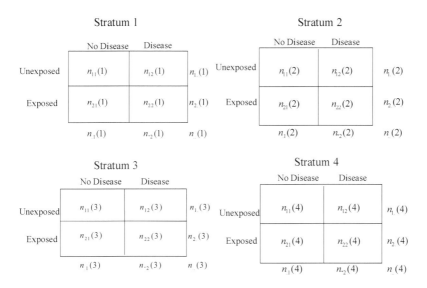

Figure 9.15   Layout and notation for four 2 by 2 tables

Each of the four tables in Figure 9.15 represents the relationship between exposure and disease in each of four categories. The categories might be age, race/ethnicity, or duration of exposure. The goal is to allow each table to contribute to the hypothesis test that examines the relationship between exposure status and disease occurrence. For the jth strata, $j = 1, 2, 3, 4$, we can compute the expected number of subjects in the cell represented by the first row and first column, i.e.,

$$e_{11}(j) = \frac{n_{1.}(j)n_{.1}(j)}{n_{..}(j)}$$

(9.70)

It has been shown that the variance of the number of subjects, $n_{11}(j)$ in this first cell is

$$Var\left[n_{11}(j)\right] = \frac{n_{.1}n_{.2}n_{1.}n_{2.}}{n_{..}^2\left(n_{..} - 1\right)}$$

(9.71)

The Maentel-Haenszel test statistic in this circumstance is

$$\frac{\left[\left|\sum_{j=1}^{4} n_{11}(j) - \sum_{j=1}^{4} e_{11}(j)\right| - \frac{1}{2}\right]^2}{\sum_{j=1}^{4} Var\left[e_{11}(j)\right]}$$

which follows a χ^2 distribution with one degree on freedom.

In general, when there are S such tables, the Maentel-Haenszel test statistic may be written as.

$$\chi^2_{MH} = \frac{\left[\left|\sum_{j=1}^{S} n_{11}(j) - \sum_{j=1}^{S} e_{11}(j)\right| - \frac{1}{2}\right]^2}{\sum_{j=1}^{S} Var\left[e_{11}(j)\right]} \sim \chi^2_1$$

9.15 Fisher's Exact Test

The contingency table is a very useful procedures in applied statistics. However the requirement of a large sample size offers an important impediment to its use. This is because large sample sizes are frequently very expensive. Hypothesis testing based on small sample sizes using the chi square test can be somewhat unreliable when there are a small number of observations in at least one of the data cells (i.e., when the expected number of observations in a data cell is less than 5). If the data set has greater than two columns and/or rows, this problem is commonly addressed by combining observations across cells. However, in a 2 by 2 table, this approach is impractical.

Fisher's exact test is the test that is most helpful in the setting of a small number of observations in a 2 by 2 table. As the name implies, this test does not rely on asymptotic theory and provides an exact result. Consider a 2 by 2 data table that provides the depiction of a relationship between two variables A and B (Figure 9.16).

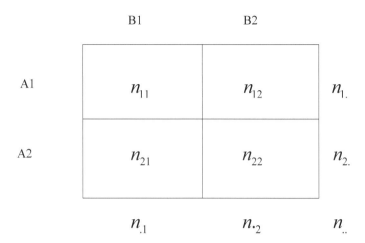

Figure 9.16 Layout and notation for a 2 by 2 table.

Fisher's exact test computes the probability of a finding that is at least as large as that observed in the data. If this probability of this event is lower than some predefined threshold, then the researcher might conclude (again, assuming that the research is well designed and well executed) that there is a relationship between the two variables). The question of independence between the two variables can actually be reduced to a sampling mechanism. If we fix the marginal values $n_{1.}$, $n_{2.}$, $n_{.1}$, and $n_{.2}$, then the probability of selecting n_{11} subjects from group $n_{1.}$ can be computed using the hypergeometric distribution that was introduced in Chapter 4. Thus

$$P\left[n_{11}, n_{12}, n_{21}, n_{22}\right] = \frac{\dbinom{n_{.1}}{n_{11}}\dbinom{n_{.2}}{n_{12}}}{\dbinom{n}{n_{1.}}} = \frac{n_{.1}!\,n_{.2}!\,n_{1.}!\,n_{2.}!}{n!\,n_{11}!\,n_{12}!\,n_{21}!\,n_{22}!}$$

Fisher's test is the computation of a result that reflects the current data from this sampling perspective, or data that is more extreme. By definition, Fisher's exact test is a one-tailed test, although it could easily be adapted to two-tailed evaluation. In its most commonly used appearance, the one-tailed computation is

$$F = \sum_{i=1}^{n_{11}} \frac{\binom{n_{.1}}{i}\binom{n_{.2}}{n_{1.} - i}}{\binom{n}{n_{1.}}} \tag{9.72}$$

This provides a type I error level that can be used to help judge the significance of the result findings.

9.16 Sample Size Computations

9.16.1 Motivation for Sample Size Computations

Good research efforts, regardless of their size, are characterized by careful planning, controlled execution, and disciplined analysis. An important component of the design of a research effort is the sample size calculation. The sample size computation is the mathematical calculation that determines how many subjects the research effort should recruit. It is based on the specialists' determinations of rates or average responses, as well as statistical considerations about the role sampling error may play in producing the research effort's results.

Since the sample size computation requires a clear set of assumptions about the primary scientific question to be addressed by the study, the expected experience of the control group, the anticipated effect of the intervention, and concerns about type I and type II errors, clearly the investigators and quantitative methodologists (i.e., , the statisticians) must be involved and agree on the estimates of these quantities. Since the researcher does not know during the design phase of the study whether the results will be positive or not, she must plan for each possibility. Thus, she should design the study in such a way that the type I error rate will be low (customarily no higher than 0.05) and the power of the experiment will be high (typically, at least 80%). Each of these considerations are part of the sample size computation.

However, the sample size computation, although composed only of mathematics, must also reflect the administrative and financial settings in which the research will be executed. These important logistical considerations, not explicitly included in the arithmetic of the sample size calculation, must nevertheless receive primary attention. The availability of an adequate number of subjects may be questionable. Alternatively, in the case where a new medication has been developed at great cost, the small number of available doses may preclude recruiting many patients. The financial cost of the research effort, and the availability of enough investigators, project managers, and skilled laboratories for executing the protocol also may limit the size of the study.

These latter, nonmathematical considerations must be factored into the final sample size determination in order for the experiment to be executable. They are blended into the plans for the research effort in   general, and the sample size in particular through several mechanisms during the design phase of the study. Among these mechanisms are (1) examination and alteration of the population from which the research sample will be recruited, (2) formulation of the primary analysis of the study, and (3) changing the duration of time over which patients will be followed. Each of these maneuvers is acceptable when considered and finalized during the design phase of the study. For this appropriate mixture of statistics, clinical science, and administration to occur, the dialogue between all involved parties should be frank, honest, and collegial. This robust research design with its recomputed sample size will be consistent with scientific, logistical, statistical, and financial considerations, making the research effort both scientifically rigorous and executable (Moyé, 2000).

9.16.2. Sample Size Computation for One-sample Test

Assume that we have a set of n i.i.d. elements from a normal distribution with unknown mean μ and known variance σ^2. We are interested in testing the hypothesis

$$H_0: \mu = \mu_0: \qquad H_a: \mu \neq \mu_0$$

Sample size computations are carried out with a specific null hypothesis in mind. In this case, we will assume that the investigators have a particular value of μ in mind under the alterative hypothesis, specifically $\mu = \mu_a$. We can assume without any loss of generality that $\mu_a > \mu_0$. Once we have identified the sample size formula for this setting, we may write the sample size formula for the hypothesis that for $\mu_a < \mu_0$

Recall from section 9.8.1 that, under the null hypothesis, the quantity $\dfrac{\left(\bar{x} - \mu_0\right)}{\sqrt{\sigma^2/n}}$

follows a standard normal distribution. Our task is to compute the sample size associated with this evaluation so that the hypothesis test has appropriate protection from type I and type II errors. All of our computations will be three phase developments as follows:

Phase I: Computation under the null hypothesis
Phase II: Computation under the alternative hypothesis
Phase III: Consolidation

Under phase I, we simply examine the test statistic under the null hypothesis. We may write at once that the half of the critical region of greatest interest to the investigators is that range where $\bar{x} > \mu_0$. This corresponds to

$$\frac{(\bar{x} - \mu_0)}{\sqrt{\sigma^2/n}} > z_{1-\alpha/2}.$$

This can easily be rewritten as

$$\bar{x} \geq z_{1-\alpha/2}\sqrt{\sigma^2/n} + \mu_0$$

This concludes phase I. Note that we have an expression that includes the variance, the mean of the distribution of the sample of x's under the null hypothesis, the type I error , and the sidedness of the test. Also note that the test represents a two-sided evaluation. Although it may be that the two-sided critical region and the assumption that the mean under the alternative hypothesis μ_a > μ_0 are contradictions they actually are quite harmonious. Recall that the assumption that μ_a > μ_0 represents only the belief (some might say, the hope) of the investigators. In actually, they do not know where μ_a is, and, in fact, μ_a may be less than μ_0. Their hope is reflected in the choice of μ_a in the sample size computation. Their uncertainty is appropriately reflected in the choice of a two-sided test.

Phase 2 incorporates the notion of statistical power. Recall from our earlier, theoretical discussion of power in this chapter that power, like the p-value, is a phenomenon of sampling error. The circumstance in which statistical power is relevant in the interpretation of a research effort is when the research results are not positive, but null; i.e., the null hypothesis is not rejected. This concept will be introduced to the sample size discussion by the following statement:

Power = P[test statistic falls in the critical region | H_a is true].

From Phase I, we know that the test statistic falls in the critical region when $\bar{x} \geq z_{1-\alpha/2}\sqrt{\sigma^2/n} + \mu_0$. We may therefore write

$$Power = P\left[\bar{x} \geq z_{1-\alpha/2}\sqrt{\sigma^2/n} + \mu_0 \mid H_a \text{ is true}\right]$$

Under the alternative hypothesis, the population mean is equal to μ_a. With this knowledge, we can write the event expressed that is formulated in the previous expression as an event that involves a standard normal random variable.

$$Power = P\left[\frac{\bar{x}_n - \mu_a}{\sqrt{\sigma^2/n}} \geq \frac{z_{1-\alpha/2}\sqrt{\sigma^2/n} - (\mu_a - \mu_0)}{\sqrt{\sigma^2/n}}\right]$$

$$= P\left[Z \geq z_{1-\alpha/2} - \frac{(\mu_a - \mu_0)}{\sqrt{\sigma^2/n}}\right]$$

where Z is a random variable from a standard normal distribution. The scientist sets the type II error level, β, before the experiment begins, just as they set the type I error. Thus

$$Power = 1 - \beta = P\left[Z \geq z_{1-\alpha/2} - \frac{(\mu_a - \mu_0)}{\sqrt{\sigma^2/n}}\right]$$

which implies that

$$z_\beta = z_{1-\alpha/2} - \frac{(\mu_a - \mu_0)}{\sqrt{\sigma^2/n}} \tag{9.73}$$

What is simply required now is to solve expression (9.73) for n to find

$$n = \left[\frac{\sigma(z_{1-\alpha/2} - z_\beta)}{\mu_a - \mu_0}\right]^2 = \left[\frac{\sigma(z_{1-\alpha/2} + z_{1-\beta})}{\mu_a - \mu_0}\right]^2 \tag{9.74}$$

where the right-hand expression in equation (9.74) is the most commonly used expression for sample size in this one-sample setting.

Example: A scientist is interested in testing whether the average diameter of field mushrooms is different than three inches. Although she does not know whether mushroom size is greater than three inches or not, she suspects that the true mushroom diameter is closer to 4 inches. Her plan is to choose mushrooms at random from a particular field. The standard deviation of the mushroom is 2 inches. She wishes to carry out a two-tailed test of this hypothesis with a type I error of 0.05 and 90% statistical power. The minimum number of mushrooms that she needs to measure is provided by the sample size from (9.74):

$$n = \left[\frac{\sigma\left(z_{1-\alpha/2} + z_{1-\beta}\right)}{\mu_a - \mu_0} \right]^2 = \left[\frac{2(1.96 + 1.28)}{4 - 3} \right]^2 = 42$$

At least 42 mushrooms have to be sampled and have their diameters measured.

Formula (9.74) reveals some interesting relationships that involve sample size. First, the sample size increases with increasing standard deviation. This is an anticipated finding, since, the larger the variance, the less precise the measurements of the mean are, the more observations are required to identify the likely location of μ. Also, the farther apart the means are, the easier it becomes to differentiate between the null and alternative hypotheses, and the greater the sample size becomes.

If the researcher was concerned about the possibility that $\mu < \mu_0$, with everything else about the problem remaining unchanged, the sample size formulation would be different. In phase 1, the half of the critical region of greatest interest to the investigators is that range where $\overline{x}_n < \mu_0$. This

corresponds to $\dfrac{\left(\overline{x}_n - \mu_0\right)}{\sqrt{\sigma^2/n}} \le z_{\alpha/2}$,

which can be rewritten as

$$\overline{x}_n \le z_{\alpha/2}\sqrt{\sigma^2/n} + \mu_0 = \mu_0 - z_{1-\alpha/2}\sqrt{\sigma^2/n}$$

This concludes phase I. The phase 2 statement that incorporates power begins with

$$Power = P\left[\overline{x}_n \le \mu_0 - z_{1-\alpha/2}\sqrt{\sigma^2/n} \mid H_a \text{ is true} \right] \tag{9.75}$$

This becomes

$$Power = P\left[\frac{\overline{x}_n - \mu_a}{\sqrt{\sigma^2/n}} \le \frac{\mu_0 - \mu_a - z_{1-\alpha/2}\sqrt{\sigma^2/n}}{\sqrt{\sigma^2/n}} \right]$$

$$= P\left[Z \le \frac{\left(\mu_0 - \mu_a\right)}{\sqrt{\sigma^2/n}} - z_{1-\alpha/2} \right]$$

The scientist sets the type II error level, β, before the experiment begins, just as they set the type I error. Thus

$$1-\beta = P\left[N(0,1) \leq \frac{\left(\mu_0 - \mu_a\right)}{\sqrt{\sigma^2/n}} - z_{1-\alpha/2} \right]$$

which implies that

$$z_{1-\beta} = z_{1-\alpha/2} - \frac{\left(\mu_a - \mu_0\right)}{\sqrt{\sigma^2/n}} \tag{9.76}$$

Producing the same sample size formula as equation (9.74).

9.16.3 Sample Size Computation for Two-Sample Test

Here we extend the development of the previous section by assuming that we have a sample of observations from each of two populations. We will assume that the populations have the same, known variance σ^2. The population means μ_x and μ_y are unknown. We are interested in testing the hypothesis

$$H_0: \mu_x = \mu_y \quad \text{versus} \quad H_a: \mu_x \neq \mu_y$$

Just as before, we will assume that the investigators have a particular value of $\mu_y - \mu_x = \Delta$ in mind for the alternative hypothesis. However, they do not actually know the value of Δ so the hypothesis test will be two-tailed. Recall from earlier in this chapter that, under the null hypothesis, the quantity $\dfrac{\left(\overline{X} - \overline{Y} - \mu_x - \mu_y\right)}{\sqrt{Var\left(\overline{X} - \overline{Y}\right)}}$ follows a standard normal distribution. Our task is to compute the sample size associated with this evaluation so that the hypothesis test has appropriate protection from type I and type II errors. We will proceed with the three phase development of the sample size computation for that was introduced in section 9.16.2 for this formulation.

Under phase I, we simply examine the test statistic under the null hypothesis. Assuming that the null hypothesis $\mu_x = \mu_y$ is true, and we may write the test statistic as $\dfrac{\overline{X} - \overline{Y}}{\sqrt{Var\left(\overline{X} - \overline{Y}\right)}}$. As before, we focus on that region of the test statistic that is of interest. This is the region where $\Delta > 0$, or $\overline{x} > \overline{y}$. This corresponds to the interval where $\dfrac{\overline{X}-\overline{Y}}{\sqrt{Var\left(\overline{X}-\overline{Y}\right)}} > z_{1-\alpha/2}$. This can easily be rewritten as

$$\overline{X}-\overline{Y} \geq z_{1-\alpha/2}\sqrt{Var\left(\overline{X}-\overline{Y}\right)},$$

concluding phase I. We now move to phase II, which examines the probability distribution of the test statistic under the alternative hypothesis. In this two-sample situation, we may write

$$Power = P\left[\overline{X}-\overline{Y} \geq z_{1-\alpha/2}Var\left(\overline{X}-\overline{Y}\right) \mid H_a \text{ is true}\right] \tag{9.77}$$

Under the alternative hypothesis, the difference in the mean populations is equal to Δ which is > 0. With this knowledge, we can write the event expressed that is formulated in expression (9.77) as an event that involves a standard normal random variable.

$$Power = P\left[\frac{\overline{X}-\overline{Y}-\Delta}{\sqrt{Var\left(\overline{x}-\overline{y}\right)}} \geq \frac{z_{1-\alpha/2}\sqrt{Var\left(\overline{X}-\overline{Y}\right)}-\Delta}{\sqrt{Var\left(\overline{X}-\overline{Y}\right)}}\right]$$

$$= P\left[Z \geq z_{1-\alpha/2} - \frac{\Delta}{\sqrt{Var\left(\overline{X}-\overline{Y}\right)}}\right]$$

Thus

$$1-\beta = P\left[Z \geq z_{1-u/2} - \frac{\Delta}{\sqrt{Var\left(\overline{X}-\overline{Y}\right)}}\right]$$

which implies that

$$Z_\beta = z_{1-\alpha/2} - \frac{\Delta}{\sqrt{Var\left(\overline{X}-\overline{Y}\right)}} \tag{9.78}$$

Recall that we assumed that $\sigma_x^2 = \sigma_y^2$. If we also assume an equal number of observations in each of the two groups, we may easily write $Var(\bar{x} - \bar{y}) = \dfrac{2\sigma^2}{n}$.

This permits us to compute

$$z_\beta = z_{1-\alpha/2} - \frac{\Delta}{\sqrt{\dfrac{2\sigma^2}{n}}}$$

What is simply required now is to solve expression (9.78) for n to find

$$n = \frac{2\sigma^2 \left(z_{1-\alpha/2} - z_\beta \right)^2}{\Delta^2}. \tag{9.79}$$

Recall that there are n observations taken from each of the two populations. Thus the total research size $N = 2n$ may be written as

$$N = \frac{4\sigma^2 \left(z_{1-\alpha/2} - z_\beta \right)^2}{\Delta^2} = \frac{4\sigma^2 \left(z_{1-\alpha/2} + z_{1-\beta} \right)^2}{\Delta^2}$$

Example: If, for this experiment, an investigator wishes to show that graduate students have higher blood pressure levels than undergraduate students, and the investigator chooses a two-sided alpha of 0.05, 90 percent power (beta = 0.10), delta = 5 mm Hg and $\sigma^2 = 18$, the trial size is

$$N = \frac{4\sigma^2 \left[z_{1-\alpha/2} - z_\beta \right]^2}{\Delta^2} = \frac{4(18)^2 \left[1.96 - (-1.28) \right]^2}{[5]^2} = 544$$

In this study, 544 subjects would be required, 272 students in each of the two groups.

Problems

1. State some arguments for and against the involvement of statisticians in the in the interpretation of the data that they collect.

2.   Two researchers draw a sample from the same population of patients with heart failure. Researcher A designs his study to compare the death rates of patients taking a new drug for heart failure with the death rate of patients on placebo therapy, and finds that there is no effect of the therapy on the death rate. Researcher B plans to examine the relationship between the same therapy and the risk of hospitalization. She finds no relationship between the use of therapy and hospitalization, but does find a statistically significant relationship between the new therapy and death. If everything else about the research efforts of these two scientists is the same (appropriate number of patients used in each study, correct use of the drug, proper ascertainment of deaths and hospitalization), what would you conclude about the relationship between use of the drug and death in patients with heart failure?

3.   A nightclub marketer hires you to determine if a practiced but relatively unknown disc jockey who plays records using a four platter record player is able to draw a larger night club crowd than a disc jockey who is more widely known but only uses an electronic collection of music. How would you design the research effort? How would you state the null and alternative hypotheses?

4.   Describe the roles of invariance, unbiasedness, and power in comparing the characteristics of a collection of statistical tests.

5.   What is the rationale for couching the question that the investigator is investigating in terms of the null hypothesis, i.e., , by disproving a negative?

6.   Show that the sum of n i.i.d. random variables following an exponential distribution with parameter 1 follows a Gamma distribution with parameters n and 1.

7.   An airline company that is interested in reducing passenger total time alters the check in process for its passengers. Rather than have passengers first purchase their ticket and check their baggage followed by going through a thorough security check, they instead choose to integrate the processes. The new procedure puts passengers through both the check in and the security procedures simultaneously once, requiring only one total line rather than two. How would you design the research effort to determine if the total time was dramatically reduced?

8.   When can the observation that a test statistic falls in its critical region mislead a researcher?

9. Describe two research circumstances in which it would be useful to have a type I error level set at a rate other than 0.05.

10. A traffic intersection has been governed by a traffic light that has only two lights, red and green. This light has been in place for 6 years, during which time there have been 113 accidents. It is replaced by a traffic signal that has three lights; red, green, and a yellow warning light. The number of traffic accidents in the one year following the installation of the new light has been 8 accidents. Using the Poisson distribution as the underlying probability distribution, carry out a two-sided hypothesis on the accident rates. Can you conclude that the introduction of the new traffic signal has reduced the number of accidents?

11. A controversial effort to change the climate has been the use of cloud seeding. In this process, airplanes seed clouds with salt crystals in the hope of inducing the clouds to produce much needed rainfall. How would you design an experiment to test the idea that cloud seeding increases rainfall?

12. Show that, in a simple hypothesis testing scenario on the mean of a normally distributed random variable, that a one-tailed test has more power than a two-tailed test.

13. If one-tailed tests are more power than two-tailed tests, why does two-tailed testing continue to be dominant statistical testing tool in some fields, e.g., medicine?

14. Of 100,000 workers, 2017 lost their jobs. In a neighboring county, out of 50,000 workers, 950 lost their jobs. Using the large sample approximation to the binomial distribution, carry out a hypothesis test to determine if the employment rate is different between the two counties.

15. A police crime lab is under scrutiny for its validity. They are provided 90 blood samples to determine the presence of a particular gene in the blood sample. In the 60 samples that contained the genome, the lab correctly identified 40 of them. In the 40 samples without the gene, the lab correctly recognized that the genome was absent in 38 cases. Are the differences between the lab's finding and the gene's presence statistically significant?

16. A marketer must determine the effect of a price reduction on the sale of CD players. For thirty days, the average number of units sold per day was 16 with a variance of four. After the price was reduced, there were 24 units sold per day with a variance of 4 over 30 days. Using the normal distribution, test the hypothesis that the number of daily units was the same before and after the price reduction.

17. A cellular phone manufacturer wishes to introduce new technology to reduce phone calls that are interrupted prematurely for technical reasons (dropped calls). Before the new technology is introduced, out of 1 million units, there are 7 dropped calls per person per week. After the institution of the new technology, out of 10,000 units, there are 6.5 dropped calls per person per week. Using the normal approximation to the Poisson distribution, is there sufficient reason to suspect the dropped call rate has significantly decreased? If so, can they conclude that the decrease in the dropped call rate is due to the new technology?

18. A major automotive company is interested in instituting methodology to improve gas mileage. Current miles per gallon is 16 mpg (miles per gallon). Testing 117 cars, the average miles per gallon was 15 mpg with a standard deviation of 4. Can the automobile manufacturer attest that they significantly altered the gas mileage for their cars?

19. An important economic indicator is the 30 year mortgage rate. In one week the average mortgage rate for a sample of 10,000 home sales is 5.21 percent with a variance of 2.75 percent. The next week the average mortgage rate is 5.20 with 2.60 percent. Is there a statistically significant difference in the mortgage rates?

20. Compute the sample size formula for the one-sample hypothesis test for the population mean of independent and identically distributed normal random variables, assuming that σ^2, μ_0, μ_a, α, and β are known. Find the ratio of the sample sizes for a one-tailed test to that of a two-tailed test.

21. Compute the sample size formula for the two-tailed hypothesis test for the two-sample test of population means from the normal distribution when $\sigma_x^2 \neq \sigma_y^2$. Assume that α, β, and $\Delta = \mu_x - u_y$ are provided.

22. Compute the sample size formula for a two-tailed hypothesis test on the equality of means from two normally distributed populations when α, β, and $\Delta = \mu_x - u_y$ are provided. Assume for this computation that

$\sigma_x^2 \neq \sigma_y^2$. and the ratio of subjects in the two groups will be k_1 to k_2 where k_1 to k_2 are known.

References

Birnbaum, A. (1962) "On the foundations of statistical inference". *Journal of the American Statistical Association* **57**: 269 – 306.

Berkson, J. (1942) "Experiences with tests of significance. A reply to R.A. Fisher". Journal *of the American Statistical Association.***37**: 242 - 246.

Berkson, J. (1942) "Tests of significance considered as evidence". *Journal of the American Statistical Association* **37**:335 - 345.

Dowdy, S., Wearden, S. *Statistics for Research* 2nd Edition. New York. Wiley. 1985.

Editorial (1988) "Feelings and frequencies: Two kinds of probability in public health research". *American Journal of Public Health* **78**:

Fisher, R. A. (1925) *Statistical methods for research workers.* Edinburg.Oliver and Boyd.

Fisher, R. A. (1926) The arrangement of field experiments. *Journal of the Ministry of Agriculture.* September 503 - 513.

Fisher, R.A. (1942) Response to Berkson. *Journal of the American Statistical Association* **37**:103 - 104.

Fisher, L.D. (1991) The use of one-sided tests in drug trials: an FDA advisory committee member's perspective. *Journal of Biopharmaceutical Statistics* **1**:151-6,

Fisher, L.D., Moyé, L.A. (1999) Carvedilol and the Food and Drug Administration Approval Process: An Introduction. *Controlled Clinical Trials* **20**:1–15.

Fisher, L.D. (1999) Carvedilol and the FDA approval process: the FDA paradigm and reflections upon hypothesis testing. *Controlled Clinical Trials* **20**:16–39.

Hoel, P.G. (1971) Introduction to Mathematical Statistics. Fourth Edition. New York. Wiley..

Knottnerus, J.A., Bouter, LM. (2001) Commentary. The ethics of sample size: two-sided testing and one-sided thinking. *Journal of Clinical Epidemiology* **54**:109-110.

Lang, J.M. Rothman, K.L. and Cann, C.L. That confounded p value. *Epidemiology.* **9**: 7 - 8.

Lehmann, E.L. (1986). *Testing Statistical Hypothesis.* 2nd Edition. Wiley. New York.

Mann, H.B., Whitney, D.R. (1947) *Annals of Mathematical Statistics* **18**:50.

Miller, R.G. (1981) *Simultaneous Statistical Inference*, 2nd ed. New York Springer.

Moyé, L.A. (2000) *Statistical Reasoning in Medicine – The Intuitive p-Value Primer.* New York. Springer.

Moyé, L.A, Tita, A. (2002) Ethics and hypothesis testing complexity. *Circulation* **105**:3062-3065.

Moyé, L.A. (1999) P–value interpretation in clinical trials. The case for discipline. *Controlled Clinical Trials* **20**:40–49.

Nester, M.R., (1996) An applied statistician's creed. *Applied Statistics* **45**: 4401–410.

Owen, D.B. (1976) *On the History of Probability and Statistics.* Marcel Dekker Inc. New York.

Pearson, E.S. (1970) *The History of Statistics in the 17th and 18th Centuries.* Macmillan Publishing, New York.

Pitt, B., Segal R., Martinez F.A., et. al. on behalf of the ELITE Study Investigators (1997) Randomized trial of losartan vs. captopril in patients over 65 with heart failure. *Lancet* **349**:747–52.

Pitt, B., Poole-Wilson P.A, Segal R. Martinez et. al (2000) Effect of losartan compared with captopril on mortality in patients with symptomatic heart failure: randomized trial─the losartan heart failure survival study. ELITE II. *Lancet* **355**:1582–87.

Poole, C. (1987) Beyond the confidence interval. *American Journal of Public Health* **77**:195–199.

Rao, C.R. *Linear Statistical Inference and Its Applications.* 2nd Edition. New York. Wiley. p 451.

Rothman, R.J. (1990) No adjustments are needed for multiple comparisons. *Epidemiology* **1**:43–46.

Simes, R.J. (1986) An improved Bonferroni procedure for multiple tests of significance. *Biometrika* **73**:819–827.

Westfall, P.H., Young S.S. (1993) *Resampling Based Multiple Testing: Examples and Methods for P-Value Adjustment.* New York. Wiley.

Westfall, P.H., Young S.S., Wright, S.P. (1993) Adjusting *p*-values for multiplicity. *Biometrics* **49**:941–945.

Westfall, P.H., Young, S. (1989) *P*-value adjustments for multiple tests in multivariate binomial models. *Journal of the American Statistical Association* **84**:780–786.

Young, A. (1771). *A Course of Experimental Agriculture.* J. Exshaw et al. Dublin.

10

Interval Estimation

10.1 Introduction

In Chapter 8, we discussed point estimation in which one random quantity was proposed to estimate the unknown parameter. Although it might have several desirable properties, the point estimator does not give any information on how close it may be to the parameter. Even for an unbiased estimator, we can not always obtain the probability that the estimator will be within a given distance of the parameter for a random sample of known size. In fact, except for some trivial cases, the probability that a point estimate is exactly equal to the parameter, is equal to 0, i.e. $P\big(T(X_1,\ldots,X_n) = \theta\big) = 0$, where $T(X_1,\ldots,X_n)$ is a point estimator. For example, if we are to study the net weight of 5-lb frozen shrimp from a packing company, we may randomly choose 10 boxes of these packages from the company and find the average net weight after we defreeze and drain the shrimp. Assuming the net weight of each box is normally distributed with unknown mean and variance, the best point estimate of the mean weight is certainly the sample mean. If the sample mean weight is 78 oz, we do not know how close the population mean net weight of this type of package really is to 5 lb. One thing we are almost certain of (with probability 1) in this case, is that the population mean net weight is not equal to the sample mean net weight (a random quantity).

In this chapter, we will first introduce the basic concept of *interval estimation*. In sections 3 and 4, some methods of constructing a confidence interval and a Bayesian credible interval will be presented. The approximate method for the construction of a confidence interval and the bootstrap method will be described in sections 5 and 6, respectively. Comparisons of confidence intervals will be discussed in section 7. The relationship between a confidence interval and the corresponding hypothesis test is summarized in section 8.

10.2 Definition

In this section, we will present the notion of interval estimation in which a random interval will be used to estimate a parameter. Using this notion, we will state the probability (called the *coverage probability*) that this random interval contains the true parameter. Therefore, for a pre-assigned probability level, we can find a random interval that has a coverage probability greater than or equal to this level. The notion of random interval introduced in this section and in section 10.3, will be different from the Bayesian credible interval that will be introduced in section 10.4. The Bayesian method may lead one to interpret that, with a fixed probability, the random parameter belongs to the established interval (or its realization).

Definition 10.1 Suppose $X_1, \ldots, X_n$ is a random sample from a population with a probability distribution depending on an unknown parameter θ. If $L(X_1, \ldots, X_n)$ and $U(X_1, \ldots, X_n)$ are two statistics of the random sample that satisfy

$$L(X_1, \ldots, X_n) \leq U(X_1, \ldots, X_n) \text{ and}$$

$$P\big(L(X_1, \ldots, X_n) \leq \theta \leq U(X_1, \ldots, X_n)\big|\theta\big) \geq 1 - \alpha \qquad (10.1)$$

for any possible value of θ, where $0 < \alpha < 1$, then the random interval $[L(X_1, \ldots, X_n), U(X_1, \ldots, X_n)]$ is called a $100(1-\alpha)\%$ confidence interval. The left hand side of the inequality in (10.1) is called the *coverage probability*. If $1 - \alpha$ is the greatest lower bound of the coverage probability for all θ, then $1 - \alpha$ is called the *confidence coefficient*. The difference $U(X_1, \ldots, X_n) - L(X_1, \ldots, X_n)$ is called the *length* of the confidence interval.

Note that, by changing the appropriate inequality from $\leq$ to $<$, we can define a $100(1-\alpha)\%$ confidence interval as an open interval or half-open interval. In this book, we will treat a confidence interval as a closed interval unless otherwise specified. For the sample from a continuous random variable, the closed, open or half-open random intervals will have the same coverage probability. For the samples from a discrete random variable, these three intervals may not have the same coverage probability.

95%CI: N=20, simulation=100, N(0,1)

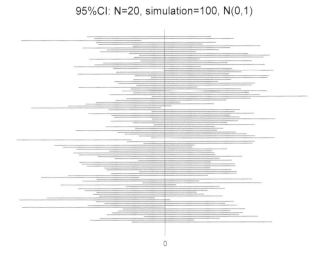

0

Figure 10.1 A collection of one hundred 95% confidence intervals for the mean as computed from repeated simulated sample of size 20.

Now, if we plan to take a random sample from the population, we can compute L and U in advance and then say that we are $100(1-\alpha)\%$ confident the interval $[L,U]$ will cover the unknown parameter. However, after a sample $x_1,\ldots,x_n$ is taken, any realization $[L(x_1,\ldots,x_n),U(x_1,\ldots,x_n)]$ of the confidence interval may or may not contain θ. We do not know if our computed interval will necessarily contain θ. However, if we take a large number of random samples of the same size and compute the realized interval $[L,U]$ for each sample, the proportion of the computed intervals that contain θ would be about $1-\alpha$. Figure 10.1 illustrates the collection of 95% confidence intervals for the unknown mean as computed from 100 simulated samples of size 20 from a $N(\mu,\sigma^2)$ population ($\mu = 0$ and $\sigma^2 = 1$). The theoretical method of constructing this confidence interval will be discussed in Example 10.1. The simulation result shows that there were 6 intervals that did not cover the true mean 0.

We caution that any numerical confidence interval can not be said to contain the parameter with probability $1-\alpha$. Since the numerical interval does not have any random component, it either contains or does not contain θ.

Definition 10.2 Suppose $X_1, \ldots, X_n$ is a random sample from a population with a probability distribution depending on an unknown parameter θ. A statistic $L(X_1, \ldots, X_n)$ is called a $100(1-\alpha)\%$ *lower confidence bound* for θ, if $P\big(L(X_1, \ldots, X_n) \le \theta \,|\, \theta\big) \ge 1 - \alpha$ for every θ.

Similarly, $U(X_1, \ldots, X_n)$ is called a $100(1-\alpha)\%$ *upper confidence bound* for θ, if $P\big(U(X_1, \ldots, X_n) \ge \theta \,|\, \theta\big) \ge 1 - \alpha$ for every θ.

10.3 Constructing Confidence Intervals

There are several methods for constructing the confidence interval for an unknown parameter. We will use the mean of a normal distribution as an example to illustrate these methods.

Example 10.1 Assume that $X_1, \ldots, X_n$ is a random sample taken from a population that is normally distributed with unknown mean μ and unknown variance σ^2. Then, as we know from Chapter 5, $\dfrac{\sqrt{n}\left(\bar{X} - \mu\right)}{S}$ has a t-distribution with $n-1$ degrees of freedom, where $\bar{X} = \dfrac{\sum\limits_{i=1}^{n} X_i}{n}$ is the sample mean and $S^2 = \dfrac{\sum\limits_{i=1}^{n}(X_i - \bar{X})^2}{n-1}$ is the sample variance. Let $t_{n-1,u}$ denote the $100u$ percentile associated with the t-distribution with $n-1$ degrees of freedom, then we can write

$$P\left(t_{n-1,\frac{\alpha}{2}} \le \frac{\sqrt{n}\left(\bar{X} - \mu\right)}{S} < t_{n-1,1-\frac{\alpha}{2}} \right) = 1 - \alpha,$$

or equivalently,

$$P\left(-t_{n-1,1-\frac{\alpha}{2}} \le \frac{\sqrt{n}\left(\bar{X} - \mu\right)}{S} \le t_{n-1,1-\frac{\alpha}{2}} \right) = 1 - \alpha \tag{10.2}$$

since $t_{n-1,\frac{\alpha}{2}} = -t_{n-1,1-\frac{\alpha}{2}}$.

For the event inside the probability statement in Equation (10.2), we multiply both sides by $\dfrac{S}{\sqrt{n}}$, and write $-\dfrac{S}{\sqrt{n}}t_{n-1,1-\frac{\alpha}{2}} \le \bar{X} - \mu \le \dfrac{S}{\sqrt{n}}t_{n-1,1-\frac{\alpha}{2}}$. By subtracting $\bar{X}$ and multiplying by -1 (this step will reverse the inequality), we can rewrite the equivalent event as

$$\bar{X} - \frac{S}{\sqrt{n}}t_{n-1,1-\frac{\alpha}{2}} \le \mu \le \bar{X} + \frac{S}{\sqrt{n}}t_{n-1,1-\frac{\alpha}{2}}. \tag{10.3}$$

Since equivalent events have the same probability, we can rewrite Equation (10.2) as

$$P\left(\bar{X} - \frac{S}{\sqrt{n}}t_{n-1,1-\frac{\alpha}{2}} \le \mu \le \bar{X} + \frac{S}{\sqrt{n}}t_{n-1,1-\frac{\alpha}{2}} \right) = 1-\alpha. \tag{10.4}$$

Therefore, if we choose

$$L(X_1,\ldots,X_n) = \bar{X} - \frac{S}{\sqrt{n}}t_{n-1,1-\frac{\alpha}{2}} \quad \text{and} \quad U(X_1,\ldots,X_n) = \bar{X} + \frac{S}{\sqrt{n}}t_{n-1,1-\frac{\alpha}{2}},$$

the $100(1-\alpha)\%$ confidence interval for μ will be $\left[L(X_1,\ldots,X_n), U(X_1,\ldots,X_n) \right]$. We can also write directly that

$$\left[\bar{X} - \frac{S}{\sqrt{n}}t_{n-1,1-\frac{\alpha}{2}}, \ \bar{X} + \frac{S}{\sqrt{n}}t_{n-1,1-\frac{\alpha}{2}} \right] \tag{10.5}$$

is the $100(1-\alpha)\%$ confidence interval.

The term $\dfrac{\sqrt{n}\left(\bar{X}-\mu\right)}{S}$ used in constructing the confidence interval in Example 10.1, is called a *pivotal quantity* (or *pivot*). In general, if a random quantity $Q(X_1,\ldots,X_n,\theta)$ associated with the distribution under consideration has a distribution that is independent of the parameter, then it is called a pivotal quantity. For a random sample taken from a population that has a location type of pdf (i.e. the pdf has the form of $f(x-\mu)$) then $\bar{X}-\mu$ is a pivotal quantity. Similarly, if the population distribution has a scale type of pdf (i.e. the pdf is of the form $\dfrac{1}{\sigma}f(\dfrac{x}{\sigma})$), then $\dfrac{\bar{X}}{\sigma}$ is a pivotal quantity. If the population distribution has a location-scale type of distribution ((i.e. the pdf has the form of $\dfrac{1}{\sigma}f(\dfrac{x-\mu}{\sigma})$), then

$\dfrac{\sqrt{n}\left(\overline{X}-\mu\right)}{S}$ is a pivotal quantity. The pivotal quantity is an important tool for constructing confidence intervals.

In Equation (10.2), we assume that the pivotal quantity belongs to an equal-tail interval that leads to an equal-tail confidence interval in (10.3). Although it is not necessary to make this assumption, this assumption will give the shortest length for the confidence interval. Although there are many $100(1-\alpha)\%$ confidence intervals for a given α, it is important to find one that has the shortest length. The reason for that will be discussed in Section 10.7.

The fact that an equal-tail confidence interval is shorter than the corresponding unequal-tail confidence interval for a unimodal distribution can be stated below: if

$$\overline{X}-\frac{S}{\sqrt{n}}t_{n-1,1-\alpha_2}\le \mu \le \overline{X}+\frac{S}{\sqrt{n}}t_{n-1,1-\alpha_1} \tag{10.6}$$

is an unequal-tail $100(1-\alpha)\%$ confidence interval, where $\alpha_1+\alpha_2=\alpha$, then the shortest length occurs when $\alpha_1=\alpha_2$. Note that the length of the confidence interval in (10.6) is $\dfrac{S}{\sqrt{n}}\left(t_{n-1,1-\alpha_1}+t_{n-1,1-\alpha_2}\right)$, where $\dfrac{S}{\sqrt{n}}$ is not a function of α_1 and α_2. To minimize the length subject to the restriction $\alpha_1+\alpha_2=\alpha$, we use Lagrange multipliers. Consider the objective function

$$F_{t_{n-1}}^{-1}(1-\alpha_1)+F_{t_{n-1}}^{-1}(1-\alpha_2)+\lambda(\alpha-\alpha_1-\alpha_2) \tag{10.7}$$

where $F_{t_{n-1}}^{-1}(\cdot)$ is the inverse function of the cdf of the t distribution with $n-1$ degrees of freedom and thus

$$F_{t_{n-1}}^{-1}(1-\alpha_1)=t_{n-1,1-\alpha_1} \text{ and } F_{t_{n-1}}^{-1}(1-\alpha_2)=t_{n-1,1-\alpha_2}.$$

Differentiating (10.7) with respect to α_1, α_2 and λ separately and setting each derivation equal to zero, we obtain the equations

$$\frac{-1}{f_{t,n-1}(1-\alpha_1)}=\lambda, \frac{-1}{f_{t,n-1}(1-\alpha_2)}=\lambda \text{ and } \alpha_1+\alpha_2=\alpha \tag{10.8}$$

where $f_{t_{n-1}}(\cdot)$ is the pdf of the t distribution with $n-1$ degrees of freedom. The closed form of $f_{t_{n-1}}(\cdot)$ was presented in Chapter 5. The solution of the above equations is clearly $\alpha_1 = \alpha_2 = \dfrac{\alpha}{2}$ since equation (10.8) implies

$$\frac{-1}{f_{t,n-1}(1-\alpha_1)} = \frac{-1}{f_{t,n-1}(1-\alpha_2)}.$$

This means that the confidence interval presented in (10.5) has the shortest length among all confidence intervals that have the same coverage probability. For most confidence intervals that have a unimodal pivotal distribution, the shortest length will be associated with equal tails. In the examples below, we will follow this principle without proving it until we revisit this issue in section 10.7.

Example 10.2 In Example 10.1, we now consider testing $H_0 : \mu = \mu_0$ versus $H_0 : \mu \neq \mu_0$. For a fixed α level, the most powerful test will reject H_0 if

$$\left| \frac{\sqrt{n}\left(\bar{X} - \mu_0\right)}{S} \right| > t_{n-1,1-\frac{\alpha}{2}}$$ (see section 9.8.2). This is equivalent to saying that, if the

sample points satisfy

$$\left| \bar{X} - \mu_0 \right| \leq \frac{S}{\sqrt{n}} t_{n-1,1-\frac{\alpha}{2}},$$

or equivalently

$$\bar{X} - \frac{S}{\sqrt{n}} t_{n-1,1-\frac{\alpha}{2}} \leq \mu_0 \leq \bar{X} + \frac{S}{\sqrt{n}} t_{n-1,1-\frac{\alpha}{2}}$$

then H_0 will be accepted. Since the level of significance for this test is α, we can state that the probability of rejecting H_0, given $\mu = \mu_0$ is α. That means the probability of accepting H_0, given $\mu = \mu_0$ is $1-\alpha$. Specifically, we have

$$P\left(\bar{X} - \frac{S}{\sqrt{n}} t_{n-1,1-\frac{\alpha}{2}} \leq \mu_0 \leq \bar{X} + \frac{S}{\sqrt{n}} t_{n-1,1-\frac{\alpha}{2}} \middle| \mu = \mu_0 \right) = 1-\alpha.$$

This probability statement is true for any μ_0. Therefore,

$$P\left(\bar{X} - \frac{S}{\sqrt{n}} t_{n-1,1-\frac{\alpha}{2}} \le \mu \le \bar{X} + \frac{S}{\sqrt{n}} t_{n-1,1-\frac{\alpha}{2}} \right) = 1 - \alpha$$

is true as it was stated in (10.4) and the $100(1-\alpha)\%$ confidence interval is the same as that in equation (10.3).

Example 10.3 For the data in Problem 8.20, we assume that the distribution of heights is normally distributed with unknown mean and unknown variance. Using the formula derived in Examples 10.1 or 10.2, the 95% confidence interval can be calculated as $[148.19, 161.81]$ and, similarly, the 90% confidence interval as $[149.48, 160.52]$. The sample mean is $\bar{x} = 155$ and the sample standard deviation is $s = 9.52$. A SAS program for calculating this 95% confidence interval is available. By changing the α value in the program, one can use it to calculate any confidence interval of this type.

Example 10.4 In Example 10.1, assume that we are interested in determining the confidence interval for the variance. We should first consider the pivotal quantity $\frac{(n-1)S^2}{\sigma^2}$ that has a χ^2 distribution with $n-1$ degrees of freedom (see Chapter 5). Let $\chi^2_{n-1,u}$ denote the $100u$ percentile of a χ^2_{n-1} distribution. Then

$$P\left(\chi^2_{n-1,\frac{\alpha}{2}} \le \frac{(n-1)S^2}{\sigma^2} \le \chi^2_{n-1,1-\frac{\alpha}{2}} \right) = 1 - \alpha.$$

It is evident that the event $\chi^2_{n-1,\frac{\alpha}{2}} \le \frac{(n-1)S^2}{\sigma^2} \le \chi^2_{n-1,1-\frac{\alpha}{2}}$ is equivalent to the event

$\frac{(n-1)S^2}{\chi^2_{n-1,1-\frac{\alpha}{2}}} \le \sigma^2 \le \frac{(n-1)S^2}{\chi^2_{n-1,\frac{\alpha}{2}}}$.   Therefore, $P\left(\frac{(n-1)S^2}{\chi^2_{n-1,1-\frac{\alpha}{2}}} \le \sigma^2 \le \frac{(n-1)S^2}{\chi^2_{n-1,\frac{\alpha}{2}}} \right) = 1 - \alpha.$

Hence

$$\left[\frac{(n-1)S^2}{\chi^2_{n-1,1-\frac{\alpha}{2}}}, \frac{(n-1)S^2}{\chi^2_{n-1,\frac{\alpha}{2}}} \right] \tag{10.9}$$

is a $100(1-\alpha)\%$ confidence interval for σ^2.

Example 10.5 For the data discussed in Example 10.3, we can calculate a 95% confidence interval for the unknown variance σ^2. The result is $[42.90, 302.18]$ while the maximum likelihood estimate is 90.67.

If we are to find the 95% confidence region for (μ, σ^2), the product of two confidence intervals calculated in Examples 10.3 and 10.5 will be too small. Since $(1-\alpha)(1-\alpha) < 1-\alpha$, the coverage probability for the product of the confidence intervals described in (10.5) and (10.9) is certainly less than $1-\alpha$. In addition, the two events may not be independent; that may further reduce the coverage probability.

The general coverage probability for the confidence intervals of two parameters θ_1 and θ_2 in the same population distribution can be described as follows. Let $[L_1, U_1]$ be a $100(1-\alpha)\%$ confidence interval for θ_1 and $[L_2, U_2]$ be a $100(1-\alpha)\%$ confidence interval for θ_2. If the event $\theta_1 \in [L_1, U_1]$ and the event $\theta_2 \in [L_2, U_2]$ are independent, then

$$\Pr\left((\theta_1, \theta_2) \in [L_1, U_1] \times [L_2, U_2]\right) = \Pr\left(\theta_1 \in [L_1, U_1], \theta_2 \in [L_2, U_2]\right) = (1-\alpha)^2.$$

It follows that $[L_1, U_1] \times [L_2, U_2]$ is a $100\left[(1-\alpha)^2\right]\%$ confidence rectangle for (θ_1, θ_2). If we have k parameters and their corresponding confidence intervals are independent, then we should first apply Bonferroni principle to find each a $100(1-\dfrac{\alpha}{k})\%$ confidence interval. The product of these confidence intervals can then be a $100(1-\alpha)\%$ confidence region. Even though the confidence coefficient associated with this confidence region will be higher than $1-\alpha$.

Example 10.6 Suppose that $X_1, \ldots, X_n$ is a random sample taken from a population that has an exponential distribution with parameter λ. We observe that a sufficient statistic $\sum_{i=1}^{n} X_i$ has a $gamma(n, \lambda)$ distribution. From Chapter 5, we know that $2\lambda \sum_{i=1}^{n} X_i$ has a χ^2 distribution with $2n$ degrees of freedom. If we are to find the confidence interval for λ, a natural choice for the pivotal quantity would be

$2\lambda \sum_{i=1}^{n} X_i$. Clearly, the probability of the event $\left\{ \chi^2_{2n,\frac{\alpha}{2}} \leq 2\lambda \sum_{i=1}^{n} X_i \leq \chi^2_{2n,1-\frac{\alpha}{2}} \right\}$ is

$1-\alpha$. By some algebraic manipulations and the information $\sum_{i=1}^{n} X_i > 0$, the

$100(1-\alpha)\%$ confidence interval can be derived as

$$\left[\frac{\chi^2_{2n,\frac{\alpha}{2}}}{2\sum_{i=1}^{n} X_i}, \frac{\chi^2_{2n,1-\frac{\alpha}{2}}}{2\sum_{i=1}^{n} X_i} \right].$$

Since the probability of the event $\left\{ \chi^2_{2n,\alpha} \leq 2\lambda \sum_{i=1}^{n} X_i \right\}$ is also $1-\alpha$, the

$100(1-\alpha)\%$ lower confidence bound for λ is $\dfrac{\chi^2_{2n,\alpha}}{2\sum_{i=1}^{n} X_i}$.

Example 10.7 Suppose that $X_{i1}, \ldots, X_{in_i}$ are independent normal random variables with mean μ_i and variance σ_i^2, $i = 1, 2$. If σ_1^2 and σ_2^2 are known, then

$\dfrac{\left(\overline{X}_1 - \overline{X}_2\right) - \left(\mu_1 - \mu_2\right)}{\sqrt{\dfrac{\sigma_1^2}{n_1} + \dfrac{\sigma_2^2}{n_2}}}$ has a standard normal distribution and therefore can be

used as a pivotal quantity, where $\overline{X}_i$ is sample mean associated with the ith population. Furthermore, we have

$$P\left(Z_{\frac{\alpha}{2}} \leq \frac{\left(\overline{X}_1 - \overline{X}_2\right) - \left(\mu_1 - \mu_2\right)}{\sqrt{\dfrac{\sigma_1^2}{n_1} + \dfrac{\sigma_2^2}{n_2}}} \leq Z_{1-\frac{\alpha}{2}} \right) = 1 - \alpha. \qquad (10.10)$$

Since

$$\left(\overline{X}_1 - \overline{X}_2\right) - Z_{1-\frac{\alpha}{2}} \sqrt{\frac{\sigma_1^2}{n_1} + \frac{\sigma_2^2}{n_2}} \leq \mu_1 - \mu_2 \leq \left(\overline{X}_1 - \overline{X}_2\right) + Z_{1-\frac{\alpha}{2}} \sqrt{\frac{\sigma_1^2}{n_1} + \frac{\sigma_2^2}{n_2}}$$

is equivalent to the event in (10.10) and has a probability $1-\alpha$, the $100(1-\alpha)\%$ confidence interval for $\mu_1 - \mu_2$ can be described as

$$\left[\left(\bar{X}_1 - \bar{X}_2\right) - Z_{1-\frac{\alpha}{2}}\sqrt{\frac{\sigma_1^2}{n_1} + \frac{\sigma_2^2}{n_2}}, \left(\bar{X}_1 - \bar{X}_2\right) + Z_{1-\frac{\alpha}{2}}\sqrt{\frac{\sigma_1^2}{n_1} + \frac{\sigma_2^2}{n_2}}\right].$$

If $\sigma_1^2 = \sigma_2^2 = \sigma^2$ is unknown, then

$$T = \frac{\left(\bar{X}_1 - \bar{X}_2\right) - \left(\mu_1 - \mu_2\right)}{S_p\sqrt{\frac{1}{n_1} + \frac{1}{n_2}}}$$

has a t distribution with $n_1 + n_2 - 2$ degrees of freedom, where

$$s_p^2 = \frac{\sum_{j=1}^{n_1}(X_{1j} - \bar{X}_1)^2 + \sum_{j=1}^{n_2}(X_{2j} - \bar{X}_2)^2}{n_1 + n_2 - 2}.$$ Therefore, T can be chosen as a pivotal

quantity. The $100(1-\alpha)\%$ confidence interval for $\mu_1 - \mu_2$ can be expressed as
$$\left[L(X_{11},\ldots,X_{1n_1},X_{21},\ldots,X_{2n_2}), U(X_{11},\ldots,X_{1n_1},X_{21},\ldots,X_{2n_2})\right],$$
where

$$L(X_{11},\ldots,X_{1n_1},X_{21},\ldots,X_{2n_2}) = \left(\bar{X}_1 - \bar{X}_2\right) - t_{n_1+n_2-2,1-\frac{\alpha}{2}}S_P\sqrt{\frac{1}{n_1} + \frac{1}{n_2}},$$

$$U(X_{11},\ldots,X_{1n_1},X_{21},\ldots,X_{2n_2}) = \left(\bar{X}_1 - \bar{X}_2\right) + t_{n_1+n_2-2,1-\frac{\alpha}{2}}S_P\sqrt{\frac{1}{n_1} + \frac{1}{n_2}}.$$

.

In Examples 10.1--10.7, we have discussed the methods of constructing confidence intervals for a random sample taken from populations that have continuous distributions. In the following examples, we will illustrate the way to construct a confidence interval using a sample from a discrete distribution.

Example 10.8 Suppose that $X_1,\ldots,X_n$ are independent random variables and that each one has a Bernoulli distribution with unknown probability of success. It has been established that the sample sum $T = \sum_{i=1}^{n} X_i$ is a sufficient statistic and that T has a $B(n, p)$ distribution. Our natural choice of pivotal quantity will be T. However, since T has a discrete distribution, we may not be able to find two numbers $l(p)$ and $u(p)$ such that $P\big(l(p) \leq T \leq u(p)\big| p\big) = 1 - \alpha$. But we can still choose two numbers such that

$$P\big(l(p) \leq T \leq u(p)\big| p\big) \geq 1 - \alpha. \tag{10.11}$$

Specifically, we can choose $l(p)$ as the largest integer among $0, 1, \ldots, n$ such that

$$P\big(T < l(p)\big| p\big) < \frac{\alpha}{2}$$

(or equivalently, $\displaystyle\sum_{i=0}^{l(p)-1} P(T = i) < \frac{\alpha}{2}$ and $\displaystyle\sum_{i=0}^{l(p)} P(T = i) \geq \frac{\alpha}{2}$) and choose $u(p)$ as the smallest integer among $0, 1, \ldots, n$ such that

$$P\big(T > u(p)\big| p\big) < \frac{\alpha}{2}$$

(or equivalently, $\displaystyle\sum_{i=u(p)}^{n} P(T = i) \geq \frac{\alpha}{2}$ and $\displaystyle\sum_{i=u(p)+1}^{n} P(T = i) < \frac{\alpha}{2}$). Observe that $l(p)$ and $u(p)$ are non-decreasing (step) functions of p (see Problem 10.10). For $T = t$, we denote $l^{-1}(t) = \inf\left\{ p : \displaystyle\sum_{i=0}^{t-1} P(T = i | p) < \frac{\alpha}{2} \right\} = \inf\{ p : l(p) \leq t \}$, then there is a unique $l^{-1}(t)$ for any t and we have $p \leq l^{-1}(t)$ if and only if $l(p) \leq t$. Similarly, if we denote

$$u^{-1}(t) = \sup\left\{ p : \sum_{i=t+1}^{n} P(T = i | p) < \frac{\alpha}{2} \right\} = \sup\{ p : t \leq u(p) \} \quad \text{then} \quad \text{the} \quad \text{unique}$$

$u^{-1}(t)$ satisfies $u^{-1}(t) \leq p$ if and only if $t \leq u(p)$. Therefore, since the event $\{ l(p) \leq T \leq u(p) \}$ $\big(= \{ T \leq u(p) \} \cap \{ l(p) \leq T \} \big)$ is equivalent to $\{ u^{-1}(T) \leq p \} \cap \{ p \leq l^{-1}(T) \}$ that is equal to the event $\{ u^{-1}(T) \leq p \leq l^{-1}(T) \}$, Equation (10.11) can be re-written as $P\big(u^{-1}(T) \leq p \leq l^{-1}(T) \big) \geq 1 - \alpha$. Hence the $100(1 - \alpha)\%$ confidence interval for p can be expressed as $\big[u^{-1}(T), l^{-1}(T) \big]$ or $\big[L(X_1, \ldots, X_n), U(X_1, \ldots, X_n) \big]$ if we define $L(X_1, \ldots, X_n) = u^{-1}\big(\displaystyle\sum_{i=1}^{n} X_i \big)$ and $U(X_1, \ldots, X_n) = l^{-1}\big(\displaystyle\sum_{i=1}^{n} X_i \big)$.

Example 10.9 A department store is interested in determining the probability p that a potential customer entering the store will actually make a purchase during their annual promotion period. A random sample of 20 adults who entered the store in that period was selected; 9 of them made purchases. To find the 95% confidence interval for p, we use the sufficient statistic $\sum_{i=1}^{20} X_i = 9$. From the derivation of Example 10.8 and $\alpha = .05$, we shall find $l^{-1}(t)$ as the smallest value of p such that

$$F(t-1; p) = \sum_{k=0}^{t-1} \binom{20}{k} p^k (1-p)^{n-k} < .025,$$

where $t = 9$. This can be accomplished by using computer software to calculate the cdf value $P\left(T \leq 8 \mid T \cdots B(20, p)\right)$ for different possibilities of p. In this example, we can start with $p = 0.1, 0.2, \ldots, 0.9$. We find that the cdf values jump from below 0.025 to above 0.025 when p changes from 0.6 to 0.7. Continuing to use this strategy, we can find $F(8, .639) = .0252$ and $F(8, .640) = .0247$. Therefore, $l^{-1}(t) = 0.640$. Similarly, using the same strategy, we can find $u^{-1}(t)$ by searching for the greatest value of p such that

$$1 - F(t; p) = \sum_{k=t+1}^{20} \binom{20}{k} p^k (1-p)^{n-k} < .025 \text{ for } t = 9.$$

We find that $1 - F(9, .271) = .0244$ and $1 - F(9, .272) = .02502$. In this example, $u^{-1}(t) = .271$. The store can now obtain the 95% confidence interval of p, based on this sample, as $[0.271, 0.640]$.

The method used in obtaining the confidence interval for the proportion discussed in Examples 10.8 and 10.9 can be generalized to broader discrete populations and to most continuous populations. The general formulas for discrete and continuous cases respectively, can be found in Section 9.2.3 of Casella and Berger (2002). Example 10.10 deals with finding a Poisson interval estimator and Example 10.11 deals with determining an interval estimator for a continuous population.

Example 10.10 Let $X_1, \ldots, X_n$ be a random sample taken from a population that has a Poisson distribution with parameter λ. Since $T(X_1, \ldots, X_n) - \sum_{i=1}^{n} X_i$ is a sufficient statistic that has a $Poisson(n\lambda)$ distribution, we can use it for purposes of interval estimation. Denoting $l(p)$ as the largest non-zero integer such that

$$P\left(T < l(p)\,\middle|\,p\right) < \frac{\alpha}{2}$$

and $u(p)$ as the smallest non-zero integer such that

$$P\left(T > u(p)\,\middle|\,p\right) < \frac{\alpha}{2},$$

then $P\left(l(p) \le T \le u(p)\,\middle|\,p\right) \ge 1 - \alpha$. Observe that $l(p)$ and $u(p)$ are non-decreasing (step) functions of p (see Problem 10.11). For any $T(X_1,\ldots,X_n) = t$, we can define $l^{-1}(t)$ and $u^{-1}(t)$ similar to those in Example 10.8 and establish the $100(1-\alpha)\%$ confidence interval as $\left[u^{-1}(T), l^{-1}(T)\right]$ just as in Example 10.8.

Using the data presented in Example 8.8, we have $\sum_{i=1}^{n} x_i = 21007$. Consequently, we can find $l^{-1}(21007) = 2661.7$ and $u^{-1}(21007) = 2590.6$. Therefore, based on this sample, the 95% confidence interval for λ is $[2590.6, 2661.7]$. The search for the upper and lower bounds is slightly different from what has been done in Example 10.8. In this problem, we start both ends of the interval with the point estimate 2625.88. We extend each end by an increment of 10, and find the cdf $F_T(t\,|\,\lambda)$ (for upper end) and $1 - F_T(t\,|\,\lambda)$ (for lower bound) until we reach the point that these values of the bounds jump from less than 0.025 to over 0.025. We then continue to perform the same search by an increment of 1, of 0.1 and so on (depending on the number of significant digits we would like to pursue). This method should be applicable for any software with a built-in cdf function. A SAS program is available for searching both bounds for this example.

Example 10.11 Assume that $X_1,\ldots,X_n$ is a random sample taken from a population that is uniformly distributed on $[0,\theta]$. It is known from Example 8.6 that the MLE for θ is $X_{(n)}$ (the largest among $X_1,\ldots,X_n$) that has a cdf defined as

$$F(t\,|\,\theta) = \begin{cases} \dfrac{t^n}{\theta^n} & \text{if } 0 \le t \le \theta \\[2mm] 0 & \text{otherwise} \end{cases}.$$

We can now find two functions $l(\theta)$ and $u(\theta)$ of θ such that

$$P\left(X_{(n)} < l(\theta)\,\middle|\,\theta\right) = \frac{\alpha}{2} \text{ and } P\left(X_{(n)} > u(\theta)\,\middle|\,\theta\right) = \frac{\alpha}{2} \qquad (10.12)$$

or $P\left(l(\theta) \le X_{(n)} \le u(\theta)\right) = 1 - \alpha$.

Using the cdf of $X_{(n)}$ (see Chapter 6), Equation (10.12) can be written as

$$\left(\frac{l(\theta)}{\theta}\right)^{n} = \frac{\alpha}{2} \text{ and } 1 - \left(\frac{u(\theta)}{\theta}\right)^{n} = \frac{\alpha}{2}.$$

Therefore, $l(\theta) = \theta\left(\frac{\alpha}{2}\right)^{1/n}$ and $u(\theta) = \theta\left(1 - \frac{\alpha}{2}\right)^{1/n}$. Hence,

$$l^{-1}(t) = t\left(\frac{\alpha}{2}\right)^{-1/n} \text{ and } u^{-1}(t) = t\left(1 - \frac{\alpha}{2}\right)^{-1/n}.$$

The $100(1-\alpha)\%$ confidence interval for θ is $\left[X_{(n)}\left(1 - \frac{\alpha}{2}\right)^{-1/n}, X_{(n)}\left(\frac{\alpha}{2}\right)^{-1/n} \right]$.

The method of constructing a confidence interval using a pivotal quantity is not limited to the population with a known distributional form. Some distribution-free nonparametric methods may also use this technique in interval estimation.

Example 10.12 Let $X_1,\ldots,X_n$ be a random sample taken from a population with a continuous distribution that has a median θ. If we define $T(X_1,\ldots,X_n)$ as the number of X_is such that $X_i > \theta$, then $T(X_1,\ldots,X_n)$ has a binomial distribution with parameters n and $p = \frac{1}{2}$. Mathematically, we can write

$T(X_1,\ldots,X_n) = \sum_{i=1}^{n} I_{(\theta,\infty)}(X_i)$. It is clear that $T(X_1,\ldots,X_n)$ can serve as a pivotal quantity. Let $b_{\alpha/2,n,1/2}$ denote the greatest integer for which

$$P\left(Y \leq b_{\alpha/2,n,1/2} \middle| Y \cdots B(n,\frac{1}{2})\right) < \frac{\alpha}{2},$$

then

$$P\left(b_{\alpha/2,n,1/2} < T(X_1,\ldots,X_n) < n - b_{\alpha/2,n,1/2}\right) \geq 1 - \alpha.$$

If we let $X_{(1)},\ldots,X_{(n)}$ be the ordered sample of observations $X_1,\ldots,X_n$, we can see that $\left\{T(X_1,\ldots,X_n) < n - b_{\alpha/2,n,1/2}\right\}$ is equivalent to $\left\{X_{(1+b_{\alpha/2,n,1/2})} \leq \theta\right\}$. The latter event indicates that at least $b_{\alpha/2,n,1/2} + 1$ of the sample observations have values less than or equal to θ and the former event means that less than $n - b_{\alpha/2,n,1/2}$ of the sample elements have values greater than θ. Similarly,

$\left\{ b_{\alpha/2,n,1/2} < T(X_1,\ldots,X_n) \right\}$ is equivalent to $\left\{ \theta \le X_{(n-b_{\alpha/2,n,1/2})} \right\}$. Therefore, the

$100(1-\alpha)\%$ confidence interval for θ is

$\left[X_{(1+b_{\alpha/2,n,1/2})}, X_{(n-b_{\alpha/2,n,1/2})} \right]$. This is a nonparametric distribution-free confidence interval.

The examples discussed in this section may not require the α error to be distributed equally between two tails. Instead, the α may be divided asymmetrically. For example, in finding the confidence interval for the normal mean (Example 10.1), we can accept an unequal-tail $100(1-\alpha)\%$ confidence interval of the

form $\left[\bar{X} - \dfrac{S}{\sqrt{n}} t_{n-1,1-\alpha_1}, \bar{X} + \dfrac{S}{\sqrt{n}} t_{n-1,1-\alpha_2} \right]$ if $\alpha_1 + \alpha_1 = \alpha$. However, for a symmet-

ric and unimodal distribution, the equal-tail one is the most desirable confidence interval since it is associated with the shortest length. Detailed criteria for a confidence interval will be discussed in section 10.7.

In general, a random set that covers the parameter may not necessarily be an interval. In that case, if the probability that a random set covers the parameter is at least $1-\alpha$, then this random set is called a $100(1-\alpha)\%$ *confidence set*. For the unimodal distribution of the pivotal quantity, the confidence interval with a good precision usually exists and is better than a non-interval random set.

10.4 Bayesian Credible Intervals

In section 10.3, we discussed the notion of the confidence interval and called attention to its interpretation. Especially, we cautioned that any realization of a $100(1-\alpha)\%$ confidence interval can not be interpreted as the probability that the true parameter will be in this confidence interval is $1-\alpha$. Instead, the true parameter either belongs to this confidence interval or does not belong to it, since the parameter in our assumption and derivations is not viewed as a random variable. For the Bayesian method, as we studied in Chapter 8, we assume that the parameter itself is random and follows a particular distribution known as the prior distribution of the parameter (see Chapter 8).

Definition 10.3 Let $X_1, X_2, \square , X_n$ be a random sample from a population with either pdf or pmf $f(x|\theta)$. If we denote the prior distribution of θ by $\pi(\theta)$ and the posterior distribution by $\pi(\theta|X_1 = x_1,\ldots,X_n = x_n)$, then the $100(1-\alpha)\%$ credible interval for θ is the interval $I \subseteq \Theta$ such that

$$1-\alpha \le P\left(\theta \in I \big| X_1 = x_1,\ldots,X_n = x_n\right)$$

$$= \begin{cases} \int_I \pi(\theta | X_1 = x_1,\ldots,X_n = x_n)d\theta \\ \sum_{\theta \in I} \pi(\theta | X_1 = x_1,\ldots,X_n = x_n). \end{cases} \tag{10.13}$$

The difference between the confidence interval and the credible interval is not only in the interpretation. The probabilities they try to measure are also different. Since the posterior distribution depends on the observations, the probability in (10.13) is a measure of final precision. That means, after taking the observations, we believe the probability that θ in I is at least $1-\alpha$. On the contrary, the coverage probability in the confidence interval is only a measure of initial precision. Hence, before observations are taken, the probability that θ belongs to that random interval is at least $1-\alpha$.

It is clear from Definition 10.3, that we have an infinite number of credible intervals. It has been suggested that we choose the one that includes the posterior mode.

Definition 10.4 If the posterior distribution is unimodal, the $100(1-\alpha)\%$ HPD (*highest posterior density*) *credible interval* for θ is the interval I of the form $I = \{\theta : \pi(\theta | x_1,\ldots,x_n) \ge k(\alpha)\}$, where $k(\alpha)$ is the largest constant such that I satisfies (10.13). If the posterior distribution is not unimodal, I may not be an interval and we call it the $100(1-\alpha)\%$ *HPD credible set* for θ.

Example 10.13 In Example 8.17, we derived the posterior distribution for a normal $N(\theta,\sigma^2)$ sample with a normal prior $N(\mu,\tau^2)$ for θ, where σ^2, μ and τ^2 are known. The posterior distribution given $X_1 = x_1,\ldots,X_n = x_n$ is normal with mean $m(x_1,\ldots,x_n) = \dfrac{1}{\varphi}\left(\dfrac{n\bar{x}}{\sigma^2} + \dfrac{\mu}{\tau^2}\right)$ and variance $\dfrac{1}{\varphi}$, where $\varphi = \dfrac{n}{\sigma^2} + \dfrac{1}{\tau^2}$.

Therefore, the $100(1-\alpha)\%$ HPD credible interval is given by

$$I = \left[m(x_1,\ldots,x_n) - z_{1-\alpha/2}\varphi^{-1/2},\; m(x_1,\ldots,x_n) + z_{1-\alpha/2}\varphi^{-1/2} \right].$$

If we use the random sample of Systolic Blood Pressure (SBP) of 10 adolescents given in Example 8.2 and assume that the variance σ^2 of the SBP of the population is known to be 40 and the prior distribution of the mean θ follows a $N(95,25)$ distribution. We can calculate

$$m(x_1,\ldots,x_n) = \frac{1}{0.29}\left(\frac{10 \times 100.85}{40} + \frac{95}{25}\right) = 100.043 \text{ and } \varphi = \frac{10}{40} + \frac{1}{25} = 0.29.$$

The 95% HPD credible interval will be $[96.4, 103.7]$. We can then believe the probability that the population mean SBP falling in the interval $[96.4, 103.7]$ is at least 95%, after observing the data. If we were to use confidence interval, we would need to perform the data collection of the same sample size a large number of times (say 100) in order to claim that there are roughly 95% of them that will cover the true population mean.

Example 10.14 Assume we observe an outcome from a binomial distribution with parameters n and θ. If n is known and θ has a beta prior with parameters α and β, the posterior distribution was calculated in Example 8.18 and is given by a $Beta(x + \alpha, n - x + \beta)$ distribution. It is easy to find a $100(1-\gamma)\%$ credible interval for θ by simply using the $100\left(\dfrac{\gamma}{2}\right)$ and $100\left(1 - \dfrac{\gamma}{2}\right)$ percentiles as the lower and upper ends of the interval. In general, it may not be easy to find a $100(1-\alpha)\%$ HPD (highest posterior density) credible interval for θ. For the beta density, it depends on the relationship of two parameters. If $a = x + \alpha$ and $b = n - x + \beta$ are equal, then the posterior density is symmetric and we can simply take the $100\left(\dfrac{\gamma}{2}\right)$ and $100\left(1 - \dfrac{\gamma}{2}\right)$ percentiles as the lower and upper ends of the HPD credible interval. If $a > 1$ and $b = 1$ ($a = 1$ and $b > 1$) then the posterior density is strictly increasing (or strictly decreasing). In these cases, the HPD credible interval is open ended. If $a < 1$ and $b < 1$, the density is U-shaped, the HPD credible set will be the union of two half-open intervals. Even when $a > 1$ and $b > 1$, the unimodality of the density does not make finding of the HPD credible interval easy. The trial and error method may be applied to find this credible interval.

   To find the HPD credible interval (set), we first find the solutions to the equation $\pi(\theta | x_1, \ldots, x_n) = c$ for any positive number c, as illustrated in Figure 10.2. If the posterior distribution is unimodal, the solutions are usually two points, say, $\theta_1(c)$ and $\theta_2(c)$. Then we solve for c the equation

$$\int_{\theta_1(c)}^{\theta_2(c)} \pi(\theta | x_1, \ldots, x_n) d\theta = 1 - \gamma.$$

For the large sample case, we can also find an approximate posterior distribution and find a HPD credible interval based on it.

   To find a numerical HPD credible interval for Example 8.18, we first choose a c and then use the Newton-Raphson method to find the solution of $\theta_1(c)$ and $\theta_2(c)$ for this c. The probability $\Pr\left(\theta \in (\theta_1(c), \theta_2(c)) | x_1, \ldots, x_n\right)$ is then calculated. If this probability is greater (or less) than $1 - \gamma$, then we increase (or decrease) the value of c until this probability reaches our desired confidence bound. A SAS program for implementing these procedures is available. A 95% HPD credible interval is found to be $(0.027, 0.122)$.

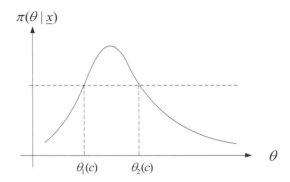

Figure 10.2 Illustration for finding a HPD credible interval

Example 10.15 In Example 10.10, we discussed the method of finding a confidence interval for the parameter λ in a Poisson sample. If we assume that the prior distribution of λ follows a gamma distribution with parameters p and α, then the posterior distribution that has been derived in Example 8.19, has gamma distribution with parameters $p + \sum_{i=1}^{n} x_i$ and $n + \alpha$. To find a $100(1-\gamma)\%$ credible interval, we can easily find an equal-tail interval $\left[g_{\gamma/2}, g_{1-\gamma/2} \right]$, where $g_{\gamma/2}$ and $g_{1-\gamma/2}$ are respectively the $100\dfrac{\gamma}{2}$ and $100\left(1 - \dfrac{\gamma}{2}\right)$ percentiles of the gamma distribution with parameters $p + \sum_{i=1}^{n} x_i$ and $n + \alpha$. Equivalently, we can also consider a transformation of the posterior distribution $2(n+\alpha)\lambda$ that has a χ^2 distribution with degree of freedom $k = 2\left(p + \sum_{i=1}^{n} x_i\right)$ (see Chapter 5). If we denote the 100γ percentile of this χ_k^2 distribution as $\chi_{k,\gamma}^2$ then the $100(1-\gamma)\%$ credible interval can also be expressed as $\left[\dfrac{1}{2(n+\alpha)} \chi_{k,\gamma/2}^2, \dfrac{1}{2(n+\alpha)} \chi_{k,1-\gamma/2}^2 \right]$.

For obtaining the $100(1-\gamma)\%$ HPD credible interval, we need to find λ_L and λ_U

such that $f_{\chi^2_{k,2\left(p+\sum\limits_{i=1}^{n}x_i\right)}}(\lambda_L)=f_{\chi^2_{k,2\left(p+\sum\limits_{i=1}^{n}x_i\right)}}(\lambda_U)$ and $\int\limits_{\lambda_L}^{\lambda_U}f_{\chi^2_{k,2\left(p+\sum\limits_{i=1}^{n}x_i\right)}}(t)dt=1-\gamma.$ The HPD

credible interval will then be $\left[\dfrac{1}{2(n+\alpha)}\lambda_L,\dfrac{1}{2(n+\alpha)}\lambda_U\right].$ The numerical method

for finding this interval is similar to the one described in Example 10.14 and will
be left as an exercise.

10.5 Approximate Confidence Intervals and MLE Pivot

In Examples 10.8 and 10.9, we obtained the exact confidence interval for the proportion. The method to obtain this confidence interval is complicated. The numerical computation even requires a trial by error approach. An alternative approach is to use the normal approximation to the binomial distribution when n is
large and $np(1-p)>5.$

Consider the independent Bernoulli sample $X_1,\ldots,X_n$ as discussed in Ex-

ample 10.8. From the Central Limit Theorem, we know that $\dfrac{\sqrt{n}\left(\bar{X}-p\right)}{\sqrt{p(1-p)}}$ con-

verges to $N(0,1)$ in distribution (Section 7.13.5). We can now choose this approximately normal statistic as a pivotal quantity and state that the random interval

$$-z_{1-\alpha/2}\le\frac{\sqrt{n}\left(\bar{X}-p\right)}{\sqrt{p(1-p)}}\le z_{1-\alpha/2} \tag{10.14}$$

has a coverage probability approximately equal to $1-\alpha$. It is obvious that (10.14)

is equivalent to $\left|\dfrac{\sqrt{n}\left(\bar{X}-p\right)}{\sqrt{p(1-p)}}\right|\le z_{1-\alpha/2}.$ By squaring both side of the inequality and

rearranging the terms, we obtain the following quadratic inequality

$$\left(1+\frac{z^2_{1-\alpha/2}}{n}\right)p^2-\left(2\bar{X}+\frac{z^2_{1-\alpha/2}}{n}\right)p+\bar{X}^2\le 0.$$

Using the solutions for the quadratic equation, we can express the above inequality
as

$$L_0(X_1,\ldots,X_n)\le p\le U_0(X_1,\ldots,X_n), \tag{10.15}$$

where

$$L_0(X_1,\ldots,X_n) = \dfrac{\bar{X} + \left(\dfrac{z_{1-\alpha/2}^2}{4n}\right) - \left(\dfrac{z_{1-\alpha/2}}{\sqrt{n}}\right)\sqrt{\bar{X}(1-\bar{X}) + \left(\dfrac{z_{1-\alpha/2}^2}{4n}\right)}}{1 + \left(\dfrac{z_{1-\alpha/2}^2}{n}\right)} \quad \text{and}$$

$$U_0(X_1,\ldots,X_n) = \dfrac{\bar{X} + \left(\dfrac{z_{1-\alpha/2}^2}{4n}\right) + \left(\dfrac{z_{1-\alpha/2}}{\sqrt{n}}\right)\sqrt{\bar{X}(1-\bar{X}) + \left(\dfrac{z_{1-\alpha/2}^2}{4n}\right)}}{1 + \left(\dfrac{z_{1-\alpha/2}^2}{n}\right)}.$$

Let us now examine the approximate distribution of $\dfrac{\sqrt{n}\left(\bar{X} - p\right)}{\sqrt{p(1-p)}}$ from another

prospective. Since $\bar{X}$ converges to p in probability, $\dfrac{\sqrt{p(1-p)}}{\sqrt{\bar{X}(1-\bar{X})}}$ converges to 1

in probability for $0 < p < 1$ and $0 < \bar{X} < 1$. According to Slutsky's theorem,

$\dfrac{\sqrt{n}\left(\bar{X} - p\right)}{\sqrt{\bar{X}(1-\bar{X})}}$ converges to $N(0,1)$ in distribution. That means $\dfrac{\sqrt{n}\left(\bar{X} - p\right)}{\sqrt{\bar{X}(1-\bar{X})}}$ has

an approximate standard normal distribution and thus can be chosen as a pivotal
quantity for estimating p. Observe that

$$P\left(-z_{1-\alpha/2} \le \dfrac{\sqrt{n}\left(\bar{X} - p\right)}{\sqrt{\bar{X}(1-\bar{X})}} \le z_{1-\alpha/2}\right) \approx 1-\alpha$$

The event inside the above probability statement is equivalent to

$$\bar{X} - z_{1-\alpha/2}\sqrt{\dfrac{\bar{X}(1-\bar{X})}{n}} \le p \le \bar{X} + z_{1-\alpha/2}\sqrt{\dfrac{\bar{X}(1-\bar{X})}{n}}. \qquad (10.16)$$

Therefore, (10.16) is the $100(1-\alpha)\%$ approximate confidence interval for p.

If we use the data given in Example 10.9, the 95% approximate confidence inter-val will be $\left[0.308, 0.784\right]$ (using (10.15)) and $\left[0.232, 0.668\right]$ (using (10.16)) compared with the exact confidence interval that is $\left[0.271, 0.640\right]$. The latter is shorter and hence more precise than the two formal ones as expected.

The method discussed above can be generalized to use the MLE as our ini-tial base for constructing an approximate confidence interval. Recall, in Chapter 8, that the MLE is consistent and asymptotically normal with asymptotic variance equal to $\dfrac{1}{I(\theta)}$, where, for any pdf or pmf $f(\cdot|\theta)$, $I(\theta) = E\left(\dfrac{\partial}{\partial\theta}\log(f(X|\theta)\right)^2$ is the Fisher information function. In other words, $\sqrt{n}(\hat{\theta} - \theta)$ converges in distribu-tion to a normal random variable with mean 0 and variance $\dfrac{1}{I(\theta)}$, as $n \to \infty$. Equivalently, $\sqrt{nI(\theta)}\left(\hat{\theta} - \theta\right)$ is approximately distributed as a standard normal random variable when n is large. It is natural now to choose this standardized MLE $\sqrt{nI(\theta)}\left(\hat{\theta} - \theta\right)$ as the pivotal quantity. Then

$P\left(-z_{1-\alpha/2} \leq \sqrt{nI(\theta)}(\hat{\theta} - \theta) \leq z_{1-\alpha/2}\right) \approx 1 - \alpha$ for large n. After replacing $I(\theta)$ by its estimate $I(\hat{\theta})$, we have

$$P\left(-z_{1-\alpha/2} \leq \sqrt{nI(\hat{\theta})}(\hat{\theta} - \theta) \leq z_{1-\alpha/2}\right) \approx 1 - \alpha.$$

After some algebraic manipulation, we establish that

$$\hat{\theta} - \frac{z_{1-\alpha/2}}{\sqrt{nI(\hat{\theta})}} \leq \theta \leq \hat{\theta} + \frac{z_{1-\alpha/2}}{\sqrt{nI(\hat{\theta})}} \tag{10.17}$$

is a $100(1 - \alpha)\%$ approximate confidence interval for θ.

Example 10.16 We will use (10.17) to establish the approximate confidence in-terval for the population proportion p using an independent Bernoulli sample. From Chapter 8, we have $\hat{p} = \bar{X}$ is the MLE for p. Using the result in Example 8.46, we can identify $I(p) = \dfrac{1}{p(1-p)}$ and thus $I(\hat{p}) = \dfrac{1}{\bar{X}(1-\bar{X})}$. Equation (10.16) can then be obtained by (10.17).

Example 10.17 Let us now revisit Example 10.10. Since the population is from a Poisson distribution, the MLE for λ is $\hat{\lambda} = \bar{X}$ (see Example 8.7). The Fisher information function can be calculated as

$$ I(\lambda) = E\left[\frac{\partial}{\partial \lambda} \log\left(\frac{\lambda^X \exp(-\lambda)}{X!} \right) \right]^2 = E\left(\frac{X}{\lambda} - 1 \right)^2 = \frac{E(X - \lambda)^2}{\lambda^2} = \frac{1}{\lambda}. $$

From (10.17) and the MLE property $I(\hat{\lambda}) = \frac{1}{\bar{X}}$, we can obtain a $100(1-\alpha)\%$ confidence interval as

$$ \bar{X} - \frac{z_{1-\alpha/2}}{\sqrt{n/\bar{X}}} \leq \lambda \leq \bar{X} + \frac{z_{1-\alpha/2}}{\sqrt{n/\bar{X}}} $$

or equivalently,

$$ \bar{X} - z_{1-\alpha/2}\sqrt{\frac{\bar{X}}{n}} \leq \lambda \leq \bar{X} + z_{1-\alpha/2}\sqrt{\frac{\bar{X}}{n}}. $$

If we use the data given in Example 8.8, the approximate 95% confidence interval will be $[2590.37, 2661.38]$, compared with the exact 95% confidence interval of $[2590.6, 2661.7]$ presented in Example 10.10.

In fact, an approximate confidence interval can be found by choosing any pivotal quantity that has any limiting distribution. The well-known Central Limit Theorem is often applied in these processes. A more general formula can also be used in constructing an approximate confidence interval. Suppose that T and S are two statistics of a random sample $X_1, \ldots, X_n$ that have a distribution that depends on a parameter θ. If $\frac{T - \theta}{S}$ converges in distribution to a standard normal random variable, then $T - z_{1-\alpha/2}S \leq \theta \leq T + z_{1-\alpha/2}S$ is a $100(1-\alpha)\%$ approximate confidence interval for θ.

Example 10.18 In Example 10.17, if we apply the Central Limit Theorem, we can see that $\frac{\bar{X} - \lambda}{\sqrt{\lambda/n}}$ converges in distribution to the standard normal. We can also use the consistency of $\bar{X}$ and Slutsky's theorem to transform the sequence of statistics to $\dfrac{\bar{X}-\lambda}{\sqrt{\bar{X}/n}}$. The approximate confidence interval will then be identical to that in Example 10.17. However, we can also directly observe the random interval

$$-\sqrt{\frac{\lambda}{n}}z_{1-\frac{\alpha}{2}} \le \bar{X}-\lambda \le \sqrt{\frac{\lambda}{n}}z_{1-\frac{\alpha}{2}}$$ that has a $100(1-\alpha)\%$ coverage probability. This

interval can also be described as the inequality $\left|\bar{X}-\lambda\right| \le \sqrt{\dfrac{\lambda}{n}}z_{1-\frac{\alpha}{2}}$, or equivalently

the quadratic inequality $\left(\bar{X}-\lambda\right)^2 \le \dfrac{\lambda}{n}(z_{1-\frac{\alpha}{2}})^2$. A simplified form for this quad-

ratic inequality is $\lambda^2 - 2\left(\bar{X}+\dfrac{\left(z_{1-\frac{\alpha}{2}}\right)^2}{2n}\right)\lambda + \bar{X}^2 \le 0$. By solving the quadratic

form, we can obtain the interval

$$\left(\bar{X}+\frac{\left(z_{1-\frac{\alpha}{2}}\right)^2}{2n}\right) - \sqrt{\left(\bar{X}+\frac{\left(z_{1-\frac{\alpha}{2}}\right)^2}{2n}\right)^2 - \bar{X}^2} \le \lambda \le \left(\bar{X}+\frac{\left(z_{1-\frac{\alpha}{2}}\right)^2}{2n}\right) + \sqrt{\left(\bar{X}+\frac{\left(z_{1-\frac{\alpha}{2}}\right)^2}{2n}\right)^2 - \bar{X}^2}$$

that is the $100(1-\alpha)\%$ approximate confidence interval for λ. The numeric answer is $\left[2590.60, 2661.63\right]$.

10.6 The Bootstrap Method*

There are several bootstrap methods for interval estimation. This section introduces one that is called the *bootstrap-t method*. Recall in Example 10.1, when the sample is taken from a normally-distributed population, we use $T = \dfrac{\sqrt{n}\left(\bar{X}-\mu\right)}{S}$ as a pivotal quantity to construct the $100(1-\alpha)\%$ confidence interval for the population mean. From the Central Limit Theorem, $\dfrac{\sqrt{n}\left(\bar{X}-\mu\right)}{\sigma}$ is approximately normally distributed. In other words, if the population is normal or the sample size is large, then the statistic T has a t-distribution. This notion can be applied to the bootstrap sample.

As in Chapter 8, we let $x_1,\ldots,x_n$ represent the original observations and $X_1^*,\ldots,X_n^*$ represent a bootstrap sample of size n that is obtained by randomly sampling with replacement from the population consisting of n objects $\{x_1,\ldots,x_n\}$. If we let $\bar{x}^* = \dfrac{\sum\limits_{i=1}^{n} x_i^*}{n}$ and $s^* = \sqrt{\dfrac{\sum\limits_{i=1}^{n}\left(x_i^* - \bar{x}^*\right)^2}{n-1}}$, then sample mean $\bar{x}^*$

and sample standard deviation s^* will play an important role in our strategy for finding a confidence interval. If we repeat the bootstrap sample B times and calculate $T* = \dfrac{\sqrt{n}\left(\bar{x}^* - \bar{x}\right)}{s^*}$ for each sample, then the B possible values of $T*$ have an

approximate t distribution. We can then use $100\dfrac{\alpha}{2}$ and $100(1-\dfrac{\alpha}{2})$ sample percentiles, of these $T*$ values to construct a $100(1-\alpha)\%$ bootstrap confidence interval. Specifically, if we let $T_{(1)}^* \le T_{(2)}^* \le \square \le T_{(B)}^*$ be the bootstrap $T*$ values in ascending order, and let $l = \left\lfloor \dfrac{\alpha}{2}B \right\rfloor$ (here $\lfloor y \rfloor$ denotes the greatest integer less

than or equal to y) and $u = B - l$, then the $100\dfrac{\alpha}{2}$ and $100(1-\dfrac{\alpha}{2})$ sample percentiles for the bootstrap $T*$ values are $T_{(l+1)}^*$ and $T_{(u)}^*$ respectively. The resulting $100(1-\alpha)\%$ confidence interval for the population mean μ is

$$\left(\bar{x} - T_{(u)}^* \frac{s}{\sqrt{n}}, \bar{x} - T_{(l+1)}^* \frac{s}{\sqrt{n}} \right). \tag{10.18}$$

Note that $T_{(l+1)}^*$ is usually negative, that is the reason it is subtracted from $\bar{x}$ in the upper bound.

Example 10.19 A school administrator is to study the mean number of absent days during a school year for an elementary school. A random sample of 20 students is chosen and the absent days are recorded as
3, 5, 7, 0, 0, 1, 0, 2, 0, 0, 3, 4, 8, 0, 0, 1, 0, 2, 0, 1.
To find a confidence interval using the bootstrap-t method, we first produce a bootstrap sample based on the above data that would generate 0 (45%), 1(15%), 2(10%), 3(10%), 4(5%), 5(5%), 7(5%) and 8(5%). Assume that we will repeatedly generate 500 bootstrap samples, and produce T* value for each sample

(using $\bar{x} = 1.85$ and $s = 2.46$ for generating $T*$ and the confidence interval). In this procedure, we dropped the first 100 simulated samples for the burn-in period that will be discussed in Chapter 11. The 95% confidence interval using equation (10.18) is obtained as (0.87, 3.42). Here we calculate $T^*_{(l+1)} = T^*_{(13)} = -2.86$ and $T^*_{(u)} = T^*_{(488)} = 1.78$. A SAS program is available for the calculation of this confidence interval.

10.7 Criteria of a Good Interval Estimator

There are several methods for constructing a confidence interval. We would prefer to choose the best one. Therefore, the criteria for assessing a good interval estimator become important. In principle, accuracy and precision are the most important criteria for interval estimation. For interval estimation, accuracy can be presented in the coverage probability (or more commonly confidence coefficient) and precision can be represented by the length or expected length of the random interval.

Similar to what we discussed in hypothesis testing, there is a tradeoff between accuracy and precision. We can not obtain perfect accuracy and perfect precision at the same time. For example, if we let the confidence interval to be $(-\infty, \infty)$, then we may achieve 100% accuracy, but this interval is associated with very poor precision. This imprecise confidence interval would not provide us with any useful information. On the other hand, a point estimator is more precise, but less accurate as we have pointed out at the beginning of this chapter.

Recall that in hypothesis testing problems, we compare the power (accuracy) when limiting the upper bound of a type I error (precision). For interval estimation, we commonly limit the confidence coefficient (or the greatest lower bound of the coverage probability) and try to find a confidence interval that has the shortest interval length (precision). The shortest confidence interval for a given coefficient is usually an equal-tail one as we demonstrated in the discussion earlier. This is true only when the density of the pivotal quantity is unimodal. The formal statement is provided in Proposition 10.1.

Proposition 10.1 Let $f(x)$ be a unimodal pdf and $F(x)$ be its associated cdf. Consider an interval $[a, b]$ that satisfies $F(b) - F(a) = 1 - \alpha$ for any α such that $0 < \alpha < 1$. Then $[a_0, b_0]$ has the shortest length among all these intervals if $f(a_0) = f(b_0) > 0$ and $a_0 \leq x^* \leq b_0$, where x^* is the mode of $f(x)$. If $f(x)$ is symmetric then $a_0 = F^{-1}\left(\dfrac{\alpha}{2}\right)$ and $b_0 = F^{-1}\left(1 - \dfrac{\alpha}{2}\right)$.

Proof. The problem for finding a_0 and b_0 is the same as that for minimizing the length $b - a$ subject to $F(b) - F(a) = 1 - \alpha$. By using Lagrange multiplier method, we will solve the partial differential equations associated with

$g(a,b,\lambda) = b - a + \lambda(1 - \alpha - F(b) + F(a))$:

$\dfrac{\partial g}{\partial b} = 1 - \lambda f(b) = 0, \quad \dfrac{\partial g}{\partial a} = 1 - \lambda f(a) = 0, \text{ and } 1 - \alpha - F(b) + F(a) = 0.$

The first two equations give us $f(a) = f(b) > 0$. If $x^* \notin [a,b]$ and $f(a) = f(b)$ then $b - a > b_0 - a_0$ since $f(x)$ is unimodal and $F(b) - F(a) = F(b_0) - F(a_0)$. This completes the proof.

Example 10.20 In Example 10.6, we have a pivotal quantity $2\lambda \sum_{i=1}^{n} X_i$ that has a χ^2 distribution with degree of freedom 2n. If $n > 1$, the pdf is unimodal. Since $\left(\chi^2_{2n,-\frac{\alpha}{2}}, \chi^2_{2n,1-\frac{\alpha}{2}} \right)$ contains the mode of the distribution for a reasonably small α,

Proposition 10.1 shows that $\left(\chi^2_{2n,-\frac{\alpha}{2}}, \chi^2_{2n,1-\frac{\alpha}{2}} \right)$ has the shortest interval that covers

$2\lambda \sum_{i=1}^{n} X_i$ with probability $1 - \alpha$. Thus the shortest $100(1 - \alpha)\%$ confidence interval is the one presented in Example 10.6.

10.8 Confidence Intervals and Hypothesis Tests

There is a close connection between hypothesis testing and the confidence intervals. Every confidence interval corresponds to a hypothesis test and every hypothesis test corresponds to a confidence interval. This correspondence is not limited to a two-sided test. For a one sided test, there is a confidence bound corresponding to it. We illustrate this relationship in Examples 10.21 and 10.22 below.

Example 10.21 In the problem described in Example 10.1, and Example 10.2, we test a hypothesis $H_0 : \mu = \mu_0$ versus $H_a : \mu \neq \mu_0$ for a random sample taken from normal population with unknown mean and variance. We used the acceptance region of H_0, given by

$$A(\mu_0) = \left\{ (x_1, \ldots, x_n) : |\bar{x} - \mu_0| \leq \frac{S}{\sqrt{n}} t_{n-1,1-\frac{\alpha}{2}} \right\}$$

to derive the $100(1 - \alpha)\%$ confidence interval for the parameter μ by

$$C(x_1, \ldots, x_n) = \left\{ \mu \left| -\frac{S}{\sqrt{n}} t_{n-1,1-\frac{\alpha}{2}} \leq \mu \leq \bar{x} + \frac{S}{\sqrt{n}} t_{n-1,1-\frac{\alpha}{2}} \right. \right\}$$

for a given sample point $(x_1,\ldots,x_n)$. The relationship of these two intervals can be described as $(x_1,\ldots,x_n) \in A(\mu_0)$ if and only if $\mu_0 \in C(x_1,\ldots,x_n)$. That means, if we take a random sample $(x_1,\ldots,x_n)$ that falls into the acceptance region for testing $H_0 : \mu = \mu_0$ with level of significance α, then μ_0 should belong to the $100(1-\alpha)\%$ confidence interval created by the method proposed using this sample point. Conversely, if we use the sample point $(x_1,\ldots,x_n)$ to construct a $100(1-\alpha)\%$ confidence interval and μ_0 belongs to it, then, using this point, we should not reject $H_0 : \mu = \mu_0$ at level α (or we should accept $H_0 : \mu = \mu_0$). This relationship brings a one to one correspondence between the confidence interval and the two-sided hypothesis test.

Example 10.22 In Example 10.4, we derived the $100(1-\alpha)\%$ confidence interval for the population variance of a normal population. If we consider hypothesis testing with $H_0 : \sigma^2 = \sigma_0^2$ and $H_a : \sigma^2 > \sigma_0^2$, the likelihood ratio test described in section 9.8.4 is $\dfrac{(n-1)S^2}{\sigma_0^2}$. The acceptance region for an α level of significance is

$A(\sigma_0^2) = \left\{ (x_1,\ldots,x_n) \left| \dfrac{(n-1)S^2}{\sigma_0^2} \le \chi_{n-1,\alpha}^2 \right. \right\}$. By similar algebraic manipulations as

in Example 10.4, we can derive the $100(1-\alpha)\%$ confidence interval for σ^2 as

$C(x_1,\ldots,x_n) = \left\{ \sigma^2 \left| \dfrac{(n-1)S^2}{\chi_{n-1,1-\alpha}^2} \le \sigma^2 \right. \right\}$. This shows that there is a one to one corre-

spondence between the acceptance regions of a one-sided test of level α and the $100(1-\alpha)\%$ lower confidence bound.

Now, we conclude that the hypothesis testing and confidence interval methods generate the same result for statistical inference. Is there any advantage for using one approach over the other? The hypothesis test can precisely tell us whether $H_0 : \theta = \theta_0$ is significant. However, the statistical significance is not always important in the scientific or practical sense. For example, we may want to test if patients taking a new drug have elevated blood pressures; the statistically significant result may not be clinically important. A confidence interval may give a range of values that the blood pressure may change after taking the new drug. The patients or their physicians may be able to use this range to judge if the new drug will clinically elevate a patient's blood pressure. On the other hand, the confidence interval does not contain all the information produced by hypothesis testing such as p-value that would have given us the degree of significance. The confidence interval only gives us the significance for a given α level.

In fact, there is a connection between the power of a hypothesis test and the length of a confidence interval derived from it. If we accept $H_0 : \theta = \theta_0$ and conclude θ_0 belongs to the $100(1-\alpha)\%$ confidence interval, then we may also want to examine other values of the θ in the interval. If values very different from θ_0 also belong to the interval, we may suspect the test does not have enough power to reject the alternative and suggest to make further investigative studies. If we reject $H_0 : \theta = \theta_0$ and find that values very close to θ_0 belong to the $100(1-\alpha)\%$ confidence interval derived from the test, we may suspect the test may have too high power to reject the alternative.

Problems:

1. Assume that $X_1, \ldots, X_n$ is a random sample of size n from a $N(\mu, 1)$ population.
    (a) Derive a $100(1-\alpha)\%$ confidence interval for μ
    (b) Show that $I = \left[\bar{x} - \dfrac{1.96}{\sqrt{n}}, \bar{x} + \dfrac{1.96}{\sqrt{n}} \right]$ is the shortest 95% confidence interval for an observed sample with sample mean $\bar{x}$.

2. In Problem 1, if an additional independent observation X_{n+1} is taken, what is the probability that X_{n+1} will fall in the interval I?

3. In Problem 3, Chapter 8, find a 90% confidence interval for p.
    (a) Using the exact method.
    (b) Using the approximate method.
    (c) Find a 90% HPD Bayesian credible interval using $Beta(2,2)$ as a prior distribution.

4. Suppose that $X_{i1}, \ldots, X_{in_i}$ are independent normal random variables with mean 0 and variance σ_i^2, $i = 1, 2$, where σ_1^2 and σ_2^2 are unknown. Find a $100(1-\alpha)\%$ confidence interval for $\theta = \dfrac{\sigma_1^2}{\sigma_2^2}$.

5. Assume T is a binomial random variable with parameter n and p. Let $a(p)$ be the greatest integer among $0, 1, \ldots, n$ such that

$P\big(T < a(p)\big) < \alpha$. Show that $a(p_1) \le a(p_2)$ if $p_1 \le p_2$. (Hint: $F_T(t\,|\,p_1) \ge F_T(t\,|\,p_2)$).

6. Suppose that $X_1,\dots,X_n$ are independent random variables, each has Bernoulli distribution with unknown probability p of success. Find a $100(1-\alpha)\%$ lower confidence bound using the acceptance region from tests of $H_0 : p = p_0$ versus $H_a : p > p_0$.

7. In Problem 8.7, if we make two observations $x_1 = 10$ and $x_2 = 15$, find the 90% confidence interval for θ.

8. Let $X_{i1},\dots,X_{in_i}$ be a random sample from a continuous population with a parameter θ. Assume that $T(X_1,\dots,X_n)$ is a sufficient statistic for θ and that the cdf of T is $F_T(t\,|\,\theta)$. If $\alpha_1 > 0$, $\alpha_2 > 0$ and $\alpha = \alpha_1 + \alpha_2$, show that an α level acceptance region for tests of $H_0 : \theta = \theta_0$ versus $H_a : \theta \ne \theta_0$ can be written as $\big\{ t \,|\, \alpha_1 \le F_T(t\,|\,\theta_0) \le \alpha_2 \big\}$ and its associated $100(1-\alpha)\%$ confidence interval is $\big\{ \theta \,|\, \alpha_1 \le F_T(t\,|\,\theta) \le \alpha_2 \big\}$.

9. In Problem 8.19, find the 95% confidence interval and calculate the coverage probability of your answer.

10. Assume T is a Poisson random variable with parameter λ. Let $a(\lambda)$ be the greatest integer such that $P\big(T < a(\lambda)\big) < \alpha$. Show that $a(\lambda_1) \le a(\lambda_2)$ if $\lambda_1 \le \lambda_2$.

11. Assume that $X_1,\dots,X_n$ is a random sample from a $N(\theta,\theta)$ population, where $\theta > 0$. Find a $100(1-\alpha)\%$ confidence interval for θ.

12. Find the expected length of the confidence interval in (10.5).

13. Find the expected length of the confidence interval in (10.9).

14. Let $X_1,\dots,X_n$ be a random sample of size n from a distribution that is $N(\mu,\sigma^2)$ distributed, where μ and σ^2 are both unknown. Assume that

$\bar{X}$ is the sample mean and S^2 is the sample variance. Find the greatest integer n such that $P\left(\bar{X} - 2\dfrac{S}{\sqrt{n}} < \mu < \bar{X} + 2\dfrac{S}{\sqrt{n}}\right) \le 0.95$.

15. In Problem 10.14, if $\bar{x} = 10$, $s^2 = 9$ and $n = 16$, find the coverage probability.

16. In Example, 10.1, find the 95% upper confidence bound for the population mean μ.

17. Assume that $X_1, \ldots, X_n$ be a random sample taken from the population that has a pdf given by $f(x|\theta) = 2x/\theta^2$ for, where $\theta > 0$. Find the $100(1 - \alpha)\%$ confidence interval for θ.

18. Find the 95% HPD credible interval for λ suing the data presented in Example 8.8 and the prior distribution discussed in Example 8.19 with $\alpha = 100$ and $\beta = 50$.

References

Berger, J.O. (1985). *Statistical Decision Theory and Bayesian Analysis.* 2nd ed. Springer-Verlag. New York.

Casella, G. and Berger, R.L. (2002). *Statistical Inference.* 2nd ed. Duxbury. Pacific Grove, CA.

Dudewicz, E.J. and Mishra, S.N. (1988). *Modern Mathematical Statistics.* Wiley. New York.

Lehmann, E.L. and Casella, G. (1998). *Theory of Point Estimation,* 2nd edition. Springer-Verlag. New York.

Wilcox, R.R. (2003). *Applying Contemporary Statistical Techniques.* Academic Press. New York.

11

Introduction to Computational Methods

Traditionally, mathematical statistics has been dominated by analytical methods using mostly algebraic operations, calculus, probability axioms and algebra, and some techniques in trigonometry and geometry. Unfortunately, the analytical approach commonly left the problem unsolved when it fails or is too complicated. The rapid development of computer technology in the past two decades, has not only made advances in data analysis and applied statistics, but also in mathematical statistics. Several computationally-intensive methods have evolved for solving problems in statistical inference. A solution to these problems would be impossible or at least extremely difficult without the use of modern computational techniques and facilities.

In this chapter, we plan to introduce the following computationally-intensive methods: Newton-Raphson's method for solving equations, EM algorithm, simulation and Markov Chain Monte Carlo methods.

11.1 The Newton-Raphson Method

In mathematical statistics, optimization techniques are often applied. When using these techniques to solve a statistical problem, we frequently face challenges to find roots of algebraically complicated functions. These functions usually do not have analytic roots. Therefore, the numeric solutions are attempted and the Newton-Raphson method is commonly selected as a tool to find the root[1].

---

[1]Sir Isaac Newton is generally credited for the development of this numeric method. It was actually Joseph Raphson who first published in 1690. He did this without previous knowledge about Newton's work on this problem. The mathematical community is now referring this method as "the Newton-Raphson method".

.

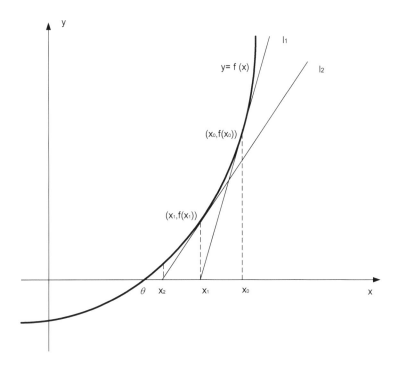

Figure 11.1 Illustration of the Newton-Raphson method

The Newton-Raphson method arises from Taylor's approximation formula $f(x) \approx f(a) + f'(a)(x-a)$, when x is close to a. If we are to obtain an approximation to the root θ of the equation $f(x) = 0$, as sketched in Figure 11.1, we consider the tangent line l_1 to the graph $y = f(x)$ at the initial point $(x_0, f(x_0))$. If x_0 is sufficiently close to θ, then the x-intercept x_1 of the line l_1 should be closer to θ. Using the point-slope form, the equation of the line l_1 can be expressed as $y = f(x_0) + f'(x_0)(x - x_0)$. Thus, the x-intercept x_1 of the line l_1 should satisfy $0 = f(x_0) + f'(x_0)(x_1 - x_0)$ or equivalently $x_1 = x_0 - \dfrac{f(x_0)}{f'(x_0)}$ if $f'(x_0) \neq 0$. If we draw another tangent line l_2 of the graph at the point $(x_1, f(x_1))$, the x-intercept x_2 of this new line l_2 will be a better approximation than x_1 and it can be expressed as $x_2 = x_1 - \dfrac{f(x_1)}{f'(x_1)}$. Continuing this process until the difference between two consecutive approximated values is small, we can find the approximated value for θ. In general, if x_n is our approximated value at the nth step, then the approximated value at the $(n+1)^{th}$ step can be expressed as $x_{n+1} = x_n - \dfrac{f(x_n)}{f'(x_n)}$ if $f'(x_n) \neq 0$. The process of successive approximation of a real root is referred to as the *Newton-Raphson method* for the function of one-dimensional domain.

Although the Newton-Raphson method is very useful in approximating the solution to an equation, its result sometimes depends on the choice of the initial approximation. Indeed, if x_0 is not chosen to be sufficiently close to θ, it is possible that x_1 is not closer to θ than x_0 (see Example 11.2). It is also important that we should avoid any x_n such that $f'(x_n)$ is close to 0. Furthermore, if the equation has more than one root, the Newton-Raphson method may not always converge to the one desired. That means, when applying the Newton-Raphson method in optimization, the solution does not guarantee convergence to the global maximum (or minimum) if there are several maxima (or minima). In other words, the Newton-Raphson method may be used to find points that have local maximum (or local minimum). It can not be used to determine that the solution maximizes (or minimizes) the function in the entire domain. A graph of the function that describes the equation should be examined before the method is applied.

The p-dimensional version of the Newton-Raphson method can be similarly described. If $\mathbf{x}$ is the root of the p-dimensional function $\mathbf{f}(\mathbf{x}) = \mathbf{0}$, the Newton-Raphson method gives us the iterative procedures $\mathbf{x}_{n+1} = \mathbf{x}_n - A^{-1}(\mathbf{x}_n)\mathbf{f}(\mathbf{x}_n)$, where

$$A(\mathbf{x}) = \begin{pmatrix} \dfrac{\partial f_1(\mathbf{x})}{\partial x_1} & \cdots & \dfrac{\partial f_p(\mathbf{x})}{\partial x_1} \\ \vdots & & \vdots \\ \dfrac{\partial f_1(\mathbf{x})}{\partial x_p} & \cdots & \dfrac{\partial f_p(\mathbf{x})}{\partial x_p} \end{pmatrix}$$

is the Jacobian matrix and $\mathbf{f}(\mathbf{x}) = (f_1(\mathbf{x}), \cdots, f_p(\mathbf{x}))^T$. Note that in the common application of the Newton-Raphson method to the maximum likelihood estimation that was discussed in Chapter 8, $f_i(\mathbf{x})$ is itself a partial derivative of the (log-) likelihood function and $A(\mathbf{x})$ is usually referred to as the *Hessian matrix*.

Example 11.1 If we try to find the positive root of the quadratic equation $f(x) = x^2 - 5 = 0$, we first use basic algebra to determine that the root θ satisfies $2 < \theta < 3$. Using the Newton-Raphson method with iterative formula

$$x_{n+1} = x_n - \frac{x_n^2 - 5}{2x_n}$$ and initial value $x_1 = 3$, we can obtain $x_2 = 2.33333$,

$x_3 = 2.23810$, $x_4 = 2.23607$ and $x_5 = 2.23607$. It is clear we can stop at x_5 and

Table 11.1 List of θ values and absolute changes at each iteration when the initial value is set to 4.

Iteration	θ value	Absolute Change
1	7.65	3.65
2	15.33	7.68
3	30.96	15.62
4	62.34	31.38
5	125.17	62.84
6	250.88	125.71
7	502.31	251.43
8	1005.18	502.87
9	2010.92	1005.74
10	4022.40	2011.49

conclude 2.23607 is the numeric root for the equation. Note that this estimate is also an approximation for $\sqrt{5}$.

Example 11.2 Consider $f(\theta) = \dfrac{1997}{2+\theta} - \dfrac{1810}{1-\theta} + \dfrac{32}{\theta} = 0$ that is an equation of the first derivative of the log-likelihood function discussed in Thisted (1988). If we choose the initial value as 0.05 and use the Newton-Raphson method, we can find the root as 0.035703 at the third iteration (by also checking the functional value of this number). However, if we choose the initial value as 4, the iterative sequence seems to be not convergent. The first 10 iterations of this process are listed in Table 1 below. At 100th iteration, the θ value becomes 4.98×10^{30}. It clearly does not converge. McLachian and Krishnan (1997) also used this example to demonstrate the dependence of the Newton-Raphson method on the choice of the initial value. A SAS program for performing this calculation is available.

Example 11.3 Consider the equation system

$$\begin{cases} f_1(x, y) = -2x^3 + 2xy + x - 1 = 0 \\ f_2(x, y) = -x^2 + y = 0 \end{cases}$$

If we try to find the numeric solution using the Newton-Raphson method, we have to calculate the Jacobian matrix of these functions

$$\begin{pmatrix} \dfrac{\partial f_1(\mathbf{x})}{\partial x} & \dfrac{\partial f_2(\mathbf{x})}{\partial x} \\[2mm] \dfrac{\partial f_1(\mathbf{x})}{\partial y} & \dfrac{\partial f_2(\mathbf{x})}{\partial y} \end{pmatrix} = \begin{pmatrix} -6x^2 + 2y + 1 & -2x \\ 2x & 1 \end{pmatrix}.$$

The determinate of this Jacobian matrix is $-2x^2 + 2y + 1$ and hence its inverse is

$$\frac{1}{-2x^2 + 2y + 1}\begin{pmatrix} 1 & 2x \\ -2x & -6x^2 + 2y + 1 \end{pmatrix}$$ assuming $-2x^2 + 2y + 1 \neq 0$. The $(n+1)^{th}$
iteration can be expressed as

$$\begin{pmatrix} x_{(n+1)} \\ y_{(n+1)} \end{pmatrix} = \begin{pmatrix} x_{(n)} \\ y_{(n)} \end{pmatrix} - \frac{1}{-2x_{(n)}^2 + 2y_{(n)} + 1}\begin{pmatrix} 1 & 2x_{(n)} \\ -2x_{(n)} & -6x_{(n)}^2 + 2y_{(n)} + 1 \end{pmatrix}$$
$$\bullet \begin{pmatrix} -2x_{(n)}^3 + 2x_{(n)}y_{(n)} + x_{(n)} - 1 \\ -x_{(n)}^2 + y_{(n)} \end{pmatrix}.$$

The numeric approximations found that $x_{(7)} = 1.0000012$ and $y_{(7)} = 1.0000024$ at the 8th iteration. The associated function values are $f_1\left(x_{(7)}, y_{(7)}\right) = 0.0000012$ and $f_2\left(x_{(7)}, y_{(7)}\right) = -1.44\exp(-12)$. A SAS program is available.

Example 11.4. In a study of youth attitude, reported by Lederman, Chan and Roberts-Gray (2003), there were 74 youth in the experimental group and 70 youth in the control group expecting to attend college and there were 15 youth in the experimental group and 10 youth in the control group not expecting to attend college. We assume that the probability that a youth in the experimental group is aspired to go to college is $\dfrac{e^\alpha}{1 + e^\alpha}$. If this youth is from the control group, that probability is assumed to be $\dfrac{e^{\alpha+\beta}}{1 + e^{\alpha+\beta}}$. This is a question in simple logistic regression. The likelihood function can be expressed as

$$\left(\frac{e^\alpha}{1 + e^\alpha}\right)^{74}\left(\frac{1}{1 + e^\alpha}\right)^{15}\left(\frac{e^{\alpha+\beta}}{1 + e^{\alpha+\beta}}\right)^{70}\left(\frac{1}{1 + e^{\alpha+\beta}}\right)^{10}.$$

The partial derivative equations of the log-likelihood function can be simplified as:

$$\begin{cases} f_1(\alpha,\beta) = 144 + 55e^{\alpha} + 64e^{\alpha+\beta} - 25e^{2\alpha+\beta} = 0 \\ f_2(\alpha,\beta) = 70 - 10e^{\alpha+\beta} = 0 \end{cases}.$$

The Jacobian matrix of this equation system is

$$\begin{pmatrix} \dfrac{\partial f_1(\alpha,\beta)}{\partial \alpha} & \dfrac{\partial f_2(\alpha,\beta)}{\partial \alpha} \\ \dfrac{\partial f_1(\alpha,\beta)}{\partial \beta} & \dfrac{\partial f_2(\alpha,\beta)}{\partial \beta} \end{pmatrix} = \begin{pmatrix} 55e^{\alpha} + 64e^{\alpha+\beta} - 50e^{2\alpha+\beta} & -10e^{\alpha+\beta} \\ 64e^{\alpha+\beta} - 25e^{2\alpha+\beta} & -10e^{\alpha+\beta} \end{pmatrix}.$$

The Jacobian (determinant) is

$$-10e^{\alpha+\beta}(55e^{\alpha} - 25e^{2\alpha+\beta}) = -550e^{2\alpha+\beta} + 250e^{3\alpha+2\beta}.$$

Therefore, the $(n+1)^{th}$ iteration in the Newton-Raphson procedure can be expressed as

$$\begin{pmatrix} \alpha_{(n+1)} \\ \beta_{(n+1)} \end{pmatrix} = \begin{pmatrix} \alpha_{(n)} \\ \beta_{(n)} \end{pmatrix} - \frac{1}{-550e^{2\alpha_{(n)}+\beta_{(n)}} + 250e^{3\alpha_{(n)}+2\beta_{(n)}}}$$

$$\bullet \begin{pmatrix} -10e^{\alpha_{(n)}+\beta_{(n)}} & 10e^{\alpha_{(n)}+\beta_{(n)}} \\ -64e^{\alpha_{(n)}+\beta_{(n)}} + 25e^{2\alpha_{(n)}+\beta_{(n)}} & 55e^{\alpha_{(n)}} + 64e^{\alpha_{(n)}+\beta_{(n)}} - 50e^{2\alpha_{(n)}+\beta_{(n)}} \end{pmatrix}$$

$$\bullet \begin{pmatrix} 144 + 55e^{\alpha_{(n)}} + 64e^{\alpha_{(n)}+\beta_{(n)}} - 25e^{2\alpha_{(n)}+\beta_{(n)}} \\ 70 - 10e^{\alpha_{(n)}+\beta_{(n)}} \end{pmatrix}.$$

If we adopt the initial value for $\alpha_{(0)}$ by using the estimate of $\dfrac{1}{1+e^{\alpha}}$ as $\dfrac{15}{89}$, then $\alpha_{(0)} = 1.60$. Similarly, we can also adopt the initial value for $\beta_{(0)}$ by setting $\dfrac{1}{1+e^{\alpha+\beta}} = \dfrac{10}{80}$ and obtain $\beta_{(0)} = 0.35$. The estimated parameters are $\hat{\alpha} = 1.5960$ and $\hat{\beta} = 0.3499$, using the Newton-Raphson. Since this set of initial values is too close to the final approximation, it is difficult to observe their changes. We now choose new initial values $\alpha_{(0)} = 1.70$ and $\beta_{(0)} = 0.45$ that are less close. Using these new initial values, we can obtain the following sequence of the numeric approximations for the solution of α and β. The estimated parameter values and the changes at each iteration are presented in Table 11.2. A SAS program that

Table 11.2 List of estimated α and β values and the changes at each iteration for Example 11.4.

Iteration	α	β	Change in α	Change in β
initial	1.70	0.45	-----	-----
1	1.6258	0.3396	-0.0742	-0.0014
2	1.5973	0.3488	-0.0285	0.0092
3	1.5960	0.3499	-0.0013	0.0011
4	1.5960	0.3499	<-0.00001	<0.00001

requests the iterations stop when the differences for both parameters are less than 0.0001 is available.

The most frequent use of the Newton-Raphson method in mathematical statistics is in maximum likelihood estimation that was presented in Chapter 8. In general, the statistical procedure is to maximize the likelihood function or the log-likelihood function by treating it as a function of the parameter or parameters in the distribution. Maximizing the log-likelihood function usually requires obtaining the roots of its partial derivatives with respect to each parameter. The Newton-Raphson method is then applied to assist in solving the equations. Examples 8.14 and 8.15 are other applications of the Newton-Raphson method to find the MLE.

11.2 The EM Algorithm

As we briefly discussed in Chapter 8, the Expectation-Maximization (EM) algorithm is an iterative procedure for finding maximum likelihood estimates of the parameter (or parameters) with partially observed data. The goal of this algorithm is to augment the observed data with hidden data that can be either missing observations, unknown parameter values or unobserved random effects, so that the log-likelihood of the augmented data conditioned on the observed can be used for maximization. In this section, we will focus on the algorithm applied to find the MLE when some observations are missing. Of course, it can also be applied to find the posterior mode (or the generalized maximum likelihood).

The EM algorithm can be divided into two steps: the expectation (E) step and the maximization (M) step. Let $\mathbf{y} = (\mathbf{y}_{obs}, \mathbf{y}_{mis})^T$ denote the combination of the observed and the missing data, where $\mathbf{y}_{obs}$ is the observed portion of the data and $\mathbf{y}_{mis}$ is the missing portion of the data. The E-step is to find the expected value denoted by $Q(\theta, \theta^{(i)})$ of the complete log-likelihood $l_C(\theta|\mathbf{y}) = \log\left(L_C(\theta|\mathbf{y})\right)$ of the parameter θ, where the expectation is taken on $\mathbf{y}_{mis}$ given $\mathbf{y}_{obs}$ and the currently estimated parameter $\hat{\theta}^{(i)}$. As we mentioned in Chapter 8, the missing mechanism is generally assumed to be Missing Completely at Random (MCAR). Mathematically, the E-step is to calculate

$$Q(\theta, \theta^{(i)}) = E\left(l_C(\theta|\mathbf{y})\big|\mathbf{y}_{obs}, \theta^{(i)}\right) = \int l_C(\theta|\mathbf{y}) f(y_{mis}|\mathbf{y}_{obs}, \theta^{(i)}) d\mathbf{y}_{mis}.$$

The M-step is then to find the $\hat{\theta}^{(i+1)}$ that maximizes $Q(\theta, \theta^{(i)})$ over all possible values of θ. This new value $\hat{\theta}^{(i+1)}$ replaces $\hat{\theta}^{(i)}$ in the E-step and $\hat{\theta}^{(i+2)}$ is then chosen to maximize $Q(\theta, \theta^{(i+1)})$ from the M-step. This procedure is repeated until the observed-data likelihood converges or practically the change of two successive observed-data likelihood values becomes very small, where the observed-data likelihood can be calculated as $L(\theta|\mathbf{y}_{obs}) = \int L_C(\theta|\mathbf{y}_{obs}, \mathbf{y}_{mis}) d\mathbf{y}_{mis}$. The EM algo-rithm is constructed such that the sequence $\left\{\hat{\theta}^{(i)}\right\}$ converges to the MLE of θ (or the posterior mode of θ when applied to Generalized MLE).

Dempster, Laird and Rubin (1977) show that, after each iteration of the EM algorithm, the estimated $\hat{\theta}^{(i)}$ does not decrease the observed-data likelihood $L(\theta|\mathbf{y}_{obs})$ of θ; that is $L(\theta^{(i)}|\mathbf{y}_{obs}) \le L(\theta^{(i+1)}|\mathbf{y}_{obs})$ for $i = 0, 1, 2, \cdots$. To examine this property, we first rewrite the pdf (or pmf) of the complete observations as

$$f(\mathbf{y}|\theta) = f(\mathbf{y}_{obs}, \mathbf{y}_{mis}|\theta) = f(\mathbf{y}_{mis}|\mathbf{y}_{obs}, \theta) f(\mathbf{y}_{obs}|\theta).$$

Consequently, the complete likelihood function of θ can be written as

$$L_C(\theta|\mathbf{y}) = L(\theta|\mathbf{y}_{obs}) f(\mathbf{y}_{mis}|\mathbf{y}_{obs}, \theta)$$

and the complete log-likelihood function as

$$\log L_C(\theta|\mathbf{y}) = l_C(\theta|\mathbf{y}) = l(\theta|\mathbf{y}_{obs}) + \log f(\mathbf{y}_{mis}|\mathbf{y}_{obs}, \theta).$$

Therefore,

$$l(\theta|\mathbf{y}_{obs}) = l_C(\theta|\mathbf{y}) - \log f(\mathbf{y}_{mis}|\mathbf{y}_{obs}, \theta) \tag{11.1}$$

Taking the expectation $E_{\mathbf{y}_{mis}|\mathbf{y}_{obs}, \theta^{(i)}}(\ )$ on both sides of equation (11.1) after the ith iteration, we obtain

$$l(\theta|\mathbf{y}_{obs}) = Q(\theta, \theta^{(i)}) - R(\theta, \theta^{(i)}), \tag{11.2}$$

where

$$R(\theta,\theta^{(i)}) = \int \left(\log f(\mathbf{y}_{mis}|\mathbf{y}_{obs},\theta)\right) f(\mathbf{y}_{mis}|\mathbf{y}_{obs},\theta^{(i)})d\mathbf{y}_{mis}$$
$$= E_{\mathbf{y}_{mis}|\mathbf{y}_{obs},\theta^{(i)}}\left(\log f(\mathbf{y}_{mis}|\mathbf{y}_{obs},\theta)\right).$$

Note that $\log f(\mathbf{y}_{mis}|\mathbf{y}_{obs},\theta)$ in the above equation is calculated for a general θ and the only random component in it is $\mathbf{y}_{mis}$. The expectation operator $E_{\mathbf{y}_{mis}|\mathbf{y}_{obs},\theta^{(i)}}(\ )$ is taken on the conditional distribution of $\mathbf{y}_{mis}$, given $\mathbf{y}_{obs}$ and $\theta^{(i)}$. Therefore, the original parameter θ will not be affected by the expectation. For example, if $\mathbf{y} = (y_{obs}, y_{mis})^T$ is a bivariate normal random vector with $E(y_{mis}) = \mu_{mis}$, $E(y_{obs}) = \mu_{obs}$, $Var(y_{mis}) = \sigma^2_{mis}$, $Var(y_{obs}) = \sigma^2_{obs}$ and $corr(y_{mis}, y_{obs}) = \rho$, then

$$\log f(y_{mis}|y_{obs},\boldsymbol{\theta}) = -\frac{1}{2}\log\left(2\pi\sigma^2_{mis}(1-\rho^2)\right) - \frac{\left(y_{mis} - \mu_{mis|obs,\theta}\right)^2}{2\sigma^2_{mis}(1-\rho^2)}$$

has y_{mis} as the only random component, where

$$\mu_{mis|obs,\theta} = \mu_{mis} + \rho\frac{\sigma_{mis}}{\sigma_{obs}}(y_{obs} - \mu_{obs}).$$

The conditional expectation can then be calculated as

$$E_{\mathbf{y}_{mis}|\mathbf{y}_{obs},\theta^{(i)}}\left(\log f(y_{mis}|y_{obs},\boldsymbol{\theta})\right)$$

$$= -\frac{1}{2}\log\left(2\pi\sigma^2_{mis}(1-\rho^2)\right) - \frac{E_{\mathbf{y}_{mis}|\mathbf{y}_{obs},\theta^{(i)}}\left(y_{mis} - \mu_{mis|obs,\theta^{(i)}} + \mu_{mis|obs,\theta^{(i)}} - \mu_{mis|obs,\theta}\right)^2}{2\sigma^2_{mis}(1-\rho^2)}$$

$$= -\frac{1}{2}\log\left(2\pi\sigma^2_{mis}(1-\rho^2)\right) - \frac{E_{\mathbf{y}_{mis}|\mathbf{y}_{obs},\theta^{(i)}}\left(y_{mis} - \mu_{mis|obs,\theta^{(i)}}\right)^2 + \left(\mu_{mis|obs,\theta^{(i)}} - \mu_{mis|obs,\theta}\right)^2}{2\sigma^2_{mis}(1-\rho^2)}$$

$$= -\frac{1}{2}\log\left(2\pi\sigma^2_{mis}(1-\rho^2)\right) - \frac{\sigma^{2(i)}_{mis}(1-\rho^{2(i)}) + \left(\mu_{mis|obs,\theta^{(i)}} - \mu_{mis|obs,\theta}\right)^2}{2\sigma^2_{mis}(1-\rho^2)}$$

that is a function of the unknown parameter $\boldsymbol{\theta} = (\mu_{obs}, \mu_{mis}, \sigma^2_{obs}, \sigma^2_{mis}, \rho)'$ and the parameter values at the ith step.

Now, we observe the difference of two successive observed-data log-likelihood in equation (11.2):

$$l(\theta^{(i+1)}|\mathbf{y}_{obs}) - l(\theta^{(i)}|\mathbf{y}_{obs})$$
$$= \left[Q(\theta^{(i+1)}, \theta^{(i)}) - Q(\theta^{(i)}, \theta^{(i)})\right] - \left[R(\theta^{(i+1)}, \theta^{(i)}) - R(\theta^{(i)}, \theta^{(i)})\right].$$

Clearly, $Q(\theta^{(i+1)}, \theta^{(i)}) \geq Q(\theta^{(i)}, \theta^{(i)})$ since $\theta^{(i+1)}$ is chosen to maximize $Q(\theta, \theta^{(i)})$ over all θ. Using the concavity of the logarithmic function and Jensen's Inequality (see Chapter 3), we can derive, for any θ

$$R(\theta, \theta^{(i)}) - R(\theta^{(i)}, \theta^{(i)})$$

$$= E_{\mathbf{y}_{mis}|\mathbf{y}_{obs}, \theta^{(i)}} \left(\log \frac{f(\mathbf{y}_{mis}|\mathbf{y}_{obs}, \theta)}{f(\mathbf{y}_{mis}|\mathbf{y}_{obs}, \theta^{(i)})} \right)$$

$$\leq \log \left(E_{\mathbf{y}_{mis}|\mathbf{y}_{obs}, \theta^{(i)}} \frac{f(\mathbf{y}_{mis}|\mathbf{y}_{obs}, \theta)}{f(\mathbf{y}_{mis}|\mathbf{y}_{obs}, \theta^{(i)})} \right)$$

$$= \log \int \frac{f(\mathbf{y}_{mis}|\mathbf{y}_{obs}, \theta)}{f(\mathbf{y}_{mis}|\mathbf{y}_{obs}, \theta^{(i)})} f(\mathbf{y}_{mis}|\mathbf{y}_{obs}, \theta^{(i)}) d\mathbf{y}_{mis}$$

$$= 0,$$

since $\int f(\mathbf{y}_{mis}|\mathbf{y}_{obs}, \theta) d\mathbf{y}_{mis} = 1$ and $\log 1 = 0$.

Therefore, $R(\theta^{(i+1)}, \theta^{(i)}) - R(\theta^{(i)}, \theta^{(i)}) \leq 0$ and hence

$l(\theta^{(i+1)}|\mathbf{y}_{obs}) - l(\theta^{(i)}|\mathbf{y}_{obs}) \geq 0.$

This establishes the property that the likelihood (or the log-likelihood) of the observed data does not decrease after each EM iteration. In fact, it will increase when $\theta^{(i+1)}$ is chosen to satisfy the strict inequality $Q(\theta^{(i+1)}, \theta^{(i)}) > Q(\theta^{(i)}, \theta^{(i)})$. Since $L(\theta^{(i)}|\mathbf{y}_{obs})$ is bounded, the sequence converges to some L^* (see Chapter 1). In most applications, L^* is a stationary value. That is $L^* = L(\theta^*|\mathbf{y}_{obs})$ for some θ^* at which $\frac{\partial}{\partial\theta} L(\theta|\mathbf{y}_{obs}) = 0$. In most of the cases, L^* will be a local maximum. If $L(\theta|\mathbf{y}_{obs})$ has several stationary points, convergence of the EM iterations to local maximum, global maximum or saddle points may depend on the initial choice of the parameter values. If the likelihood is unimodal, any EM iterations converge to the unique MLE, regardless of the initial choice of the parameter values.

McLachlan and Krishnan (1997) provided a list of the advantages and disadvantages of the EM algorithm by comparing with other iterative procedures such as Newton-Raphson and Fisher's scoring methods. The attractive properties of the EM algorithm include that it is numerically stable, easy to implement, easy to program, and easy to monitor its convergence. It also requires small storage space in computing and simpler analytical work (only conditional expectation). The major shortcoming of the EM algorithm similar to the Newton-type method is that it does not guarantee convergence to the global maximum when there are multiple maxima. In this case, the estimate often depends on the initial choice of the parameter value.

Example 11.5 The yearly dental cost of an individual is usually assumed to follow a normal distribution. A study is to estimate the mean μ and variance σ^2 of the yearly dental cost for the employees of a large company; all of them have the same health insurance policy through the same insurance company. A random sample of 30 employees is taken. Their last year's dental expenses filed through this insurance company were retrieved. Under the policy, the upper limit for each individual's yearly dental expenses is set at \$1000. Those who exceeded this amount would not have accurate information and were considered as censored observations. Their actual dental costs were not recorded. The record only indicated that they had spent at least \$1000. There were 5 individuals who exceeded the cap. The yearly expenses for the other 25 individuals in the sample were (in dollars):
458, 701, 623, 902, 923, 457, 767, 822, 955, 718, 948, 695, 481, 720, 368, 386, 854, 724, 696, 896, 794, 878, 879, 377 and 959.

Let $y_{1,obs}, y_{2,obs}, \cdots, y_{m,obs}$, $m = 25$ denote the actual dental expenses for those who did not exceed the cap $c = 1000$, and let the censored information be $y_{m+1,mis}, y_{m+2,mis}, \cdots, y_{n,mis}$, $n = 30$. Therefore, $y_{k,mis} \geq c, k = m+1,\ldots,n$. The complete log-likelihood can be described as

$$l(\mu,\sigma^2) = -\frac{n}{2}\log(2\pi\sigma^2) - \sum_{k=1}^{m}\frac{(y_{k,obs}-\mu)^2}{2\sigma^2} - \sum_{k=m+1}^{n}\frac{(y_{k,mis}-\mu)^2}{2\sigma^2}$$
$$= -\frac{n}{2}\log(2\pi\sigma^2) - \sum_{k=1}^{m}\frac{(y_{k,obs}-\mu)^2}{2\sigma^2} - \sum_{k=m+1}^{n}\frac{y_{k,mis}^2-2\mu y_{k,mis}+\mu^2}{2\sigma^2}.$$

The distribution of $y_{k,mis}$, given $y_{k,mis} \geq c$, is a truncated normal distribution. Its pdf can be found in Chapter 5.

Using this pdf, we can now compute the conditional expectation of the missing observations as follows:

$$E(y_{k,mis} \mid \mathbf{y}_{obs}, \mu^{(i)}, \sigma^{2(i)}, y_{k,mis} \geq c)$$

$$= \frac{1}{P(y_{k,mis} > c)} \int_c^\infty \frac{w}{\sqrt{2\pi\sigma^{2(i)}}} \exp\left(-\frac{(w-\mu^{(i)})^2}{2\sigma^{2(i)}}\right) dw$$

$$= \frac{1}{1 - \Phi\left(\dfrac{c-\mu^{(i)}}{\sigma^{(i)}}\right)} \int_{\frac{c-\mu^{(i)}}{\sigma^{(i)}}}^\infty \frac{\mu^{(i)} + \sigma^{(i)} z}{\sqrt{2\pi}} \exp\left(-\frac{z^2}{2}\right) dz.$$

The integral in the above formula can be simplified as

$$\int_{\frac{c-\mu^{(i)}}{\sigma^{(i)}}}^\infty \frac{\mu^{(i)} + \sigma^{(i)} z}{\sqrt{2\pi}} \exp\left(-\frac{z^2}{2}\right) dz$$

$$= \mu^{(i)} \left[1 - \Phi\left(\frac{c-\mu^{(i)}}{\sigma^{(i)}}\right)\right] + \sigma^{(i)} \int_{\frac{c-\mu^{(i)}}{\sigma^{(i)}}}^\infty \frac{z}{\sqrt{2\pi}} \exp\left(-\frac{z^2}{2}\right) dz$$

$$= \mu^{(i)} \left[1 - \Phi\left(\frac{c-\mu^{(i)}}{\sigma^{(i)}}\right)\right] + \sigma^{(i)} \frac{-1}{\sqrt{2\pi}} \exp\left(-\frac{z^2}{2}\right)\Bigg|_{\frac{c-\mu^{(i)}}{\sigma^{(i)}}}^\infty$$

$$= \mu^{(i)} \left[1 - \Phi\left(\frac{c-\mu^{(i)}}{\sigma^{(i)}}\right)\right] + \sigma^{(i)} \frac{1}{\sqrt{2\pi}} \exp\left(-\frac{1}{2}\left(\frac{c-\mu^{(i)}}{\sigma^{(i)}}\right)^2\right).$$

Hence we have

$$E(y_{k,mis} \mid \mathbf{y}_{obs}, \mu^{(i)}, \sigma^{2(i)}, y_{k,mis} \geq c) = \mu^{(i)} + \frac{\sigma^{(i)} \phi(\dfrac{c-\mu^{(i)}}{\sigma^{(i)}})}{\left[1 - \Phi\left(\dfrac{c-\mu^{(i)}}{\sigma^{(i)}}\right)\right]}, \qquad (11.3)$$

where $\phi(\ )$ and $\Phi(\ )$ are respectively the pdf and the cdf of a standard normal random variable. References to them may be found in Chapter 5.

It is left as an exercise to show that

$$E(y^2_{k,mis} \mid \mathbf{y}_{obs}, \mu^{(i)}, \sigma^{2(i)}, y_{k,mis} \geq c)$$

$$= \left(\mu^{(i)}\right)^2 + \sigma^{2(i)} + \frac{\sigma^{(i)}(c + \mu^{(i)})\phi(\frac{c - \mu^{(i)}}{\sigma^{(i)}})}{\left[1 - \Phi\left(\frac{c - \mu^{(i)}}{\sigma^{(i)}}\right)\right]}. \tag{11.4}$$

For the conditional expectation of the complete-data log-likelihood, we can then proceed as follows:

$$Q(\mu, \sigma^2, \mu^{(i)}, \sigma^{2(i)})$$

$$= -\frac{n}{2}\log(2\pi\sigma^2) - \sum_{k=1}^{m}\frac{(y_{k,obs} - \mu)^2}{2\sigma^2} - (n - m)\frac{\mu^2}{2\sigma^2}$$

$$- \sum_{k=m+1}^{n}\frac{E(y^2_{k,mis} \mid \mathbf{y}_{obs}, \mu^{(i)}, \sigma^{2(i)}, y_{k,mis} \geq c) - 2\mu E(y_{k,mis} \mid \mathbf{y}_{obs}, \mu^{(i)}, \sigma^{2(i)}, y_{k,mis} \geq c)}{2\sigma^2},$$

where $E(y^2_{k,mis} \mid \mathbf{y}_{obs}, \mu^{(i)}, \sigma^{2(i)}, y_{k,mis} \geq c)$ and $E(y_{k,mis} \mid \mathbf{y}_{obs}, \mu^{(i)}, \sigma^{2(i)}, y_{k,mis} \geq c)$ are presented in equations (11.4) and (11.3) respectively.

For the M-step, we differentiate $Q(\ )$ with respect to μ and σ^2 and set the derivatives to 0. The resulting equations are

$$0 = \frac{\partial}{\partial\mu}Q(\mu, \sigma^2, \mu^{(i)}, \sigma^{2(i)})$$

$$= \sum_{k=1}^{m}\frac{(y_{k,obs} - \mu)}{\sigma^2} - (n - m)\frac{\mu}{\sigma^2} + \sum_{k=m+1}^{n}\frac{E(y_{k,mis} \mid \mathbf{y}_{obs}, \mu^{(i)}, \sigma^{2(i)}, y_{k,mis} \geq c)}{\sigma^2} \tag{11.5}$$

and

$$0 = \frac{\partial}{\partial \sigma^2} Q(\mu, \sigma^2, \mu^{(i)}, \sigma^{2(i)})$$

$$= -\frac{n}{2\sigma^2} + \sum_{k=1}^{m} \frac{(y_{k,obs} - \mu)^2}{2\sigma^4} + (n-m)\frac{\mu^2}{2\sigma^4}$$

$$+ \sum_{k=m+1}^{n} \frac{E(y_{k,mis}^2 \mid \mathbf{y}_{obs}, \mu^{(i)}, \sigma^{2(i)}, y_{k,mis} \geq c)}{2\sigma^4}$$

$$- \sum_{k=m+1}^{n} \frac{2\mu E(y_{k,mis} \mid \mathbf{y}_{obs}, \mu^{(i)}, \sigma^{2(i)}, y_{k,mis} \geq c)}{2\sigma^4}.$$

$$(11.6)$$

Solving equation (11.5) and substituting the values of the expectations by the result in equation (11.3), we obtain the next iteration of μ as

$$\mu^{(i+1)} = \frac{1}{n}\left\{ \sum_{k=1}^{m} y_{k,obs} + (n-m)\mu^{(i)} + \frac{(n-m)\sigma^{(i)}\phi(\frac{c-\mu^{(i)}}{\sigma^{(i)}})}{\left[1 - \Phi\left(\frac{c-\mu^{(i)}}{\sigma^{(i)}}\right)\right]} \right\}.$$

Note that the solution of the unknown parameter μ in equation (11.5) will become $\mu^{(i+1)}$ and that the original i^{th} step parameter values $\mu^{(i)}$ and $\sigma^{2(i)}$ carried from equation (11.3) will remain the same and be used for calculating the new (revised) value of μ.

However, the unknown value of μ in equation (11.6) will be replaced by $\mu^{(i+1)}$ for calculation of $\sigma^{2(i+1)}$ while the coefficients in equation (11.6) that include the parameter values of $\mu^{(i)}$ and $\sigma^{2(i)}$ will remain the same. By using this rule and the results in equation (11.3) and equation (11.4), we obtain the next iteration of υ^2 as

$$\sigma^{2(i+1)} = \frac{1}{n} \{ \sum_{k=1}^{m} (y_{k,obs} - \mu^{(i+1)})^2 + (n-m)\left(\mu^{(i+1)}\right)^2 - 2(n-m)\mu^{(i+1)}\mu^{(i)}$$

$$-\frac{2(n-m)\mu^{(i+1)}\sigma^{(i)}\phi(\dfrac{c-\mu^{(i)}}{\sigma^{(i)}})}{\left[1-\Phi\left(\dfrac{c-\mu^{(i)}}{\sigma^{(i)}}\right)\right]} + (n-m)\left(\mu^{(i)}\right)^2 + (n-m)\sigma^{2(i)}$$

$$+\frac{(n-m)\sigma^{(i)}(c+\mu^{(i)})\phi(\dfrac{c-\mu^{(i)}}{\sigma^{(i)}})}{\left[1-\Phi\left(\dfrac{c-\mu^{(i)}}{\sigma^{(i)}}\right)\right]} \}.$$

A SAS program that can perform the EM algorithm for finding the MLE of these parameters is provided. The result shows that $\hat{\mu} = 842.76$, $\hat{\sigma}^2 = 414218.82$ (or $\hat{\sigma} = 643.60$) and the log-likelihood at these estimates is -236.58.

Example 11.6 Research on human behavior usually depends on interviews with the subjects under study. The response from a subject may not always be truthful, especially pertaining to unethical or confidential activities. In an HIV-related behavior study, a survey was conducted on a randomly chosen sample regarding the number of sexual partners in the previous month. It is assumed that this number follows a Poisson distribution with parameter λ. However, from past experience, the researchers also assume that a subject may either tell the truth (with probability p) by giving the exact number of sexual encounters in the previous month, or simply give the number 0 regardless of the actual number of his/her sexual encounters in the previous months. In a pilot study, the reported numbers of sexual partners for 20 subjects are
5, 5, 3, 2, 2, 0, 0, 8, 0, 2, 3, 0, 0, 2, 0, 3, 1, 0, 0 and 1.
Let $x_1, \ldots, x_n$ denote the reported numbers of sexual partners in the previous month for the n sampled subjects. Assume that $Y_1, \ldots, Y_n$ are the random variables that represent the true numbers of sexual partners in the previous month and $Z_1, \ldots, Z_n$ denote the dichotomous random variables that represent the subject's character on the response. We assume that "$Z_i = 1$" means subject i tells the truth by reporting the true number of sexual encounters and "$Z_i = 0$" means subject i simply reported the number 0, regardless of his/her actual number of sexual encounters. It is noteworthy to emphasize that neither Y_i nor Z_i is observable. For notational convenience, we will use y_i and z_i to represent the realization. However, if $x_i > 0$ then $Y_i = x_i$ and $Z_i = 1$.

Assuming that Y_i and Z_i are independent, the complete-data likelihood function for λ and p can be written as

$$L_C(\lambda, p) = \prod_{i=1}^{n} \frac{\lambda^{y_i} e^{-\lambda}}{y_i!} p^{z_i} (1-p)^{1-z_i}.$$

The complete-data log-likelihood is

$$l_C(\lambda, p) = C(\mathbf{y}) + \left(\sum_{i=1}^{n} y_i\right)\log\lambda - n\lambda + \left(\sum_{i=1}^{n} z_i\right)\log p + \left(n - \sum_{i=1}^{n} z_i\right)\log(1-p),$$

where $C(\mathbf{y}) = -\sum_{i=1}^{n}\log(y_i!)$.

If $x_i > 0$, then $E(Y_i|x_i > 0) = x_i$ and $E(Z_i|x_i > 0) = 1$. If $x_i = 0$, it is left as an exercise for the readers to show that

$$E\left(y_i|x_i = 0\right) = \frac{(1-p)\lambda}{(1-p+pe^{-\lambda})} \text{ and } E\left(z_i|x_i = 0\right) = \frac{pe^{-\lambda}}{(1-p+pe^{-\lambda})}. \quad (11.7)$$

Therefore, for a given set of parameters $p^{(k)}$ and $\lambda^{(k)}$, we can obtain

$$E\left(\sum_{i=1}^{n} y_i \,\middle|\, x_1 = 0,\dots,x_{n_0} = 0, x_{n_0+1} > 0,\dots,x_n > 0\right) = \sum_{i=n_0+1}^{n} x_i + \frac{n_0(1-p^{(k)})\lambda^{(k)}}{(1-p^{(k)}+p^{(k)}e^{-\lambda^{(k)}})}$$

and

$$E\left(\sum_{i=1}^{n} z_i \,\middle|\, x_1 = 0,\dots,x_{n_0} = 0, x_{n_0+1} > 0,\dots,x_n > 0\right) = n - n_0 + \frac{n_0 p^{(k)} e^{-\lambda^{(k)}}}{(1-p^{(k)}+p^{(k)}e^{-\lambda^{(k)}})},$$

where n_0 denote the number of individuals who reported 0.

The conditional expectation of $l_C(\lambda, p)$ conditioned on x_i at $(k+1)^{th}$ iteration when $\lambda = \lambda^{(k)}$ and $p = p^{(k)}$ have been calculated at the kth iteration, is

$$Q(\lambda, p, \lambda^{(k)}, p^{(k)})$$

$$= E(C(\mathbf{y})|\mathbf{x}) + \left(\sum_{i=n_0+1}^{n} x_i + \frac{n_0(1-p^{(k)})\lambda^{(k)}}{(1-p^{(k)}+p^{(k)}e^{-\lambda^{(k)}})} \right) \log \lambda$$

$$+ \left(n - n_0 + \frac{n_0 p^{(k)}e^{-\lambda^{(k)}}}{(1-p^{(k)}+p^{(k)}e^{-\lambda^{(k)}})} \right) \log p - n\lambda$$

$$+ \left(n_0 - \frac{n_0 p^{(k)}e^{-\lambda^{(k)}}}{(1-p^{(k)}+p^{(k)}e^{-\lambda^{(k)}})} \right) \log(1-p),$$

where $\mathbf{x} = (x_1, \ldots, x_n)$.

For the M-step, we can set up the equations for the first derivatives of $Q(\lambda, p, \lambda^{(k)}, p^{(k)})$ as $\frac{\partial}{\partial \lambda} Q(\lambda, p, \lambda^{(k)}, p^{(k)}) = 0$ and $\frac{\partial}{\partial p} Q(\lambda, p, \lambda^{(k)}, p^{(k)}) = 0$.

The roots for the these two equations can be solved as

$$\lambda^{(k+1)} = \frac{1}{n} \left(\sum_{i=n_0+1}^{n} x_i + \frac{n_0(1-p^{(k)})\lambda^{(k)}}{(1-p^{(k)}+p^{(k)}e^{-\lambda^{(k)}})} \right)$$

and

$$p^{(k+1)} = \frac{1}{n} \left(n - n_0 + \frac{n_0 p^{(k)}e^{-\lambda^{(k)}}}{(1-p^{(k)}+p^{(k)}e^{-\lambda^{(k)}})} \right).$$

The MLE for λ and p using this procedure can be found as $\hat{\lambda} = 2.92$ and $\hat{p} = 0.63$. A SAS program for the implementation of this procedure is provided.

11.3 Simulation

All simulations build on the generation of random numbers, that represents the value (or a realization) of a $U(0,1)$ random variable. Traditionally, random numbers were generated by drawing a card, flipping a coin or rolling a dice. After a couple decades of using the table of random digits, a new approach uses a computer to generate pseudo-random numbers. The pseudo-random numbers are obtained by iterating a deterministic formula starting from a *seed*. The function of the seed is to initiate the sequence of the pseudo-random numbers resulting from the deterministic formula using the *multiplicative congruential method*. Detailed discussion of this method is beyond the scope of this book. Readers are referred to a simulation book for further study (see for example, Ross (2002)). For the generation of pseudo-random numbers, different seeds will initiate different sequences. Most of the computer software packages allow the user to choose a seed for generation of a random number or any random variable with a well known distribution. If the seed is not specified, the computer will assign one usually related to the time the simulation procedure is run. Pseudo-random numbers have been rigorously tested and can be treated as random. However, this method still has the potential disadvantages for producing irregular results.

11.3.1 The Inverse Transformation Method

For the generation of random variables other than continuous uniform variable, the *inverse transformation method* is most commonly used. In generating a discrete random variable X, the unit interval is divided into several segments. Each segment is associated with a possible outcome and has a length equal to $P(X = i)$, where i is a possible value of the random variable. If a random number is generated and the result falls into a particular segment that is associated with $P(X = i)$, then $X = i$ is declared a realization of X. Example 11.7 below demonstrates the implementation of this algorithm. When the random variable has a large (or even an infinite) number of possible outcomes, the procedure may be more complicated. Example 11.8 will be used to display this scenario. In generating a continuous random variable, the following result from the probability integral transformation discussed in chapter 6 is often applied. Recall that, if Y is a continuous random variable with cdf $F(\ )$, then $F(Y)$ has a $U(0,1)$ distribution. Conversely, if U is a $U(0,1)$ random variable and $F(\ )$ is a legitimate cdf, then $F^{-1}(U)$ is a random variable with cdf $F(\ )$. Note that, when a function $F(\ )$ is referred as a legitimate cdf, $F(\ )$ must be (1) monotonely non-decreasing, (2) $\lim_{x \to -\infty} F(x) = 0$, (3) $\lim_{x \to \infty} F(x) = 1$, and (4) continuous from the right, as discussed in Section 3.3. In practice, if we try to generate a value from a known cdf $F(\ )$, we can first generate a $U(0,1)$ random number u and then calculate $F^{-1}(u)$ that will become a generated value from the cdf $F(\ )$.

Example 11.7 Let X be a discrete random variable with the following pdf

$$p_X(x) = \begin{cases} 0.1 & if \ x = 1 \\ 0.3 & if \ x = 2 \\ 0.4 & if \ x = 3 \\ 0.2 & if \ x = 4 \end{cases}.$$

If we are to simulate this random variable, we first divide the unit interval as the following four segments $[0, 0.1)$, $[0.1, 0.4)$, $[0.4, 0.8)$ and $[0.8, 1)$. We can set the X value as 1 if the generated random number U is less than 0.1, as 2 if $0.1 \leq U < 0.4$, as 3 if $0.4 \leq U < 0.8$ and as 4 if $0.8 \leq U < 1$. A program using random numbers to generate X is available. If we simulate X 1000 times, the results showed that 106 of them obtained value 1, 295 of them obtained value 2, 397 of them obtained value 3 and 202 of them obtained value 4.

Example 11.8 The inverse transformation method can also be applied to simulate a binomial random variable, say $B(5, 0.2)$. As we did in example 11.7, we can divide the unit interval into 6 segments, each has a length $\binom{5}{i}(0.2)^i (0.8)^{5-i}$, $i = 0, 1, \cdots, 5$. However, the calculation of the boundaries of these subintervals may be cumbersome, especially when n is large. A better way is to use the property of the binomial distribution: $p_X(i+1) = \left(\frac{n-i}{i+1}\right)\left(\frac{p}{1-p}\right) p_X(i)$ (see section 4.3, Chapter 4). Since $p_X(0) = (1-p)^n = (0.8)^5$, we can first check to see if the generated random number U is less than $(0.8)^5$. If it is, then the generated X value can be declared as 0. Otherwise, we can check if U is less than $p_X(0) + p_X(1)$ that is equal to $(0.8)^5 + \left(\frac{5-0}{0+1}\right)\left(\frac{0.2}{1-0.2}\right)(0.8)^5$. If it is, we set X to be 1. Otherwise, we continue this iterative process until we reach $i = n$. An algorithm can be easily written using this iterative procedure. A computer program is provided. Table 11.3 presents the frequency tables of $B(5, 0.2)$ using two different simulation methods. One took the built-in function from SAS (Frequency 1) and the other used the iterative method presented in this example (Frequency 2). Each contains 1000 samples. The theoretical probabilities of all possible values are also presented for comparison.

Table 11.3 Proportions of simulated data in each value of $B(5, 0.2)$ compared with the corresponding theoretical probabilities.

i	$\Pr(X = i)$	Frequency 1	Frequency 2
0	0.3277	0.3350	0.3330
1	0.4096	0.4000	0.3880
2	0.2048	0.1880	0.2170
3	0.0512	0.0690	0.0530
4	0.0064	0.0070	0.0090
5	0.0003	0.0010	0

Example 11.9 Let X be an exponential random variable with rate $\lambda = 1$. It is well known that the cdf of X is $F(x) = 1 - e^{-x}$. As it was discussed in Section 6.3, if U represents the random number (or the $U(0,1)$ random variable), then the random variable $F^{-1}(U) = -\log(1-U)$ has the distribution the same as that of X. We can generate X by first generating a random number and then applying this formula to calculate $F^{-1}(U)$. Since $1-U$ is also uniformly distributed in $(0,1)$, we can use a slightly more efficient procedure by using $-\log U$ to generate X. A computer program using random numbers to simulate the exponential random variable with unit rate is available. The empirical frequency of 1000 replications is used to compare the theoretical pdf of the distribution. The q-q plots of the theoretical and empirical quantiles is presented in Figure 2.

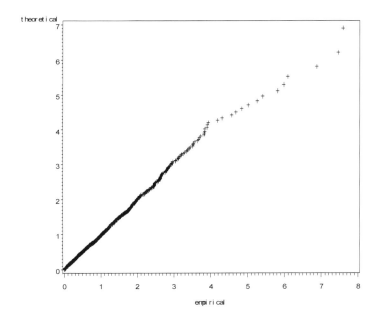

Figure 11.2 Q-Q plots of the simulated quantiles versus the corresponding theoretical quantiles

If $\lambda \neq 1$, then $F^{-1}(u) = -\frac{1}{\lambda}\log(1-u)$. Therefore, we can generate the exponential random variable by the formula $-\frac{1}{\lambda}\log U$ for any $\lambda > 0$.

Note that, due to the complexity of the cdf of the gamma distribution, it is very difficult or even impossible to find its inverse function in a closed form. Direct application of the inverse transformation method to generate a gamma random variable is not likely. From Chapter 5, we know that the sum of n independent exponential random variables with rate λ has a gamma distribution with parameters (n, λ). Therefore, we can generate a $gamma(n, \lambda)$ variable by

$$-\frac{1}{\lambda}\log U_1 - \frac{1}{\lambda}\log U_2 - \ldots - \frac{1}{\lambda}\log U_n,$$

where $U_i, i = 1, 2, \ldots, n$ are independent random numbers. Or, we may equivalently use $-\frac{1}{\lambda}\log(U_1 U_2 \cdots U_n)$ to generate a $gamma(n, \lambda)$. This more efficient procedure can be justified by some simple algebra.

11.3.2 The Rejection Method*

Although the inverse transformation method is convenient, it is not always applicable. When the inverse of the cdf of the random variable to be simulated is difficult to find or does not exist in an explicit form, the rejection method is another choice for simulating a random variable. The rejection method uses a random variable Y that has a known pdf $g(y)$ (or pmf $q(y)$) and can be efficiently simulated, as the basis for generating values of our target random variable X with pdf $f(x)$ (or pmf $p(x)$). This method first generates a y from the distribution of Y, then generates a random number U. If $U \leq \frac{f(y)}{cg(y)}$ (or $U \leq \frac{p(y)}{cq(y)}$), then we accept this simulated value and set $X = y$. Otherwise, we restart the procedure by generating another y and U and making the similar comparison. Here c is chosen so that $\frac{f(y)}{g(y)} \leq c$ (or $\frac{p(y)}{q(y)} \leq c$) for all possible values of Y. This process is iteratively executed until a valid X value is generated.

We will show that the generated value has the same distribution as that of X in the discrete case. It can be similarly shown for the continuous case.

At any iteration, observe that the probability that a $Y = j$ is accepted can be calculated as

$$P(Y = j \text{ and } j \text{ is accepted})$$

$$= P(j \text{ is accepted}|Y = j)P(Y = j)$$

$$= P\left(U \le \frac{p(j)}{cq(j)}\right)q(j)$$

$$= \frac{p(j)}{cq(j)}q(j)$$

$$= \frac{p(j)}{c}.$$

It is clear that, by summing over all j, the probability of accepting a y value at any iteration is $\frac{1}{c}$. The number of iterations needed to obtain a valid X value follows a geometric distribution. Therefore,

$$P(X = j)$$

$$= \sum_{n=1}^{\infty} P(j \text{ is accepted at the } n^{th} \text{ iteration})$$

$$= \sum_{n=1}^{\infty} P(\text{no acceptance at iterations 1 to n-1})P(\text{acceptance of } j \text{ at the } n^{th} \text{ iteration})$$

$$= \sum_{n=1}^{\infty} \left(1 - \frac{1}{c}\right)^{n-1}\frac{p(j)}{c}$$

$$= p(j).$$

Example 11.10 We can now use the unit exponential distribution as a basis to generate a standard normal random variable Z. First note that the pdf of $X = |Z|$ can be written as $f(x) = \frac{2}{\sqrt{2\pi}}\exp(-\frac{x^2}{2})$ for $0 < x < \infty$. The pdf of the unit exponential random viable Y is $g(y) = \exp(-y)$ for $0 < y < \infty$. Now, observe that $\frac{f(x)}{g(x)} = \sqrt{\frac{2}{\pi}}\exp\left(x - \frac{x^2}{2}\right)$ is bounded by $\sqrt{\frac{2e}{\pi}}$, since the ratio of two pdfs has a maximum at $x = 1$. We can generate a y from Y and a random number U. If

$$U \le \frac{f(y)}{\sqrt{\frac{2e}{\pi}}g(y)} = \exp\left(-\frac{(y-1)^2}{2}\right),$$ we then set $X=y$. Otherwise, we continue this

procedure. After we generate an X value x, we can obtain a Z value as either $Z = x$ or $Z = -x$ with equal probability. That means we generate another random number U^*. If $U^* \leq 0.5$ then we set $Z = x$, otherwise, we set $Z = -x$. A computer program using this method to generate the standard normal random variable is available. This program generates 10,000 numbers from standard normal distribution and the generated data have a sample mean of -0.0000342 and a sample standard deviation of 0.9951. The test for normality produces a p-value 0.15 (using Kolmogorov-Smirnov test) or 0.25 (using Anderson-Darling test).

Note that the common method for generating a normal random variable is using Box-Muller algorithm. This algorithm uses a mathematical property that $X = \cos(2\pi U_1)\sqrt{-2\log(U_2)}$ and $Y = \sin(2\pi U_1)\sqrt{-2\log(U_2)}$ are independent standard normal random variables, if U_1 and U_2 are two random numbers.

11.3.3 Basic Methods on Generation of Multivariate Distributions*

For generation of samples from multivariate distributions, the standard rejection method is usually difficult to apply. In addition, the standard rejection method is not very efficient for a large dimensional distribution.

The common device is to make a transformation on i.i.d. univariate random variables. This method arises from the theoretical framework especially when the distribution is normal. It can be proved by the Jacobian method discussed in Chapter 5, that, if $\mathbf{y}$ is a multivariate normal random vector with mean vector $\boldsymbol{\mu}$ and variance-covariance matrix Σ, then $A\mathbf{y}$ is also a multivariate normal random vector with mean vector $A\boldsymbol{\mu}$ and variance-covariance matrix $A\Sigma A'$, where A is a non-singular matrix (that means the matrix has an inverse) to guarantee that $A\mathbf{y}$ is not degenerate. Therefore, if we are to generate samples from an n-dimensional normal distribution with mean vector $\boldsymbol{\mu}_0$ and variance-covariance matrix Σ_0, we first choose a non-singular matrix A such that $AA' = \Sigma_0$ and make transformation on the n-dimensional standard normal vector $\mathbf{z}$ formed by n i.i.d. standard normal samples. This transformation was carried by pre-multiplying $\mathbf{z}$ by A and adding the product to $\boldsymbol{\mu}_0$. The resulting $\mathbf{y} = A\mathbf{z} + \boldsymbol{\mu}_0$ will be the desired vector sample. Note that we assume that Σ_0 is positive definite matrix so the existence of A is guaranteed by the spectral decomposition theorem and its related results. Detailed discussion on the related linear algebra can be found in, for example, Searle (1982). The theory on the distribution of the transformation of a normal random vector can be found in any multivariate statistics book (see, for example, Anderson (2003)).

The other method for generation of a multivariate distribution is by generating a sequence of random samples from conditional univariate distributions that together can yield the multivariate distribution. If we are to simulate a multivariate normal vector with mean vector $\boldsymbol{\mu} = (\mu_1, \mu_2, \ldots, \mu_n)'$ and variance covariance matrix $\Sigma_0 = (\sigma_{ij})_{n \times n}$, then we can generate x_1 from $X_1 \sim N(\mu_1, \sigma_{11})$, and generate x_2 from $X_2 | X_1 = x_1 \sim N \left(\mu_2 + \dfrac{\sigma_{12}(x_1 - \mu_1)}{\sigma_{11}}, \sigma_{22} - \dfrac{\sigma_{12}^2}{\sigma_{11}} \right)$, and x_3 from the

conditional normal distribution of $X_3 | X_1 = x_1, X_2 = x_2$, and so on. The theoretical justification is based on the property that

$$
\begin{aligned}
f(x_1, \ldots, x_n) &= f(x_n | x_1, \ldots, x_{n-1}) f(x_1, \ldots, x_{n-1}) \\
&= f(x_n | x_1, \ldots, x_{n-1}) f(x_{n-1} | x_1, \ldots, x_{n-2}) f(x_1, \ldots, x_{n-2}) \\
&= f(x_n | x_1, \ldots, x_{n-1}) f(x_{n-1} | x_1, \ldots, x_{n-2}) f(x_{n-2} | x_1, \ldots, x_{n-3}) \cdots f(x_1 | x_2) f(x_1).
\end{aligned}
$$

We can then successively generate each term and obtain a multivariate pdf.

Example 11.11   For the generation of the three-dimensional normal samples $(X, Y, W)'$ with means, variances and covariances specified as

$\mu_X = 1, \mu_Y = 2, \mu_W = 3, \sigma_X^2 = 2, \sigma_Y^2 = 6, \sigma_W^2 = 6$,

$COV(X, Y) = 3$, $COV(Y, W) = -1$, and $COV(X, W) = -2$,

we can first generate three i.i.d. standard normal random samples $(z_1, z_2, z_3)'$ and then make the following transformations

$$
\begin{aligned}
x &= 1 + z_1 - z_2 \\
y &= 2 + 2z_1 - z_2 + z_3 \\
w &= 3 - z_1 + z_2 + 2z_3
\end{aligned}
$$

It is left for the readers to check the means, the variances and the covariances of $(X, Y, W)'$. A SAS program that generates 1000 vector samples, is available. The sample means for X, Y and W are 1.05, 2.04 and 2.94 respectively. The sample variances are 2.10, 6.14 and 6.57 respectively. The sample covariances for (X, Y), (X, W) and (Y, W) are 3.09, -2.21 and -1.10 respectively. Another SAS program using generation of conditional distributions is also available. In this program, we first generate x from $N(1, 2)$, then generate y from

$$
N \left(2 + \frac{3}{2}(x - 1), 6 - \frac{3^2}{2} \right).
$$

After x and y are generated, we generate w from

$$N\left(3+\begin{pmatrix}-3 & \frac{4}{3}\end{pmatrix}\begin{pmatrix}x-1\\y-2\end{pmatrix}, \frac{4}{3}\right),$$

where the mean is calculated from the formula

$$3+\begin{pmatrix}-2 & -1\end{pmatrix}\begin{pmatrix}2 & 3\\3 & 6\end{pmatrix}^{-1}\left(\begin{pmatrix}x\\y\end{pmatrix}-\begin{pmatrix}1\\2\end{pmatrix}\right)$$

and the variance is from

$$6-\begin{pmatrix}-2 & -1\end{pmatrix}\begin{pmatrix}2 & 3\\3 & 6\end{pmatrix}^{-1}\begin{pmatrix}-2\\-1\end{pmatrix}=\frac{4}{3}.$$

From this method, the sample means for X, Y and W are 1.01, 1.96 and 2.92 respectively. The sample variances are 2.06, 6.15 and 6.51 respectively. The sample covariances for (X,Y), (X,W) and (Y,W) are 3.08, -2.11 and -1.01 respectively. The sample statistics are very close to the one generated by the first method.

Although these methods are available and theoretically justifiable, they are usually slow and less efficient, especially when matrix operations are involved. A type of more efficient methods, called the *Markov chain Monte Carlo* methods has been increasingly popular. These methods can generate multi-dimensional samples using only the univariate conditional distributions of each component, given the rest of the components. The iterative procedures of these methods, along with their theoretical justifications will be introduced in section 11.5.

11.4 Markov Chains

Markov chains is a subject in applied probability and is used to study a sequence of random variables. It has been used in several areas of practical applications such as inventory, genetics and queuing theory. This section is to focus on the definition and some basic properties of Markov chains that are to be used in presenting Markov chain Monte Carlo methods to be discussed in section 11.5.

Consider a sequence of random variables $X_1,\ldots,X_n,\ldots$, where X_i represents the state of a system being studied at time i. The set of all possible values of all states is called the *state space*. For discrete state Markov chains, the state space is either equivalent to the set of non-negative integers or a finite set $\{1,2,\ldots,N\}$. The section will focus on the case that the Markov chain has a finite state space.

If $\{X_n\}$ has the property that, given the value of X_k, the value of X_m, for any integer $m > k$, does not depend on the value of X_i, for $i < k$, then $\{X_n\}$ is called a (discrete-time) *Markov chain*. Mathematically, a Markov chain satisfies

$$P\left(X_{n+1} = j \big| X_1 = i_0, X_1 = i_1, \ldots, X_{n-1} = i_{n-1}, X_n = i\right) = P\left(X_{n+1} = j \big| X_n = i\right)$$

for all n and all possible values of $i_0, i_1, \ldots, i_{n-1}, i, j$. If the probability that X_{n+1} being in state j, given that X_n is in state i is independent of the time n, then the Markov chain is referred to as a *stationary Markov chain*. In a stationary Markov chain, the probability $P\left(X_{n+1} = j \big| X_n = i\right)$ is called *the (one-step) transition probability* and is denoted by p_{ij}. Since a Markov chain must be in a certain state after it leaves the state i, these transition probabilities satisfy the relationship $\sum_{j=1}^{N} p_{ij} = 1$. It is convenient to use a matrix to express the transition probabilities and call it *the transition probability matrix*, denoted as $P = \left(p_{ij}\right)$. In this section, we will only focus on the stationary Markov chain and will use the simpler term Markov chain to refer to the stationary Markov chain. Therefore, each Markov chain we refer to will have a transition probability matrix associated with it.

Note that, a sequence of continuous random variables $\{X_n\}$ is also called a Markov chain if the distribution of the future (X_m, $m > k$) does not depend on the past (X_i, $i < k$), given the present value X_k. Similar definition for the stationary Markov chain can also be defined. For a stationary Markov chain of continuous random variables, the conditional probability density $p_{xy} = f_{X_{n+1}|X_n}(y|x)$ is called the *transition kernel*. It shares many good properties with the transition probability matrix for the discrete random variables. In this section, we will only focus on the Markov chain of discrete random variables and their associate transition probability matrices. For the application of these results in the remaining sections, we will assume similar properties derived in discrete cases, can also be applied for the continuous cases with slightly different regularity assumptions.

Now, consider the probability that a Markov chain moves from state i to state j in k steps, denoted by $p_{ij}^{(k)} = P\left(X_{n+k} = j \big| X_n = i\right) = P\left(X_k = j \big| X_0 = i\right)$. Due to the stationary property, this probability does not depend on n and can be defined by either one of the probabilities above. For notional convenience, we also use $p_{ij}^{(0)} = 1$ or 0, depending on whether or not $j = i$. The quantity $p_{ij}^{(k)}$ is called the k-step transition probability and satisfies the following relations

$$p_{ij}^{(m+n)} = \sum_{\text{all possible state } k} p_{ik}^{(m)} p_{kj}^{(n)}, \tag{11.8}$$

for all $m, n \geq 0$.

Equation (11.8) can be proved by using the law of total probability described in equation (2.5) and mathematical induction. In matrix notation, equation (11.8) can be expressed as $\mathrm{P}^{(m+n)} = \mathrm{P}^{m+n}$, where $\mathrm{P}^{(m+n)}$ is the matrix of $m+n$ step transition probabilities and P^{m+n} is the $(m+n)^{th}$ power of the one-step transition probability matrix P defined above.

Example 11.12 Let $\{X_n\}$ be a stationary Markov chain that has a state space $\{0,1,2,3\}$. Assume that $\{X_n\}$ has a transition probability matrix P such that

$$\mathrm{P} = \begin{pmatrix} \frac{1}{2} & \frac{1}{6} & \frac{1}{3} & 0 \\ \frac{1}{4} & \frac{1}{4} & \frac{1}{4} & \frac{1}{4} \\ \frac{1}{6} & \frac{1}{2} & \frac{1}{6} & \frac{1}{6} \\ \frac{1}{3} & 0 & \frac{1}{2} & \frac{1}{6} \end{pmatrix}.$$

For this Markov chain, if you start with state 0, then, in the next time point you will have a probability of $\frac{1}{6}$ that the process moves to state 1, a probability of $\frac{1}{3}$ that this process moves to state 2 and a probability of $\frac{1}{2}$ that it stays in state 0. It will be impossible for you to move to state 3 in the next time point, if you start with state 0. However, it is not impossible for you to move to state 3 later in time. In fact, one can move to state 3 from state 0 through state 1 or state 2. If you are in state 0 at time $t = 0$, your probability of moving to state 3 at time $t = 2$ will be the sum of $\left(\frac{1}{6}\right)\left(\frac{1}{4}\right)$ (through state 1) and $\left(\frac{1}{3}\right)\left(\frac{1}{6}\right)$ (through state 2), that is $\frac{7}{72}$. This number is also the $(1,4)$ element of the 2-step transition matrix P^2. Using similar computations, we can find that this Markov chain allows us to move from one state to another in a finite number of steps. A Markov chain having this characteristic is called an *irreducible* Markov chain.

Definition 11.1 A Markov chain is said to be *irreducible* if for any two states i and j, there are positive integers m and n such that $p_{ij}^{(m)} > 0$ and $p_{ji}^{(n)} > 0$. This means that, with positive probability starting with state i, the chain will reach state j in a finite number of steps.

Definition 11.2 An irreducible Markov chain is said to be *aperiodic* if there is a state j and a non-negative integer n such that the probabilities of returning to j at n^{th} step and at $(n+1)^{th}$ step, given the initial state j, are positive. Mathematically, for some $n \geq 0$ and some state j, $p_{jj}^n > 0$ and $p_{jj}^{n+1} > 0$.

Theorem 11.1. Let $P = \left(p_{ij} \right)_{N \times N}$ be the transition probability matrix of an irreducible and aperiodic Markov chain with finite number of states. Then the sequence $\left\{ p_{ij}^{(n)} \right\}_{n=1}^{\infty}$ converges to a positive number π_j for every j regardless of the initial state i. In other words,

$$\lim_{n \to \infty} p_{ij}^{(n)} = \pi_j > 0 \tag{11.9}$$

Moreover, π_js satisfy the following linear equation system:

$$\sum_{j=1}^{N} \pi_j P_{jm} = \pi_m, \, m = 1, 2, \ldots, M \tag{11.10}$$

and

$$\sum_{m=1}^{N} \pi_m = 1 \tag{11.11}$$

Proof. For the proof of the convergence of the sequence, the readers are referred to a book on stochastic processes such as Taylor and Karlin (1998). Assuming the limit in (11.9) exists, we first rewrite the sequence as $p_{im}^{(n+1)} = \sum_{k=1}^{N} p_{ik}^{(n)} P_{km}$. Taking the limit on both sides, we obtain

$$\pi_m = \lim_{n \to \infty} \sum_{k=1}^{N} p_{ik}^{(n)} P_{km} = \sum_{k=1}^{N} \lim_{n \to \infty} p_{ik}^{(n)} P_{km} = \sum_{k=1}^{N} \pi_k P_{km} \text{, for every } m = 1, 2, \ldots, M.$$

To show (11.11), we rewrite the π_js as a limit and then exchange the finite sum with the limit, i.e. $\sum_{m=1}^{N} \pi_m = \sum_{m=1}^{N} \lim_{n\to\infty} p_{im}^{(n)} = \lim_{n\to\infty} \sum_{m=1}^{N} p_{im}^{(n)} = 1$, where we use the prop-

erty $\sum_{m=1}^{N} p_{im}^{(n)} = 1$.

In this case, $\{\pi_1, \pi_2, \dots, \pi_N\}$ is unique and often called the *stationary probabilities*. A Markov chain satisfying equation (11.9) is called an *ergodic Markov chain*. In addition, if a Markov chain has a stationary distribution, equations (11.10) and (11.11) can be used to find the unique solution of the stationary distribution.

Now consider the reversed process of a Markov chain $X_n, X_{n-1} \dots, X_{n-k}, \dots$, (for any n). This process also forms a Markov chain with transition probabilities defined by

$$
\begin{aligned}
p_{ij}^* &= P\left(X_{k-1} = j \middle| X_k = i\right) \\
&= \frac{P\left(X_k = i \middle| X_{k-1} = j\right)P\left(X_{k-1} = j\right)}{P\left(X_k = i\right)} \\
&= \frac{\pi_j p_{ji}}{\pi_i}.
\end{aligned} \tag{11.12}
$$

Furthermore, the reverse process is finite, aperiodic and irreducible. It satisfies Equation (11.10) and hence we have $\sum_{i=1}^{N} \pi_i p_{ij}^* = \sum_{i=1}^{N} \pi_j p_{ji} = \pi_j$.

Suppose that the sequence $\gamma_1, \gamma_2, \dots \gamma_N$ satisfies $\sum_{i=1}^{N} \gamma_i = 1$ and $\gamma_i p_{ij} = \gamma_j p_{ji}$ for all

i, j. We have $\sum_{i=1}^{N} \gamma_i p_{ij} = \sum_{i=1}^{N} \gamma_j p_{ji} = \gamma_j$. But this is the unique solution for the sta-

tionary distribution and hence $\gamma_i = \pi_i$ for all i. In this case we have $p_{ij}^* = p_{ij}$ and thus the original Markov chain and its reverse process the same properties. This type of Markov chain is called the time-reversible Markov chain.

11.5 Markov Chain Monte Carlo Methods

As we discussed in Chapter 8, the two most important point estimation methods are the Maximum Likelihood method and the Bayesian method using posterior distribution. Due to analytical and algebraic difficulty, most of the practical Bayesian estimates can not be obtained without numerical computations. Recent developments in Bayesian estimation methods have focused on techniques of simulation from the posterior distribution. Traditional simulation practice such as the inverse transformation method or the rejection method, that were discussed in section 11.3, would not be a practical tool when the posterior distribution is complicated. The *Markov Chain Monte Carlo* (MCMC) method is a class of more efficient methods for simulation. It has proved to be of great value. The MCMC refers to a type of simulation method that generates samples of an ergodic Markov chain and use the Markov chain convergence property such as the one presented in theorem 11.1, to make statistical inference from the generated samples. Uses of the MCMC method are not limited in Bayesian estimation. They have tremendous potential in application to a variety of statistical problems (see Gelfand and Smith (1990)). Almost any maximum likelihood computation can also use some MCMC approaches (see Geyer and Thompson (1992)).

The most popular MCMC method is the Gibbs sampler that is a special case of the Metropolis-Hastings algorithm. In this section, we will first describe the Gibbs sampler and then introduce the Metropolis-Hastings algorithm. We will also use the result in section 4 to show how the Metropolis-Hastings algorithm can converge and reach the desired distribution.

11.5.1 The Gibbs Sampler

The Gibbs sampler was first introduced by Geman and Geman (1984) and further developed by Gelfand and Smith (1990). It uses the conditional distribution of each component variable given the tentative values of the remaining random variables to construct the multivariate samples. It has been shown that the samples converge in distribution to the joint distribution of these variables. The main function of the Gibbs sampler is to generate samples from a multivariate probability density function (or probability mass function) $f(x_1,\ldots,x_n)$ when we know the form (up to a constant) of each of the *full conditional distributions*
$f(x_i | x_1,x_2,\cdots,x_{i-1},x_{i+1}\cdots,x_n)$, $i=1,2,\cdots,n$. This method is especially usefully when the multivariate distribution has a non-standard form. In this case, it will often be practical to simulate scalar quantity from a univariate conditional distribution.

For a given set of initial values $x_1^{(0)},x_2^{(0)},\cdots,x_n^{(0)}$, the Gibbs sampler algorithm simulates the next cycle of the values (in order) following the distributions below:

$$x_1^{(1)} \sim f\left(x_1 \big| x_2^{(0)},\cdots,x_n^{(0)}\right),$$
$$x_2^{(1)} \sim f\left(x_2 \big| x_1^{(1)},x_3^{(0)},\cdots,x_n^{(0)}\right),$$
$$x_3^{(1)} \sim f\left(x_2 \big| x_1^{(1)},x_2^{(1)},x_4^{(0)},\cdots,x_n^{(0)}\right),$$
$$\vdots$$

$$x_n^{(1)} \sim f\left(x_n \,\middle|\, x_1^{(1)}, x_2^{(1)}, \cdots, x_{n-1}^{(1)}\right).$$

After i cycles of iterations, the simulated values become $\left(x_1^{(i)}, x_2^{(i)}, \cdots, x_n^{(i)}\right)$. Under some mild conditions, the generated values can be shown to have a distribution approximately equal to $f\left(x_1, \ldots, x_n\right)$ when i is large.

The advantage of the Gibbs Sampler is that we trade off n doable simulations of univariate random variables for a usually-impossible simulation of a n dimensional random vector. In practical applications, the conditional distributions described above have standard forms when the joint distribution does not. For the case that direct simulation of the joint distribution is possible, the Gibbs sampler is still easier or more efficient as demonstrated in Example 11.13 below. If some of the conditional distributions do not have a standard form, it might be easier to use the rejection method described in section 11.3.2 or other simulation methods for univariate distributions to complete the Gibbs sampler.

Some theoretical issues that have arisen from the implementation of the Gibbs sampler were briefly summarized by Armitage, Berry and Matthews (2002). The first is whether the joint distributions that are constructed by the conditional distributions described above are unique. Brook (1964) and Besag (1974) have proved that these conditional distributions can characterize the joint distribution. No two different joint distributions will give the same set of full conditional distributions. The second issue is the choice of initial values. The underlying theory only states that, when i is large, $\left(x_1^{(i)}, x_2^{(i)}, \cdots, x_n^{(i)}\right)$ has a distribution approximately equal to $f\left(x_1, \ldots, x_n\right)$, i.e. $\left(x_1^{(i)}, x_2^{(i)}, \cdots, x_n^{(i)}\right)$ converges to a random vector that has a joint distribution equal to $f\left(x_1, \ldots, x_n\right)$. It does not exclude the effect of initial values. In practice, this problem can be minimized by discarding the first few runs of the simulations, called the *burn-in period*. Of course, the choice of the initial values and the length of the burn-in period are still dependent on each individual problem. The third concern is the dependence of the successive sampling values. In the simulation procedures discussed in section 3, most of them generate independent samples. From the construction of the Gibbs sampler, it is clear that the vectors generated from the successive cycles are not independent. Therefore, applications of the Gibbs sampler in statistical inferences require extra caution. Some will not be affected by the dependence of the samples and some will. For example, if we try to use the sample mean from the generated samples to estimate the distributional mean, this dependence will not affect our result. If we try to use the sample variance from the Gibbs samples to estimate our distributional variance, the traditional sample variance may not be suitable due to the inter-dependence of the variables.

Example 11.13 Consider generating a random sample $(X,Y)'$ from a bivariate normal distribution with both means equal to 0, $Var(X) = \sigma_X^2$, $Var(Y) = \sigma_Y^2$ and the correlation between X and Y equal to ρ. It is well known that we can first simulate two independent standard normal random samples Z_1 and Z_2, and then make an appropriate transformation so that $X = aZ_1 + bZ_2$ and $Y = cZ_1 + dZ_2$, for some constants a, b, c and d. However, calculations of a, b, c and d may require trial and error on the following equations:

$$Var(X) = a^2 + b^2 = \sigma_X^2,$$

$$Var(Y) = c^2 + d^2 = \sigma_Y^2$$

and

$$Cov(X,Y) = Cov(aZ_1 + bZ_2, cZ_1 + dZ_2) = ac\sigma_X^2 + bd\sigma_Y^2 = \rho\sigma_X\sigma_Y.$$

If we use the Gibbs sampler, we can iteratively generate samples $x^{(i+1)}$ and $y^{(i+1)}$ based on the conditional distributions, given the values of the other, that are univariate normal:

$$x^{(i+1)}\left|y^{(i)} \sim N\left(\rho\frac{\sigma_X}{\sigma_Y}y^{(i)}, \sigma_X^2(1-\rho^2)\right)\right.$$

and

$$y^{(i+1)}\left|x^{(i+1)} \sim N\left(\rho\frac{\sigma_Y}{\sigma_X}x^{(i+1)}, \sigma_Y^2(1-\rho^2)\right).\right.$$

Note that the conditional distribution of one given the other in bivariate normal random variables can be found in Chapter 5 or Chapter 8. This method does not require trial-and-error computations. A SAS program for simulating 10,000 bivariate normal samples with $\sigma_X^2 = 1$, $\sigma_Y^2 = 2$ and $\rho = 0.3$ is provided. A burn-in period of 1000 is used for this implementation. The estimated variances and correlation are $\hat{\sigma}_X^2 = 0.998$, $\hat{\sigma}_Y^2 = 1.996$ and $\hat{\rho} = 0.307$. The plot of these samples on the two-dimensional plane is attached in figure 11.3. If means of X and Y are not zero, additional constants will add to the samples.

Example 11.14 If we are to use Gibbs sampling method to generate the trivariate normal samples discussed in example 11.11, we have to calculate the full conditional distributions for all three components, given the other two. The three conditional distributions are given by

$$x^{(i+1)} \sim N\left(1+\frac{16}{35}\left(y^{(i)}-2\right)+\frac{-9}{35}\left(w^{(i)}-3\right),2,-\frac{66}{35}\right),$$

$$y^{(i+1)} \sim N\left(2+2\left(x^{(i)}-1\right)+0.5\left(w^{(i)}-3\right),0.5\right)$$

and

$$w^{(i+1)} \sim N\left(3+(-3)\left(x^{(i)}-1\right)+\left(\frac{4}{3}\right)\left(y^{(i)}-2\right),\frac{4}{3}\right).$$

Note that the above conditional distributions were derived using the property in multivariate normal distributions. The version for a trivariate normal distribution can be stated below:

Let $(X,Y,W)'$ be a trivariate normal random vector with mean vector $(\mu_X,\mu_Y,\mu_W)'$ and variance covariance matrix

$$\Sigma = \begin{pmatrix} \sigma_X^2 & COV(X,Y) & COV(X,W) \\ COV(X,Y) & \sigma_Y^2 & COV(Y,W) \\ COV(X,W) & COV(Y,W) & \sigma_W^2 \end{pmatrix}.$$

Then the conditional distribution of X, given $Y = y$ and $W = w$ is normal distributed with mean $\mu_{X|y,w} =$

$$\mu_X +\left(COV(X,Y) \quad COV(X,W)\right)\begin{pmatrix} \sigma_Y^2 & COV(Y,W) \\ COV(Y,W) & \sigma_W^2 \end{pmatrix}^{-1}\begin{pmatrix} y-\mu_Y \\ w-\mu_W \end{pmatrix}$$

and variance $\sigma_{X|y,w}^2 =$

$$\sigma_X^2 -\left(COV(X,Y) \quad COV(X,W)\right)\begin{pmatrix} \sigma_Y^2 & COV(Y,W) \\ COV(Y,W) & \sigma_W^2 \end{pmatrix}^{-1}\begin{pmatrix} COV(X,Y) \\ COV(X,W) \end{pmatrix}.$$

Similarly, we can derive the conditional distribution for the other two components. The proof of the above derivations can be obtained by writing down the ratio of the joint distribution and the appropriate marginal distribution.

A SAS program implemented shows that the sample means for X, Y and W are 0.96, 1.96 and 3.09 respectively. The sample variances are 2.13, 6.31 and 6.12 respectively. The sample covariances for (X,Y), (X,W) and (Y,W) are 3.20, -2.12 and -1.18 respectively. This program created 10,500 trivariate samples and excluded the first 500 trivariate samples for the burn in period.

Example 11.15 Consider the simulation of the following joint distribution discussed in Casella and George (1992):

$$f(x,y,n) = C\binom{n}{x} y^{x+\alpha-1}\left(1-y\right)^{nx+\beta-1} e^{-\lambda}\frac{\lambda^n}{n!},$$

for $x = 0,1,\ldots,n$, $0 \le y \le 1$ and $n = 1,2,\ldots$.

We can generate samples from this trivariate distribution using the Gibbs sampling method. Observe that the full conditional distributions

$$f(x|y,n) \sim B(n,y),$$

$$f(y|x,n) \sim Beta(x+\alpha, n-x+\beta)$$

and

$$f(n|x,y) = \frac{\left[(1-y)\lambda\right]^{n-x}}{(n-x)!}\exp\left(-(1-y)\lambda\right), \text{ for } n = x, x+1,\ldots.$$

Note that the conditional distribution of $n-x$ given x and y, follows a Poisson distribution with parameter $(1-y)\lambda$. The Gibbs samples obtained can be used for making statistical inference on the joint distribution. For example, we can estimate the means of X, Y and n. A SAS program for implementing this procedure is available with $\alpha = 3$, $\beta = 2$ and $\lambda = 20$. To avoid an early termination of the iterative process due to some numeric problems, we chose relatively large initial values: $y = 0.8$ and $n = 50$. After excluding 1000 burn-in trivariate samples, the 10,000 samples that were kept, gave the mean estimates: $\hat{\mu}_X = 11.98$, $\hat{\mu}_Y = 0.60$ and $\hat{n} = 20.02$.

11.5.2 The Metropolis-Hastings Sampler*

From an initial value, the Gibbs sampler specifies how to generate the next vector sample $\mathbf{x}^{(i+1)}$ from the current vector sample $\mathbf{x}^{(i)}$ by use of the full conditional distributions. If any full conditional distribution in Gibbs sampler is difficult to simulate, there is another MCMC method, called the *Metropolis- Hastings algorithm*. This algorithm was originally proposed by Metropolis et al. (1953) and later extended by Hastings (1970). It chooses a different transition matrix (kernel) that is time reversible, to simulate a candidate sample and then uses a special acceptance method to screen this candidate sample.

The Metropolis-Hastings algorithm is used to simulate a sequence of samples that form a Markov chain and have limiting distribution proportional to a known functional form. This algorithm accomplishes the task by constructing a time reversible Markov chain with the desired limiting distribution.

Let $\frac{1}{c} f(\ )$ be the target probability density function we try to simulate samples from and $\mathbf{x}^{(0)}$ be the initial point. The Metropolis- Hastings algorithm first chooses a conditional density $q(\cdot|\mathbf{x})$ (or transition kernel) that is easy to simulate from. Note that this transitional kernel must satisfy the time reversibility condition

$$f(\mathbf{y})q(\mathbf{y}|\mathbf{x}) = f(\mathbf{x})q(\mathbf{x}|\mathbf{y}).$$

The algorithm generates a candidate sample $\mathbf{y}_n \sim q(\mathbf{y}|\mathbf{x}^{(n)})$ at iteration n $n = 1,2,\ldots$. It then take $\mathbf{x}^{(n+1)} = \mathbf{y}^{(n)}$ with probability $\alpha(\mathbf{x}^{(n)},\mathbf{y}_n)$ and $\mathbf{x}^{(n+1)} = \mathbf{x}^{(n)}$ with probability $1 - \alpha(\mathbf{x}^{(n)},\mathbf{y}_n)$, where $\alpha(\mathbf{x},\mathbf{y}) = \min\left\{\dfrac{f(\mathbf{y})}{f(\mathbf{x})} \cdot \dfrac{q(\mathbf{x}|\mathbf{y})}{q(\mathbf{y}|\mathbf{x})}, 1\right\}$.

For the discrete distribution with finite number of states, $f(\ )$ can be expressed as $\boldsymbol{\pi} = (\pi_1, \pi_2, \ldots, \pi_N)'$ (here, $\sum_{i=1}^{N} \pi_i = \frac{1}{c}$) and $q(\cdot|\mathbf{x})$ can be expressed as a transition probability matrix $Q = \left(q_{ij}\right)_{N \times N}$. At time n, if $x^{(n)} = i$, then the candidate sample $y = j$ will be generated according to the transition probability q_{ij}, $j = 1,2,\ldots,N$. The acceptance probability of $y = j$ can be expressed as $\alpha_{ij} = \min\left\{\dfrac{\pi_j}{\pi_i} \cdot \dfrac{q_{ji}}{q_{ij}}, 1\right\}$. If $y = j$ is accepted then $x^{(n+1)} = j$, otherwise $x^{(n+1)} = i$.

In the discrete distributional case, we can easily see that the samples $\left\{x^{(n)}\right\}$ constructed above constitutes a Markov chain with transition probability matrix denoted by $\left\{p_{ij}\right\}$. From the above description and the definitions of the notation, the transition probabilities satisfy

$$p_{ij} = q_{ij}\alpha_{ij}, \text{ if } j \neq i \qquad (11.13)$$

and

$$p_{ii} = q_{ii} + \sum_{k \neq i} q_{ik}(1 - \alpha_{ik}).$$

Furthermore, if $\alpha_{ij} = \dfrac{\pi_j q_{ji}}{\pi_i q_{ij}}$, then $\dfrac{\pi_j q_{ji}}{\pi_i q_{ij}} \leq 1$ and thus $\dfrac{\pi_i q_{ij}}{\pi_j q_{ji}} \geq 1$, i.e. $\alpha_{ji} = 1$. Conversely, if $\alpha_{ji} = 1$, then $\dfrac{\pi_i q_{ij}}{\pi_j q_{ji}} \geq 1$ and thus $\dfrac{\pi_j q_{ji}}{\pi_i q_{ij}} \leq 1$, i.e. $\alpha_{ij} = \dfrac{\pi_j q_{ji}}{\pi_i q_{ij}}$. In both cases, we can derive $\pi_i q_{ij}\alpha_{ij} = \pi_j q_{ji}\alpha_{ji}$, for $i \neq j$, using the definition of α_{ij}. This equation implies that $\pi_i p_{ij} = \pi_j p_{ji}$ using equation (11.13). That means the Markov chain $\{x^{(n)}\}$ constructed is time reversible and hence has limiting probabilities equal to $\boldsymbol{\pi} = (\pi_1, \pi_2, \ldots, \pi_N)'$ as discussed at the end of section 11.4.

Now we can conclude that the samples generated by the Metropolis-Hastings algorithm have a distribution approximately equal to $(\pi_1, \pi_2, \ldots, \pi_N)$. This is true for the continuous random variables using transition kernel. The readers are referred to Robert and Casella (1999) for the proof.

Note that Gibbs sampler is a special case of the Metropolis-Hastings algorithm. Gibbs sampler takes each full conditional distribution as the candidate sample's transition kernel and set $\alpha(\mathbf{x}, \mathbf{y})$ equal to 1.

Problems:

1. Generate 1000 values from a Poisson random variable with mean=3.
2. Prove equation (11.4).
3. Prove equation (11.7).
4. Verify that the means vector and the variances and covariance matrix of $(X, Y, W)'$ in Example 11.11 are as they are specified in the problem.

References

Anderson, T.W. (2003). *An Introduction to Multivariate Statistical Analysis.* 3rd ed. Wiley-Interscience. New York.

Armitage, P., Berry, G., and Matthews, J.N.S. (2002). *Statistical Methods in Medical Research.* 4th ed. Blackwell Science. New York.

Gelfand, A.E. and Smith, A.F.M. (1990). Sampling-based approaches to calculating marginal densities. *Journal of the American Statistical Association.* 85: 398-409.

Geman, S. and Geman, D. (1984). Stochastic relaxation, Gibbs distributions, and the Bayesian restoration of images. *IEEE Transactions on Pattern Analysis and Machine Intelligence.* 6:721-741.

Hastings, W.K. (1970) Monte Carlo sampling methods using Markov chains and their applications. *Biometrika* 57:97-109.

Lederman, R., Chan, W. and Roberts-Gray, C. (2004). Sexual risk attitudes and intentions of youth age 12-14 years: survey comparisons of parent-teen prevention and control groups. *Behavioral Medicine.* 29:155-163.

Metropolis, N., Rosenbluth, AW, and Rosenbluth, MN (1953). Equations of state calculations by fast computing machine. *Journal of Chemical Physics.* 21:1087-1091.

Robert, C.P. and Casella, G. (1999). *Monte Carlo Statistical Methods.* Springer. New York.

Ross, S.M. (2002). *Simulation.* 3rd ed.. Academic Press. Orlando.

Searle, S.R. (1982). *Matrix Algebra Useful for Statistics.* Wiley. New York.

Taylor, H. and Karlin, S. (1998). *An Introduction to Stochastic Modeling.* 3rd ed.. Academic Press. Orlando.

Thisted, R.A. (1988). *Elements of Statistical Computing: Numerical Computation.* Chapman & Hall. London.

Index

Printed and bound by CPI Group (UK) Ltd, Croydon, CR0 4YY

24/10/2024

01778493-0001